Actinomycetes

Actinomycetes

The Antibiotics Producers

Editors

Evi Stegmann
Yvonne Mast

MDPI • Basel • Beijing • Wuhan • Barcelona • Belgrade • Manchester • Tokyo • Cluj • Tianjin

Editors
Evi Stegmann
Interfaculty Institute of Microbiology and Infection Medicine, Department of Microbial Bioactive Compounds, University of Tübingen
Germany

Yvonne Mast
Leibniz Institute DSMZ - German Collection of Microorganisms and Cell Cultures GmbH, Department Bioresources for Bioeconomy and Health Research
Germany

Editorial Office
MDPI
St. Alban-Anlage 66
4052 Basel, Switzerland

This is a reprint of articles from the Special Issue published online in the open access journal *Antibiotics* (ISSN 2079-6382) (available at: https://www.mdpi.com/journal/antibiotics/special_issues/actinomycetes).

For citation purposes, cite each article independently as indicated on the article page online and as indicated below:

LastName, A.A.; LastName, B.B.; LastName, C.C. Article Title. *Journal Name* **Year**, *Article Number*, Page Range.

ISBN 978-3-03936-910-2 (Hbk)
ISBN 978-3-03936-911-9 (PDF)

Cover image courtesy of Dr. Imen Nouioui; Leibniz Institute DSMZ - German Collection of Microorganisms and Cell Cultures GmbH, Braunschweig, Germany.

Contents

About the Editors

Evi Stegmann (Dr.) is a senior group leader in the Department of Microbial Bioactive Compounds at the University of Tübingen. She obtained her doctoral degree and Venia Legendi in microbiology from Eberhard Karls University in Tübingen, Germany. Her research is focused on the discovery and investigation of interesting secondary metabolites, as well as the understanding of their regulation. Presently, she is particularly focused on the biosynthesis of glycopeptide antibiotics, their mode of action, and the resistance mechanism against those medically important antibiotics. A further research area of interest is the optimization of both the properties and yields of molecules.

Yvonne Mast (Prof. Dr.) is the head of the Department of Bioresources for Bioeconomy and Health Research at the Leibniz Institute DSMZ - German Collection of Microorganisms and Cell Cultures GmbH and a Professor of Applied Microbiology at the Technical University of Braunschweig, Germany. She obtained her doctoral degree and Venia Legendi in microbiology from Eberhard Karls University in Tübingen, Germany. Her research focuses on the regulation of antibiotic biosynthesis in actinomycetes, as well as genetic engineering and genome mining of antibiotic-producing actinobacteria.

antibiotics

Editorial

Actinomycetes: The Antibiotics Producers

Yvonne Mast [1,2,3,4,*] and Evi Stegmann [3,4,*]

1 Department of Bioresources for Bioeconomy and Health Research, Leibniz Institute DSMZ-German Culture Collection for Microorganisms and Cell Cultures, 38124 Braunschweig, Germany
2 Institute for Microbiology, Technical University of Braunschweig, 38106 Braunschweig, Germany
3 Department of Microbiology/Biotechnology, Interfaculty Institute of Microbiology and Infection Medicine, Faculty of Science, Eberhard Karls University of Tübingen, Auf der Morgenstelle 28, D-72076 Tübingen, Germany
4 German Center for Infection Research (DZIF), Partner Site Tübingen, D-72076 Tübingen, Germany
* Correspondence: yvonne.mast@dsmz.de (Y.M.); evi.stegmann@biotech.uni-tuebingen.de (E.S.); Tel.: +49-(0)-531-2616-418 (Y.M.); +49-(0)-7071-2978-840 (E.S.)

Received: 12 July 2019; Accepted: 25 July 2019; Published: 29 July 2019

Actinomycetes are well known as an inexhaustible source for antibiotics. Most of the known antimicrobials today were originally isolated from actinomycetes, especially from the genus *Streptomyces*. The produced substances include all important drug classes used in clinics today, such as β-lactams, tetracyclines, macrolides, aminoglycosides, or glycopeptides. However, in the past years the effectiveness of these impressive weapons have become endangered by the rise of resistances of live-threatening pathogenic bacteria. In addition to that, the golden era of antibiotic discovery has been over for quite a while and the antibiotic pipeline was feared to run dry. Next-generation sequencing techniques in combination with genome mining approaches revolutionized the field of antibiotic research and might reboot the pipeline in the near future. In 2002, the first *Streptomyces* genome sequence was published [1]. This was the genome sequence of the model actinomycete *Streptomyces coelicolor*. Mining this sequence revealed that *S. coelicolor* harbors 22 secondary metabolite gene clusters but indeed produces only four of the encoded metabolites under standard lab conditions. Currently, more than 625 genome sequences are available only of the genus *Streptomyces* [2]. Genome mining analyses suggest that less than 10% of the genetic potential of antibiotic producers is currently being used [3], which implicates that there is a huge untapped genetic reservoir waiting to be exploited for drug discovery. In addition to that, metagenomic data indicate that there are far more potential antibiotic producers in nature awaiting isolation and investigation. Thus, today, almost 80 years after Selman Waksman introduced the genus of *Streptomyces* and, with actinomycin being the first antibiotic that was isolated from an actinomycete, these bacteria still represent a treasure chest for the identification of novel antibiotics.

Novel cultivation strategies, elaborated screening techniques, new genetic manipulation tools, further insights into physiological aspects of actinobacterial live style, and also knowledge on new secondary metabolite biosynthetic pathways may open up a new era of antibiotic drug discovery.

In the Special Issue of "Actinomycetes: The Antibiotics Producers" we will highlight the latest research findings in the field. The current issue brings together 19 publications (9 research articles and 10 reviews) from expert groups, which address scientific topics associated with the biology of actinomycetes as well as their antibiotic biosynthesis capacity.

Two research articles deal with the isolation and characterization of actinobacteria from unique habitats. Using a high-throughput sequencing approach, Maciejewska et al. were able to show that there is a quite complex actinobacterial community present in cave moonmilk deposits obtained from the cave "Grotte des Collemboles", Belgium [4]. By applying several strategies to allow growth and isolation of hard-to-culture actinobacteria, Adam et al. were able to isolate new actinobacterial representatives from these cave moonmilk deposits and demonstrated that some of them exert antibacterial activity [5]. Another extraordinary source for novel actinobacteria producing unique chemistry are certain insects. Benndorf et al. isolated 97 actinobacteria from the fungus-growing termite *Macrotermes natalensis* and analyzed them in terms of their antibiotic production properties. Thereby, the authors isolated several new actinobacterial species and identified a dichlorinated diketopiperazine derivative and two tetracyclic lanthipeptides (rubrominins A and B) as new secondary metabolites [6]. However, not just new strain isolates but also old strain isolates can be promising sources for novel natural compounds, when investigated with state-of the art analytical technologies. This has been effectively highlighted in the review of Takahashi and Nakashima, which gives an update on novel compounds from old strains of the Kitasato Microbial Library [7]. Indeed, the authors describe the identification of 36 novel compounds from 11 actinomycetes, most of them more than 35 years old. Thereby, the list of producers resembles a quite diverse set of actinomycetes, which illustrates that also non-*Streptomyces* actinomycetes represent good sources for novel natural compounds. This is also underpinned by the study of Iorio et al., which specifically focusses on the underexplored genus *Actinoallumurus*, for which two novel polyether substances (α-770 and α-823) were described [8]. This raises the question if and to what extent gene cluster composition differs between different types of actinobacterial antibiotic producers. Two of the Journal's papers deal with that issue in more detail: Choudoir et al. and Vicente et al. focus on comparative biosynthetic gene cluster (BGC) distributions in similar streptomycetes and thereby concentrate on the evolutionary dynamics of BGCs within these organisms [9,10]. These manuscripts already depict the importance of omics approaches in antibiotic research, which in general is shortly highlighted in the review of Genilloud [11]. Finally, a very comprehensive review on novel strategies and innovative methods to get access to novel natural compounds from actinomycetes is given by Hug et al., which also offers a very reasonable outlook to future perspectives in antibiotic research [12].

Besides the studies that aim to identify novel antibiotics, the Special Issue also contains four papers dealing with specific antibiotics and their biosyntheses: Schwarzer et al. report on the biosynthesis of the polyketide rishirilide B [13], whereas Hofeditz et al. describe the production and structure elucidation of the novel polyphenolic tridecaketide lysoquinone-TH1 [14]. Both compounds are polyketides, and biosynthesis has successful been made possible after heterologous expression of the respective gene clusters in *Streptomyces albus*. Viana Marques et al. provide an update on β-lactam inhibitors and their applications [15], whereas Musiol-Kroll and Wohlleben focus on the biochemical function and usage of acyltransferases during polyketide biosynthesis for engineering approaches [16].

Especially when it comes to engineering and production attempts of the respective compounds, regulation and self-resistance of the producer organisms are important factors to be targeted in order to increase yields. Thus, aspects of regulation and self-resistance are also addressed in the Special Issue: Ordóñez-Robles et al. summarize the current knowledge on regulation of tacrolimus biosynthesis in *Streptomyces tsukubaensis* [17]. Alduina et al. report on the regulatory network influencing the biosynthesis of the glycopeptide antibiotic A40926 from *Nonomuraea* sp. [18], whereas Binda et al. concentrate on the self-resistance mechanisms and characteristics of glycopeptide producers in general [19].

Besides their overwhelming importance as antibiotic producers, actinomycetes undoubtedly have further imposing characteristics, and some of them are handled in the Special Issue. For example, actinomycetes undergo a complex life cycle, whereby morphological development and secondary metabolism often coincide. Manteca and Yagüe review strategies to improve *Streptomyces* differentiation in liquid cultures with the aim to elicit antibiotic production [20]. Furthermore, actinomycetes are well known as potent producers of extracellular hydrolytic enzymes, which is relevant for industrial applications. Here, Gullón and Mellado give an overview focusing on the secretion machinery of the most widely used host strains for exoenzyme production—*Streptomyces lividans* [21]. Finally, Jones reports on known and novel aspects of the polynucleotide phosphorylase (PNPase) in *Streptomyces*, which exerts a dual function as 3′-5′-exoribonuclease and RNA 3′-polyribonucleotide polymerase [22].

We are happy to bring together so many valuable works from respected experts in the field. It was our great pleasure to be Guest Editors of this special issue of *Antibiotics*.

Author Contributions: Y.M. and E.S. equally contributed in writing the article.

Funding: This research received funding from the German Center for Infection research (DZIF), TTU 09.819.

Conflicts of Interest: The authors declare no conflicts of interest."

References

1. Bentley, S.D.; Chater, K.F.; Cerdeño-Tárraga, A.M.; Challis, G.L.; Thomson, N.R.; James, K.D.; Harris, D.E.; Quail, M.A.; Kieser, H.; Harper, D.; et al. Complete genome sequence of the model actinomycete *Streptomyces coelicolor* A3(2). *Nature* **2002**, *417*, 141–147. [CrossRef] [PubMed]
2. Blin, K.; Pascal Andreu, V.; de Los Santos, E.L.C.; Del Carratore, F.; Lee, S.Y.; Medema, M.H.; Weber, T. The antiSMASH database version 2: A comprehensive resource on secondary metabolite biosynthetic gene clusters. *Nucleic Acids Res.* **2019**, *47*, D625–D630. [CrossRef] [PubMed]
3. Nett, M.; Ikeda, H.; Moore, B.S. Genomic basis for natural product biosynthetic diversity in the actinomycetes. *Nat. Prod. Rep.* **2009**, *26*, 1362–1384. [CrossRef] [PubMed]
4. Maciejewska, M.; Całusińska, M.; Cornet, L.; Adam, D.; Pessi, I.S.; Malchair, S.; Delfosse, P.; Baurain, D.; Barton, H.A.; Carnol, M.; et al. High-Throughput Sequencing Analysis of the Actinobacterial Spatial Diversity in Moonmilk Deposits. *Antibiotics* **2018**, *7*, 27. [CrossRef] [PubMed]
5. Adam, D.; Maciejewska, M.; Naômé, A.; Martinet, L.; Coppieters, W.; Karim, L.; Baurain, D.; Rigali, S. Isolation, Characterization, and Antibacterial Activity of Hard-to-Culture Actinobacteria from Cave Moonmilk Deposits. *Antibiotics* **2018**, *7*, 28. [CrossRef] [PubMed]
6. Benndorf, R.; Guo, H.; Sommerwerk, E.; Weigel, C.; Garcia-Altares, M.; Martin, K.; Hu, H.; Küfner, M.; de Beer, Z.W.; Poulsen, M.; et al. Natural Products from Actinobacteria Associated with Fungus-Growing Termites. *Antibiotics* **2018**, *7*, 83. [CrossRef] [PubMed]
7. Takahashi, Y.; Nakashima, T. Actinomycetes, an Inexhaustible Source of Naturally Occurring Antibiotics. *Antibiotics* **2018**, *7*, 45. [CrossRef] [PubMed]
8. Iorio, M.; Tocchetti, A.; Santos Cruz, J.C.; Del Gatto, G.; Brunati, C.; Maffioli, S.I.; Sosio, M.; Donadio, S. Novel Polyethers from Screening *Actinoallomurus* spp. *Antibiotics* **2018**, *7*, 47. [CrossRef] [PubMed]
9. Choudoir, M.J.; Pepe-Ranney, C.; Buckley, D.H. Diversification of Secondary Metabolite Biosynthetic Gene Clusters Coincides with Lineage Divergence in *Streptomyces*. *Antibiotics* **2018**, *7*, 12. [CrossRef]
10. Vicente, C.M.; Thibessard, A.; Lorenzi, J.N.; Benhadj, M.; Hôtel, L.; Gacemi-Kirane, D.; Lespinet, O.; Leblond, P.; Aigle, B. Comparative Genomics among Closely Related *Streptomyces* Strains Revealed Specialized Metabolite Biosynthetic Gene Cluster Diversity. *Antibiotics* **2018**, *7*, 86. [CrossRef]
11. Genilloud, O. Mining Actinomycetes for Novel Antibiotics in the Omics Era: Are We Ready to Exploit This New Paradigm? *Antibiotics* **2018**, *7*, 85. [CrossRef] [PubMed]
12. Hug, J.J.; Bader, C.D.; Remškar, M.; Cirnski, K.; Müller, R. Concepts and Methods to Access Novel Antibiotics from Actinomycetes. *Antibiotics* **2018**, *7*, 44. [CrossRef] [PubMed]
13. Schwarzer, P.; Wunsch-Palasis, J.; Bechthold, A.; Paululat, T. Biosynthesis of Rishirilide, B. *Antibiotics* **2018**, *7*, 20. [CrossRef] [PubMed]
14. Hofeditz, T.; Unsin, C.E.; Wiese, J.; Imhoff, J.F.; Wohlleben, W.; Grond, S.; Weber, T. Lysoquinone-TH1, a New Polyphenolic Tridecaketide Produced by Expressing the Lysolipin Minimal PKS II in *Streptomyces albus*. *Antibiotics* **2018**, *7*, 53. [CrossRef] [PubMed]
15. Viana Marques, D.A.; Machado, S.E.F.; Ebinuma, V.C.S.; Duarte, C.A.L.; Converti, A.; Porto, A.L.F. Production of β-Lactamase Inhibitors by *Streptomyces* Species. *Antibiotics* **2018**, *7*, 61. [CrossRef] [PubMed]
16. Musiol-Kroll, E.M.; Wohlleben, W. Acyltransferases as Tools for Polyketide Synthase Engineering. *Antibiotics* **2018**, *7*, 62. [CrossRef] [PubMed]
17. Ordóñez-Robles, M.; Santos-Beneit, F.; Martín, J.F. Unraveling Nutritional Regulation of Tacrolimus Biosynthesis in *Streptomyces tsukubaensis* through *omic* Approaches. *Antibiotics* **2018**, *7*, 39. [CrossRef] [PubMed]
18. Alduina, R.; Sosio, M.; Donadio, S. Complex Regulatory Networks Governing Production of the Glycopeptide A40926. *Antibiotics* **2018**, *7*, 30. [CrossRef]

19. Binda, E.; Cappelletti, P.; Marinelli, F.; Marcone, G.L. Specificity of Induction of Glycopeptide Antibiotic Resistance in the Producing Actinomycetes. *Antibiotics* **2018**, *7*, 36. [CrossRef]
20. Manteca, Á.; Yagüe, P. *Streptomyces* Differentiation in Liquid Cultures as a Trigger of Secondary Metabolism. *Antibiotics* **2018**, *7*, 41. [CrossRef]
21. Gullón, S.; Mellado, R.P. The Cellular Mechanisms that Ensure an Efficient Secretion in *Streptomyces*. *Antibiotics* **2018**, *7*, 33. [CrossRef] [PubMed]
22. Jones, G.H. Novel Aspects of Polynucleotide Phosphorylase Function in *Streptomyces*. *Antibiotics* **2018**, *7*, 25. [CrossRef] [PubMed]

antibiotics

MDPI

Communication

Comparative Genomics among Closely Related *Streptomyces* Strains Revealed Specialized Metabolite Biosynthetic Gene Cluster Diversity

Cláudia M. Vicente [1], Annabelle Thibessard [1], Jean-Noël Lorenzi [2], Mabrouka Benhadj [1,3], Laurence Hôtel [1], Djamila Gacemi-Kirane [4], Olivier Lespinet [2], Pierre Leblond [1,*] and Bertrand Aigle [1,*]

1 Université de Lorraine, INRA, DynAMic, F-54000 Nancy, France; claudia.morgado-vicente@univ-lorraine.fr (C.M.V.); annabelle.thibessard@univ-lorraine.fr (A.T.); mabrouka.benhadj@gmail.com (M.B.); laurence.hotel@univ-lorraine.fr (L.H.)

2 Institute for Integrative Biology of the Cell (I2BC), CEA, CNRS, University Paris-Sud, University Paris-Saclay, Gif-sur-Yvette CEDEX, France; jean-noel.lorenzi@i2bc.paris-saclay.fr (J.-N.L.); olivier.lespinet@i2bc.paris-saclay.fr (O.L.)

3 Biomolecules and Application Laboratory, Faculty of Exact Sciences and Natural and Life Sciences, University of Tebessa, Tebessa 12002, Algeria

4 Department of Biochemistry, Faculty of Science, University Badji Mokhtar Annaba, Annaba 23000, Algeria; dj_gacemi@yahoo.fr

* Correspondence: pierre.leblond@univ-lorraine.fr (P.L.); bertrand.aigle@univ-lorraine.fr (B.A.); Tel.: +33-372-745-143 (P.L.); +33-372-745-141 (B.A.)

Received: 12 September 2018; Accepted: 28 September 2018; Published: 2 October 2018

check for updates

Abstract: Specialized metabolites are of great interest due to their possible industrial and clinical applications. The increasing number of antimicrobial resistant infectious agents is a major health threat and therefore, the discovery of chemical diversity and new antimicrobials is crucial. Extensive genomic data from *Streptomyces* spp. confirm their production potential and great importance. Genome sequencing of the same species strains indicates that specialized metabolite biosynthetic gene cluster (SMBGC) diversity is not exhausted, and instead, a pool of novel specialized metabolites still exists. Here, we analyze the genome sequence data from six phylogenetically close *Streptomyces* strains. The results reveal that the closer strains are phylogenetically, the number of shared gene clusters is higher. Eight specialized metabolites comprise the core metabolome, although some strains have only six core gene clusters. The number of conserved gene clusters common between the isolated strains and their closest phylogenetic counterparts varies from nine to 23 SMBGCs. However, the analysis of these phylogenetic relationships is not affected by the acquisition of gene clusters, probably by horizontal gene transfer events, as each strain also harbors strain-specific SMBGCs. Between one and 15 strain-specific gene clusters were identified, of which up to six gene clusters in a single strain are unknown and have no identifiable orthologs in other species, attesting to the existing SMBGC novelty at the strain level.

Keywords: *Streptomyces*; strain; specialized metabolites; biosynthetic gene cluster

1. Introduction

Microorganisms are unmatched in their capacity to produce chemically diverse specialized metabolites [1]. *Streptomyces* in particular, which are ubiquitous filamentous bacteria with a differentiating mycelial life cycle, have a significant role in biotechnology industry. They are employed in the production of more than half of all antibiotics used in human and veterinary medicine, as well as a large assortment of other high-value molecules such as immune suppressants, anticancer, and anti-parasitic molecules [2,3]. The increasing incidence of antibiotic resistant infections and the lack of new antimicrobial agents have driven the race for natural product discovery. The significant advances made in high-throughput DNA sequencing, along with the relatively low cost of whole genome sequencing, have greatly contributed to this effort [4]. Strategies to address the antimicrobial shortage include both the intensive screening of uncultivable microorganisms through metagenomics and environmental DNA cloning for heterologous expression [5,6], as well as further genome mining of cultivable isolates. Consequently, correlation data between species and specialized metabolomes are accumulating progressively.

Streptomyces spp. are amongst the best-characterized bacteria, and the vastly increasing genomic data that is available has revealed the presence of thousands of specialized metabolite biosynthetic gene clusters (SMBGCs), highlighting their enormous potential for natural product biosynthesis [7,8]. Estimates say that no more than 10% have been identified to date, and furthermore, the majority of these SMBGCs appear to be "silent" [9]. A recent bioinformatic analysis of six *Streptomyces albus* strains led to the identification of strain-specific SMBGCs, speaking to the chemical diversity at the strain level [10]. It has also been shown that *Streptomyces* species with identical 16S rRNA sequences can have distinct specialized metabolomes [11]. Moreover, studies in actinobacteria *Salinispora* demonstrated an important biosynthetic pathway diversity in marine populations, providing insight into new chemical diversity-generating mechanisms [12].

A *Streptomyces* strain collection that included isolates with interesting biological activities against both Gram-positive and Gram-negative indicator strains was recently characterized [13]. Here, we report the genomic analysis of a small group of these environmental *Streptomyces* strains, including those antimicrobial-producing isolates and the associated SMBGC diversity that was encountered. This analysis highlights that the use of closely related strains to identity new SMBGCs in *Streptomyces* constitutes a promising approach for the identification of novel specialized biosynthetic pathways.

2. Results and Discussion

2.1. Analyzed Strains are Phylogenetically Diverse

The genomes of five environmental strains from a previously constructed *Streptomyces* collection [13], isolates E5N91, E2N166, E2N171, E1N211, E5N298, were sequenced and examined. Draft genome sequences have 7.0–8.5 Mb with an average G+C content of 71.8%, which is in good accordance with the genus (Table 1).

Table 1. Strains analyzed in this work.

Strain	G + C Content (%)	Genome Size (Mb)
E5N91	71.9	8.51
E2N166	70.2	7.91
E2N171	72.3	7.00
E1N211	72.1	7.32
E5N298	71.9	7.87
Streptomyces sp. ETH9427	72.1	7.75

A comparative genomics analysis of the nucleotide sequences, including the taxonomically close *Streptomyces* sp. ETH9427 recently sequenced [14], was carried out through a multilocus sequence typing (MLST) using five genes: *atpD* (ATP synthase, β subunit), *gyrB* (DNA gyrase, subunit B), *recA* (recombination protein), *rpoB* (RNA polymerase, β subunit), and *trpB* (tryptophan synthase, β subunit). It highlighted the taxonomic relationships between the strains and with other model *Streptomyces* species (Figure 1). Four clades are distinguishable in the constructed phylogenetic tree: the first clade with strain E5N91, *Streptomyces coelicolor* A3(2) [7], and *S. lividans* TK24 [15], the second containing strain E5N298 and the recently described *Streptomyces* sp. M1013 [16], clade 3, including strain E2N166 and *S. ambofaciens* strains ATCC 23877 [17] and DSM 40697 [18], and finally strains E2N171, E1N211, and ETH9427 grouped with *Streptomyces* sp. 4F (accession number CP013142.1) in clade 4 (Figure 1).

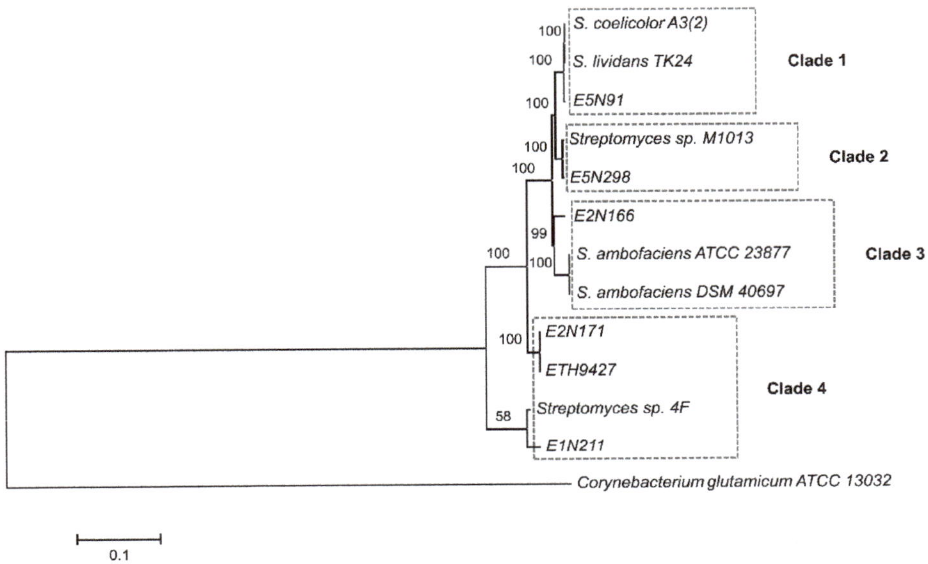

Figure 1. Phylogenetic tree of sequenced isolated and reference strains. Constructed based on the concatenated sequence alignment of five loci (*atpB-gyrB-recA-rpoB-trpB*), using inference by neighbor-joining with 100 bootstrap replicates. Only positions with a minimum 92% site coverage were used and the final dataset has 9141 positions in total. The scale bar indicates 10% estimated sequence divergence. The tree is rooted on *Corynebacterium glutamicum* ATCC 13032.

Taxonomic affiliations were further assessed by calculating the average nucleotide identity (ANI) using JSpeciesWS [19], and are resumed in Table 2. The ANI in clade 1 between strain E5N91 and *S. coelicolor* and with *S. lividans* is 97.32% (with 75.5% of the genome nucleotides aligned) and 97.65% (77.0% of nucleotides aligned), respectively. In clade 2, the ANI value is 98.19% (80.9% nucleotides aligned) between strain E5N298 and *Streptomyces* sp. M1013, and in clade 3 between strain E2N166, *S. ambofaciens* ATCC 23877, and *S. ambofaciens* DSM 40697, the ANI values are 88.86% (65.3% alignment) and 88.75% (with 66.0% nucleotides aligned), respectively. The values in clade 4 are as follows: the E2N171 strain's ANI value is 98.98% with strain E1N211 (83.5% of aligned nucleotides), 99.14% (84.8% alignment) with strain ETH9427, and 94.56% (76.3% of nucleotides aligned) with *Streptomyces* sp. 4F; the ANI value of strain E1N211 is 99.90% (86.4% of aligned nucleotides) with strain ETH9427, and 94.41% (71.1% alignment) with *Streptomyces* sp. 4F; and the ETH9427 strain's ANI value is 94.13% (70.7% of nucleotides aligned) with *Streptomyces* sp. 4F. These results suggest that strain E5N91 belongs to a species that is taxonomically closely related to *S. coelicolor* and *S. lividans* in clade 1, as do strain E5N298 and *Streptomyces* sp. M1013 in clade 2. Moreover, the ANI value among the six studied strains varies between 81.40–99.90%, except in those from clade 4, where the ANI value is 99.00–99.90%, well above the species boundary [20] and indicating that strains E2N171, E1N211, and ETH9427 belong to the same species. The closest relative to these strains is *Streptomyces* sp. 4F, and as such, all are clustered in clade 4. However, these strains and *Streptomyces* sp. 4F belong to distinct but related species, as seen by the ANI values between 94.13–94.56% and a low branch bootstrap in the phylogenetic tree (Figure 1). Still, these strains have highly similar 16S ribosomal DNA sequences (dissimilarities between 0.1–1.1%), which would mistakenly suggest that most of the strains belong to the same species. These results indicate that 16S single-gene analysis is not the most accurate tool for streptomycetes phylogeny analysis and inferring intraspecific genetic diversity, as already seen previously [21,22]. Instead, taxonomic relationships are better resolved using genome comparison indexes, such as ANI, and multi-gene analysis, such as MLST [23,24], which in some cases can even match whole-genome comparisons [25].

Table 2. Average nucleotide identity values of the compared strains.

	SLI	*E5N91*	*M1013*	*E5N298*	*ATCC*	*DSM*	*E2N166*	*E2N171*	*E1N211*	*ETH9427*	*4F*
SCO	99.0	97.3	91.5	91.3	86.9	86.9	87.5	81.7	81.8	81.9	81.8
SLI		97.7	91.8	91.6	87.0	87.0	87.7	81.6	81.7	81.8	81.7
E5N91			91.2	91.0	86.6	86.6	87.4	81.4	81.4	81.6	81.3
M1013				98.2	87.1	87.1	87.9	81.5	81.5	81.7	81.4
E5N298					87.2	87.2	87.9	81.7	81.7	81.8	81.6
ATCC						99.0	88.9	81.8	81.7	81.8	81.9
DSM							88.8	81.7	81.6	81.7	81.9
E2N166								82.1	82.1	82.2	82.0
E2N171									99.0	99.1	94.6
E1N211										99.9	94.4
ETH9427											94.1

2.2. Specialized Metabolism Diversity

The identification of SMBGCs was performed using antiSMASH 3.0 [26] and then edited to take into account previously published experimental data. A total of 28–33 SMBGCs was predicted in the isolated strains (Figure 2), with strain E2N171 harboring the least (28) and strain E5N91 harboring the most (33) gene clusters, expectantly coinciding with the smallest and largest genome, respectively. All six strains harbor SMBGCs belonging to the principal classes such as polyketide synthases (PKS), non-ribosomal synthetases (NRPS), and PKS–NRPS hybrids. Interestingly, the analysis revealed that a single strain within its clade can contain up to 48% strain-specific SMBGCs, suggesting that strain-level search for new SMBGCs diversity is not exhausted, and genome sequencing of several strains from a single species can lead to the identification of new gene clusters, as already seen previously [10,12].

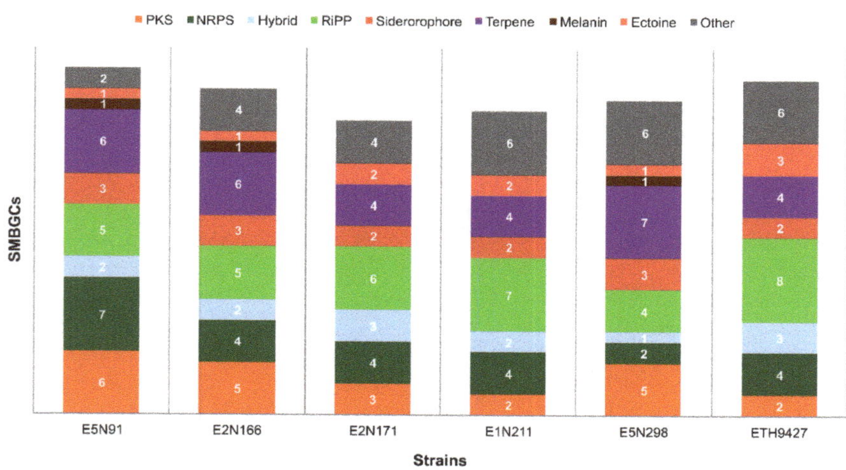

Figure 2. Specialized metabolite biosynthetic gene clusters identified in analyzed strains. Genome sequences were analyzed with antiSMASH 3.0 for gene cluster identification. The "Hybrid" category refers only to PKS-NRPS hybrid gene clusters. Other hybrid clusters that were detected were separated into their constituent parts to better visualize cluster diversity. Clusters encoding bacteriocins, lantipeptides, and lassopeptides are grouped in the category "RiPP", and the category "Other" includes other KS pathways and less common clusters such as phenazines and butyrolactones, among others.

2.3. The Core and Conserved Specialized Metabolites

Some of the identified gene clusters are well known and usually found in *Streptomyces* spp. The core SMBGCs are gene clusters that are present in the majority of streptomycetes and in the examined strains were selected based on conservation, with at least 58% of the strains sharing all of the core SMBGCs that encode for: melanin [27], ectoine osmolytes [28], spore pigment [29], the siderophore desferrioxamine B [30], the terpenes albaflavenone [31], hopene [32], isorenieratene [33], and the volatile geosmin [34]. This set of eight biosynthetic gene clusters is present in all of the strains from clades 1, 2, and 3, although no isorenieratene gene cluster was found in strain E2N166 (Figure 3). Interestingly, none of the strains in clade 4 possess either melanin or isorenieratene encoding gene clusters. Furthermore, strains E2N171, E1N211, and ETH9427 have additional ectoine gene clusters: two in ETH9427, and one in each of the other two strains. These have different gene contents (Figure 3, yellow and green E clusters), and as expected, the phylogenetically closer E2N171 and ETH9427 strains have the same extra ectoine encoding gene cluster, that in the case of ETH9427 is doubled due to its location on the chromosomal terminal inverted repeats. This result is consistent with strains from the same species, which usually share the same core metabolome [35].

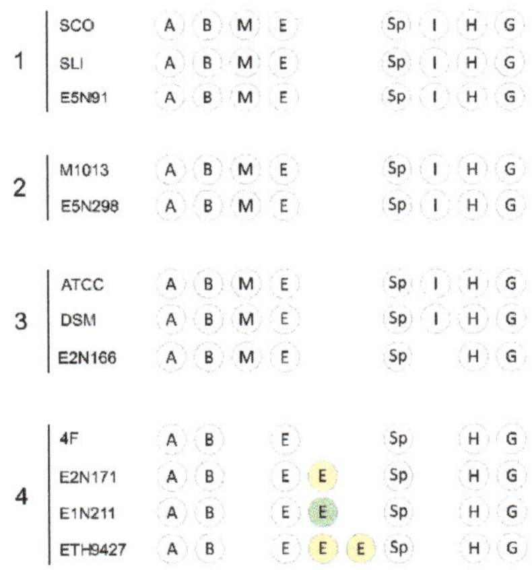

Figure 3. The core specialized metabolites produced by the strains. From the total number of SMBGCs that were identified, the core metabolome constitutes a set of gene clusters frequently conserved across *Streptomyces* species. The letters stand for the specialized metabolites albaflavenone (A), desferrioxamine B (B), melanin (M), ectoine (E), spore pigment (Sp), isorenieratene (I), hopene (H), and geosmin (G). Cluster disposition is not related to chromosomal position. The ectoine clusters in yellow and green represent different gene clusters.

In addition to the core metabolome, other SMBGCs are conserved within each clade (Figure 4). Strain E5N91 contains 18 conserved gene clusters out of 33 in total, which are also present in *S. coelicolor* in clade 1. Surprisingly, biosynthetic gene clusters for the characteristic metabolites coelichelin [36], coelibactin [37], calcium-dependent antibiotic [38], and coelimycin [39] are conserved in strain E5N91, whereas the actinorhodin [40] and undecylprodigiosin [41] gene clusters are not, even if they are present in phylogenetically closely related species. In clade 2, two-thirds of the gene clusters identified in strain E5N298 are conserved in *Streptomyces* sp. M1013 (21 out of 30 in total). Conversely, in clade 3, strain E2N166 has only nine gene clusters out of 31 in total conserved with *S. ambofaciens* strains ATCC 23877 and DSM 40697. These include the species-distinctive Type II PKS gene cluster encoding kinamycin [42], the NRPS biosynthetic gene cluster for coelichelin, two siderophore and two bacteriocin gene clusters, and a terpene-encoding gene cluster [43]. The E2N171, E1N211, and ETH9427 strains in clade 4 have a remarkably high number of conserved SMBGCs, with each strain showing 54 to

72% of the identified gene clusters that are present in one or both of the other two strains. However, only seven gene clusters are conserved in *Streptomyces* sp. 4F, which is consistent with its taxonomic relationship in clade 4. These results indicate that phylogenetic relationships are accompanied by specialized gene cluster maintenance.

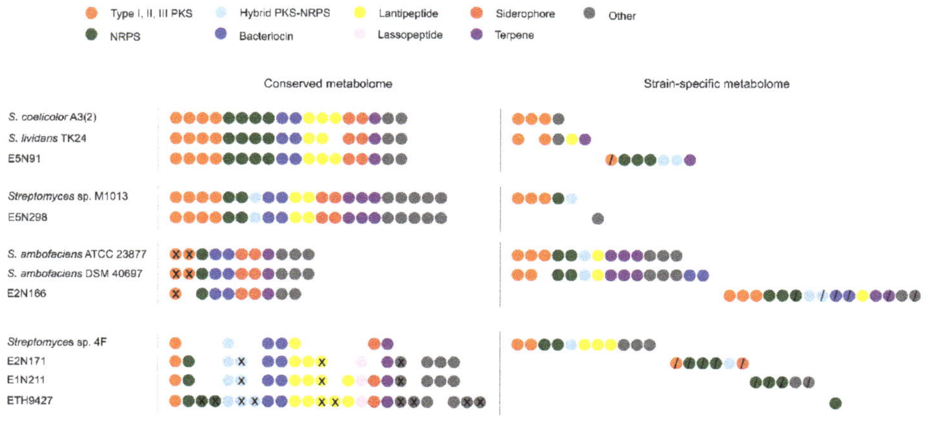

Figure 4. Conserved and strain-specific biosynthetic gene clusters. Within each clade, conserved and strain-specific gene clusters (circles) were identified. Original chromosomal organization is not considered, and clusters with the same position are identical. Colors represent the major classes of specialized metabolites: Type I, II, and III PKS (orange), NRPS (green), hybrid PKS–NRPS (light blue), bacteriocin (dark blue), lantipeptide (yellow), lassopeptide (pink), siderophore (red), terpene (purple), and others that include other KS pathways and less common metabolites such as phenazines, butyrolactones, and indoles (grey). Unknown clusters are indicated with "/" and duplicated clusters with "X".

2.4. Strain-Specific Biosynthetic Gene Clusters

Other gene clusters exist in each of the six examined strains beyond the core and conserved metabolome, the strain-specific SMBGCs. These were determined as gene clusters that are exclusive to a given strain and absent from its phylogenetically closest relatives, based on the nucleotide identity of putative biosynthetic genes. The putative products of these strain-specific biosynthetic pathways comprise all the major classes of specialized metabolites (Figure 4). Each strain harbors at least one strain-specific gene cluster. The fewest strain-specific gene clusters were encountered in strains E5N298 and ETH9427, with only one each. Strains E5N91 and E5N298, which are taxonomically close to their clade's counterparts, dedicate almost opposite amounts of their specialized metabolome to strain-specific products (21% and 3%, respectively), perhaps as a result of horizontal gene transfer events and reflecting different levels of adaptation [44,45]. Conversely, other strains show a significantly higher number of strain-specific gene clusters. Strains E2N171 and E2N166 harbor six and 15 strain-specific gene clusters that represent 21% and 48% of their specialized metabolome, respectively.

The preliminary results of genome alignment and sequence comparison of the relatively close E2N166 and *S. ambofaciens* strains reveals large chromosome reorganizations events, such as a pericentral inversion, and most of the acquired strain-specific gene clusters are unexpectedly located in the central region of the chromosome. Moreover, genome alignment and comparison of the ETH9427 strain with its closest related strain *Streptomyces* sp. 4F and the reference species *S. ambofaciens* ATCC 23477 shows a relatively conserved chromosome organization, particularly between the first two strains (Figure 5). The identified SMBGCs also reflect the shared chromosomal regions, with seven gene clusters shared between ETH9427 and *Streptomyces* sp. 4F, most of which are also maintaining relative positions, and only three gene clusters between ETH9427 and *S. ambofaciens* ATCC 23477 (Figure 5), supporting the notion that the distribution of SMBGCs is coherent with phylogenetic relationships [46].

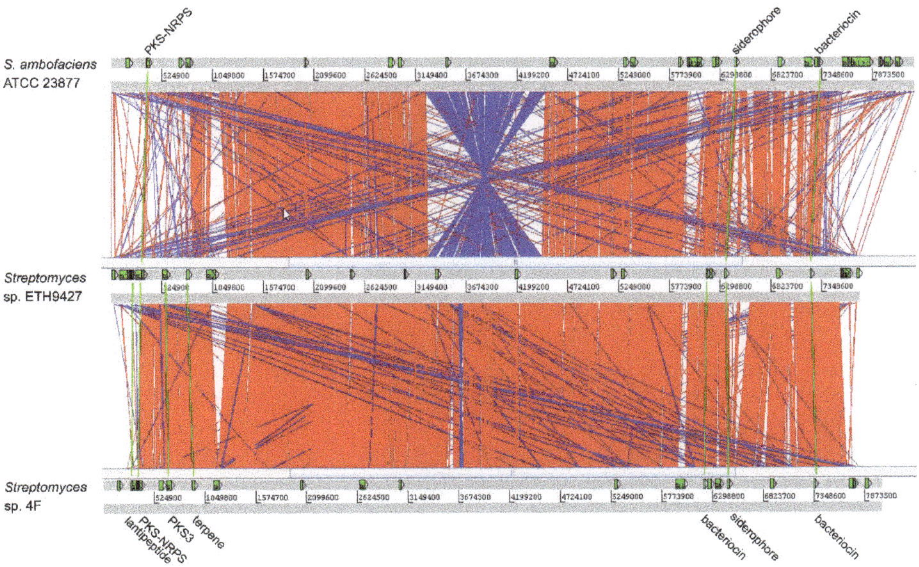

Figure 5. Genome alignment and specialized metabolome comparison of *Streptomyces* sp. ETH9427. Alignment was performed with the Artemis Comparison Tool software (ACT) using an identity threshold of 77% and a score threshold of 500. Synteny regions are represented by red lines, inversion events are represented in blue, and breaks in synteny are seen as blank spaces. SMBGCs are marked in the horizontal panels (green arrows), and conserved gene clusters are linked (green lines).

While some of the identified strain-specific gene clusters have orthologues in other species, four out of the six strains that are used in this work (strains E5N91, E2N166, E2N171, and E1N211) have between one and six unknown biosynthetic gene clusters. These represent 3% to 19% of the metabolome, leading to a total of 16 SMBGCs that could possibly lead to novel chemistry. The strain-specific gene clusters identified in E2N171 and E1N211 (six NRPS, a type I PKS and a siderophore-like and another gene cluster classified as "other") are of particular interest, as they can conceivably be responsible for the antimicrobial activities previously observed in Benhadj et al., namely against nosocomial infections responsible bacterial strains of *Escherichia coli*, *Streptococcus aureus* and *Pseudomonas aeruginosa*. These bioactivities are exclusive to E2N171 and E1N211 amongst the entire collection, and are absent from strains such as E5N298; therefore, the identified strain-specific SMBGCs in these two strains are good candidates for further detailed characterization [13].

3. Materials and Methods

3.1. Strains and DNA Extraction

Five strains from a previously built *Streptomyces* collection were used in this work [13]. In brief, isolates were obtained from water samples of Lake Fetzara (Annaba, Algeria) on different agar media supplemented with antifungals. Additionally, a strain from the ETH collection (Eidgenosische Technische Hochschule Kultursammlungen, Zurich, Switzerland), *Streptomyces* sp. ETH9427 (accession numbers CP029624, CP029625, CP029626), was also analyzed. Genomic DNA was extracted from liquid cultures using a standard "*salting out*" protocol [47].

3.2. Whole-Genome Sequencing and Annotation

Draft genomes were constructed by assembling reads from an Illumina Genome Analyzer (Illumina, San Diego, CA, USA), with pair-ends (c.300 bp) and/or mate-pairs (c.8 kb) DNA fragments libraries, and were processed as described in Thibessard et al. [14]. Sequences were trimmed to eliminate adaptor sequences, and contigs with a size smaller than 1000 pb were eliminated. Draft genomes of the strains E5N91, E2N166, E2N171, E1N211, and E5N298 were submitted into the NCBI GenBank database and given the following accession numbers: RAIE00000000, RAIF00000000, RAIH00000000, RAII00000000, and RAIG00000000, respectively. Whenever possible, sequence assembly and contig ordering were achieved using the CLC Main Workbench (Qiagen, Hilden, Germany), and coding sequence prediction and annotation were automatically performed using the NCBI Prokaryotic Genome Annotation Pipeline (PGAP) [48]. Genome alignments were performed with Artemis Comparison Tool (ACT) [49]. Average nucleotide identity (ANI) was determined using the JSpeciesWS server with the BLAST algorithm (http://jspecies.ribohost.com/jspeciesws/) [19].

3.3. Phylogenetic Analysis and Biosynthetic Gene Cluster Identification

The phylogenetic relationships between strains were reconstructed with an MLST analysis [50], using DNA sequences from *atpB*, *gyrB*, *recA*, *rpoB*, and *trpB* genes. Several reference *Streptomyces* species were included in the analysis: *S. ambofaciens* ATCC 23877 (accession number CP012382.1), *S. ambofaciens* DSM 40697 (accession number CP012949.1), *S. coelicolor* A3(2) (accession number NC_003888.3), *S. lividans* TK24 (accession number CP009124.1), *Streptomyces* sp. M1013 (accession number MQUH00000000.1), and *Streptomyces* sp. 4F (accession number CP013142.1). The actinobacteria *Corynebacterium glutamicum* ATCC 13032 (accession number BA000036.3) was used to root the phylogenetic tree. Nucleotide sequence alignments were performed with SeaView software [51] using the muscle algorithm, and a concatenated gene sequence (20592 nt in total) was used to construct a neighbor-joining phylogenetic tree with Kimura's 2-parameter distance correction and 100 bootstrap replicates with MEGA7 software [52]. Overall, there was a good separation and statistical support for most of the branches in the tree. Biosynthetic gene clusters were predicted using the antiSMASH 3.0 server (https://antismash.secondarymetabolites.org/#!/start) without ClusterFinder algorithm [26].

4. Conclusions

The efforts for *Streptomyces* environmental strains genome sequencing continue to show the genus SMBGC's diversity. Strain de-replication of isolates can be achieved using multi-gene sequence comparison instead of the traditional 16S rRNA, and the analysis of isolated strains reveals a pool for novel specialized metabolites in taxonomically closely related strains. Encountered gene cluster diversity supports phylogenetic relationships, where closely related strains share a higher number of SMBGCs. Horizontal gene transfer as a source of genetic diversity does not distort phylogenetic examination, as these are rare events, and the presence or absence of a SMBGC is not enough to blur phylogenetic relationships. Strain-level analysis has already been demonstrated to lead to the identification of new biosynthetic pathways. Though the strain-specific gene clusters identified here require further characterization, they nonetheless reflect that chemical richness can be found at this level, and imply that the deep sequencing of strains belonging to same species will continue to produce originality.

Author Contributions: Conceptualization, C.M.V., P.L. and B.A.; methodology, C.M.V.; validation, P.L. and B.A.; formal analysis, C.M.V.; investigation, C.M.V., L.H. and J.-N.L.; resources, M.H., D.K.-G.; data curation, C.M.V., P.L., A.T. and O.L.; writing—original draft preparation, C.M.V.; writing—review and editing, B.A. and P.L.; visualization, all authors; supervision, P.L. and B.A.; project administration, B.A.; funding acquisition, B.A. and P.L.

Funding: This work was founded by the French National Research Agency through two programs (ANR Streptoflux ANR-07-BLAN-0096, ANR MiGenIs ANR-13-BSV6-0006), the French National Institute for Agricultural

Research (INRA), Région Lorraine and the Laboratory Excellence ARBRE (ANR-11-LABX-0002-01). CMV was also supported by the AgreenSkillsPlus Program (FP7-609398.0000) and the Région Grand Est.

Acknowledgments: The authors acknowledge the support of the DynAMic laboratory by the Impact Biomolecules project of the "Lorraine Université d'Excellence (Investissements d'avenir—ANR)".

Conflicts of Interest: The authors declare no conflict of interest. The funders had no role in the design of the study; in the collection, analyses, or interpretation of data; in the writing of the manuscript, or in the decision to publish the results.

References

1. Bérdy, J. Bioactive microbial metabolites. *J. Antibiot.* **2005**, *58*, 1–26. [CrossRef] [PubMed]
2. Worrall, J.A.R.; Vijgenboom, E. Copper mining in *Streptomyces*: Enzymes, natural products and development. *Nat. Prod. Rep.* **2010**, *27*, 742–756. [CrossRef] [PubMed]
3. Newman, D.J.; Cragg, G.M. Natural Products as Sources of New Drugs from 1981 to 2014. *J. Nat. Prod.* **2016**, *79*, 629–661. [CrossRef] [PubMed]
4. Corre, C.; Challis, G.L. New natural product biosynthetic chemistry discovered by genome mining. *Nat. Prod. Rep.* **2009**, *26*, 977–986. [CrossRef] [PubMed]
5. Lewis, K. Platforms for antibiotic discovery. *Nat. Rev. Drug. Discov.* **2013**, *12*, 371–387. [CrossRef] [PubMed]
6. Rebets, Y.; Kormanec, J.; Luzhetskyy, A.; Bernaerts, K.; Anné, J. Cloning and Expression of Metagenomic DNA in *Streptomyces* lividans and Subsequent Fermentation for Optimized Production. *Methods Mol. Biol.* **2017**, *1539*, 99–144. [CrossRef] [PubMed]
7. Bentley, S.D.; Chater, K.F.; Cerdeño-Tárraga, A.-M.; Challis, G.L.; Thomson, N.R.; James, K.D.; Harris, D.E.; Quail, M.A.; Kieser, H.; Harper, D.; et al. Complete genome sequence of the model actinomycete *Streptomyces* coelicolor A3(2). *Nature* **2002**, *417*, 141–147. [CrossRef] [PubMed]
8. Medema, M.H.; Fischbach, M.A. Computational approaches to natural product discovery. *Nat. Chem. Biol.* **2015**, *11*, 639–648. [CrossRef] [PubMed]
9. Nett, M.; Ikeda, H.; Moore, B.S. Genomic basis for natural product biosynthetic diversity in the actinomycetes. *Nat. Prod. Rep.* **2009**, *26*, 1362–1384. [CrossRef] [PubMed]
10. Seipke, R.F. Strain-level diversity of secondary metabolism in *Streptomyces* albus. *PLoS ONE* **2015**, *10*, e0116457. [CrossRef] [PubMed]
11. Antony-Babu, S.; Stien, D.; Eparvier, V.; Parrot, D.; Tomasi, S.; Suzuki, M.T. Multiple *Streptomyces* species with distinct secondary metabolomes have identical 16S rRNA gene sequences. *Sci. Rep.* **2017**, *7*, 11089. [CrossRef] [PubMed]
12. Ziemert, N.; Lechner, A.; Wietz, M.; Millán-Aguiñaga, N.; Chavarria, K.L.; Jensen, P.R. Diversity and evolution of secondary metabolism in the marine actinomycete genus Salinispora. *Proc. Natl. Acad. Sci. USA* **2014**, *111*, E1130–E1139. [CrossRef] [PubMed]
13. Benhadj, M.; Gacemi-Kirane, D.; Toussaint, M.; Hotel, L.; Bontemps, C.; Duval, R.E.; Aigle, B.; Leblond, P. Diversity and antimicrobial activities of *Streptomyces* isolates from Fetzara Lake, north eastern Algeria. *Ann. Biol. Clin.* **2018**, *76*, 81–95. [CrossRef] [PubMed]
14. Thibessard, A.; Vicente, C.M.; Bertrand, C.; Aigle, B.; Leblond, P. Draft genome of *Streptomyces* sp. ETH9427, whole genome shotgun sequence. *Genome Announc.* **2018**, in press.
15. Rückert, C.; Albersmeier, A.; Busche, T.; Jaenicke, S.; Winkler, A.; Friðjónsson, Ó.H.; Hreggviðsson, G.Ó.; Lambert, C.; Badcock, D.; Bernaerts, K.; et al. Complete genome sequence of *Streptomyces* lividans TK24. *J. Biotechnol.* **2015**, *199*, 21–22. [CrossRef] [PubMed]
16. Haas, D.; Gerbaud, C.; Sahin, N.; Pernodet, J.-L.; Lautru, S. Draft Genome Sequence of *Streptomyces* sp. M1013, a Close Relative of Streptomyces ambofaciens and *Streptomyces* coelicolor. *Genome Announc.* **2017**, *5*. [CrossRef] [PubMed]
17. Thibessard, A.; Haas, D.; Gerbaud, C.; Aigle, B.; Lautru, S.; Pernodet, J.-L.; Leblond, P. Complete genome sequence of *Streptomyces* ambofaciens ATCC 23877, the spiramycin producer. *J. Biotechnol.* **2015**, *214*, 117–118. [CrossRef] [PubMed]
18. Thibessard, A.; Leblond, P. Complete Genome Sequence of *Streptomyces* ambofaciens DSM 40697, a Paradigm for Genome Plasticity Studies. *Genome Announc.* **2016**, *4*. [CrossRef] [PubMed]

19. Richter, M.; Rosselló-Móra, R.; Oliver Glöckner, F.; Peplies, J. JSpeciesWS: A web server for prokaryotic species circumscription based on pairwise genome comparison. *Bioinformatics* **2016**, *32*, 929–931. [CrossRef] [PubMed]
20. Rosselló-Móra, R.; Amann, R. Past and future species definitions for Bacteria and Archaea. *Syst. Appl. Microbiol.* **2015**, *38*, 209–216. [CrossRef] [PubMed]
21. Alam, M.T.; Merlo, M.E.; Takano, E.; Breitling, R. Genome-based phylogenetic analysis of *Streptomyces* and its relatives. *Mol. Phylogenet. Evol.* **2010**, *54*, 763–772. [CrossRef] [PubMed]
22. Rong, X.; Liu, N.; Ruan, J.; Huang, Y. Multilocus sequence analysis of *Streptomyces* griseus isolates delineating intraspecific diversity in terms of both taxonomy and biosynthetic potential. *Antonie. Van. Leeuwenhoek* **2010**, *98*, 237–248. [CrossRef] [PubMed]
23. Rong, X.; Huang, Y. Taxonomic evaluation of the *Streptomyces* hygroscopicus clade using multilocus sequence analysis and DNA-DNA hybridization, validating the MLSA scheme for systematics of the whole genus. *Syst. Appl. Microbiol.* **2012**, *35*, 7–18. [CrossRef] [PubMed]
24. Labeda, D.P.; Dunlap, C.A.; Rong, X.; Huang, Y.; Doroghazi, J.R.; Ju, K.-S.; Metcalf, W.W. Phylogenetic relationships in the family Streptomycetaceae using multi-locus sequence analysis. *Antonie Van Leeuwenhoek* **2017**, *110*, 563–583. [CrossRef] [PubMed]
25. Jiménez, G.; Urdiain, M.; Cifuentes, A.; López-López, A.; Blanch, A.R.; Tamames, J.; Kämpfer, P.; Kolstø, A.-B.; Ramón, D.; Martínez, J.F.; et al. Description of Bacillus toyonensis sp. nov., a novel species of the Bacillus cereus group, and pairwise genome comparisons of the species of the group by means of ANI calculations. *Syst. Appl. Microbiol.* **2013**, *36*, 383–391. [CrossRef] [PubMed]
26. Weber, T.; Blin, K.; Duddela, S.; Krug, D.; Kim, H.U.; Bruccoleri, R.; Lee, S.Y.; Fischbach, M.A.; Müller, R.; Wohlleben, W.; et al. AntiSMASH 3.0-a comprehensive resource for the genome mining of biosynthetic gene clusters. *Nucleic Acids Res.* **2015**, *43*, W237–W243. [CrossRef] [PubMed]
27. Bernan, V.; Filpula, D.; Herber, W.; Bibb, M.; Katz, E. The nucleotide sequence of the tyrosinase gene from Streptomyces antibioticus and characterization of the gene product. *Gene* **1985**, *37*, 101–110. [CrossRef]
28. Malin, G.; Lapidot, A. Induction of synthesis of tetrahydropyrimidine derivatives in *Streptomyces* strains and their effect on Escherichia coli in response to osmotic and heat stress. *J. Bacteriol.* **1996**, *178*, 385–395. [CrossRef] [PubMed]
29. Davis, N.K.; Chater, K.F. Spore colour in Streptomyces coelicolor A3(2) involves the developmentally regulated synthesis of a compound biosynthetically related to polyketide antibiotics. *Mol. Microbiol.* **1990**, *4*, 1679–1691. [CrossRef] [PubMed]
30. Neilands, J.B. Microbial iron compounds. *Annu. Rev. Biochem.* **1981**, *50*, 715–731. [CrossRef] [PubMed]
31. Gürtler, H.; Pedersen, R.; Anthoni, U.; Christophersen, C.; Nielsen, P.H.; Wellington, E.M.; Pedersen, C.; Bock, K. Albaflavenone, a sesquiterpene ketone with a zizaene skeleton produced by a streptomycete with a new rope morphology. *J. Antibiot.* **1994**, *47*, 434–439. [CrossRef] [PubMed]
32. Poralla, K.; Muth, G.; Härtner, T. Hopanoids are formed during transition from substrate to aerial hyphae in Streptomyces coelicolor A3(2). *FEMS Microbiol. Lett.* **2000**, *189*, 93–95. [CrossRef] [PubMed]
33. Krügel, H.; Krubasik, P.; Weber, K.; Saluz, H.P.; Sandmann, G. Functional analysis of genes from Streptomyces griseus involved in the synthesis of isorenieratene, a carotenoid with aromatic end groups, revealed a novel type of carotenoid desaturase. *Biochim. Biophys. Acta* **1999**, *1439*, 57–64. [CrossRef]
34. Gerber, N.N.; Lechevalier, H.A. Geosmin, an earthly-smelling substance isolated from actinomycetes. *Appl. Microbiol.* **1965**, *13*, 935–938. [PubMed]
35. Choudoir, M.J.; Pepe-Ranney, C.; Buckley, D.H. Diversification of Secondary Metabolite Biosynthetic Gene Clusters Coincides with Lineage Divergence in Streptomyces. *Antibiotics* **2018**, *7*, 12. [CrossRef]
36. Lautru, S.; Deeth, R.J.; Bailey, L.M.; Challis, G.L. Discovery of a new peptide natural product by Streptomyces coelicolor genome mining. *Nat. Chem. Biol.* **2005**, *1*, 265–269. [CrossRef] [PubMed]
37. Hesketh, A.; Kock, H.; Mootien, S.; Bibb, M. The role of absC, a novel regulatory gene for secondary metabolism, in zinc-dependent antibiotic production in Streptomyces coelicolor A3(2). *Mol. Microbiol.* **2009**, *74*, 1427–1444. [CrossRef] [PubMed]
38. Chong, P.P.; Podmore, S.M.; Kieser, H.M.; Redenbach, M.; Turgay, K.; Marahiel, M.; Hopwood, D.A.; Smith, C.P. Physical identification of a chromosomal locus encoding biosynthetic genes for the lipopeptide calcium-dependent antibiotic (CDA) of *Streptomyces* coelicolor A3(2). *Microbiology* **1998**, *144 Pt 1*, 193–199. [CrossRef] [PubMed]

39. Takano, E.; Kinoshita, H.; Mersinias, V.; Bucca, G.; Hotchkiss, G.; Nihira, T.; Smith, C.P.; Bibb, M.; Wohlleben, W.; Chater, K. A bacterial hormone (the SCB1) directly controls the expression of a pathway-specific regulatory gene in the cryptic type I polyketide biosynthetic gene cluster of *Streptomyces* coelicolor. *Mol. Microbiol.* **2005**, *56*, 465–479. [CrossRef] [PubMed]
40. Wright, L.F.; Hopwood, D.A. Actinorhodin is a chromosomally-determined antibiotic in *Streptomyces* coelicolar A3(2). *J. Gen. Microbiol.* **1976**, *96*, 289–297. [CrossRef] [PubMed]
41. Rudd, B.A.; Hopwood, D.A. A pigmented mycelial antibiotic in *Streptomyces* coelicolor: Control by a chromosomal gene cluster. *J. Gen. Microbiol.* **1980**, *119*, 333–340. [CrossRef] [PubMed]
42. Aigle, B.; Pang, X.; Decaris, B.; Leblond, P. Involvement of AlpV, a new member of the *Streptomyces* antibiotic regulatory protein family, in regulation of the duplicated type II polyketide synthase alp gene cluster in *Streptomyces* ambofaciens. *J. Bacteriol.* **2005**, *187*, 2491–2500. [CrossRef] [PubMed]
43. Aigle, B.; Lautru, S.; Spiteller, D.; Dickschat, J.S.; Challis, G.L.; Leblond, P.; Pernodet, J.-L. Genome mining of *Streptomyces* ambofaciens. *J. Ind. Microbiol. Biotechnol.* **2014**, *41*, 251–263. [CrossRef] [PubMed]
44. Jensen, P.R.; Williams, P.G.; Oh, D.-C.; Zeigler, L.; Fenical, W. Species-specific secondary metabolite production in marine actinomycetes of the genus Salinispora. *Appl. Environ. Microbiol.* **2007**, *73*, 1146–1152. [CrossRef] [PubMed]
45. Penn, K.; Jenkins, C.; Nett, M.; Udwary, D.W.; Gontang, E.A.; McGlinchey, R.P.; Foster, B.; Lapidus, A.; Podell, S.; Allen, E.E.; et al. Genomic islands link secondary metabolism to functional adaptation in marine actinobacteria. *ISME J.* **2009**, *3*, 1193–1203. [CrossRef] [PubMed]
46. Joynt, R.; Seipke, R.F. A phylogenetic and evolutionary analysis of antimycin biosynthesis. *Microbiology* **2018**, *164*, 28–39. [CrossRef] [PubMed]
47. Kieser, T.; Bibb, M.; Buttner, M.; Chater, K.; Hopwood, D.A. *Practical Streptomyces Genetics*; The John Innes Foundation: Norwich, UK, 2000.
48. Tatusova, T.; DiCuccio, M.; Badretdin, A.; Chetvernin, V.; Nawrocki, E.P.; Zaslavsky, L.; Lomsadze, A.; Pruitt, K.D.; Borodovsky, M.; Ostell, J. NCBI prokaryotic genome annotation pipeline. *Nucleic. Acids. Res.* **2016**, *44*, 6614–6624. [CrossRef] [PubMed]
49. Carver, T.; Berriman, M.; Tivey, A.; Patel, C.; Böhme, U.; Barrell, B.G.; Parkhill, J.; Rajandream, M.-A. Artemis and ACT: Viewing, annotating and comparing sequences stored in a relational database. *Bioinformatics* **2008**, *24*, 2672–2676. [CrossRef] [PubMed]
50. Guo, Y.; Zheng, W.; Rong, X.; Huang, Y. A multilocus phylogeny of the *Streptomyces* griseus 16S rRNA gene clade: Use of multilocus sequence analysis for streptomycete systematics. *Int. J. Syst. Evol. Microbiol.* **2008**, *58*, 149–159. [CrossRef] [PubMed]
51. Gouy, M.; Guindon, S.; Gascuel, O. SeaView version 4: A multiplatform graphical user interface for sequence alignment and phylogenetic tree building. *Mol. Biol. Evol.* **2010**, *27*, 221–224. [CrossRef] [PubMed]
52. Tamura, K.; Peterson, D.; Peterson, N.; Stecher, G.; Nei, M.; Kumar, S. MEGA5: Molecular evolutionary genetics analysis using maximum likelihood, evolutionary distance, and maximum parsimony methods. *Mol. Biol. Evol.* **2011**, *28*, 2731–2739. [CrossRef] [PubMed]

antibiotics

MDPI

Article

Natural Products from Actinobacteria Associated with Fungus-Growing Termites

René Benndorf [1], Huijuan Guo [1], Elisabeth Sommerwerk [1], Christiane Weigel [1], Maria Garcia-Altares [1], Karin Martin [1], Haofu Hu [2], Michelle Küfner [1], Z. Wilhelm de Beer [3], Michael Poulsen [2] and Christine Beemelmanns [1,*]

1 Leibniz Institute for Natural Product Research and Infection Biology—Hans-Knöll-Institute, Beutenbergstraße 11a, 07745 Jena, Germany; rene.benndorf@hki-jena.de (R.B.); huijuan.guo@hki-jena.de (H.G.); e.sommerwerk@gmail.com (E.S.); christiane.weigel@leibniz-hki.de (C.W.); maria.garcia-altares@leibniz-hki.de (M.G.-A.); karin.martin@leibniz-hki.de (K.M.); michelle.kuefner@gmx.de (M.K.)

2 Section for Ecology and Evolution, Department of Biology, University of Copenhagen, 2100 Copenhagen East, Denmark; Haofu.Hu@bio.ku.dk (H.H.); MPoulsen@bio.ku.dk (M.P.)

3 Department of Microbiology and Plant Pathology, Forestry and Agriculture Biotechnology Institute, University of Pretoria, Pretoria 0001, South Africa; wilhelm.debeer@fabi.up.ac.za

* Correspondence: christine.beemelmanns@hki-jena.de; Tel.: +49-3641-532-1525

Received: 13 August 2018; Accepted: 3 September 2018; Published: 13 September 2018

check for updates

Abstract: The chemical analysis of insect-associated Actinobacteria has attracted the interest of natural product chemists in the past years as bacterial-produced metabolites are sought to be crucial for sustaining and protecting the insect host. The objective of our study was to evaluate the phylogeny and bioprospecting of Actinobacteria associated with fungus-growing termites. We characterized 97 Actinobacteria from the gut, exoskeleton, and fungus garden (comb) of the fungus-growing termite *Macrotermes natalensis* and used two different bioassays to assess their general antimicrobial activity. We selected two strains for chemical analysis and investigated the culture broth of the axenic strains and fungus-actinobacterium co-cultures. From these studies, we identified the previously-reported PKS-derived barceloneic acid A and the PKS-derived rubterolones. Analysis of culture broth yielded a new dichlorinated diketopiperazine derivative and two new tetracyclic lanthipeptides, named rubrominins A and B. The discussed natural products highlight that insect-associated Actinobacteria are highly prolific natural product producers yielding important chemical scaffolds urgently needed for future drug development programs.

Keywords: actinobacteria; symbiosis; secondary metabolites; drug discovery; chemical ecology

1. Introduction

Historically, natural products of microbial origin have been a rich source of drug-like lead structures and until today, almost 35% of all drugs are based on structures of naturally occurring small molecules [1–3]. Despite this prevalence, natural product chemistry has faced declining enthusiasm and dwindling investments for decades as bioactivity-guided screening programs resulted mostly in the rediscovery of already known compounds. The low success rates of industrial antibiotic drug discovery programs worldwide resulted in todays' eminent lack of new antibiotic drug leads. At the same time, increasing numbers of multiresistant human-pathogenic microbes causing non-treatable infections in clinics are reported. These eminent health threats led to the recent realization that new natural product derived scaffolds are urgently needed to combat the life threating infections caused by multidrug resistant pathogens.

The revolutionary developments in genome sequencing and analytical technologies in the last decade have dramatically changed the field of natural product discovery [4]. Particularly, ecology-driven natural product discovery approaches including the chemical analyses of symbiotic microorganisms, in combination with omics-based dereplication strategies, have become highly efficient approaches to identify new natural products with unique chemical scaffolds and bioactivities [5–8]. Most notably, the analysis of insect-microbe symbioses, and more specifically insect-Actinobacteria interactions, have been the focus of a series of recent natural product discovery studies, as bacterial symbionts are required to communicate with the host or participate in host defense using small molecules. The importance of defensive secondary metabolites in insect-Actinobacteria symbiosis is evident, as exemplified in firebugs (Pyrrhocoridae: *Pyrrhocoris apterus*) [9] or the European beewolf (Crabronidae: *Philanthus*) that harbors antibiotic-producing *Streptomyces* in their antennae to help protect wasp larvae from fungal infections [10]. Similarly, fungus-growing ants (*Attini* species) carry symbiotic *Pseudonocardia* that help protect the ants' fungal gardens against specialized parasites [11–13]. Insect-associated Actinobacteria have also been reported from other insects, including Ambrosia beetles [14], dung beetles (Scarabaeidae: *Copris tripartitus*) [15,16], and fungus-growing termites [17,18].

We have recently focused efforts on the chemical analyses of the delicate interplay between fungus-growing termites (Termitidae: Macrotermitinae), their fungal mutualist *Termitomyces* (Basidiomycota: Agaricales: Lyophyllaceae), and bacteria residing within termite guts and fungus gardens (fungus combs). Fungus-growing termites cultivate the fungal mutualist in subterranean monoculture fungus gardens as their main food source [19]. The maintenance of such a monoculture in a nutritionally-rich environment is expected to make the fungal garden prone to exploitation by competitors and disease, such as mites, nematodes, and co-occurring fungi. In addition to antimicrobial and behavioral defense mechanisms of the termites themselves [20], it has been hypothesized that bacteria are employed as defensive symbionts [17,21].

Using bacteria-fungus interaction assays, we have demonstrated that Actinobacteria associated with *Macrotermes natalensis* secrete secondary metabolite mixtures that are active against co-occurring fungi and the weed fungus *Pseudoxylaria* sp. (Ascomycota: Xylariales: Xylariaceae) [22]. Subsequent analysis of single species resulted in the isolation of a new geldanamycin derivative, named natalamycin (**4**), from *Streptomyces* sp. M56 (Figure 1) [23]. In a follow-up study, activity-based analysis of *Amycolatopsis* sp. M39 identified several new macrolactams named macrotermycins (macrotermycin A, **5**) [24], and comparative genome and metabolomic analysis of *Streptomyces* sp. M41 yielded, amongst others, the novel depsipeptide dentigerumycin B (**2**) [25]. Recently, activity and NMR-guided analysis of *Streptomyces* sp. RB1 led to the isolation of termisoflavones A–C (termisoflavone A, **1**) [26], and co-cultivation studies of *Actinomadura* sp. RB29 yielded a new group of tropolone derivatives named rubterolones (e.g., rubterolone D, **6**) [27].

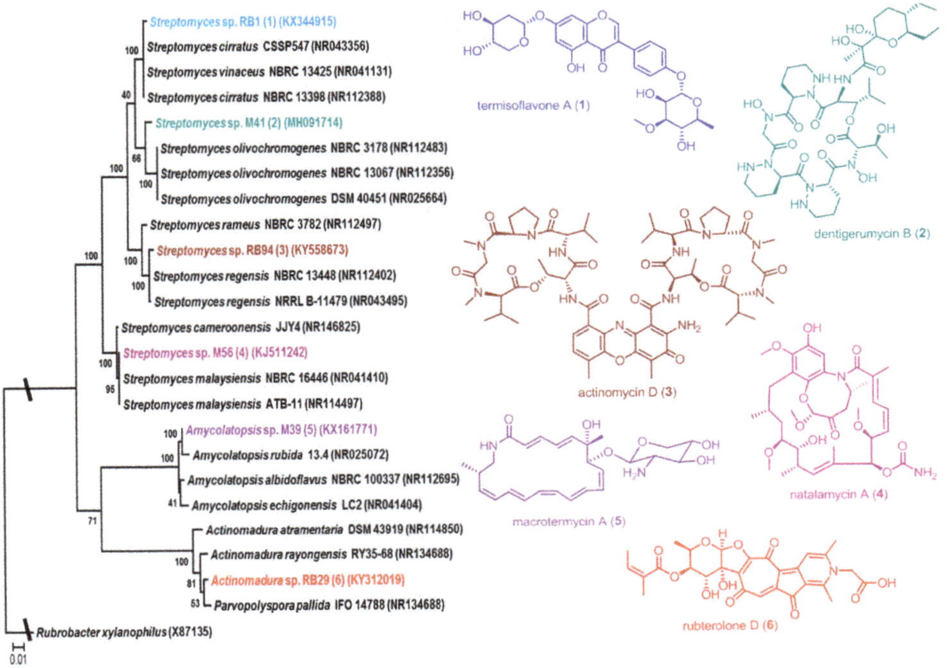

Figure 1. Phylogenetic placement Actinobacteria that have previously been reported from fungus-growing termites and isolated natural products: termisoflavone A (**1**) from *Streptomyces* sp. RB1, dentigerumycin B (**2**) from *Streptomyces* sp. M41, actinomycin D (**3**) from *Streptomyces* sp. RB94, natalamycin A (**4**) from *Streptomyces* sp. M56, macrotermycin A (**5**) from *Amycolatopsis* sp. M39, and rubterolone D (**6**) from *Actinomadura* sp. RB29.

Here, we present a comprehensive phylogenetic and bioactivity survey of Actinobacteria associated with the fungus-growing termite *M. natalensis*. In summary, our study shows that bioprospecting for secondary metabolites in Actinobacteria associated with this termite species leads to the identification of novel natural with unique chemical scaffolds and provides a foundation for gaining a better understanding of the general sanitary role of Actinobacteria in fungus-growing termites.

2. Results

2.1. Phylogenetic Diversity

To assess the culturable actinobacterial diversity, we chose three different sample origins (fresh fungus comb material and the termite worker exoskeleton and gut content) from eleven different *M. natalensis* termite colonies collected in South Africa (Table S1). We focused on the isolation of Actinobacteria capable of living on cellulose or chitin as a sole C-source (Figure 2), as these bacterial isolates are likely adapted to living within the cellulose-rich comb material [28]. Actinobacteria (97) with unique morphotypes were isolated from the termite gut (68), termite abdomen (13), and fungus comb material (16) (Table 1, Table S2). Subsequent phylogenetic analysis of all isolates using 16S rRNA sequencing was conducted, showing that the characterized isolates did not form a monophyletic group (Figure 3, Figure S1) but were interspersed within the Actinobacterium phylum [17,29]. Interestingly, 73 isolates belonged to the genus *Streptomyces* and covered most of the reported phylogenetic diversity of this widespread genus. The remaining 24 isolates belonged to 12 genera within the Actinobacteria

(Table 1). For species delineation a threshold of <98.65% sequence similarity was applied revealing seven putative new Actinobacteria species (Figure 4, Table S3) [30,31].

Figure 2. (**A**) Royal chamber of *Macrotermes natalensis* containing the queen, the king, and workers. (**B**) Fungus comb with a major and a minor soldier. (**C**) A plate exemplifying the diversity of culturable bacteria that can be isolated from the gut of a fungus-growing termite worker.

Table 1. The number of Actinobacteria isolated within this study and by Visser et al. [17], and their origin of isolation by genera and family.

Genus	Family	This Study: Origin of Isolation (Number of Isolates)			Visser et al. [17] Origin of Isolation (Number of Isolates)	
		Termite Gut	Termite Exoskeleton	Fungus Comb	Termite Exoskeleton	Fungus Comb
Streptomyces	Streptomycetaceae	46	14	13	13	2
Kitasatospora	Streptomycetaceae	0	0	0	2	0
Actinomadura	Thermomonosporaceae	4	1	0	1	0
Leifsonia	Microbacteriaceae	3	0	0	0	0
Curtobacterium	Microbacteriaceae	1	0	0	0	0
Arthrobacter	Micrococcaceae	0	1	0	0	0
Micromonospora	Micromonosporaceae	3	0	0	1	1
Nocardia	Nocardiaceae	3	0	0	0	0
Aeromicrobium	Nocardioidaceae	1	0	0	0	0
Cellulosimicrobium	Promicromonosporaceae	1	0	0	0	0
Mycobacterium	Mycobacteriaceae	3	0	0	0	0
Sphaerisporangium	Streptosporangiaceae	1	0	0	0	0
Microbispora	Streptosporangiaceae	1	0	0	0	0
Luteimicrobium	family of order *Micrococcales*	1	0	0	0	0

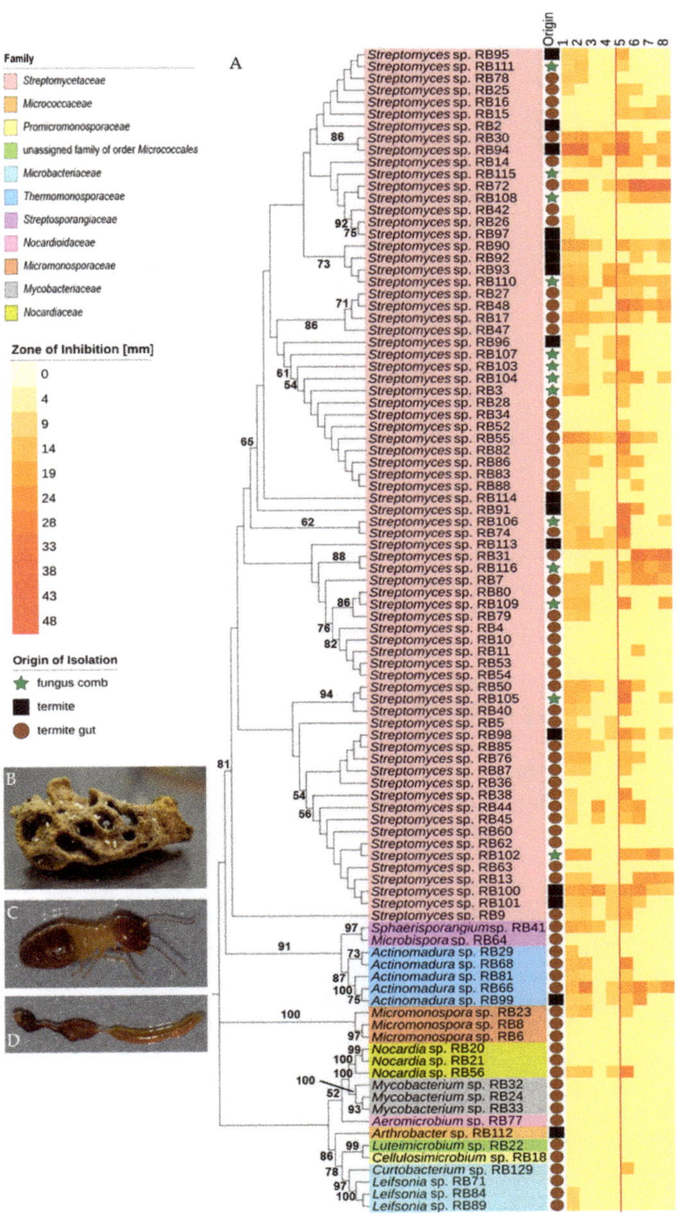

Figure 3. Phylogeny and antimicrobial activity of newly isolated Actinobacteria: (**A**) Phylogenetic analysis based on near full-length 16S rRNA sequences of isolated Actinobacteria including phylogenetic placement to the family level. An unrooted neighbor-joining distance tree is shown with branch values indicating bootstrap support (>50 are given) of 1000 pseudoreplicates, tree was constructed with Mega 7.0 and edited with iTOL v3. Middle: Origin of isolation: termite abdomen: black box, termite gut: brown circle, fungus comb: green star. Right: activity heatmap against test strains *Bacillus subtilis* ATCC 6633 (1), *Staphylococcus aureus* IMET 10760 (2), *Escherichia coli* SG 458 (3), *Pseudomonas aeruginosa* K799/61 (4), *Mycobacterium vaccae* IMET 10670 (5), *Sporobolomyces salmonicolor* SBUG 549 (6), *Candida albicans* BMSY 212 (7) and *Penicillium notatum* JP36 (8). Representative picture of (**B**) fungus comb, (**C**) major worker, (**D**) dissected gut of major worker.

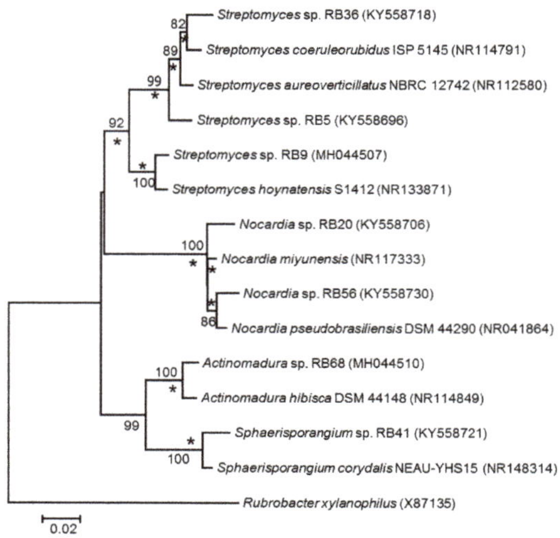

Figure 4. Rooted neighbor-joining tree based on near-complete 16S rRNA gene sequences showing relationship between putative new Actinobacteria strains (based on 98.65% similarity threshold) and closest relatives. Stars (*) indicate branches that were also recovered in maximum-likelihood tree. Only bootstrap values above 50% (based on 1000 pseudoreplicates) are shown. *Rubrobacter xylanophilus* was used as an outgroup. The scale bar indicates 0.02 substitutions per nucleotide position.

2.2. Antimicrobial Activities Against Test Strains

We first assessed the bioactivities of standardized culture extracts (1 mg/mL) of all 97 isolates against a panel of test strains, including human-pathogens (the Gram-positive bacterium *Staphylococcus aureus*, the Gram-negative bacterium *Pseudomonas aeruginosa*, and the fungus *Candida albicans*) (Figure 3, Table S5). While most *Streptomyces* strains produced compounds with antibacterial and antifungal properties, we observed varying intensities of activity. On average, *Streptomyces* extracts (73) inhibited four test strains, but individual strains varied substantially in the number of test strains they suppressed. Several *Streptomyces* extracts showed only antibacterial activity (e.g., *Streptomyces* sp. RB74, *Streptomyces* sp. RB106 and *Streptomyces* sp. RB113), while others exhibited only antifungal activity (e.g., *Streptomyces* sp. RB31). Most notably, *Streptomyces* sp. RB94 and RB100 showed strong antibacterial activity. In contrast, 11 out of 73 *Streptomyces* strains (15%) inhibited none of the tested strains. Isolates belonging to the genera *Actinomadura*, *Sphaerisporangium* and *Micromonospora* inhibited on average three bacterial test strains and only two strains showed antifungal activity (*Actinomadura* sp. RB66 and *Actinomadura* sp. RB99). In contrast, extracts obtained from isolates belonging to *Arthrobacter*, *Cellulosimicrobium*, *Aeromicrobium*, *Luteimicrobium* and *Mycobacterium* showed almost no inhibitory activity against any of the investigated test strains. Thirty-seven extracts with moderate to strong antifungal activity were subjected to a second antifungal assay against ecologically-relevant co-occurring fungi derived from termite nests, four different fungal cultivar isolates (two species) and two entomopathogenic fungi (*Beauvaria bassiana* ST 17960 and *Metarhizium anisopliae* ATCC 24942) [32,33]. As depicted in Figure 5, the majority of culture extracts inhibited none of the representative competing or mutualistic fungi. Extracts of four strains (*Streptomyces* sp. RB7, RB72, RB13, and RB31) did, however, inhibit on average ten of the ecologically-relevant fungal test strains, including the entomopathogenic fungi and *Termitomyces* (Table S6). Interestingly *Streptomyces* sp. RB116 and *Streptomyces* sp. RB31 have the same closest type strains (Table S3) in the blast search in NCBI (https://blast.ncbi.nlm.nih.gov/Blast.cgi, last visit 26th of July, 2018, 00:58 AM) [34], but show variation in antifungal activity (Figure 5, Table S8).

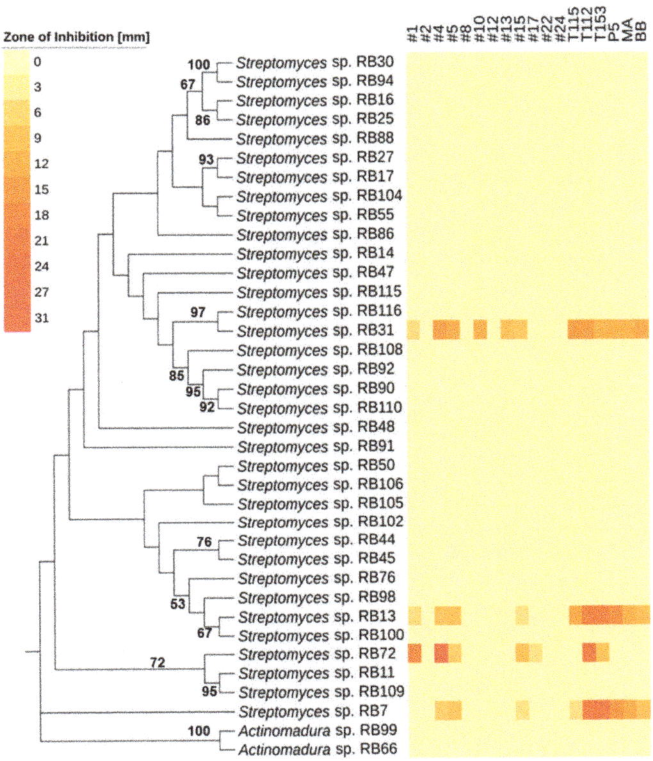

Figure 5. Phylogeny and antifungal activity of isolated Actinobacteria. Left: Phylogenetic analysis based on near full-length 16S rRNA sequences of isolated Actinobacteria. An unrooted maximum-likelihood distance tree is shown with branch values indicating bootstrap support (>50 are given) of 1000 pseudoreplicates; tree was constructed with Mega 7.0 and edited with iTOL v3. Right: Antifungal activity assays of 37 different culture extracts against ecologically relevant fungi: #1: *Cladosporium* sp., #2: *Cladosporium* sp., #4: *Pleosporales* sp., #5: *Fusarium* sp., #8: *Coriolopsis* sp., #10: *Fusarium* sp., #12: *Cunninghamella* sp., #13: *Cladosporium* sp., #15: *Alternaria* sp., #17: *Trichoderma* sp., #22: *Trichoderma* sp., #24: *Hypocrea* sp., T115: *Termitomyces* sp., T112: *Termitomyces* sp., T153: *Termitomyces* sp., P5: *Termitomyces* sp., MA: *Metarhizium anisopliae* ATCC 24942, and BB: *Beauveria bassiana* ST 17960.

2.3. Chemical Analysis

As gene expression of important biosynthetic clusters often is under the control of promoters that respond to certain external factors, cultivation using standard laboratory conditions is likely to lead to limited amount of secondary metabolite production. To activate so-called "cryptic" gene clusters, it is often necessary to mimic natural key stress factors such as limited nutrient availability or the presence of other potentially competing organisms, as exemplified in a recent study of *Amycolatopsis* sp. M39 [24]. We therefore selected *Actinomadura* sp. RB29 and *Streptomyces* sp. RB108, and subjected both strains to co-cultivation set-ups against co-isolated fungi to stimulate the production of cryptic metabolites.

First, we tested strain RB108, as comparative phenotypical and phylogenetic analyses indicated it to be a novel *Streptomyces* species (Figure S1, Table S3). Although standardized extracts of RB108 showed only moderate antifungal activity (Figure 3, Table S5), co-cultivation induced a strong antifungal activity against almost all tested fungal strains (Table S9) and an increased brownish pigment production. Due to the large inhibition zone, we selected co-cultivation set-up RB108/*Pleosporales* sp.

#4 for in-depth analysis. We performed comparative ultra-performance-liquid chromatography-mass spectrometry (UHPLC-MS, Shimadzu, Japan) analysis of concentrated extracts obtained from the zone of inhibition (ZOI). In addition to several other upregulated signals of minor intensity, a distinct UV-detectable metabolite (m/z = 303.1/321.1) was found to be only produced in co-cultivation with *Pleosporales* sp. #4. Mass spectrometry (MS)-guided high pressure liquid chromatography (HPLC, Shimadzu, Japan) purification resulted in the isolation of barceloneic acid A (**7**) (Figure 6), a fungal metabolite acting as a farnesyl-protein transferase inhibitor [35]. We then performed Matrix Assisted Laser Desorption Ionization Imaging MS (MALDI Imaging MS, Bruker Daltonics) to resolve the spatial distribution of **7** and to identify possible antifungal candidates from RB108. However, due to low ionization capacity of this compound class, a clear spatial location of barceloneic acid A (**7**) was not observable. Instead, the detailed analysis of the co-culture assay revealed a cluster of ions with *m/z* values between 2000 Da and 2500 Da (*m/z* 2188.15 and *m/z* 2134.71 being the most intense) that were upregulated and accumulated in the center and on the edges of the colonies facing the fungus *Pleosporales* sp. #4 (Figure 7). This *m/z* range is typical for ribosomally-synthesized peptides (RiPPs) often associated with high antimicrobial activities. We currently hypothesize that *Pleosporales* sp. #4 modulates the interaction with strain RB108 using barceloneic acid A (**7**) and stimulates the production of RiPPs of yet unknown composition.

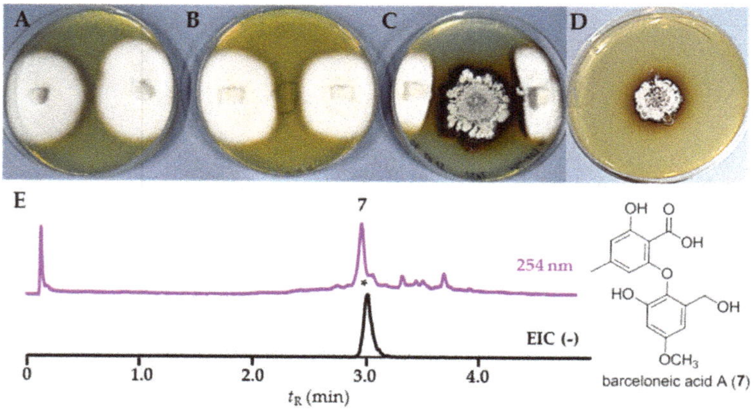

Figure 6. Co-cultivation studies of strain RB108 with *Pleosporales* sp. #4: (**A**) negative control: axenic *Pleosporales* sp. #4; (**B**) positive control: *Pleosporales* sp. #4 in the presence of amphotericin B (8 mg/mL, middle); (**C**) interaction assay of *Streptomyces* sp. RB108 (middle) with *Pleosporales* sp. #4 (edge of the plate); (**D**) axenic *Streptomyces* sp. RB108; (**E**) structure of barceloneic acid A (**7**); and representative UHPLC-MS analysis (254 nm) of zone of inhibition extracts: EIC (–) of barceloneic acid A (**7**) at *m/z* 319.0.

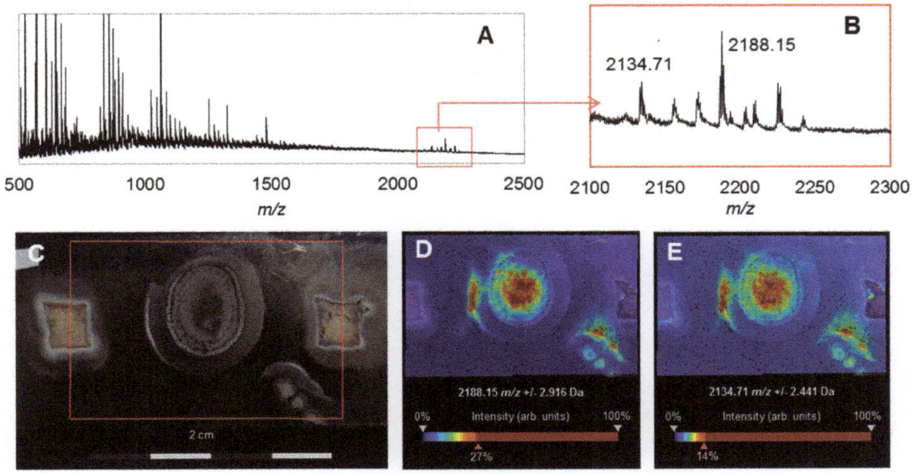

Figure 7. MALDI imaging of a co-cultivation study of *Streptomyces* sp. RB108 with *Pleosporales* sp. #4: (**A**) average MS spectra of the MALDI Imaging MS analysis (TIC normalization) and red rectangle defines the region of the spectra zoomed in; (**B**) extended view of the region from 2100 to 2300 m/z showing the upregulated RiPPs; (**C**) photograph of co-culture set up: RB108 grown in the middle and right lower side; *Pleosporales* sp. #4 grown on agar plugs on the right and left edge of the plate. Area covered by the MALDI Imaging MS analysis defined in red; (**D**) visualization of the most intense peak ion m/z 2188.15 (TIC normalization, weak denoising); and (**E**) visualization of the second most intense peak ion m/z 2134.71 (TIC normalization, weak denoising).

In a second study, we pursued a comparative UHPLC-MS analysis of *Actinomadura* sp. RB29 as co-cultivation against, for example, *Trichoderma* sp. #22, which induced strong antifungal activities (Figure 8). In contrast, no antifungal properties were originally observed in standard culture extracts (Figure 3).

By tracing back the antifungal agents, comparative analysis of culture extracts resulted in the detection of predominant rubterolone derivatives (**6**, **8**, and **9**) [27] and a unique *m/z* signal pattern indicative for a dichlorinated natural product (**11**). Subsequent purification of co-cultivation and liquid culture extracts by semipreparative HPLC led to the isolation of the natural product banegasine (**10**) and the chlorinated natural product cyclo(*N*Me-L-3,5-dichlorotyrosine-Dhb) (**11**) (Table S7) [36].

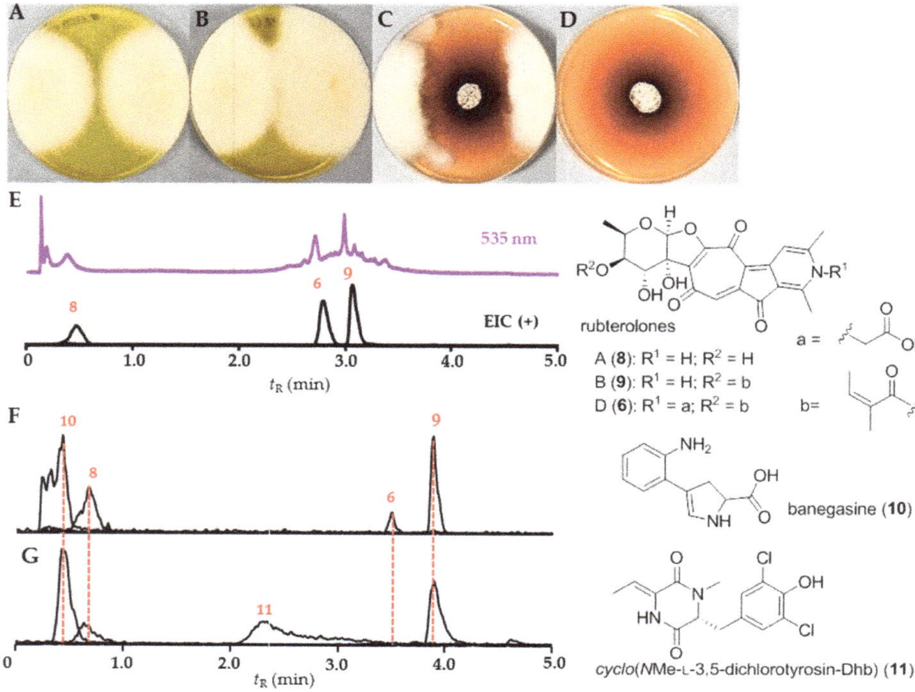

Figure 8. Co-cultivation studies of strain RB29 with *Trichoderma* sp. #22: (**A**) negative control: axenic *Trichoderma* sp. #22; (**B**) positive control: *Trichoderma* sp. #22 in the presence of AmpB; (**C**) co-cultivation of *Actinomadura* sp. RB29 (middle) and *Trichoderma* sp. #22 (edge of the plate); (**D**) axenic culture of *Actinomadura* sp. RB29; (**E**) Representative UHPLC-MS analysis (535 nm) of zone of inhibition extracts: EIC (+) of rubterolone A, B, and D at m/z 414.1, 496.1, and 554.1 and structures of isolated rubterolone A (**8**), B (**9**), and D (**6**); (**F**) production of banegasine (**10**) on ISP2 agar plate and extract with 1% AcOH containing MeOH; and (**G**) production of cyclo(*N*Me-L-3,5-dichlorotyrosine-Dhb) (**11**) in Soya Broth (here represent from XAD16 80% MeOH eluate).

We also analyzed standard culture extracts using LC-HRMS and detected, in addition to several of the previously reported rubterolone derivatives, a RiPP-type MS^2 pattern of two parent ions m/z at 957.8 ($[M + 2H]^{2+}$) and 993.5 ($[M + 2H]^{2+}$) within the SPE 70% and 80% MeOH C18-SPE eluates. Purification by Sephadex LH20 resin and semipreparative HPLC purification resulted in the identification of two tetracyclic lanthipeptides rubrominin A (**12**) and B (**13**).

The molecular formulas of **12** and **13** were established to be $C_{80}H_{115}N_{21}O_{28}S_3$ and $C_{83}H_{120}N_{22}O_{29}S_3$ based on the exact mass analysis of protonated ion **12** (m/z 1914.73938 $[M + H]^+$, calcd. 1914.7 $\Delta = -3.21$ ppm) and **13** (m/z 1985.77539 $[M + H]^+$, calcd. 1985.77141 $\Delta = 2.01$ ppm). The mass difference of 71.03601 between **12** and **13** suggested an additional alanine residue from the *N*-terminus. The MS^2 spectra were recorded, submitted to Global Natural Product Social Molecular Networking (GNPS) and processed by RiPPquest [37]. The combined analysis led to the identification of a putative candidate peptide ([A]CSSTCTSGPFTFACDGTTKG), which is presumably modified by dehydrations and oxidation reactions. However, the estimate p-value of PSM (peptide-spectrum matches) did not allow an assignment of the modified positions.

Genome analysis of *Actinomadura* sp. RB29 using antiSMASH [38] and Blast resulted in the identification of a cinnamycin-homolog gene cluster [39,40], which we named *rum*. It contains 21 open reading frames (ORF) and homolog genes to *cinA*, *cinM*, *cinX*, and *cinorf7* from *Streptomyces cinnamoneous* (Figure S6, Table S11). The candidate peptide sequence ([A]CSSTCTSGPFTFACDGTTKG) is located at the *N*-terminus of RumA. Furthermore, a precursor peptide sequence and an AYA (AXA motif) between the *C*-terminus leader sequence and the core region of RumA was identified; both of which are likely be recognized by a type I signal peptidases of the general secretory (sec) pathway. The RumM sequence shows high homology with other class II LanM (lanthipeptide synthetase) enzymes and is likely to catalyze the dehydration of Thr4, Ser7, Thr11, and Thr18 in the precursor peptide and the subsequent addition/cyclization reaction of three Cys residues to form the three methyllanthionine bridges. In addition, RumX, a homolog Nif11 family protein, might hydroxylate Asp15 of the precursor peptide. The Cinorf7 homolog named RumN is presumably involved in the formation of the cross-link between Lys19 and dehydroalanine7 to form a lysinoalanine bridge. We then performed isotope labeling experiments using L-serine-2,3,3-D_3 and DL-cysteine-3,3-D_2, which showed the incorporation of both, serine and cysteine, into the core peptide. Subsequent Marfey's derivatization of both compounds revealed the partial amino acid composition in the matured peptide as L-Phe, L-Ala, Gly, L-Ser, L-Thr, and L-Pro. ^{1}H NMR spectra and COSY correlation of **12** and **13** recorded in D_2O revealed amino acid-like chemical shifts, and 10% D_2O/90% H_2O deduced the amide chemical shifts from peptide bonds, in particular phenylalanine and alanine spin systems. Due to the dominant rubterolone formation and low production titers of **12** and **13**, full NMR assignment and evaluation of the bioactivities is the topic of current investigations. Based on the acquired genomic data, MS^2-analysis and Marfey analysis, we are confident propose the tetracyclic structure of rubrominins A (**12**) and B (**13**) as depicted in Figure 9.

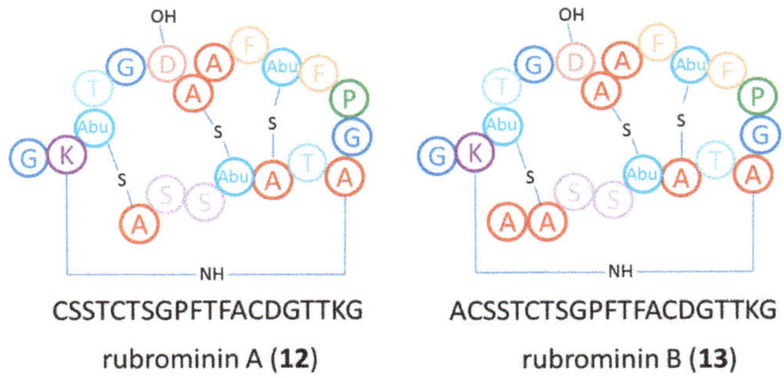

Figure 9. Proposed structures of lanthipeptides rubrominin A (**12**) and B (**13**).

3. Discussion

3.1. Phylogenetic and Ecological Relevance of the Actinobacteria

Termites forage for a broad range of organic material to manure the mutualistic food fungus and this substrate is expected to contain a broad diversity of fungal and bacterial isolates. After predigestion of the harvested material during a first gut passage [20], the termites deposit the resulting feces as fresh comb material for fungal growth. However, it is a conundrum that active combs lack any signs of the presence of fungal contaminates or diseases. This suggests that the gut and comb environment, including the microbial community, provides effective defenses against any incoming potentially parasitic and competitive species. To test the hypothesis that Actinobacteria provide have the potential to be a line of defense against invading fungal species, and to identify novel antimicrobial metabolites produced by these bacteria, we investigated the culturable actinobacterial diversity of guts, combs, and exoskeletons of the fungus-growing termite species *M. natalensis*. Here, we acknowledge the fact that due to the limited numbers of tested culture conditions, we miss "unculturable" and extremophilic members.

Overall, Actinobacteria were found throughout all samples and colonies of *M. natalensis* with 97 representatives, covering twelve genera, ten families, and two orders of the known actinobacterial diversity [29]. We found a dominance of diverse *Streptomyces* species (75%) and a low abundances of other genera, such as *Actinomadura, Microbispora, Micromonospora*, and *Nocardia* consistent with previous findings [17,29,41]. Here, it is interesting to note that *Actinomadura* are frequently isolated from soil [42] and from other social insects such as bee hives [43], similar to members of the genera *Microbispora, Micromonospora*, and *Nocardia* [44,45].

Based on the threshold of 98.65% sequence similarity of the 16S rRNA gene for species delimination, we defined seven isolates as putative novel species (Figure 4), and which are currently investigated for their physiological properties and biosynthetic potential.

The dominant isolation rates of *Streptomyces* from all biological samples and the lack of phylogenetic specificity between Actinobacteria genera and the different biological samples (exoskeleton, gut fluids, and fungus comb), may suggest that Actinobacteria are transient microbes. They are presumably taken up as spores or vegetative mycelium present within digested soil particles and then incorporated as part of the fecal deposits within the comb material; a similar strategy allows the introduction and propagation of the fungal mutualist *Termitomyces*.

In particular, the high isolation rate of Actinobacteria from anaerobic gut fluids was intriguing as previous metagenomic studies indicate that the microbial gut community is dominated by members of Firmicutes, Bacteroidetes, Spirochaetes, Proteobacteria, and Synergistetes, with the most abundant genera being anaerobic or microaerophilic, such as *Alistipes, Treponema, Desulfovibrio, Paludibacter*, and a member of the Synergistaceae [45]. 16S rRNA sequences affiliated with Actinobacteria appear to account for only a minor component of the gut bacterial microbiota. Similar results were obtained in a related study of the fungus-growing termite *Odontotermes formosanus*, with four phylogenetic groups, Firmicutes, the Bacteroidetes/Chlorobi group, Proteobacteria, and Actinobacteria dominating [46]. Considering the anaerobic or microaerophilic environment of the gut compartments, it is likely that only a small fraction of Actinobacteria are actively growing (Table S10) [47,48], and most of our isolates might originate from germination of spores present within the gut fluids [49].

In contrast to gut microbial communities, taxonomic analyses of the comb microbiota of different termite genera showed a clear shift in microbial composition to a more dynamic microbiota of about 33 different phyla, with Firmicutes, Bacteroidetes, Proteobacteria, and Actinobacteria (47 families of four different classes) as the most abundant phyla [50]. It has been hypothesized that this shift would allow for a second microbe-assisted aerobic decomposition, detoxification of plant substrates, and defense against invading and potentially non-beneficial microbes. In particular, Actinobacteria have a multitude of enzymatic capabilities to break down polysaccharides (cellulose, chitin, xylan, and agar) [29,41,51], and to detoxify microbial metabolites, metabolic capacities that most likely contribute to the optimal growth conditions for *Termitomyces*.

3.2. Bioactivities and Natural Products

Using two standardized bioassays, we explored the antimicrobial activities of associated Actinobacteria. More than three quarters of the actinobacterial isolates produced compounds with antimicrobial activity against one or more test strains (human-pathogenic), but only four of the generated extracts revealed antifungal activity against fungal garden weeds, in addition to inhibitory activity against the fungal cultivar. Here, we acknowledge the possibility that metabolite secretion of isolated bacterial strains is strongly depended on the culture environment. We also noticed that even very closely related strains showed a high variability amongst activities. In subsequent actinobacterium-fungus co-cultivation studies we showed that metabolite production was stimulated in the presence of a fungal species and that strains which previously secreted no antifungal metabolites were stimulated and produced compounds with strong antifungal activity. We decided to analyze two co-culture case studies in detail. In the first case study, we analyzed the interaction zone of a co-culture between *Streptomyces* sp. RB108 and *Pleosporales* sp. #4 using LC-MS and MALDI Imaging to visualize the production of potential "cryptic" metabolites. UHPLC-MS-based analysis of the co-culture between *Streptomyces* sp. RB108 and *Pleosporales* sp. #4 revealed that the fungal metabolite barceloneic acid A (**7**), a known farnesyl-protein transferase inhibitor, was strongly upregulated. Although barceloneic acid A did not reveal any antimicrobial activity, it is likely that the molecule modulates the bacterium-fungal interactions using yet unknown mechanisms. We then used a MALDI imaging approach to identify the origin of the antifungal activities and found an increased production of RiPP-like metabolites, which are known for antimicrobial activities and their detailed structural analysis is topic of recent investigations. Upregulations of RiPP-like metabolites having different *m/z* ranges were also observed in other co-cultivation studies; revealing a glimpse into the plethora of metabolites present within complex multipartner interactions.

In a second case study, we analyzed the metabolome of *Actinomadura* sp. RB 29 in more detail, as growth studies on different media and co-culture assay induced strong metabolomic shifts and inducible antifungal activity. Comparative analysis revealed that in addition to the previously reported rubterolones, the natural product banegasine (**10**) and dichlorinated diketopiperazine derivative **11** were produced. Banegasine (**10**) displayed moderate antimicrobial activities against Gram-positive bacteria, including *Mycobacteria*, and is known to potentiate the antimicrobial activity of, for example, pyrrolnitrin [52]. Although the dichlorinated diketopiperazine derivative **11** has been previously reported as part of a screening library, its origin has been undisclosed [36]. In general, diketopiperazine derivatives are common secondary metabolites from bacteria and fungi and the combination of natural and modified amino acids produces diverse structural and bioactivity diversity. The structure of compound **11** is unique as it contains two modified amino acids. First, it contains a 2,3-dehydro-2-aminobutyric acid (Dhb), which presumably originates from the dehydration of threonine [53], and is frequently found in bioactive natural products family like nonribosomal peptides [54] and lanthipeptides [55]. Secondly, it contains a 3,5-dichlorotyrosine moiety, which is a building block of several natural products like chloropeptin [56] from *Streptomyces lavendulae*, and cyclo(13,15-dichloro-L-Pro-L-Tyr) from fungi *Leptoxyphium* sp. In addition, the modified amino acid has been detected in cuticles from several insect species, where they might play important roles in the sclerotization process [57]. A similar diketopiperazine *cyclo*(L-*N*-MePhe-Dhb) was identified from *Streptomyces globisporus*, exhibiting interesting morphogenic and biosynthetic regulator effects [58]. Additionally, we also identified two ribosomally synthesized and post-translationally modified peptides, named rubrominins A (**12**) and B (**13**). The post-translational modification reactions include the dehydration of Ser residue to dehydroalanine and a cyclization step that includes the addition of Cys residues to the dehydrated Ser residues yielding the lanthionine and thioether cross-link. The resulting polycyclic peptides, named lanthipeptides, have constrained conformations that often confer their biological activities. Due to the very low production titers of rubrominins A (**12**) and B (**13**) in co-cultures and axenic cultures, their biological activities have not yet been elucidated and is a topic of current investigations.

Overall, the chemical analysis of *Actinomadura* sp. RB 29 revealed growth-condition dependent metabolite production, and although the origin of the antifungal activity within co-cultures remains to be fully elucidated, the identified natural products (rubterolones, banegasine a chlorinated diketopiperazine derivative **11**, and two lanthipeptides **12** and **13**) exhibit interesting chemical features that are key in many pharmacologically important compound classes.

4. Materials and Methods

General procedures: NMR measurements were performed on a Bruker AVANCE III 500 MHz and 600 MHz spectrometer, equipped with a Bruker Cryoplatform. The chemical shifts are reported in parts per million (ppm) relative to the solvent residual peak of DMSO-d_6 (^{1}H: 2.50 ppm, quintet; ^{13}C: 39.52 ppm, heptet). LC-ESI-HRMS measurements were carried out on an Accela UPLC system (Thermo Scientific) coupled with an Accucore C18 column (100 mm × 2.1 mm, particle size 2.6 μm) combined with a Q-Exactive mass spectrometer (Thermo Scientific) equipped with an electrospray ion

(ESI) source. UHPLC-MS measurements were performed on a Shimadzu LCMS-2020 system equipped with single quadrupole mass spectrometer using a Phenomenex Kinetex C18 column (50 mm × 2.1 mm, particle size 1.7 µm, pore diameter 100 Å). The column oven was set to 40 °C; scan range of MS was set to m/z 150–2000 with a scan speed of 10,000 u/s and event time of 0.25 s under positive and negative mode. The DL temperature was set to 250 °C with an interface temperature of 350 °C and a heat block temperature of 400 °C. The nebulizing gas flow was set to 1.5 L/min and dry gas flow to 15 L/min. Semipreparative HPLC was performed on a Shimadzu HPLC system using a Phenomenex Luna C18(2) 250 mm × 10 mm column (particle size 5 µm, pore diameter 100 Å). IR spectra were recorded on an FT/IR-4100 ATR spectrometer (JASCO). Optical rotations were recorded in methanol on a P-1020 polarimeter (JASCO). Solid phase extraction was carried out using Chromabond C18ec cartridges filled with 2 g and 10 g of octadecyl-modified silica gel (Macherey-Nagel, Düren, Germany). Open column chromatography was performed on Sephadex LH20 (GE Healthcare, Hamburg, Deutschland). Chemicals: Methanol and acetonitrile LC-MS grade (VWR International GmbH, Dresden); water for analytical and preparative HPLC (Millipore, Darmstadt, Germany); formic acid (Carl Roth, Karlsruhe, Germany); acetonitrile (VWR as LC-MS grade); media ingredients (Carl Roth, Karlsruhe, Germany).

Sample collections and isolation procedures: Biological material (soldiers, workers, and fungus comb) was collected from eleven *M. natalensis* nests (stored in 50% glycerol) and one *M. natalensis* and one *Odontotermes* sp. nest for transcriptomic analysis (Table S1) (stored in RNAlater@, Sigma Aldrich, St. Louis, MO, USA). Samples were kept on ice immediately after collection and stored at −80 °C within one day. Frozen termite workers (gut and cuticle) and fresh fungus comb material were used for the isolation of Actinobacteria and each sample was processed separately. First, termites and fungus comb samples were individually washed with ddH_2O (250 µL) and the wash water collected separately for subsequent isolation procedure. Then, major termite workers were surface sterilized with 70% ethanol and washed in sterile Ringer solution (7.5 g/L NaCl, 0.35 g/L KCl, 0.21 g/L $CaCl_2$). Termites were dissected using sterile, fine tipped forceps and intact guts were immediately removed and stored in 500 µL PBS on ice until further use (5 guts per sample). Dissected guts were crushed using a sterile pestle and a series of dilution (up to 10^{-6} in PBS) was produced. Bacteria from each sample were isolated by plating 100 µL of each dilution series (10^{-4}–10^{-6}) on two different selective low-nutrient media: chitin and microcrystalline medium supplemented with 0.05 g/L cycloheximide (Table 2) [17]. Isolates with Actinobacteria-like morphology were transferred to the nutrient-richer medium ISP2 and subcultured. A total of 68 isolates were obtained from gut compartment, 16 isolates from termite cuticle, and 13 isolates from fungus comb (Table S2).

Table 2. Media compositions used for initial isolations and subsequent growth assays and large-scale cultivation.

Medium (Abbreviation)	Content Per L
Potato Dextrose Broth (PDB)	26.5 g potato extract glucose (6.5 g potato extract, 20 g glucose)
Potato Dextrose Agar (PDA)	26.5 g potato extract glucose, 20.0 g agar
ISP2 Broth	4.0 g yeast extract, 10.0 g malt extract, 4.0 g glucose
ISP2 Agar	4.0 g yeast extract, 10.0 g malt extract, 4.0 g glucose, 20.0 g agar
Chitin Agar	4.0 g chitin, 0.7 g K_2HPO_4, 0.3 g KH_2PO_4, 0.57 g $MgSO_4{\cdot}7H_2O$, 0.01 g $FeSO_4{\cdot}7H_2O$, 0.0018 g $ZnSO_4{\cdot}7H_2O$, 0.0016 g $MnCl_2{\cdot}4H_2O$
Minimal Media	2.0 g Na-acetate, 2.0 g NH_4Cl 0.7 g K_2HPO_4, 0.3 g KH_2PO_4, 0.57 g $MgSO_4{\cdot}7H_2O$, 0.01 g $FeSO_4{\cdot}7H_2O$, 0.0018 g $ZnSO_4{\cdot}7H_2O$, 0.0016 g $MnCl_2{\cdot}4H_2O$
MC Agar	5.0 g microcrystalline cellulose, 20.0 g agar
Soya Broth	20.0 g soya flour, 20.0 g glucose, 5.0 g NaCl, 3.0 g $CaCO_3$, 0.25 mL desmophen

DNA extraction, PCR amplification, pairwise sequence similarities and phylogenetic analysis: Actinobacteria were grown in nutrient-rich ISP2 broth for 5 to 7 days at 30 °C (150 rpm). Cells were harvested, and genomic DNA was extracted using the GenJet Genomic DNA Purification Kit (Thermo Scientific, Waltham, MA, USA, #K0721) following the manufacturer's instructions with slight changes (lysozyme incubation time 40 min, protein kinase K treatment 40 min). DNA was quantified spectrophotometrically using Nanodrop (Thermo Scientific, Waltham, MA, USA). 16S rRNA gene was amplified using the primers pair 27F/1492R [17]. The amplification reaction was prepared in 25 μL final volume containing: 7.25 μL dH_2O, 5.0 μL HF buffer, 5.0 μL of each primer (2.5 μM), 0.5 μL dNTPs (10 μM), and 0.25 μL Phusion High Fidelity DNA Polymerase (New England Biolabs). PCR was performed with the conditions: 98 °C for 38 s, followed by 32 cycles of 98 °C for 30 s, 52 °C for 45 s, 72 °C for 1 min 20 s, and a final extension of 72 °C for 8 min. PCR products were visualized by agarose gel electrophoresis. PCR reactions were purified using the PCR Purification Kit (Thermo Scientific, Waltham, MA, USA, #K0702) and sequenced at GATC (Konstanz).

Sequences were checked for purity and mismatches using BioEdit [59]. Forward and reverse sequences of each sequence were assembled with BioEdit and tested for chimeras using DECIPHER [60]. For strains RB9, RB54, RB74, RB85, and RB129 only reverse or forward sequences were generated (Table S3). Resulting sequences were used for a BLASTn search in GenBank using "refseq_rna" database [61]. Pairwise sequence similarities were calculated using the method recommended by Meier-Kolthoff [30] for 16S rRNA gene available via the GGDC web server available at http://ggdc.dsmz.de/. Sequence similarities were calculated for all strains with first three hits (Table S3). A phylogenetic analysis was done with the 16S rRNA sequences (GenBank accession numbers KX344916-KX344918, KY312017-KY312022, KY558669-KY558746, and MH044507-MH044516) and the first hit from the BLASTn search (Figure S1, Table S3). Sequences were aligned with muscle [62] and trimmed using MEGA 7.0.26 [63]. Two different phylogenetic trees were reconstructed with neighbor-joining [64] and maximum likelihood algorithms (Figure S1) [65].

The evolutionary distance model of Kimura or Tamura and Nei was used to generate evolutionary distance matrices for the maximum likelihood [66,67], and neighbor joining algorithm with deletion of complete gaps and missing data. For the maximum-likelihood algorithm, a discrete Gamma distribution was used (+G) and the rate variation model allowed for some sites to be evolutionarily invariable (+I). For the neighbor-joining algorithm rate variation among sites was modeled with a gamma distribution. For all constructed trees the confidence values of nodes were evaluated by bootstrap analysis based on 1000 resamplings [68]. For graphic design, iTOL v3 (https://itol.embl.de/, 31st of July, 2018) was used with the following settings: leaf sorting = none, branch length = ignore, scaling factors: Hor = 0.3, Vert: = 0.8) [69].

Phylogenetic comparison of strain RB108: Near-complete 16S rRNA sequences (1365 bp, GenBank accession number KY558675) were used for a search in NCBI database (reference RNA sequences). The first three hits were *Streptomyces pulveraceus* NBRC 3855, *Streptomyces atratus* NRRL B-16927, and *Streptomyces gelaticus* NRRL B-2928 with an Ident value of 99%. All three hits were phenotypically compared using the Wink compendium [70]. Strain RB108 exhibited a different phenotype compared to the above listed *Streptomyces*. Therefore, the full-length 16S rRNA sequence (1514 bp, GenBank accession number MH828334) was extracted of the genome of strain RB108 and used for comparison. The first three hits of the reference RNA sequence database were *Streptomyces fulvissimus* DSM 40593, *Streptomyces caviscabies* ATCC 51928, and *Streptomyces luridiscabiei* S63. Virtual DDH estimation of strain RB108 and *Streptomyces fulvissimus* DSM 40593 was performed, resulting in a value of 26.20% (23.8–28.7%). According to the DDH threshold of <70% both strains can be regarded as distinct two separate strains [30,31].

Culture extracts: Actinobacteria were cultivated in 25 mL ISP2 or PDB for 4 days at 30 °C at 150 rpm, after 4 days additional 25 mL ISP2 or PDB broth were added and cultivation was continued for another 3 to 5 days. Cultures were centrifuged (6000 rcf, 10 min) and the resulting cell pellets were lysed using MeOH (9 mL). The resulting methanolic cell extracts were combined with the culture supernatant to yield a 20% MeOH culture supernatant. Metabolites from the supernatant were concentrated using an activated (20% MeOH) Chromabond C18ec cartridges filled with 500 mg of octadecyl-modified silica gel (Macherey-Nagel, Düren, Germany). (Unless stated otherwise: %MeOH refers to a mixture of MeOH and dH_2O). Metabolites were eluted using 100% MeOH (5 mL) and 100% acetone (2 mL) and pooled and concentrated in vacuo. The resulting extracts (E) were adjusted with MeOH to 1 mg/mL and used for bioactivity assays.

Antimicrobial activities against test strains: Antimicrobial assays against *Bacillus subtilis* ATCC 6633, *Staphylococcus aureus* IMET 10760, *Escherichia coli* SG 458, *Pseudomonas aeruginosa* K799/61, *Mycobacterium vaccae* IMET 10670, *Sporobolomyces salmonicolor* SBUG 549, *Candida albicans* BMSY 212, and *Penicillium notatum* JP36 were done using the broth dilution method according to the NCCLS (National Committee for Clinical Laboratory Standards) (Table S5).

Antifungal activity assay against co-isolated fungi: All fungal isolates (Table S4) were cultivated on PDA plates for a maximum of six weeks (23 °C) and subcultured by plating mycelium-containing agar pieces (1 cm × 1 cm) onto fresh PDA. To evaluate antifungal activity, a filter paper disk (d = 6 mm) was soaked with 10 µL extract (1 mg/mL) and dried (sterile air flow). Depending on the growth behavior of each fungus, two different assays were applied. Method A (fast-growing fungi), mycelium covered agar pieces were placed in the middle of a PDA plate (standard petri dish 92 mm × 16 mm) and sterile filter paper discs were placed at a distance of 1–2 cm from each agar plug. Method B (slow to medium-fast growing fungi): Fungi were grown in 25 mL PDB for 10 to 14 days at 30 °C (150 rpm) and 500 µL of actively growing culture or a spore solution (*M. anisopliae*) was used to inoculate a PDA plate (standard petri dish 92 mm × 16 mm). Distribution of mycelium or spores on plates was performed using sterile glass beads. Plates were dried and filter discs soaked with equal amounts (10 µL) of extracts were put onto the PDA plate. Plates were checked daily and the diameter of the zone of inhibition (ZOI) (no growth, mycelium free) was recorded as a measure of inhibition (Tables S6 and S8). Amphotericin B (8 mg/mL in DMSO) and cycloheximide (50 mg/mL in MeOH) were used as positive controls, 100% MeOH was used as a negative control. All combinations were prepared in duplicate.

Co-cultivation studies: *Streptomyces* sp. RB108 was grown for 14 days at 30 °C in ISP2, 25 µL of liquid culture were used to inoculate ISP2 and PDA plates centrally. Plates were incubated for 7 days at 30 °C until a clear colony (1 cm in diameter) was apparent. Then, plates were inoculated at the edge of the agar plate with two agar pieces covered with fungal mycelium (Table S9). All combinations were prepared in triplicate. Amphotericin B (8 mg/mL in MeOH) was used as positive control; ddH_2O was used as negative control. Plates were incubated for 10 days at room temperature Plates were checked daily for the formation of a zone of inhibition (ZOI). When a clear, stable ZOI was detectable (normally after 10 days), the ZOI was cut out and extracted with 100% methanol (overnight). Methanol extracts were dried and stored at −20 °C until further use and then subjected to comparative LC-MS and HPLC analysis. Due to significant upregulation of metabolites during co-cultivation, the combination of *Streptomyces* sp. RB108 and fungus #4 was selected for subsequent experiments.

Isolation and structural elucidation of barceloneic acid A (7): For isolation of upregulated metabolites, large scale co-cultivation of *Streptomyces* sp. RB108 and *Pleosporales* sp. #4 was accomplished by inoculation of 30 PDA agar plates (92 mm × 16 mm, 20 mL agar/plate) as described above. Each ZOI was excised from the plate then pooled and extracted with MeOH overnight and solvent removed by evaporation. The resulting crude extract (adjusted to 10% MeOH) was loaded on an activated and equilibrated SPE C18 column and fractionated by step-gradient from 10% MeOH to 100% MeOH (100 mL each). The 50% MeOH eluate was separated by semipreparative HPLC to yield pure compound 7 (m/z 319.19 $[M - H]^-$; 302.95 $[M + H - H_2O]^+$). The molecular formula of barceloneic acid A (**7**) was assigned as $C_{16}H_{16}O_7$ based on ESI-HRMS (m/z 321.0964 $[M + H]^+$, calcd. 321.0969 $\Delta = -1.46$ ppm) (Figure S11). The ^1H NMR analysis (Table S12) revealed 13 protons as sharp signals, suggesting additional three invisible exchangeable protons. Two *meta*-substituted aromatic moieties could be deduced according to the ^1H chemical shifts and multiplicity of protons at δ_H 6.23 ppm (1H, doublet, J = 2.7 Hz, H-11) and δ_H 6.44 ppm (1H, doublet, J = 2.7 Hz, H-13); δ_H 5.95 ppm (1H, broad singlet, H-4) and δ_H 6.24 ppm (1H, broad singlet, H-6), respectively. The methyl group at $\delta_{H3\text{-}8}$ 2.07 ppm/$\delta_{C\text{-}8}$ 21.2 ppm was deduced to attach on aromatic ring based on the HMBC correlations of H_3-8 to C-4/C-5/C-6 and H-4 to C-1/C-3/C-5/C-6/C-8 and H-6 to C-1/C-4/C-7/C-8. The oxygenated methyl moiety at $\delta_{H3\text{-}16}$ 3.68 ppm/$\delta_{C\text{-}16}$ 54.9 ppm was suggested to connect on second aromatic ring based on the HMBC correlation of H_3-16 to C-12. The oxygenated methylene moiety at $\delta_{H\text{-}15}$ 4.51 ppm (2H, doublet, J = 1.2 Hz, H-15)/$\delta_{C\text{-}15}$ 58.3 ppm was deduced to connect on the same aromatic ring based on the HMBC correlations of H_2-15 to C-9/C-13/C-14, H-13 to C-12/C-14/C-15, and H-11 to C-9/C-10/C-12/C-13. Finally, the proposed structure of barceloneic acid A (**7**) matched the literature reported HRMS and 2D NMR data (Table S13) [35]. NMR spectra are shown in Figure S7–S10.

MS Imaging: An indium tin oxide (ITO) coated glass slide was used for Imaging MS (Bruker Daltonics, Billerica, MA, USA) and covered with 1 mL of ISP2 and PDA medium, respectively. The dried agar glass slide was then inoculated with 10 µL of a *Streptomyces* sp. RB108 liquid culture (middle of the cover slide) and incubated for 7 d at 30 °C. Two square agar pieces (0.3 cm × 0.3 cm) covered with mycelium of fungus # 4 were arranged at a distance of 1cm on the cover slide (Figure S2). After an incubation period of 7 days at RT, the slides were dried for an additional half hour next to an open flame and then sprayed with a saturated solution (7 g/L) of universal MALDI matrix (1:1 mixture of 2,5-dihydroxybenzoic acid and α-cyano-4-hydroxy-cinnamic acid; Sigma Aldrich) in acetonitrile HPLC grade, using the automatic system ImagePrep device 2.0 (Bruker Daltonics, Bremen, Germany). The sample was analyzed in an UltrafleXtreme MALDI TOF/TOF (Bruker Daltonics, Bremen, Germany), which was operated in positive reflector mode using flexControl 3.0 (Bruker, Bremen, Germany). The analysis was performed in the 100–3000 and 400–4000 Da ranges, with 40% laser intensity (laser type 3), accumulating 1000 shots by taking 50 random shots at every raster position. Raster width was set at 150 µm. Spectra were processed with baseline subtraction in flexAnalysis 3.3 (Bruker, Bremen, Germany). Processed spectra were uploaded in flexImaging 3.0 for visualization and SCILS Lab 2015b for analysis and representation. Chemical images were obtained after peak alignment on the dataset using Median normalization and weak denoising.

Isolation and structure elucidation of banegasine (**10**): Large-scale cultivation of *Actinomadura* sp. RB29 was performed by inoculation of 100 ISP2 agar plates (standard 150 mm × 20 mm, 40 mL ISP2 agar/plate) at 30 °C for 10 days. Whole agar plates were cut into pieces and extracted twice with 2 L MeOH (1% AcOH) at 4 °C overnight. MeOH/AcOH extracts were filtered and concentrated under reduced pressure. The crude extract was dissolved using 10% MeOH and loaded on an activated and equilibrated SPE C18 column (10 g) and fractionated by step-gradient from 10% MeOH to 100% MeOH (100 mL each). The eluent using 30% MeOH was first purified by Sephadex LH20 using 50% MeOH to obtain subfractions Fr.3.1–Fr.3.6 (20 mL/Fr.). Concentrated violet band Fr.3.6 was further separated by semipreparative HPLC to yield pure banegasine (**10**, 2.0 mg, t_R = 6.74 min) using the following gradient: 0–5 min, 50% B; 5–20 min, 50–100% B; 20–25 min, 100% B (A: dd H_2O + 0.1% formic acid; B: MeOH) with a flow rate of 2.0 mL/min. The molecular formula of banegasine (**10**) was assigned as $C_{11}H_{12}O_2N_2$ based on ESI-HRMS (m/z 205.0972 $[M + H]^+$, calcd. 205.0972 Δ = 0.13 ppm) and λ_{max} 217,.278 nm (MeCN/H_2O/FA) (Figure S13). The ^1H NMR analysis (Table S14) revealed nine protons as sharp signals, suggesting additional three invisible exchangeable protons. The *ortho*-substituted aromatic moiety was deduced from the ^1H chemical shifts and multiplicity of protons at δ_H 7.35 ppm (1H, doublet, J = 8.1 Hz, H-3′); δ_H 7.06 ppm (1H, triplet, J = 7.8 Hz, H-4′); δ_H 6.97 ppm (1H, triplet, J = 7.2 Hz, H-5′); δ_H 7.56 ppm (1H, doublet, J = 7.9 Hz, H-6′). The dihydropyrrole moiety was deduced from the aliphatic protons at δ_H 3.48 ppm (1H, doublet of doublet, J = 8.6, 3.9 Hz, H-2) and δ_H 2.99 ppm (1H, doublet of doublet, J = 15.1, 8.9 Hz, H-3a) and δ_H 3.31 ppm (1H, doublet of doublet, J = 15.1, 3.7 Hz, H-3b), and olefinic proton at δ_H 7.22 ppm (1H, doublet, J = 1.2 Hz, H-5). ^1H NMR spectrum of banegasine is shown in Figure S12.

Isolation and structure elucidation of *cyclo*(*N*Me-L-3,5-dichlorotyrosine-Dhb) (**11**): Large-scale liquid cultivation of *Actinomadura* sp. RB29 was performed in a 50 L fermenter (20 L Soya Broth liquid media, pH 6.8) for 5 days at 28 °C (stirring). The culture supernatant was separated from biomass by separator and collected and loaded onto activated XAD16 resin (1 kg). The resin was first washed by water (2 L) and then eluted by MeOH/H_2O mixture in a step gradient manner, with 1 L 10% MeOH, 30% MeOH, 50% MeOH, 80% MeOH, 100% MeOH, respectively. The corresponding eluates were concentrated under reduced pressure and redissolved into 50% MeOH or 100% MeOH as 5 mg/mL for standard metabolomic LCMS analysis. The interesting ion with m/z at 328.9 ([M+H]$^+$) was redetected from 80% MeOH eluate and allowed us to submit this 80% MeOH eluate onto Sephadex LH20 resin and eluted by 100% MeOH for further purification. Subfractions containing the ion m/z at 328.9 ([M+H]$^+$) were concentrated under reduced pressure and finally purified by semipreparative HPLC (Nucleodur C18 250 mm × 10 mm) to yield compound **11** (2.0 mg, t_R = 13.33 min) using the following gradient: 0–5 min, 30% B; 5–25 min, 30–83% B; 25–30 min, 100% B (A: dd H_2O + 0.1% formic acid; B: MeCN) with a flow rate of 2.0 mL/min. The molecular formula of compound (**11**) was determined as $C_{14}H_{14}O_3N_2Cl_2$ based on ESI-HRMS analysis (m/z 329.04520 [M + H]$^+$, calcd. 329.04542 $\Delta = -0.68$ ppm) and the isotope abundance of chlorine, and further confirmed by the observation of 14 carbon atoms from ^{13}C-NMR and DEPT135 spectra analysis (Figure S15 and S16). Detailed analysis of 1D and 2D NMR spectra (Figure S14, S17–S20) indicated two carbonyl groups, one four-substituted benzyl moiety, one olefinic moiety, one methylene and one methine group, two methyl groups, and one visible OH/NH signal. The presence of two amide groups ($\delta_{C\text{-}1}$ 165.5 ppm/$\delta_{C\text{-}9}$ 159.0 ppm) suggested the amino acids origin. Methylated olefin moiety was deduced from the ^{1}H-^{1}H COSY correlation of H-11 (δ_H 5.41 ppm/δ_C 112.0 ppm) to H_3-12 (δ_H 1.46 ppm/δ_C 10.7 ppm) and the HMBC correlations of H-11 to C-12, and H_3-12 to C-10 and C-11. The further HMBC correlations of H-11 and H_3-12 to C-9 (δ_C 159.0 ppm), OH/NH (δ_H 9.92 ppm) to C-9 and C-10 suggested the possible dehydroaminobutyric acid (dhAbu or Dhb) moiety. Second spin system C-2–C-3 from COSY correlation of H-2 to H_2-3 and the aromatic protons ($\delta_{H\text{-}5}$ 6.94 ppm/$\delta_{C\text{-}5}$ 129.7 ppm), and *N*-methyl group ($\delta_{H3\text{-}8}$ 2.94 ppm/$\delta_{C3\text{-}8}$ 32.1 ppm) were deduced to belong to the substituted tyrosine skeleton based on the deeper observation of HMBC correlations of H-2 to C-1/-3/-4/-8, and H_2-3 to C-1/-2/-4/-5, and H-5 to C-2/-3/-6/-7, and H_3-8 to C-2. Based on the ^{13}C chemical shift compared with the literature [70] and demand of elemental composition, two chlorine atoms were suggested to attach on the C-6 individually, leading to the build-up of *N*-methyl 3,5-dichlorotyrosine moiety. The diketopiperazine structure condensed from Dhb and *N*-methyl 3,5-dichlorotyrosine residue was deduced by the HMBC correlations of *N*H to C-1/-2, and H-2 to C-9, and H_3-8 to C-9. Furthermore this conclusion was confirmed by the observation of MS fragmentation of m/z 329.04520 ([M+H]$^+$) (Figure S21) and prediction by Mass Frontier 7.0 (Thermo, Figure S22). The fragment ion at m/z 301.05118 ($C_{13}H_{15}N_2O_2Cl_2^+$) might originate from the diketopiperazine ring opening and loss of carbonyl group; ion at m/z 218.01393 ($C_9H_{10}NOCl_2^+$) might derive from the diketopiperazine ring opening and loss of carbonyl group and Dhb moiety; and ion at m/z 174.97168 ($C_7H_5OCl_2^+$) might represent the (3,5-dichloro-4-hydroxyphenyl)methylium and the ion at m/z 127.08707 ($C_6H_{11}N_2O^+$) might deduce from the Dhb moiety and *N*-methyl group with rearrangement (e.g., 1-(2-iminobutanoyl)aziridin-1-ium). The NOESY correlation of H_3-12 to NH suggested the configuration of double bond from Dhb moiety as *Z* form. Finally the planar structure of **11** was suggested as *cyclo*(dehydroaminobutyric acid- *N*-methyl 3,5-dichlorotyrosine). Comparison of the specific optical rotation of **11** with reported tyrosine or phenylalanine containing diketopiperazine [71,72], the stereochemistry of *N*-methyl 3,5-dichlorotyrosine moiety (–104°) was deduced to be L-configurated.

cyclo(*N*-Me-L-3,5-dichlorotyrosine-Dhb) (**11**): light yellow solid; $[\alpha]_D^{25}$ –104.0 (c 0.1 *w*/*v*%, MeOH); UV (MeCN/H_2O/FA) λ_{max} 230 nm; IR (ATR) ν_{max} 3857, 3738, 3650, 2925, 2855, 1745, 1683, 1650, 1539, 1512, 1460, 674 cm^{-1}; NMR spectral data, see Table S15; ESI-HRMS [M+H]$^+$ m/z 329.04520 (calcd for $C_{14}H_{15}O_3N_2Cl_2$, 329.04542). A SCIFinder search indicated the commercial availability from Aurora Screening Library (CAS 1798281-02-3) however the natural origin of the compound is unassigned.

Antimicrobial activities of *cyclo*(*N*-Me-L-3,5-dichlorotyrosine-Dhb) against test strains: Antimicrobial assays against *Bacillus subtilis* ATCC 6633, *Staphylococcus aureus* IMET 10760, *Escherichia coli* SG 458, *Pseudomonas aeruginosa* SG137, *Pseudomonas aeruginosa* K799/61, MRSA *Staphylococcus aureus* 134/93, VRSA *Enterococcus faecalis* 1528, *Mycobacterium vaccae* IMET 10670, *Sporobolomyces salmonicolor* SBUG 549, *Candida albicans* BMSY 212, and *Penicillium notatum* JP36 were done using the broth dilution method according to the NCCLS (National Committee for Clinical Laboratory Standards) (Table S7).

Isolation and structure elucidation of rubromidin A (**12**) and B (**13**): Large-scale liquid cultivation was performed as mentioned above. SPE 70% and 80% MeOH C18-SPE eluates, which containing ion with m/z at 957.8 ($[M+2H]^{2+}$) and 993.5 ($[M+2H]^{2+}$).

Those two fractions were concentrated under reduced pressure and pooled and resubmitted to Sephadex LH20 resin and eluted by 100% MeOH for further purification. The subfractions containing ion with m/z at 957.8 ($[M + 2H]^{2+}$) and 993.5 ($[M + 2H]^{2+}$) were concentrated under reduced pressure and finally purified by semipreparative HPLC (Nucleodur C18 250 mm × 10 mm) to yield rubromidin A (**12**, 0.5 mg, t_R = 10.02 min) and rubromidin B (**13**, 0.5 mg, t_R = 10.80 min) using the following gradient: 0–20 min, 30% B (A: dd H_2O + 0.1% formic acid; B: MeCN) with a flow rate of 2.0 mL/min. ESI-HRMS analysis of purified lanthipeptide **12** revealed the protonated molecular ion at m/z 1914.73938, as well as the doubly protonated ion at m/z 957.87408 (Figure S23). The MS^2 spectra at m/z 1914.73938 and 957.87408 under positive mode were recorded and submitted to GNPS and processed by RiPPquest, a tandem mass spectrometry database search tool for identification of microbial RiPPs [73]. The exact mass of protonated ion of compound **13** was assigned with m/z 1985.77539, by the observation of the doubly protonated ion at m/z 993.39203 (Figure S24). The mass difference of 71.03601 between **12** and **13** suggested an additional alanine residue from the *N*-terminus. Similarly, the MS^2 spectra at m/z 1985.77539 and 993.39203 were recorded and submitted to GNPS and processed by RiPPquest, and led to the identification of the candidate peptide (ACSSTCTSGPFTFACDGTTKG), including an additional alanine which might due to the alternative cleavage of RumM.

Marfey's reaction [74]: Compounds **12** and **13** (0.1 mg each) were hydrolyzed separately by 6 N HCl (1.0 mL) at 110 °C for 15 h. Then HCl was removed using a SpeedVac and 20 μL FDAA (1-fluoro-2,4-dinitrophenyl-5-L-alanine amide, 10 mg/mL in acetone) and 100 μL $NaHCO_3$ (1 N aqueous solution) were added. The reaction was heated at 80 °C for 10 min, and the reaction quenched by addition of 50 μL 2 N HCl. L- and D-phenylalanine and L- and D-alanine, glycine, L- and D-serine, L- and D-theorine, and L- and D-proline were converted accordingly. After centrifugation for 10 min at 13,000 rpm, the reaction mixture was analyzed by UHPLC-MS (Figures S29 and S30). Five μL of the reaction mixture were injected and analyzed using the following gradient: 0–1 min, 10% B; 1–7 min, 10–100% B; 7.1–10 min, 100% B (A: dd H_2O with 0.1% formic acid; B: MeCN with 0.1% formic acid) at a flow rate of 0.7 mL/min.

Ser/Cys labeling: *Actinomadura* sp. RB29 was first incubated in 50 mL ISP2 at 30 °C (150 rpm shaking) for one week. Then, biomass was collected by centrifugation (4000 rpm, 10 min, rt), washed twice using autoclaved minimal media and transferred into 100 mL minimal media containing L-serine-2,3,3-D_3 (100 mg) and DL-cystein-3,3-D_2 (100 mg), respectively. The cultures were incubated for one week at 30 °C (150 rpm). The culture broth was collected by filtration and extracted with activated HP20 resin (20 g/L) at 4 °C overnight. The resin was washed by H_2O firstly then eluted by 100% MeOH. The MeOH eluate was concentrated under reduced pressure and resuspended into MeOH for LCMS analysis and ESI-HRMS analysis (Figure S24–27).

Gene cluster identification: The putative biosynthetic gene cluster of rubromidin (*rum*) was predicted using antiSMASH and compared with already described lantipeptide gene cluster *cin* (from *Streptomyces cinnamoneus*) and putative biosynthetic gene cluster of cinnamycin B from *Actinomadura atramentaria* [39,40,75] (Figure S6, Table S11). The peptide sequence of lantibiotic cinnamycin precursor (predicted with antiSMASH) was used for al BLASTp search in GenBank using "refseq protein" or "nonredundant protein sequence" database. A phylogenetic analysis was performed using lantibiotic cinnamycin precursor peptide sequence (GenBank accession number WP 103565569) and LanM lanthipeptide synthetase (GenBank accession number WP 103565568) and related hits from the

BLASTp search. Phylogentic trees are shown in Figures S3 and S4. Comparative sequence alignment of precursor peptide sequence is shown in Figure S5. Sequences were aligned with Muscle [62]. Two different phylogenetic trees were reconstructed with neighbor-joining and maximum-likelihood algorithms [63,64]. The evolutionary distance model of Jones, Taylor, and Thorton [76] was used to generate evolutionary distance matrices for the maximum likelihood and neighbor joining algorithm with deletion of complete gaps and missing data. For the maximum-likelihood algorithm, a discrete Gamma distribution was used (+G). For the neighbor-joining algorithm, the rate variation among sites was modeled with a gamma distribution. For all constructed trees the confidence values of nodes were evaluated by bootstrap analysis based on 1000 resamplings.

5. Conclusions

This study provides an extended taxonomical and chemical analysis of Actinobacteria isolated from the fungus-growing termite *M. natalensis*. Our findings clearly verify that a high diversity of Actinobacteria can be found with these termites, and most notably the termite gut. The high recovery rates from gut fluids suggest that most species survive the relatively short passage and are inoculated together with the fungal mutualists into the fresh fungus comb. Although the high phylogenetic diversity denotes a certain lack of specificity between bacteria and origin of isolation, it simultaneously ensures the presence of diverse chemistry and biochemical capacity, which may be necessary for protection against alien species and the assisted breakdown of complex plant material. It also needs to be noted that antibiotics at higher concentrations might be deployed as weapons against competing microbes; they also have been found to elicit changes in the global bacterial transcription patterns and metabolism at sub-inhibitory concentrations [77,78]. Therefore, it could be speculated that natural products, including antibiotics, serve as signaling molecules between different microorganisms within the termite symbiosis and contribute to the overall unique stability of the system. Consequently, the structural diversity of metabolites produced within the mutualistic agricultural system might provide a net benefit for the fungal mutualist and farming termites; however, documenting the individual compounds and costs and benefits involved in this potential mutualism are still required.

An intriguing feature of studying the defensive symbiosis paradigm is its parallel to human medicine as it both deploys mediating or antagonistic molecules to suppress pathogens. Insect defensive symbioses, such as the fungus-growing termite systems, probably offer the clearest window into antibiotic use in nature, and the presence of natural product factories, such as Actinobacteria, represent an untapped chemical treasure trove of novel chemical scaffolds, a fact that is underlined by the impressive amount of new natural product scaffolds derived from termite-associated bacteria reported in the last decade. However, key to the discovery of new natural products is the ability to mimic the natural environment and the dynamics present within the system. As exemplified in our two case studies, co-culture assays are a first step towards mimicking natural habitats and identifying key secondary metabolites from both bacteria and fungi. However, more efforts are clearly needed to steadily increase the levels of complexity and more holistically describe the metabolic network within the symbiosis.

Supplementary Materials: The following are available online at http://www.mdpi.com/2079-6382/7/3/83/s1: Table S1: Colonies information of *M. natalensis* for Actinobacteria isolation and metatranscriptomic analysis, Table S2: Actinobacteria IDs of isolates from fungus-growing termites, Table S3: Identities of isolated Actinobacteria strains, Table S4: Identities of ecologically-relevant fungal strains used as targets in the bioactivity tests, Table S5: Antimicrobial assay results of extracts of isolated bacteria against eight medically relevant bacteria and fungi, Table S6: Antifungal assay results of extracts of isolated bacteria against ecologically-relevant fungi, Table S7: Antimicrobial assay results of *cyclo*(*N*-Me-L-3,5-dichlorotyrosine-Dhb) (**11**), Table S8: Representative images of antifungal assay of extract against ecologically relevant fungi, Table S9: Representative interactions between *Streptomyces* sp. RB108 and co-cultivated fungi, Table S10: Presence/absence of transcripts identified in gut microbiome metatranscriptome data from major old worker guts of *M. natalensis* Mn156 and Odontotermes sp. Od127, Table S11: Rubromidin biosynthetic protein annotations based on sequence homology, Tables S12–S15: NMR tables, Figure S1: Unrooted Neighbor-joining tree based on near-complete 16S rRNA gene sequences showing relationships between isolated Actinobacteria and closest relatives, Figure S2: MALDI-TOF imaging of co-cultivation of RB108 and *Pleosporales* sp. #4, Figure S3: Neighbor-joining tree based on peptide sequence of precursor peptide sequence (RumA), Figure S4: Neighbor-joining tree based on type 2 lantipeptide synthetase RumM, Figure S5: Comparative sequence alignment of precursor peptide sequence, Figure S6: Comparative gene maps of *rum* and *cin* and proposed cinnamycin B biosynthetic clusters, Figures S7–S30: NMR spectra and LCMS of Marfey's reaction.

Author Contributions: R.B., M.P., and C.B. conceived and designed the experiments; R.B., E.S., H.G., C.W., M.G.-A., K.M., H.H., and M.K. performed the experiments and analyzed the data; C.W., Z.W.d.B., and C.H. contributed reagents/materials/analyses tools; and R.B., M.P., and C.B. wrote the paper.

Funding: Rene Benndorf was funded by the International Leibniz Research School for Microbial and Biomolecular Interactions (ILRS) and Jena School for Microbial Communication (JSMC, DFG). Financial support of the Boehringer Ingelheim Foundation, the Daimler Benz foundation, and the German Research Foundation (CRC 1127 (ChemBioSys) and BE-4799/3-1) to Christine Beemelmanns is greatly acknowledged. This work was performed with financial support from the Villum Kann Rasmussen Foundation Young Investigator Fellowship (VKR10101) to Michael Poulsen.

Acknowledgments: We thank the Oerlemans family (Mookgophong) for permission to sample colonies at their farm.

Conflicts of Interest: The authors declare no conflicts of interest.

References

1. Newman, D.J.; Cragg, G.M. Natural products as sources of new drugs over the 30 years from 1981 to 2010. *J. Nat. Prod.* **2012**, *75*, 311–335. [CrossRef] [PubMed]
2. Mishra, B.B.; Tiwari, V.K. Natural products: An evolving role in future drug discovery. *Eur. J. Med. Chem.* **2011**, *46*, 4769–4807. [CrossRef] [PubMed]
3. Dias, D.A.; Urban, S.; Roessner, U. A historical overview of natural products in drug discovery. *Metabolites* **2012**, *2*, 303–336. [CrossRef] [PubMed]
4. Weber, T.; Kim, H.U. The secondary metabolite bioinformatics portal: Computational tools to facilitate synthetic biology of secondary metabolite production. *Synth. Syst. Biotechnol.* **2016**, *5*, 69–79. [CrossRef] [PubMed]
5. Molloy, E.M.; Hertweck, C. Antimicrobial discovery inspired by ecological interactions. *Curr. Opin. Microbiol.* **2017**, *39*, 121–127. [CrossRef] [PubMed]
6. Machado, H.; Tuttle, R.N.; Jensen, P.R. Omics-based natural product discovery and the lexicon of genome mining. *Curr. Opin. Microbiol.* **2017**, *39*, 136–142. [CrossRef] [PubMed]
7. Beemelmanns, C.; Guo, H.; Rischer, M.; Poulsen, M. Natural products from microbes associated with insects. *Beilstein J. Org. Chem.* **2016**, *12*, 314–327. [CrossRef] [PubMed]
8. Genilloud, O. Actinomycetes: Still a source of novel antibiotics. *Nat. Prod. Rep.* **2017**, *34*, 1203–1232. [CrossRef] [PubMed]
9. Salem, H.; Kreutzer, E.; Sudakaran, S.; Kaltenpoth, M. Actinobacteria as essential symbionts in firebugs and cotton stainers (Hemiptera, Pyrrhocoridae). *Environ. Microbiol.* **2013**, *15*, 1956–1968. [CrossRef] [PubMed]
10. Kaltenpoth, M.; Yildirim, E.; Gürbüz, M.F.; Herzner, G.; Strohm, E. Refining the roots of the beewolf-*Streptomyces* symbiosis: Antennal symbionts in the rare genus *Philanthinus* (Hymenoptera, Crabronidae). *Appl. Environ. Microbiol.* **2012**, *78*, 822–827. [CrossRef] [PubMed]
11. Currie, C.R.; Mueller, U.G.; Malloch, D. The agricultural pathology of ant fungus gardens. *Proc. Natl. Acad. Sci. USA* **1999**, *96*, 7998–8002. [CrossRef] [PubMed]
12. Poulsen, M.; Cafaro, M.J.; Erhardt, D.P.; Little, A.E.; Gerardo, N.M.; Tebbets, B.; Klein, B.S.; Currie, C.R. Variation in *Pseudonocardia* antibiotic defence helps govern parasite-induced morbidity in *Acromyrmex* leaf-cutting ants. *Environ. Microbiol. Rep.* **2010**, *2*, 534–540. [CrossRef] [PubMed]
13. Oh, D.C.; Poulsen, M.; Currie, C.R.; Clardy, J. Dentigerumycin: A bacterial mediator of an ant-fungus symbiosis. *Nat. Chem. Biol.* **2009**, *5*, 391–393. [CrossRef] [PubMed]
14. Peleg, B.; Norris, D.M. Symbiotic interrelationships between microbes and ambrosia beetles. *J. Invertebr. Pathol.* **1972**, *20*, 59–65. [CrossRef]
15. Shukla, S.P.; Sanders, J.G.; Byrne, M.J.; Pierce, N.E. Gut microbiota of dung beetles correspond to dietary specializations of adults and larvae. *Mol. Ecol.* **2016**, *25*, 6092–6106. [CrossRef] [PubMed]

16. Um, S.; Park, S.H.; Kim, J.; Park, H.J.; Ko, K.; Bang, H.S.; Lee, S.K.; Shin, J.; Oh, D.C. Coprisamides A and B, new branched cyclic peptides from a gut bacterium of the Dung Beetle *Copris tripartitus*. *Org. Lett.* **2017**, *5*, 1272–1275. [CrossRef] [PubMed]
17. Visser, A.A.; Nobre, T.; Currie, C.R.; Aanen, D.K.; Poulsen, M. Exploring the potential for actinobacteria as defensive symbionts in fungus-growing termites. *Microb. Ecol.* **2012**, *63*, 975–985. [CrossRef] [PubMed]
18. Kurtböke, D.I.; French, J.R.; Hayes, R.A.; Quinn, R.J. Eco-taxonomic insights into actinomycete symbionts of termites for discovery of novel bioactive compounds. *Adv. Biochem. Eng. Biotechnol.* **2015**, *147*, 111–135. [PubMed]
19. Aanen, D.K.; Eggleton, P.; Rouland-Lefevre, C.; Guldberg-Froslev, T.; Rosendahl, S.; Boomsma, J.J. The evolution of fungus-growing termites and their mutualistic fungal symbionts. *Proc. Natl. Acad. Sci. USA* **2002**, *99*, 14887–14892. [CrossRef] [PubMed]
20. Nobre, T.; Rouland-Lefevre, C.; Aanen, D.K. Comparative biology of fungus cultivation in termites and ants. In *Biology of Termites: A Modern Sysnthesis*; Bignell, D., Roisin, Y.L.N., Eds.; Springer: Dordrecht, The Netherlands, 2010; pp. 193–210.
21. Um, S.; Fraimout, A.; Sapountzis, P.; Oh, D.C.; Poulsen, M. The fungus-growing termite *Macrotermes natalensis* harbors bacillaene-producing *Bacillus* sp. that inhibit potentially antagonistic fungi. *Sci. Rep.* **2013**, *3*, 3250. [CrossRef] [PubMed]
22. Visser, A.A.; Kooij, P.W.; Debets, A.J.M.; Kuyper, T.W.; Aanen, D.K. Pseudoxylaria as stowaway of the fungus-growing termite nest: Interaction asymmetry between *Pseudoxylaria*, *Termitomyces* and free-living relatives. *Fungal Ecol.* **2011**, *4*, 322–332. [CrossRef]
23. Kim, K.H.; Ramadhar, T.R.; Beemelmanns, C.; Cao, S.; Poulsen, M.; Currie, C.R.; Clardy, J. Natalamycin a, an ansamycin from a termite-associated *Streptomyces* sp. *Chem. Sci.* **2014**, *5*, 4333–4338. [CrossRef] [PubMed]
24. Beemelmanns, C.; Ramadhar, T.R.; Kim, K.H.; Klassen, J.L.; Cao, S.; Wyche, T.P.; Hou, Y.; Poulsen, M.; Bugni, T.S.; Currie, C.R.; et al. Macrotermycins A-D, glycosylated macrolactams from a termite-associated *Amycolatopsis* sp. M39. *Org. Lett.* **2017**, *19*, 1000–1003. [CrossRef] [PubMed]
25. Wyche, T.P.; Ruzzini, A.C.; Beemelmanns, C.; Kim, K.H.; Klassen, J.L.; Cao, S.; Poulsen, M.; Bugni, T.S.; Currie, C.R.; Clardy, J. Linear peptides are the major products of a biosynthetic pathway that encodes for cyclic depsipeptides. *Org. Lett.* **2017**, *19*, 1772–1775. [CrossRef] [PubMed]
26. Kang, H.R.; Lee, D.; Benndorf, R.; Jung, W.H.; Beemelmanns, C.; Kang, K.S.; Kim, K.H. Termisoflavones A-C, isoflavonoid glycosides from termite-associated *Streptomyces* sp. RB1. *J. Nat. Prod.* **2016**, *79*, 3072–3078. [CrossRef] [PubMed]
27. Guo, H.; Benndorf, R.; Leichnitz, D.; Klassen, J.L.; Vollmers, J.; Görls, H.; Steinacker, M.; Weigel, C.; Dahse, H.-M.; Kaster, A.K.; et al. Isolation, biosynthesis and chemical modifications of rubterolones A-F: Rare tropolone alkaloids from *Actinomadura* sp. 5–2. *Chem. Eur. J.* **2017**, *23*, 9338–9345. [CrossRef] [PubMed]
28. Da Costa, R.R.; Hu, H.; Pilgaard, B.; Vreeburg, S.M.E.; Schückel, J.; Pedersen, K.S.K.; Kračun, S.K.; Busk, P.K.; Harholt, J.; Sapountzis, P.; et al. Enzyme activities at different stages of plant biomass decomposition in three species of fungus-growing termites. *Appl. Environ. Microbiol.* **2018**, *14*, e01815-17. [CrossRef] [PubMed]
29. Gao, B.; Gupta, R.S. Phylogenetic framework and molecular signatures for the main clades of the phylum Actinobacteria. *Microbiol. Mol. Biol. Rev.* **2012**, *76*, 66–112. [CrossRef] [PubMed]
30. Meier-Kolthoff, J.P.; Goker, M.; Sproer, C.; Klenk, H.P. When should a DDH experiment be mandatory in microbial taxonomy? *Arch. Microbiol.* **2013**, *195*, 413–418. [CrossRef] [PubMed]
31. Kim, M.; Oh, H.S.; Park, S.C.; Chun, J. Towards a taxonomic coherence between average nucleotide identity and 16S rRNA gene sequence similarity for species demarcation of prokaryotes. *Int. J. Syst. Evol. Microbiol.* **2014**, *64*, 346–351. [CrossRef] [PubMed]
32. Wang, C.; Wang, S. Insect pathogenic fungi: Genomics, molecular interactions, and genetic improvements. *Annu. Rev. Entomol.* **2017**, *62*, 73–90. [CrossRef] [PubMed]
33. Sbaraini, N.; Guedes, R.L.; Andreis, F.C.; Junges, Â.; de Morais, G.L.; Vainstein, M.; de Vasconcelos, A.T.; Schrank, A. Secondary metabolite gene clusters in the entomopathogen fungus *Metarhizium anisopliae*: Genome identification and patterns of expression in a cuticle infection model. *BMC Genom.* **2016**, *17*, 736. [CrossRef] [PubMed]
34. BLAST: Basic Local Alignment Search Tool—NCBI-NIH. Available online: https://blast.ncbi.nlm.nih.gov/Blast.cgi (accessed on 26 July 2018).

35. Jayasuriya, H.; Ball, R.G.; Zink, D.L.; Smith, J.L.; Goetz, M.A.; Jenkins, R.G.; Nallin-Omstead, M.; Silverman, K.C.; Bills, G.F.; Lingham, R.B.; et al. Barceloneic acid A, a new farnesyl-protein transferase inhibitor from a *Phoma* species. *J. Nat. Prod.* **1995**, *58*, 986–991. [CrossRef] [PubMed]
36. Aurora Screening Library; Order Number: K13.438.893; CAS Registry Number: 1798281-02-3. Available online: http://www.aurorafinechemicals.com (accessed on 3 July 2018).
37. GNPS: Global Natural Products Social Molecular Networking. Available online: https://gnps.ucsd.edu/ProteoSAFe/static/gnps-splash.jsp (accessed on 29 June 2018).
38. Blin, K.; Wolf, T.; Chevrette, M.G.; Lu, X.; Schwalen, C.J.; Kautsar, S.A.; Duran, H.G.S.; de los Santos, E.L.C.; Kim, H.U.; Nave, M.; et al. AntiSMASH4.0-improvments in chemistry prediction and gene cluster boundary identification. *Nucleic Acids Res.* **2017**, *45*, W36–W41. [CrossRef] [PubMed]
39. Repka, L.M.; Chekan, J.R.; Nair, S.K.; van der Donk, W.A. Mechanistic understanding of lanthipeptide biosynthetic enzymes. *Chem. Rev.* **2017**, *117*, 5457–5520. [CrossRef] [PubMed]
40. Ökesli, A.; Cooper, L.E.; Fogle, E.J.; van der Donk, W.A. Nine post-translational modifications during biosynthesis of cinnamycin. *J. Am. Chem. Soc.* **2011**, *133*, 13753–13760. [CrossRef] [PubMed]
41. Barka, E.A.; Vatsa, P.; Sanchez, L.; Gaveau-Vaillant, N.; Jacquard, C.; Klenk, H.P.; Clement, C.; Ouhdouch, Y.; van Wezel, G.P. Taxonomy, physiology, and natural products of actinobacteria. *Microbiol. Mol. Biol. Rev.* **2016**, *80*, 1–43. [CrossRef] [PubMed]
42. Songsumanus, A.; Kudo, T.; Ohkuma, M.; Phongsopitanun, W.; Tanasupawat, S. *Actinomadura montaniterrae* sp. nov., isolated from mountain soil. *Int. J. Syst. Evol. Microbiol.* **2016**, *66*, 3310–3316. [CrossRef] [PubMed]
43. Promnuan, Y.; Kudo, T.; Ohkuma, M.; Chantawannakul, P. *Actinomadura apis* sp. nov., isolated from a honey bee (*Apis mellifera*) hive, and the reclassification of *Actinomadura cremea* subsp. Rifamycini gauze et al. 1987 as *Actinomadura rifamycini* (gauze et al. 1987) sp. nov., comb. nov. *Int. J. Syst. Evol. Microbiol.* **2011**, *61*, 2271–2277. [CrossRef] [PubMed]
44. Han, C.; Liu, C.; Zhao, J.; Guo, L.; Lu, C.; Li, J.; Jia, F.; Wang, X.; Xiang, W. *Microbispora camponoti* sp. nov., a novel actinomycete isolated from the cuticle of *Camponotus japonicus mayr*. *Antonie Van Leeuwenhoek* **2016**, *109*, 215–223. [CrossRef] [PubMed]
45. Otani, S.; Mikaelyan, A.; Nobre, T.; Hansen, L.H.; Kone, N.A.; Sorensen, S.J.; Aanen, D.K.; Boomsma, J.J.; Brune, A.; Poulsen, M. Identifying the core microbial community in the gut of fungus-growing termites. *Mol. Ecol.* **2014**, *23*, 4631–4644. [CrossRef] [PubMed]
46. Shinzato, N.; Muramatsu, M.; Matsui, T.; Watanabe, Y. Phylogenetic analysis of the gut bacterial microflora of the fungus-growing termite *Odontotermes formosanus*. *Biosci. Biotechnol. Biochem.* **2007**, *71*, 906–915. [CrossRef] [PubMed]
47. Mikaelyan, A.; Dietrich, C.; Köhler, T.; Poulsen, M.; Sillam-Dussès, D.; Brune, A. Diet is the primary determinant of bacterial community structure in the guts of higher termites. *Mol. Ecol.* **2015**, *24*, 5284–5295. [CrossRef] [PubMed]
48. Long, Y.H.; Xie, L.; Liu, N.; Yan, X.; Li, M.H.; Fan, M.Z.; Wang, Q. Comparison of gut-associated and nest-associated microbial communities of a fungus-growing termite (*Odontotermes yunnanensis*). *Insect Sci.* **2010**, *17*, 265–276. [CrossRef]
49. Haofu, H. Gut Bacterial Contributions to Antimicrobial Defense and Biomass Decomposition in Fungus-Farming Termites. Ph.D. Thesis, University of Copenhagen, Copenhagen, Denmark, 2018.
50. Otani, S.; Hansen, L.H.; Sorensen, S.J.; Poulsen, M. Bacterial communities in termite fungus combs are comprised of consistent gut deposits and contributions from the environment. *Microb. Ecol.* **2016**, *71*, 207–220. [CrossRef] [PubMed]
51. Liu, C.; Jiang, Y.; Wang, X.; Chen, D.; Chen, X.; Wang, L.; Han, L.; Huang, X.; Jiang, C. Diversity, antimicrobial activity, and biosynthetic potential of cultivable actinomycetes associated with lichen symbiosis. *Microb. Ecol.* **2017**, *74*, 570–584. [CrossRef] [PubMed]
52. Cain, C.C.; Lee, D.; Waldo Iii, R.H.; Henry, A.T.; Casida, E.J.; Wani, M.C.; Wall, M.E.; Oberlies, N.H.; Falkinham, J.O., III. Synergistic antimicrobial activity of metabolites produced by a nonobligate bacterial predator. *Antimicrob. Agents Chemother.* **2003**, *47*, 2113–2117. [CrossRef] [PubMed]
53. Grgurina, I.; Mariotti, F. Biosynthetic origin of syringomycin and syringopeptin 22, toxic secondary metabolites of the phytopathogenic bacterium *Pseudomonas syringae* pv. *Syringae*. *FEBS Lett.* **1999**, *462*, 151–154. [CrossRef]

54. Cheng, Y.; Yang, M.; Matter, A.M. Characterization of a gene cluster responsible for the biosynthesis of anticancer agent FK228 in *Chromobacterium violaceum* No. 968. *Appl. Environ. Microbiol.* **2007**, *73*, 3460–3469. [CrossRef] [PubMed]
55. Ortega, M.A.; van der Donk, W.A. New insights into the biosynthetic logic of ribosomally synthesized and post-translationally modified peptide natural products. *Cell Chem. Biol.* **2016**, *23*, 31–44. [CrossRef] [PubMed]
56. Matsuzaki, K.; Ogino, T.; Sunazuka, T.; Tanaka, H.; Omura, S. Chloropeptins, new anti-HIV antibiotics inhibiting gp120-CD4 binding from *Streptomyces* sp. II. Structure elucidation of chloropeptin I. *J. Antibiot.* **1997**, *50*, 66–69. [CrossRef] [PubMed]
57. Andersen, S.O. Chlorinated tyrosine derivatives in insect cuticle. *Insect Biochem. Mol. Biol.* **2004**, *34*, 1079–1087. [CrossRef] [PubMed]
58. Matselyukh, B.; Mohammadipanah, F.; Laatsch, H.; Rohr, J.; Efremenkova, O.; Khilya, V. *N*-methylphenylalanyl-dehydrobutyrine diketopiperazine, an A-factor mimic that restores antibiotic biosynthesis and morphogenesis in *Streptomyces globisporus* 1912-B2 and *Streptomyces griseus* 1439. *J. Antibiot.* **2015**, *68*, 9–14. [CrossRef] [PubMed]
59. Hall, T.A. Bioedit: A user-friendly biological sequences alignment editor and analysis program for windows 95/98/nt. *Nucleic Acids Symp. Ser.* **1999**, *41*, 95–98.
60. Wright, E.S.; Yilmaz, L.S.; Noguera, D.R. Decipher, a search-based approach to chimera identification for 16S rRNA sequences. *Appl. Environ. Microbiol.* **2012**, *78*, 717–725. [CrossRef] [PubMed]
61. Meier-Kolthoff, J.P.; Auch, A.F.; Klenk, H.-P.; Göker, M. Genome sequence-based species delimitation with confidence intervals and improved distance functions. *BMC BioInform.* **2013**, *14*, 60. [CrossRef] [PubMed]
62. Edgar, R.C. Muscle: Multiple sequence alignment with high accuracy and high throughput. *Nucleic Acids Res.* **2004**, *32*, 1792–1797. [CrossRef] [PubMed]
63. Kumar, S.; Stecher, G.; Tamura, K. Mega7: Molecular evolutionary genetics analysis version 7.0 for bigger datasets. *Mol. Biol. Evol.* **2016**, *33*, 1870–1874. [CrossRef] [PubMed]
64. Saitou, N.; Nei, M. The neighbor-joining method: A new method for reconstructing phylogenetic trees. *Mol. Biol. Evol.* **1987**, *4*, 406–425. [PubMed]
65. Felsenstein, J. Evolutionary trees from DNA sequences: A maximum likelihood approach. *J. Mol. Evol.* **1981**, *17*, 368–376. [CrossRef] [PubMed]
66. Kimura, M. A simple method for estimating evolutionary rates of base substitutions through comparative studies of nucleotide sequences. *J. Mol. Evol.* **1980**, *16*, 111–120. [CrossRef] [PubMed]
67. Tamura, K.; Nei, M. Estimation of the number of nucleotide substitutions in the control region of mitochondrial DNA in humans and chimpanzees. *Mol. Biol. Evol.* **1993**, *10*, 512–526. [PubMed]
68. Felsenstein, J. Confidence limits on phylogenies: An approach using the bootstrap. *Evolution* **1985**, *39*, 783–791. [CrossRef] [PubMed]
69. iTOL v3. Available online: https://itol.embl.de/ (accessed on 31 July 2018).
70. Leibniz Institute DSMZ-German Collection of Microorganisms and Cell Cultures. Available online: https://www.dsmz.de/ (accessed on 24 July 2018).
71. Fdhila, F.; Vazquez, V.; Sanchez, J.L.; Riguera, R. dd-Diketopiperazines: Antibiotics active against *Vibrio anguillarum* isolated from marine bacteria associated with cultures of *Pecten maximus*. *J. Nat. Prod.* **2003**, *66*, 1299–1301. [CrossRef] [PubMed]
72. Stierle, A.C.; Cardellina, J.H., II; Strobel, G.A. Maculosin, a host-specific phytotoxin for spotted knapweed from *Alternaria alternata*. *Proc. Natl. Acad. Sci. USA* **1988**, *85*, 8008–8011. [CrossRef] [PubMed]
73. Mohimani, H.; Kersten, R.D.; Liu, W.T.; Wang, M.; Purvine, S.O.; Wu, S.; Brewer, H.M.; Pasa-Tolic, L.; Bandeira, N.; Moore, B.S.; et al. Automated genome mining of ribosomal peptide natural products. *ACS Chem. Biol.* **2014**, *9*, 1545–1551. [CrossRef] [PubMed]
74. Marfey, P. Determination of D-amino acids. II. Use of a bifunctional reagent, 1,5-difluoro-2,4-dinitrobenzene. *Carlsberg Res. Commun.* **1984**, *49*, 591. [CrossRef]
75. Kodani, S.; Komaki, H.; Ishimura, S.; Hemmi, H.; Ohnishi-Kameyama, M. Isolation and structure determination of a new lantibiotic cinnamycin B from *Actinomadura atramentaria* based on genome mining. *J. Ind. Microbiol. Biotechnol.* **2016**, *43*, 1159–1165. [CrossRef] [PubMed]
76. Jones, D.T.; Taylor, W.R.; Thornton, J.M. The rapid generation of mutation data matrices from protein sequences. *Comput. Appl. Biosci.* **1992**, *8*, 275–282. [CrossRef] [PubMed]

77. Goh, E.B.; Yim, G.; Tsui, W.; McClure, J.; Surette, M.G.; Davies, J. Transcriptional modulation of bacterial gene expression by subinhibitory concentrations of antibiotics. *Proc. Natl. Acad. Sci. USA* **2002**, *99*, 17025–31700. [CrossRef] [PubMed]
78. Davies, J.; Spiegelman, G.B.; Yim, G. The world of subinhibitory antibiotic concentrations. *Curr. Opin. Microbiol.* **2006**, *9*, 445–453. [CrossRef] [PubMed]

antibiotics

MDPI

Article

Lysoquinone-TH1, a New Polyphenolic Tridecaketide Produced by Expressing the Lysolipin Minimal PKS II in *Streptomyces albus*

Torben Hofeditz [1], Claudia Eva-Maria Unsin [2], Jutta Wiese [3], Johannes F. Imhoff [3], Wolfgang Wohlleben [2,4], Stephanie Grond [1,*] and Tilmann Weber [2,5,*]

1. Institut für Organische Chemie, Eberhard Karls Universität Tübingen, Auf der Morgenstelle 18, 72076 Tübingen, Germany; torben.hofeditz@gmx.de
2. Interfakultäres Institut für Mikrobiologie und Infektionsmedizin, Eberhard Karls Universität Tübingen, Auf der Morgenstelle 28, 72076 Tübingen, Germany; claudia_unsin@yahoo.de (C.E.-M.U.); wolfgang.wohlleben@biotech.uni-tuebingen.de (W.W.)
3. GEOMAR, Helmholtz-Zentrum für Ozeanforschung Kiel, Düsternbrooker Weg 20, 24105 Kiel, Germany; jwiese@geomar.de (J.W.); jimhoff@geomar.de (J.F.I.)
4. German Center for Infection Research (DZIF), partner site Tübingen, Auf der Morgenstelle 28, 72076 Tübingen, Germany
5. The Novo Nordisk Foundation Center for Biosustainability, Technical University of Denmark, Kemitorvet bygning 220, 2800 Kongens Lyngby, Denmark

* Correspondence: stephanie.grond@uni-tuebingen.de (S.G.); tiwe@biosustain.dtu.dk (T.W.)

Received: 4 May 2018; Accepted: 22 June 2018; Published: 28 June 2018

Abstract: The structural repertoire of bioactive naphthacene quinones is expanded by engineering *Streptomyces albus* to express the lysolipin minimal polyketide synthase II (PKS II) genes from *Streptomyces tendae* Tü 4042 (*llpD-F*) with the corresponding cyclase genes *llpCI-CIII*. Fermentation of the recombinant strain revealed the two new polyaromatic tridecaketides lysoquinone-TH1 (**7**, identified) and TH2 (**8**, postulated structure) as engineered congeners of the dodecaketide lysolipin (**1**). The chemical structure of **7**, a benzo[a]naphthacene-8,13-dione, was elucidated by NMR and HR-MS and confirmed by feeding experiments with [1,2-$^{13}C_2$]-labeled acetate. Lysoquinone-TH1 (**7**) is a pentangular polyphenol and one example of such rare extended polyaromatic systems of the benz[a]napthacene quinone type produced by the expression of a minimal PKS II in combination with cyclases in an artificial system. While the natural product lysolipin (**1**) has antimicrobial activity in nM-range, lysoquinone-TH1 (**7**) showed only minor potency as inhibitor of Gram-positive microorganisms. The bioactivity profiling of lysoquinone-TH1 (**7**) revealed inhibitory activity towards phosphodiesterase 4 (PDE4), an important target for the treatment in human health like asthma or chronic obstructive pulmonary disease (COPD). These results underline the availability of pentangular polyphenolic structural skeletons from biosynthetic engineering in the search of new chemical entities in drug discovery.

Keywords: lysolipin; minimal PKS II; cyclases; benz[a]naphthacene quinone; tridecaketide; aromatic polyketide; pentacyclic angular polyphenol; extended polyketide chain

1. Introduction

Polyketides are a large family of structurally-diverse natural products, mainly produced by bacteria, fungi, and plants. Their biosynthesis is catalyzed by distinct enzymes, termed type I polyketide synthases (PKS), type II PKS, type III PKS or variants thereof [1]. They exhibit broad ranges of pharmacological properties for use in clinical applications [2].

Many bacterial aromatic polyketides, such as the clinically used tetracycline, are biosynthesized by type II polyketide synthases (PKS II). Each PKS II contains a minimal set of enzymes (minPKS) that is required to synthesize a polyketide chain of defined length usually primed by acetyl-CoA and extended with malonyl-CoA to polycyclic products [3]. A minimal PKS II consists of two β-ketoacyl synthases (KS_α and KS_β) and one acyl carrier protein (ACP). KS_α is responsible for loading malonyl-CoA extender onto the PKS II system and also for the iterative Claisen condensations to extend the polyketide chain. KS_β, also referred to as chain length factor (CLF), is contributing to control the polyketide chain length [4]. Diverse cyclization patterns convert the polyketones to an enormous variety of structural polycyclic skeletons [5].

The final bioactive polycyclic PKS II natural products are aromatic compounds and often arise from additional enzymatic conversions, among them cyclases (CYC) [6], reductases (KR) [7], oxidases [8] and decarboxylating enzymes [2,9] or amidotransferases [10] for insertion of nitrogen or glycosyl transferases [11].

Lysolipin I (**1**) from *Streptomyces* Tü 4042 [12] and other members of microbial aromatic polyketides, such as pradimicin A (**2**) [13], fredericamycin A (**3**) [14], benastatin A (**4**) [15], bequinostatin C (**4a**) [16] and xantholipin (**5**) [17], are among the largest type II PKS products that have been described. They differentiate from the important groups of the smaller angucyclinones and anthracyclines (A-type, resp. B in Figure 1). They have distinct pentangular polycyclic aromatic core structures with pyridone, piperidone or lactone rings, respectively, added to a hexacyclic core structure as in xantholipin (**5**), lysolipin (**1**) or fredericamycins (**3**, **3a**) [18] (Figure 1). This additional ring F varies in δ-position with different substituents next to the amide. Several lysolipin derivatives have been described. While lysolipin I (**1**) carries a methoxy substituent at C-24, derivatives of lysolipin have been engineered which have a methyl group at this position (Patent WO2007079715 (A3), Combinature Biopharm, Berlin, Germany).

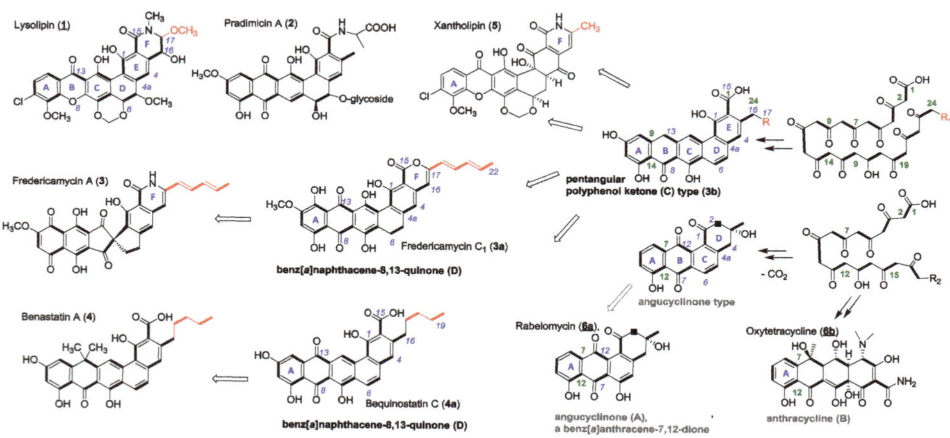

Figure 1. Chemical structures of PKS II products (**1–6**) and examples of biosynthetic congeners (**3a**, **b**, **4a**). PKS chains (acetate units bold) of angucyclinone (**A**) and anthracycline (**B**) type structures start to cyclize with ring A at C-7/C-12. Benz[a]naphthacene (**C**) and benz[a]naphthacene quinone (**D**) type structures cyclize starting with C-9/C-14 according to polyketide chain numbering (green). Variations of δ-substituents of ring-F highlighted in red. R = H or alkyl-, allylic carbon chains (usual chemical nomenclature in blue).

Several biosynthetic gene clusters of these large PKS-II antibiotics are known and have been subject to extensive genetic and biochemical characterization [18–23]. In the biosynthetic pathways of pentangular polyaromatic polyketides, e.g., **1–5**, the respective cyclized polyphenols (**3b**) and polyphenolic quinones (**3a**) have been demonstrated as pathway intermediates; they have also been obtained as products of genetic engineering efforts (Figure 2). Therefore, we regard the class of pentangular quinones as the metabolic hub of the important aromatic polyketide products.

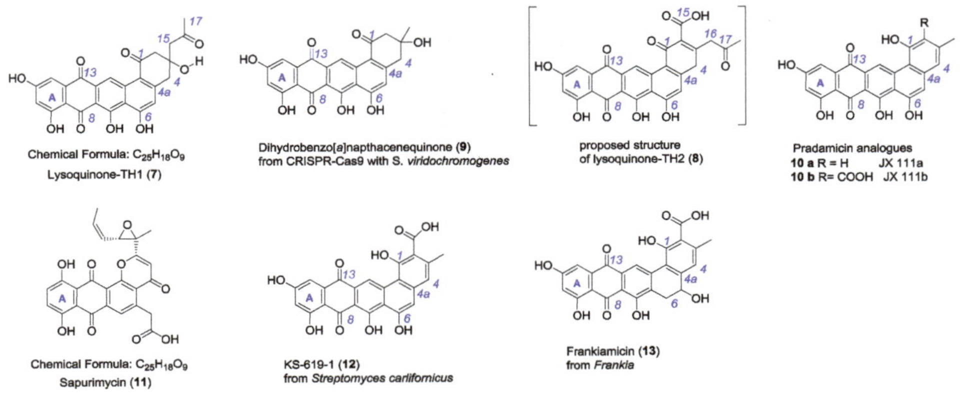

Figure 2. Chemical structure of lysoquinone-TH1 (**7**, identified), lysoquinone-TH2 (**8**, proposed), the constitutional isomer sapurimycin (**11**) and structurally related pentangular polyketides **9–13**.

Here, we report that the heterologous expression of the lysolipin minimal PKS II genes, which comprise *llpF*, coding for the ketosynthase α (KSα), *llpE*, coding for the ketosynthase β (KSβ), and *llpD*, coding for the ACP, in combination with genes encoding the cyclases (*llpCI-CIII*) in the host *S. albus* J1074 resulted in the production of novel metabolites (Figure 2). The structure elucidation of the novel bioactive lysoquinone-TH1 (**7**), strong evidence for its tridecaketide backbone from the doubly labeled [1,2-$^{13}C_2$]-acetate-feeding experiments, and the potent bioactivity as potent phosphodiesterase inhibitor are discussed.

2. Results and Discussion

2.1. Heterologous Production of Lysoquinone-TH 1 (7) in S. albus

The lysolipin gene cluster has been identified on a 42-kb genomic region in *Streptomyces tendae* Tü 4042 and analyzed by sequence comparison and heterologous expression in *S. albus* [21]. For this study, the genes coding for the minimal PKS II (*llpD-F*) and cyclases (*llpCI-CIII*) were amplified by PCR from the cosmid 4H04 encoding the complete lysolipin (**1**) gene cluster. The 4.1-kb PCR fragment was cloned into the vector pSET152*ermE*p* [24] under control of the constitutive promoter *ermE*p* yielding plasmid pCU1 (Figure S1). pCU1 was introduced into *S. albus* J1074 by intergeneric conjugation and the minimal PKS II and cyclase genes were heterologously expressed in culture. The recombinant strain expressing the minimal PKS and cyclases changed the color of the nutrient media (R5) and colonies on plates from yellowish to dark brown. In comparison, *S. albus* strain containing a pSET152*ermE*p* plasmid without insert does not show the phenotype.

Thus, this strain gained the ability to produce new substances. Thin layer chromatography (TLC) analysis revealed a strong red fraction proved to be not identical to lysolipin (**1**).

2.2. Isolation and Structure Elucidation of Lysoquinone-TH1 (7)

Starting from the newly observed colored product, we developed a work-up procedure towards the isolation of the pure compound for structure elucidation and biological profiling from a four liter fermentation broth in M65 medium. Acetone and methanol extractions removed water soluble and other unwanted compounds. Filtration with RP-18 material preceded the preparative HPLC purification using a C_4 column and H_2O (Ammonium formate, TFA): acetonitrile as solvents (see Supplemental Materials). Thereby, 3.4 mg of purified red compound (lysoquinone-TH1, **7**) were obtained from a four-liter fermentation broth from medium supplemented with sterile adsorber resin XAD-16 and subsequently characterized by TLC, LC-mass spectrometry (MS) and NMR-analysis (Figures S2–S5).

HR-ESI-MS data for the red metabolite **7** revealed a m/z = 461.087756 [M−H]$^-$ ((Δppm = 0.11 ppm) and suggested the molecular formula for **7** to be $C_{25}H_{18}O_9$ (M_R = 462,4). Consistently low resolution ESI-MS monitoring of the crude extracts exhibited m/z = 461.1 [M−H]$^-$ and 463.2 [M+H]$^+$) ions with stronger ionization in the negative mode. Thus, the sum formula for compound **7** proposed 17 degrees of unsaturation and implied a large aromatic ring system.

The LC-MS/MS fragmentation analysis of lysoquinone-TH1 (**7**) gave evidence for a mass difference of m/z 58 with m/z = 403.1 [M−C_3H_6O-H]$^-$ as the only fragmentation product, which was assigned to a McLafferty rearrangement and a neutral loss of acetone (Figure S2). These results indicated a stable substance in MS-fragmentation.

One- and two-dimensional NMR-data (^{1}H-NMR, ^{13}C-NMR, HMBC, HSQC, COSY) yielded ten proton signals; one aliphatic methyl-, one methylene group, four protons with methylene character and four aromatic protons (Figures S3–S5). The carbon to proton correlation (HSQC experiments) pointed to two isolated methylene groups with diasterotopic protons. These doublet protons (δ_H = 2.67, 2.91 and δ_H = 3.01, 3.25 ppm) underlined that they only couple within the methylene group, each (J = 15.9 Hz), and suggest a ring structure. Furthermore, a methyl group singlet (δ_H = 2.18 ppm) next to an aliphatic carbonyl group (δ_C = 207.8) implied no other direct substituents. However, this carbonyl group is adjacent to a methylene group (δ_H = 2.70, δ_C = 53.3 ppm), attached to a quaternary carbon (δ_C = 71.0 ppm) as part of the aliphatic ring system (C-1-C-4, C-4a, C-14b). The proton at δ_H = 9.48 indicated an aromatic proton with exceptional low field shift from two carbonyl groups (C-1, C-13) in spatial proximity. In consistence with the recorded ^{13}C-NMR- and ^{1}H-NMR-data, the HMBC-experiment unambiguously delivered assigned key correlations due to well separated signals which established the connectivity to the whole scaffold of lysoquinone-TH1 (**7**) and a full structural assignment (Table S1) to a pentangular polyphenolic core (Figure S4). Lysoquinone-TH1 (**7**) is a 3-(2-oxo-propyl)-decorated dihydrobenz[a]napthacene-8,13-quinone, a novel compound to the best of our knowledge.

Additionally, the purification protocol revealed another violet fraction with obviously a second compound (proposed as lysoquinone-TH2, **8**) not identical to lysolipin (**1**). However, only minute amounts were observed in production cultures, and presumably a distinct instability did not allow for purification and full structure elucidation. HR-ESI-MS data from extract fractions of the violet metabolite **8** revealed a m/z = 487.0670 [M−H]$^-$ (Δppm = 0.1 ppm) and suggested the molecular formula for **8** to be $C_{26}H_{16}O_{10}$ (M_R = 488.4). In comparison to **7**, 19 instead of 17 degrees of unsaturation were calculated. A large aromatic ring system is also concluded. The isolation of **8** was not achieved to allow for NMR studies. Thus, the analytical MS/MS and the UV-data of lysoquinone-TH1 (**7**) in addition to the knowledge of the mentioned McLafferty rearrangement are pointing to structure **8**. This proposed compound was named lysoquinone-TH2 (**8**) and resembles the C-2 carboxyl analogue of **7** as a yet unknown structure.

In comparison to **7**, the only known constitutional isomer sapurimycin (**11**, $C_{25}H_{18}O_9$) [25] is an annealed tetracyclic ring system. The rare moiety of a reduced ring E of angular naphthacene quinone of **7** is only known from metabolite **9**, generated via CRISPR-Cas9 technology with *S. viridochromogenes* [26]. JX111a (**10a**), JX111b (**10b**), and further precursors of pradimicin A (**2**) [27], KS-619-1 (**12**) [28], and frankiamicin (**13**) from *Frankia* [29] display a similar structural skeleton to lysoquinone-TH1 (**7**) and to the proposed structure lysoquinone-TH2 (**8**) (Figure 2). It could be anticipated that the benz[a]napthacene carbon skeleton originates from a native tridecaketide polyketide chain for the lysoquinones **7** and **8** from the min PKS while the parent natural product of strain S. Tü4042 lysolipin I (**1**) has a dodecaketide backbone.

2.3. Feeding Experiment with [1,2-$^{13}C_2$]-Labeled Acetate

For validating the biogenesis of lysoquinone-TH1 (**7**) a feeding experiment with the doubly ^{13}C-labeled [1,2-$^{13}C_2$] acetate was carried out, and lysoquinone-TH1 (**7**) was purified from the culture and subjected to NMR spectroscopy. All carbon atoms turned out to be enriched [30] with highly specific incorporation rates between 2.0 and 7.3 (Figure 3, Figure S4 and S6, Table S1). The specific coupling constants from NMR analysis again confirmed the structure of lysoquinone-TH1 (**7**) and suggest a polyketide origin assembled from 13 acetate extender units (tridecaketide, Figure 2). It is therefore among the largest polyketides formed by a minimal PKS II (LlpD, E, F) with cyclases (LlpCI–CIII) via heterologous expression. It could be anticipated that the proposed structure of lysoquinone-TH2 (**8**) further corroborates the native tridecaketide precursor and carries the additional carboxyl group (C-15) of the final extender acetate unit. The continuous labeled-acetate chain of **7** corresponds with the pattern of the further post-PKS-processed antibiotic lysolipin I (**1**) [31].

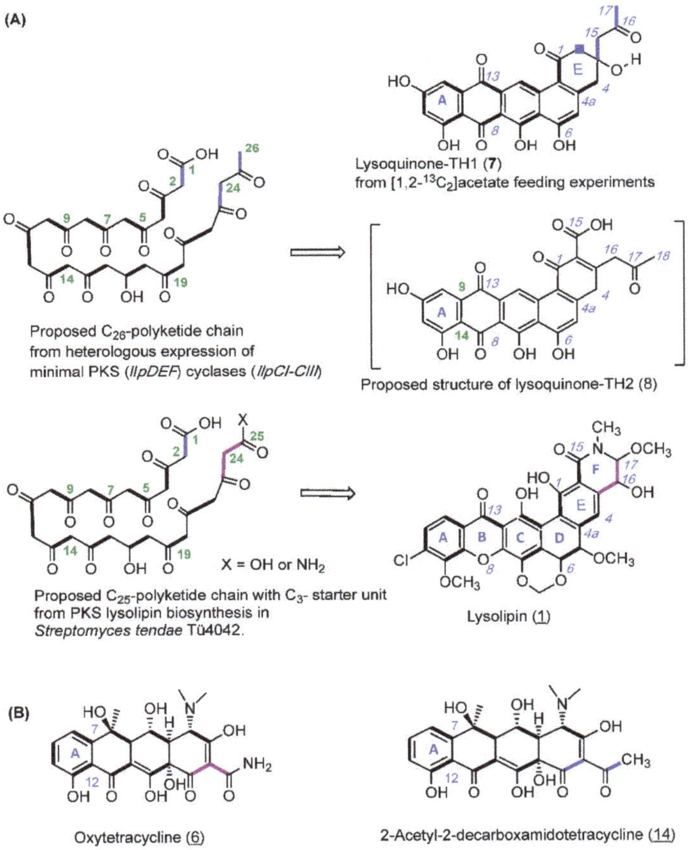

Figure 3. (**A**) Biosynthetic origin of lysoquinone-TH1 (**7**) and the proposed structure of lysoquinone-TH2 (**8**): Biosynthesis hypothesis from feeding experiments with doubly labeled [1,2-$^{13}C_2$] acetate with the *S. albus* host strain and heterologous expression of the lysolipin minimal PKS genes (*llpD-F*) and cyclase genes (*llpCI-CIII*). (**B**) Oxytetracycline (**6**) from a biosynthesis primed with malonamide in the wild-type producer, and with two acetate units in the mutant producer.

A specific feature of the lysolipin polyketide biosynthesis is the priming by a malonate-derived starter unit [31]. However, in the identified lysoquinone-TH1 (**7**) as well as in the proposed structure of lysoquinone-TH2 (**8**), two acetate units replace the original C_3-malonyl/malonamide starter unit. Respective observations were also made for oxytetracycline (**6b**) with a malonamide starter unit since heterologous expression of the oxytetracycline minimal PKS of *Streptomyces rimosus* in *S. coelicolor* yielded tetracycline analogue **14** with two acetate units priming the polyketide backbone instead [32] (Figure 3). Co-expressing the *oxyD* gene with the oxytetracyline minimal PKS in heterologous expression experiments has shown that it is presumably responsible for generating the characteristic malonamate starter unit. OxyD codes for an amidotransferase and is homologous to the putative amidotransferase LlpA of lysolipin biosynthesis. Therefore, priming of the PKS biosynthesis with malonamate and and formation of a N-heterocyclic product does not require additional enzymes [10].

Studies on hybrid PKS pathways with benastatin A (**4**) revealed isolated hybrid PKS-II products with the number of extender units increased if shorter starter units were used. In analogy, based on the observations on the identified lysoquinone-TH 1 (**7**) and the proposed structure of lysoquinone-TH2 (**8**) it can be anticipated that the number of elongation steps is dependent on the length of the polyketide chain [33] fitting into the substrate pocket of the KSα/KSβ-complex.

2.4. *Biological Activity of Lysoquinone-TH1 (7)*

Lysolipin (**1**) is highly active against various Gram-positive bacteria and shows antifungal activity. While the molecular target of lysolipin (**1**) still remains elusive, there is strong indication that this xanthone antibiotic targets the bacterial cell envelope [12,34].

Because of the high antibiotic activity of lysolipin (**1**) in nM-range, biological assays with lysoquinone-TH1 (**7**) were performed. Only weak antibiotic activities at 100 μM concentration of lysoquinone-TH1 (**7**) were observed against the Gram-positive strains *Staphylococcus lentus*, *Staphylococcus epidermidis* and *Propionibacterium acnes* with inhibition of 39%, 66% and 74%, respectively, in comparison to the positive control chloramphenicol [35]. Therefore, no IC_{50} values were determined.

KS-619-1 (**12**) and K-259-2 are representing inhibitors of the cyclic nucleotide phosphodiesterase (PDE4) [28,36–38]. The enzyme PDE4 is a very attractive target for the treatment of asthma, chronic obstructive pulmonary disease (COPD), psoriasis, schizophrenia, diet induced obesity, glucose intolerance and multiple sclerosis [39–44]. PDE4 addresses cyclic nucleotides like cAMP and cGMP and degrades these cellular messengers. However, these messengers possess regulatory functions in almost all cells so it is regarded an important target [45], therefore, detailed studies with different inhibitors are needed to evaluate an ideally selective bioactivity.

For profiling lysoquinone-TH1 (**7**), an enzyme assay using PDE-4B2 was carried out according to Schulz et al. [39]. A clearly defined IC_{50} value could not be determined in this assay due to an additional luminescence signal derived from the chromophore of lysoquinone-TH1 (**7**), an extensive polyaromatic skeleton. However, an IC_{50} range of 10–20 μM could be inferred from the assays against the PDE-4B2 enzyme. The MIC (minimal inhibition concentration) of lysoquinone-TH1 (**7**) was determined to a value of 2.33 μM (±0.04) corresponding to 3.368 log(nM) (±0.007) (Figure S7). The standard used in these assays was rolipram (IC_{50} = 0.8 μM (±0.1)) [39], an optimized and approved drug. Lysoquinone-TH1 (**7**), a completely new substance, is only 10-fold less active as this well-known PDE4 inhibitor rolipram. When lysolipin (**1**) was tested in the same assay, no PDE4 inhibition was detected up to a concentration of 50 μM.

3. Materials and Methods

3.1. *Cloning of the Lysolipin Minimal PKS*

The genes of the lysolipin minimal PKS II (*llpD*, *llpE*, *llpF*), which are surrounded by the cyclases *llpCI*, *llpCII*, and *llpCIII*, were amplified by PCR using the primers minPKScyc-fw-Hind (aaa gct tga gta gcc aaa cgg gtt c) and minPKScyc-revSs (aag aat tca ata ttg tgc cca cca gta cac) and the template cosmid 4H04 [21] using ProofStart PCR polymerase kit (*Qiagen*). The PCR Program used in an PTC-100 thermocycler (MJ Research, Waltham, MA, USA) was: 95 °C—5 min; 30 cycles with (94 °C—90 s, 62 °C—90 s, 72 °C—4 min); 72 °C—10 min. The PCR product was then cloned into pSETermE*p [24] (Combinature Biopharm AG, Berlin, Germany) via the EcoRI/HindIII restriction sites. The resulting plasmid pCU1 was checked by restriction and DNA sequencing.

The plasmid pCU1 was introduced into *Streptomyces albus* J1074 via a standard intergeneric conjugation protocol as, for example, described in [46].

3.2. Culture Conditions

A pre-culture with medium G20 (600 mL) in six 300 mL Erlenmeyer flasks was inoculated with *S. albus* J1074 and apramycin (50 µg/mL) for 48 h at 28 °C and 180 rpm (B. Braun Certomat HK with shaker B. Braun Certomat U, B. Braun, Melsungen, Germany). The main culture was inoculated with 400 mL of the pre-culture and was grown up in medium M65 (3.6 L) under selection with apramycin (50 µg/mL) in a fermenter (B. Braun Biostat B, B. Braun, Melsungen, Germany) for 96 h at 28 °C and 300 rpm. After 24 h, 15 g/300 mL of sterile XAD-16 was added to the culture. Nutrient solutions: G20 (Glycerol (20 g), malt extract (10 g), yeast extract (4 g) in 1 L of tab water. pH = 7.2). M65 (Malt extract (10 g), yeast extract (4 g), D-glucose (4 g), $CaCO_3$ (2 g) in 1 L of tab water. pH = 7.2).

3.3. Extraction and Isolation

For initial detection, agar plates from *S. albus* incubation were extracted with ethyl acetate, the organic phases evaporated and the extract applied to silica gel TLC analysis (solvent cyclohexane/ethylacetate/methanol 6:8:1, with 1% of trifluoric acetate acid added). For purification, 4 L of culture broth with XAD-16 was filtered over Celite to separate the mycelia from the liquid culture. The filtrate was autoclaved and discarded. The mycelia were extracted two times with acetone/methanol 7:3 and then in acetone/methanol 1:1 in an ultrasonic bath. After filtration, the organic phases were combined, evaporated, water was added, and the pH adjusted to 4–5 with 1 M HCl. Extraction with ethyl acetate and evaporation gave the crude extract. RP silica gel was pretreated with 3–4 column volumes (CV) of pyridine and washed with 3–4 CV of water as a basic activation of the RP phase. After column conditioning with 1–2 CV of the solvent acetone/methanol 1:1 the extract was loaded, and red and violet fractions were selected, accompanied by TLC analysis on silica gel (see above) and LC-ESI-MS analysis (HPLC, Agilent 1100 series. Ion trap, Bruker Daltonic Esquire 3000+, He as reactant gas, with Data Analysis software, Bruker Daltonik, Bremen, Germany). Combined red fractions were dissolved in DMSO and purified with HPLC (Thermo Ultimate 3000 Thermo Scientific, Dreieich, Germany); Column: Dr. Maisch, Ammerbuch-Entringen, Germany, Reprosil 120 C-4, 5 µm, 250 × 20 mm id, pre-colum: Dr. Maisch, standard guard Reprosil 120 C-4, 30 × 20 mm id, flow rate: 13.0 mL/min.; program: 20 min at 45% B, in 5 min to 75% B, 8 min at 75% B, in 2 min to 45% B, 8 min at 45% B; solvent: A = ammonium formate (20 mM) + 0.1% TFA in water; B = acetonitrile). A retention time from 12 to 13 min was observed. After evaporating a Sephadex LH-20 (2 × 2 cm, methanol) and following extraction three times with ethyl acetate and three times with diisopropyl ether was necessary for desalting the sample. This work up procedure gives 3.4 mg lysoquinone-TH1 (**7**) from four liters of culture.

3.4. Feeding Experiment with [1,2-$^{13}C_2$]-Labeled Acetate

For the feeding experiment 2.0 g of the doubly labeled [1,2-$^{13}C_2$] acetate (99% enrichment; Cambridge Isotope Laboratories, Inc., Tewksbury, MA, USA), which corresponds to 5.95 mM end concentration in the fermenter (4 L, B. Braun Biostat B) were added to the culture broth after 32 h of cultivation. The isotope labeled lysoquinone-TH1 (**7**) was purified with same protocol as described above and subjected to NMR-analysis (^{13}C-NMR, ^{1}H-NMR, HMBC, HSQC).

3.5. Biological Activity Assays

Antibacterial assays were carried out with the test strains *Staphylococcus lentus* DSM 6672, *Staphylococcus epidermidis* DSM 20044 and *Propionibacterium acnes* DSM 1897 using a cell viability test based on the reduction of resazurin to resorufin. Details on the cultivation conditions of the strains *S. epidermidis* and *P. acnes*, as well as on the evaluation of cell viability are described by Silber et al. [34]. The experiments with *S. lentus* were performed in the same manner as *S. epidermidis*. The positive control chloramphenicol was applied in a concentration of 10 µM for *S. lentus* and *S. epidermidis* and of 1 µM for *P. acnes*.

The effect of lysoquinone-TH1 (**7**) on PDE-4B2, a human recombinant cyclic adenosine monophosphate (cAMP) specific phosphodiesterase (BPS Bioscience no. 60042, San Diego, CA, USA) was determined in 96 well plates using the PDELight HTS cAMP Phosphodiesterase Kit (Lonza, LT07-600, Wuppertal, Germany). Lysoquinone-TH1 (**7**) was diluted in 50 mM Tris-HCl buffer (pH 7.5) containing 8.3 mM $MgCl_2$ and 1.7 mM EGTA. 10 µL of each dilution was transferred to a well. 20 µL PDE-4B2 solution (0.25 U/µL) were added. The reaction was started by adding 10 µL of 12 mM cAMP (Sigma A9501, Taufkirchen, Germany) dissolved in 50 mM Tris-HCl buffer (pH 7.5) containing 8.3 mM $MgCl_2$ and 1.7 mM EGTA to each well of the microtiter plate. PDE-4B2 hydrolysed cAMP to adenosine monophosphate (AMP). After an incubation at 30 °C for 30 min, the reaction was stopped by transferring 30 µL solution containing 10 µL PDELight Stop Solution and 20 µL PDELight AMP detection reagent. The detection reagent converted AMP to ATP and luciferase catalyzed the formation of light from ATP and luciferin. The emitted light is proportional to the level of AMP produced. AMP was quantified after incubation at 30 °C for 10 min by measuring the luminescence using the microtiter plate reader Infinite M200 (Tecan, Crailsheim, Germany) with 0.1 s integration time. The assays were performed in duplicates. Rolipram (4-[3-(cyclopentyloxy)-4-methoxyphenyl]-2-pyrrolidinone) was used as positive control.

4. Conclusions

Lysoquinone-TH1 (**7**) is a "non-natural natural product", a pentangular aromatic polyketide derived from engineering of the lysolipin biosynthetic pathway. It was produced with *Streptomyces albus* as host expressing the minimal PKS II genes (*llpD-F*) in combination with three cyclases (*llpCI-CIII*) of the lysolipin gene cluster. In a bioactivity profiling study, it was shown that lysoquinone-TH1 (**7**) only has weak antibacterial activity, but instead is an inhibitor of phosphodiesterase 4 (PDE4) which is a target for treatment of pulmonary diseases. The lysoquinone-TH1 (**7**) biosynthetic pathway, which was deduced based on NMR data and supported by the new but postulated analogue lysoquinone-TH2 (**8**) also provides insights on the biosynthesis of lysolipin I (**1**). Evidence is given for the acetate-derived tridecaketide backbone of **7** in contrast to the dodecaketide malonyl-derived (malonyl- or malonamide CoA) precursor chain of the parent compound lysolipin I (**1**).

Supplementary Materials: The following are available online at http://www.mdpi.com/2079-6382/7/3/53/s1. Figure S1: Plasmid map of pCU1; Figure S2: HPLC and LC-MS of lysoquinone-TH1 (7) and proposed lysoquinone-TH2 (8); Figure S3: NMR spectra of lysoquinone-TH1 (7); Figure S4: NMR spectra of 13C-enriched lysoquinone-TH1 (7); Figure S5: 2D-NMR data of lysoquinone-TH1 (7); Figure S6: ^{13}C enrichment of 7; Figure S7: PDE-4B2 inhibition assay with lysoquinone-TH1; Table S1: Level of enrichment and specific incorporation of lysoquinone-TH1 (7), resulted from the feeding experiment with doubly labeled [1,2-$^{13}C_2$] acetate.

Author Contributions: T.H. performed strain cultivation and feeding experiments, designed and performed chemical preparative isolation and chemical analyses, wrote the draft manuscript with contributions of all authors. C.E.-M.U. designed and performed cloning of the minimal PKS. J.W. and J.F.I. designed and performed bioactivity studies. W.W., S.G. and T.W. conceived the study, all authors contributed and approved the manuscript.

Funding: This work was supported by the German Ministry of Education and Research [GenBioCom 0315585A] to S.G., T.W. and W.W. T.W. is supported by a grant of the Novo Nordisk Foundation [CFB, grant NNF10CC1016517] and W.W. by a grant from the German Center for Infection Research [DZIF TTU 09.912, FKZ 8020809912].

Acknowledgments: We thank Bruker Daltonics, Bremen, Germany for valuable discussions.

Conflicts of Interest: The authors declare no conflict of interest.

References

1. Hertweck, C. The biosynthetic logic of polyketide diversity. *Angew. Chem. Int. Ed.* **2009**, *48*, 4688–4716. [CrossRef] [PubMed]
2. Kharel, M.K.; Pahari, P.; Shepherd, M.D.; Tibrewal, N.; Nybo, S.E.; Shaaban, K.A.; Rohr, J. Angucyclines: Biosynthesis, mode-of-action, new natural products, and synthesis. *Nat. Prod. Rep.* **2012**, *29*, 264–325. [CrossRef] [PubMed]
3. Zhang, Z.; Pan, H.-X.; Tang, G.-L. New insights into bacterial type II polyketide biosynthesis. *F1000Research* **2017**, *6*, 172. [CrossRef] [PubMed]
4. Tang, Y.; Tsai, S.C.; Khosla, C. Polyketide chain length control by chain length factor. *J. Am. Chem. Soc.* **2003**, *125*, 12708–12709. [CrossRef] [PubMed]
5. Zhou, H.; Li, Y.; Tang, Y. Cyclization of aromatic polyketides from bacteria and fungi. *Nat. Prod. Rep.* **2010**, *27*, 839–868. [CrossRef] [PubMed]
6. Zhang, W.; Watanabe, K.; Wang, C.C.C.; Tang, Y. Investigation of early tailoring reactions in the oxytetracycline biosynthetic pathway. *J. Biol. Chem.* **2007**, *282*, 25717–25725. [CrossRef] [PubMed]
7. Valentic, T.R.; Jackson, D.R.; Brady, S.F.; Tsai, S.C. Comprehensive Analysis of a Novel Ketoreductase for Pentangular Polyphenol Biosynthesis. *ACS Chem. Biol.* **2016**, *11*, 3421–3430. [CrossRef] [PubMed]
8. Kong, L.; Zhang, W.; Chooi, Y.H.; Wang, L.; Cao, B.; Deng, Z.; Chu, Y.; You, D. A Multifunctional monooxygenase XanO4 catalyzes xanthone Formation in xantholipin biosynthesis via a cryptic demethoxylation. *Cell Chem. Biol.* **2016**, *23*, 508–516. [CrossRef] [PubMed]
9. Gullon, S.; Olano, C.; Abdelfattah, M.S.; Brana, A.F.; Rohr, J.; Mendez, C.; Salas, J.A. Isolation, characterization, and heterologous expression of the biosynthesis gene cluster for the antitumor anthracycline steffimycin. *Appl. Environ. Microbiol.* **2006**, *72*, 4172–4183. [CrossRef] [PubMed]
10. Zhang, W.; Ames, B.D.; Tsai, S.C.; Tang, Y. Engineered biosynthesis of a novel amidated polyketide, using the malonamyl-specific initiation module from the oxytetracycline polyketide synthase. *Appl. Environ. Microbiol.* **2006**, *72*, 2573–2580. [CrossRef] [PubMed]
11. Luzhetskyy, A.; Vente, A.; Bechthold, A. Glycosyltransferases involved in the biosynthesis of biologically active natural products that contain oligosaccharides. *Mol. Biosyst.* **2005**, *1*, 117–126. [CrossRef] [PubMed]
12. Drautz, H.; Keller-Schierlein, W.; Zähner, H. Metabolic products of microorganisms, 149. Lysolipin I, a new antibiotic from *Streptomyces violaceoniger* (author's transl). *Arch. Microbiol.* **1975**, *106*, 175–190. [CrossRef] [PubMed]
13. Oki, T.; Konishi, M.; Tomatsu, K.; Tomita, K.; Saitoh, K.; Tsunakawa, M.; Nishio, M.; Miyaki, T.; Kawaguchi, H. Pradimicin, a novel class of potent antifungal antibiotics. *J. Antibiot.* **1988**, *41*, 1701–1704. [CrossRef] [PubMed]
14. Misra, R.; Pandey, R.C.; Silverton, J.V. Fredericamycin A, an Antitumor antibiotic of a novel skeletal type. *J. Am. Chem. Soc.* **1982**, *104*, 4478–4479. [CrossRef]
15. Aoyagi, T.; Aoyama, T.; Kojima, F.; Matsuda, N.; Maruyama, M.; Hamada, M.; Takeuchi, T. Benastatins A and B, new inhibitors of glutathione S-transferase, produced by *Streptomyces* sp. MI384-DF12. I. Taxonomy, production, isolation, physico-chemical properties and biological activities. *J. Antibiot.* **1992**, *45*, 1385–1390. [CrossRef] [PubMed]
16. Yamazaki, T.; Tatee, T.; Aoyama; Kojima, F.; Takeuchi, T.; Aoyagi, T. Bequinostatins C and D, new inhibitors of glutathione S-transferase, produced by *Streptomyces* sp. MI384-DF12. *J. Antibiot.* **1993**, *46*, 1309–1311. [CrossRef] [PubMed]
17. Terui, Y.; Yiwen, C.; Jun-Ying, L.; Ando, T.; Yamamoto, H.; Kawamura, Y.; Tomishima, Y.; Uchida, S.; Okazaki, T.; Munetomo, E.; et al. Xantholipin, a novel inhibitor of HSP47 gene expression produced by *Streptomyces* sp. *Tetrahedron Lett.* **2003**, *44*, 5427–5430. [CrossRef]
18. Chen, Y.; Wendt-Pienkowski, E.; Ju, J.; Lin, S.; Rajski, S.R.; Shen, B. Characterization of FdmV as an amide synthetase for fredericamycin A biosynthesis in *Streptomyces griseus* ATCC 43944. *J. Biol. Chem.* **2010**, *285*, 38853–38860. [CrossRef] [PubMed]
19. Kim, B.C.; Lee, J.M.; Ahn, J.S.; Kim, B.S. Cloning, sequencing, and characterization of the pradimicin biosynthetic gene cluster of *Actinomadura hibisca* P157-2. *J. Microbiol. Biotechnol.* **2007**, *17*, 830–839. [PubMed]

20. Das, A.; Szu, P.-H.; Fitzgerald, J.T.; Khosla, C. Mechanism and engineering of polyketide chain initiation in fredericamycin biosynthesis. *J. Am. Chem. Soc.* **2010**, *132*, 8831–8833. [CrossRef] [PubMed]
21. Lopez, P.; Hornung, A.; Welzel, K.; Unsin, C.; Wohlleben, W.; Weber, T.; Pelzer, S. Isolation of the lysolipin gene cluster of *Streptomyces tendae* Tü 4042. *Gene* **2010**, *461*, 5–14. [CrossRef] [PubMed]
22. Zhang, W.; Wang, L.; Kong, L.; Wang, T.; Chu, Y.; Deng, Z.; You, D. Unveiling the post-PKS redox tailoring steps in biosynthesis of the type II polyketide antitumor antibiotic xantholipin. *Chem. Biol.* **2012**, *19*, 422–432. [CrossRef] [PubMed]
23. Lackner, G.; Schenk, A.; Xu, Z.; Reinhardt, K.; Yunt, Z.S.; Piel, J.; Hertweck, C. Biosynthesis of pentangular polyphenols: Deductions from the benastatin and griseorhodin pathways. *J. Am. Chem. Soc.* **2007**, *129*, 9306–9312. [CrossRef] [PubMed]
24. Menges, R.; Muth, G.; Wohlleben, W.; Stegmann, E. The ABC transporter Tba of *Amycolatopsis balhimycina* is required for efficient export of the glycopeptide antibiotic balhimycin. *Appl. Microbiol. Biotechnol.* **2007**, *77*, 125–134. [CrossRef] [PubMed]
25. Uosaki, Y.; Yasuzawa, T.; Hara, M.; Saitoh, Y.; Sano, H. Sapurimycin, new antitumor antibiotic produced by *Streptomyces*. Structure determination. *J. Antibiot.* **1991**, *44*, 40–44. [CrossRef] [PubMed]
26. Zhang, M.M.; Wong, F.T.; Wang, Y.; Luo, S.; Lim, Y.H.; Heng, E.; Yeo, W.L.; Cobb, R.E.; Enghiad, B.; Ang, E.L.; et al. CRISPR-Cas9 strategy for activation of silent *Streptomyces* biosynthetic gene clusters. *Nat. Chem. Biol.* **2017**, *13*, 607–609. [CrossRef] [PubMed]
27. Zhan, J.; Watanabe, K.; Tang, Y. Synergistic actions of a monooxygenase and cyclases in aromatic polyketide biosynthesis. *Chembiochem* **2008**, *9*, 1710–1715. [CrossRef] [PubMed]
28. Yasuzawa, T.; Yoshida, M.; Shirahata, K.; Sano, H. Structure of a novel Ca^{2+} and calmodulin-dependent cyclic nucleotide phosphodiesterase inhibitor KS-619-1. *J. Antibiot.* **1987**, *40*, 1111–1114. [CrossRef] [PubMed]
29. Ogasawara, Y.; Yackley, B.J.; Greenberg, J.A.; Rogelj, S.; Melançon, C.E. Expanding our understanding of sequence-function relationships of Type II polyketide biosynthetic gene clusters: Bioinformatics-guided identification of frankiamicin a from *Frankia* sp. EAN1pec. *PLoS ONE* **2015**, *10*, 1–25. [CrossRef] [PubMed]
30. Scott, A.I.; Townsend, C.A.; Okada, K.; Kajiwara, M.; Cushley, R.J.; Whitman, P.J. Biosynthesis of corrins. II. Incorporation of 13C-labeled substrate in vitamins B12. *J. Am. Chem. Soc.* **1974**, *96*, 8069–8080. [CrossRef] [PubMed]
31. Bockholt, H.; Udvarnoki, G.; Rohr, J.; Mocek, U.; Beale, J.M.; Floss, H.G. Biosynthetic studies on the xanthone antibiotics lysolipins X and I. *J. Org. Chem.* **1994**, *59*, 2064–2069. [CrossRef]
32. Fu, H.; Ebert-Khosla, S.; Khosla, C.; Hopwood, D.A. Relaxed Specificity of the oxytetracycline polyketide synthase for an acetate primer in the absence of a malonamyl primer. *J. Am. Chem. Soc.* **1994**, *116*, 6443–6444. [CrossRef]
33. Xu, Z.; Schenk, A.; Hertweck, C. Molecular analysis of the benastatin biosynthetic pathway and genetic engineering of altered fatty acid-polyketide hybrids. *J. Am. Chem. Soc.* **2007**, *129*, 6022–6030. [CrossRef] [PubMed]
34. Winter, D.K.; Sloman, D.L.; Porco, J., Jr. A. Polycyclic xanthone natural products: Structure, biological activity and chemical synthesis. *Nat. Prod. Rep.* **2013**, *30*, 382–391. [CrossRef] [PubMed]
35. Silber, J.; Ohlendorf, B.; Labes, A.; Erhard, A.; Imhoff, J.F. Calcarides A–E, antibacterial macrocyclic and linear polyesters from a calcarisporium strain. *Mar. Drugs* **2013**, *11*, 3309–3323. [CrossRef] [PubMed]
36. Matsuda, Y.; Kase, H. KS-619-1, a new inhibitor of Ca^{2+} and calmodulin-dependent cyclic nucleotide phosphodiesterase from *Streptomyces californicus*. *J. Antibiot.* **1987**, *40*, 1104–1110. [CrossRef] [PubMed]
37. Matsuda, Y.; Asano, K.; Kawamoto, I.; Kase, H. K-259-2, a new inhibitor of Ca^{2+} and calmodulin-dependent cyclic nucleotide phosphodiesterase from *Micromonospora olivasterospora*. *J. Antibiot.* **1987**, *40*, 1092–1100. [CrossRef] [PubMed]
38. Yasuzawa, T.; Yoshida, M.; Shirahata, K.; Sano, H. Structure of a novel Ca^{2+} and calmodulin-dependent cyclic nucleotide phosphodiesterase inhibitor K-259-2. *J. Antibiot.* **1987**, *40*, 1101–1103. [CrossRef] [PubMed]
39. Schulz, D.; Beese, P.; Ohlendorf, B.; Erhard, A.; Zinecker, H.; Dorador, C.; Imhoff, J.F. Abenquines A–D: Aminoquinone derivatives produced by *Streptomyces* sp. strain DB634. *J. Antibiot.* **2011**, *64*, 763–768. [CrossRef] [PubMed]
40. Houslay, M.D.; Schafer, P.; Zhang, K.Y.J. Keynote review: Phosphodiesterase-4 as a therapeutic target. *Drug Discov. Today* **2005**, *10*, 1503–1519. [CrossRef]

41. Boswell-Smith, V.; Spina, D.; Page, C.P. Phosphodiesterase inhibitors. *Br. J. Pharmacol.* **2009**, *147*, S252–S257. [CrossRef] [PubMed]
42. Millar, J.K.; Pickard, B.S.; Mackie, S.; James, R.; Christie, S.; Buchanan, S.R.; Malloy, M.P.; Chubb, J.E.; Huston, E.; Baillie, G.S.; et al. DISC1 and PDE4B are interacting genetic factors in schizophrenia that regulate cAMP signaling. *Science* **2005**, *310*, 1187–1191. [CrossRef] [PubMed]
43. Park, S.-J.; Ahmad, F.; Philp, A.; Baar, K.; Williams, T.; Luo, H.; Ke, H.; Rehmann, H.; Taussig, R.; Brown, A.L.; et al. Resveratrol ameliorates aging-related metabolic phenotypes by inhibiting cAMP phosphodiesterases. *Cell* **2012**, *148*, 421–433. [CrossRef] [PubMed]
44. Kanes, S.J.; Tokarczyk, J.; Siegel, S.J.; Bilker, W.; Abel, T.; Kelly, M.P. Rolipram: A specific phosphodiesterase 4 inhibitor with potential antipsychotic activity. *Neuroscience* **2007**, *144*, 239–246. [CrossRef] [PubMed]
45. Houslay, M.D.; Baillie, G.S.; Maurice, D.H. cAMP-Specific phosphodiesterase-4 enzymes in the cardiovascular system: A molecular toolbox for generating compartmentalized cAMP signaling. *Circ. Res.* **2007**, *100*, 950–966. [CrossRef] [PubMed]
46. Musiol, E.M.; Härtner, T.; Kulik, A.; Moldenhauer, J.; Piel, J.; Wohlleben, W.; Weber, T. Supramolecular templating in kirromycin biosynthesis: The acyltransferase KirCII loads ethylmalonyl-CoA extender onto a specific ACP of the trans-AT PKS. *Chem. Biol.* **2011**, *18*, 438–444. [CrossRef] [PubMed]

antibiotics

MDPI

Article

Novel Polyethers from Screening *Actinoallomurus* spp.

Marianna Iorio [1], Arianna Tocchetti [1], Joao Carlos Santos Cruz [2], Giancarlo Del Gatto [1], Cristina Brunati [2], Sonia Ilaria Maffioli [1], Margherita Sosio [1,2] and Stefano Donadio [1,2,*]

1 NAICONS Srl, Viale Ortles 22/4, 20139 Milano, Italy; miorio@naicons.com (M.I.); atocchetti@naicons.com (A.T.); giancarlo_delgatto@hotmail.it (G.D.G.); smaffioli@naicons.com (S.I.M.); msosio@naicons.com (M.S.)

2 KtedoGen Srl, Viale Ortles 22/4, 20139 Milano, Italy; maildocruz@gmail.com (J.C.S.C.); cbrunati@naicons.com (C.B.)

* Correspondence: sdonadio@naicons.com

Received: 1 May 2018; Accepted: 13 June 2018; Published: 14 June 2018

Abstract: In screening for novel antibiotics, an attractive element of novelty can be represented by screening previously underexplored groups of microorganisms. We report the results of screening 200 strains belonging to the actinobacterial genus *Actinoallomurus* for their production of antibacterial compounds. When grown under just one condition, about half of the strains produced an extract that was able to inhibit growth of *Staphylococcus aureus*. We report here on the metabolites produced by 37 strains. In addition to previously reported aminocoumarins, lantibiotics and aromatic polyketides, we described two novel and structurally unrelated polyethers, designated α-770 and α-823. While we identified only one producer strain of the former polyether, 10 independent *Actinoallomurus* isolates were found to produce α-823, with the same molecule as main congener. Remarkably, production of α-823 was associated with a common lineage within *Actinoallomurus*, which includes *A. fulvus* and *A. amamiensis*. All polyether producers were isolated from soil samples collected in tropical parts of the world.

Keywords: *Actinoallomurus*; antibiotics polyethers; screening

---

1. Introduction

Antimicrobial resistance among bacterial pathogens is becoming a major threat to human health and well-being. While different approaches can be deployed to mitigate and delay the insurgence and spread of antibiotic resistance, it is also clear that we will need a constant supply of new antibiotics, especially new chemical classes not affected by current resistance mechanisms. However, new chemical classes of antibiotics have been extremely difficult to discover from combinatorial and chemical libraries and microbial products still represent a major source of drug leads as antibiotics [1].

One of the main issues with antibiotic discovery based on microbial products is the probability of rediscovering known metabolites. This requires introducing one or more elements of novelty in the screening with respect to past efforts [2,3]. An attractive element of novelty can be represented by using novel strains, for example a taxonomic group that has not witnessed extensive analyses of its secondary metabolites, since taxonomic diversity can be seen as a surrogate for chemical diversity [4]. The main idea behind this concept is that organisms that have been subjected to different evolutionary pressures have developed unique biology to survive and, for some taxa, secondary metabolites are an important part of their biology. However, since production of secondary metabolites is not distributed equally among all species, it is important to select a taxon with a high potential to produce bioactive compounds in order to increase the probability of finding new compounds with a reasonable screening effort. Following this rationale, we initiated over a decade ago a project aimed at finding taxonomically divergent filamentous *Actinobacteria*, which led to the discovery of several novel taxa, including new suborders, families and genera [5–8]. One of the new taxa, originally designated as "alpha" [5], turned out to coincide with the genus *Actinoallomurus* (family *Thermomonosporaceae*), formally described in 2009 [9] with new entries added since [10–15]. With proper isolation methods, strains belonging to the genus *Actinoallomurus* could be effectively retrieved from a variety of soil samples, enabling the creation of a consistent collection of about 1000 isolates [2,12].

Strains belonging to the genus *Actinoallomurus* have been shown to produce a variety of metabolites [12,15–19] originating from different biosynthetic pathways. In this study, we explored 200 randomly picked *Actinoallomurus* isolates from the NAICONS strain collection. Together with the metabolites previously described [16–19], we analyzed the antibacterial compounds produced by 37 strains. This set of compounds includes two novel polyethers, as described here.

2. Results and Discussion

2.1. The Screened Set

The selected *Actinoallomurus* strains were isolated from a variety of samples collected in different continents, and representing diverse environments such as densely vegetated areas, sulfur-enriched craters of volcanic origin, and plant rhizosphere. The geographic distribution of the screened strains is listed in Table 1.

Table 1. Geographic origin of the analyzed strains.

Continent	Analyzed Strains	Active Strains	Active (%)
Europe	124	59	48%
Africa [a]	24	17	71%
Asia [a]	12	3	25%
Americas [b]	40	25	62%

[a] All from tropical countries; [b] All from tropical countries, except four strains from continental USA.

Three extract types were prepared from the strains—see Material and Methods—and evaluated for their ability to inhibit growth of *Staphylococcus aureus* and of a Δ*tolC* mutant of *Escherichia coli*. Overall, 104 and 17 strains produced at least one extract with activity against *S. aureus* and *E. coli*, respectively. All extracts with activity against *E. coli* were also active against *S. aureus*. The highest activity was observed in the mycelium and the ethyl acetate extracts at comparable frequency (57 and 46, respectively), and only in one case was the ethyl acetate exhaust extract more active. Except perhaps for an under-representation of active strains isolated from Asian samples, there was no apparent effect of the continent of origin on the frequency of anti-staphylococcal activity (Table 1). Positive extracts were analyzed as described under Material and Methods, leading to preliminary information on the chemical identity of the identified compounds. We report below the characterization of the molecules identified from 37 of the active strains.

2.2. Coumermycins, Spirotetronates, Lantibiotics and Diketopiperazines

Coumarin antibiotics target bacterial DNA gyrase and one member of this family, novobiocin, has been used to treat bacterial infections in humans caused by Gram-positive bacteria [20]. Coumermycins are other member of this family with higher antibacterial activity than novobiocin [20]. We have previously reported that the *Actinoallomurus* sp. K275, belonging to the Alp18 phylotype, produced several members of the coumermycin complex [12]. In the course of screening the 200 strains, two additional coumermycin producers were identified: strains ID145250 and ID145519, belonging to the Alp22 phylotype. The main coumermycin congeners produced by these three strains were A2, D1 and A1, respectively. Relevant data are shown in Supplementary Materials Figure S1. Coumermycins have been reported mostly from *Streptomyces* spp. [20].

Tetronate-containing polyketide natural products represent a large and diversified family of microbial metabolites with different bioactivities [21]. Halogenated spirotetronates designated NAI-414 A and B were previously described as the main products of *Actinoallomurus* sp. ID145414 [16]. During our screening, two strains (ID145260 and ID145814) were found to produce a molecule with *m/z* $[M-H]^-$ 839, an isotopic pattern compatible with the presence of two chlorines and a UV spectrum with maxima at 228 and 268 nm (see Supplementary Materials Figure S2). These properties closely resemble those reported for NAI-414 and actually match those reported for pyrrolosporin, a compound structurally related to NAI-414 but containing an additional unsaturation in the polyketide backbone. Pyrrolosporin was previously reported as a metabolite from a *Micromonospora* sp. [22,23].

Ribosomally synthesized and post-translationally modified peptides represent a rapidly expanding family of microbial metabolites, with lantibiotics as one of better known representatives [24]. *Actinoallomurus* sp. ID145699 was previously reported to produce the chlorinated lantibiotic NAI-107 and its brominated variant in Br-supplemented medium [19]. In the course of our screening, we identified strain ID145640 as an additional NAI-107 producer, with the known variations at Trp4 (hydrogen or chlorine) but just zero or one hydroxylation at Pro14 (Supplementary Materials Figures S3 and S4). Previously, NAI-107 was reported as the product of two independent *Microbispora* isolates [25,26].

Diketopiperazines represent a broad family of cyclized dipeptides produced by a large variety of microorganisms [27]. It is thus no surprise that one of the strains in the present work, *Actinoallomurus* sp. ID145219, produced two compounds with activity against *S. aureus* that were identified as cyclo-Phe-Leu and cyclo-Phe-Phe (see Supplementary Materials Figure S5). The structures of the metabolites mentioned in Section 2.2 are illustrated in Figure 1.

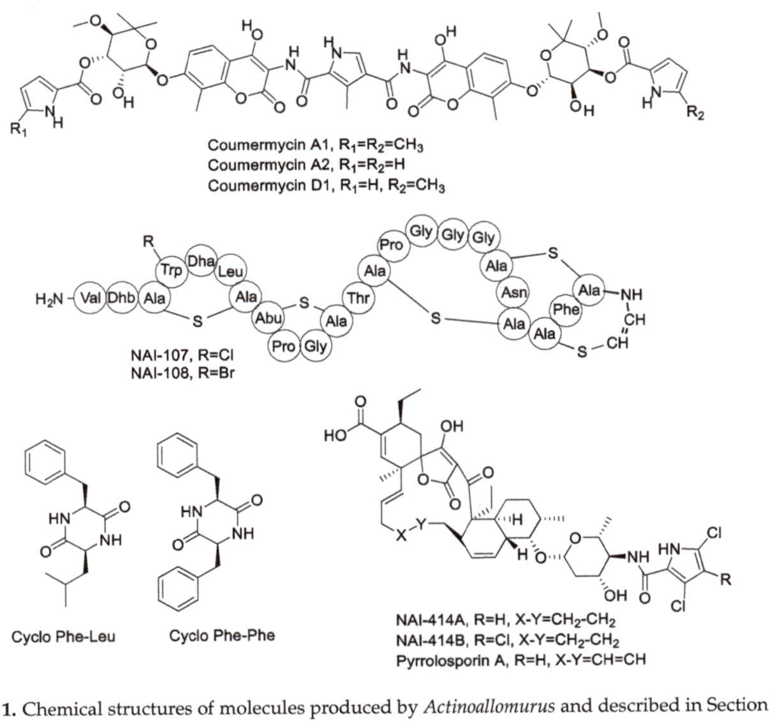

Figure 1. Chemical structures of molecules produced by *Actinoallomurus* and described in Section 2.2.

2.3. Aromatic Polyketides

During the course of our screening, we frequently encountered strains producing aromatic polyketides. Identified products included the allocyclinones, hyper-halogenated angucyclinones detected from twelve independent strains belonging to three different phylotypes [17]; the paramagnetoquinones, highly paramagnetic tetracenes produced by three independent strains belonging to three different phylotypes [18]; three producers of the related dihydrobenzo-(alpha)-naphthacenequinones pradimicin (strains ID145114 and ID145318) and benanomicin (strain ID145226) (see Supplementary Materials Figure S6). Pradimicin and benanomicin present a common polyketide core decorated with a disaccharide unit and differ for the presence/absence of an *N*-methyl on the aminated-sugar. Both compounds were previously reported as products of *Actinomadura* spp., with pradimicin produced by a confirmed species of the genus, *Actinomadura hibisca* [28]. Benanomicin had already been reported as a metabolite of *Actinoallomurus* strain K15 [12].

Overall, this brief survey of aromatic polyketides indicates that *Actinoallomurus* spp. are capable of producing decaketides (i.e., paramagnetoquinones), undecaketides (i.e., allocyclinones, presumably undergoing oxidative ring cleavage after polyketide formation) and dodecaketides (i.e., pradimicin and benanomicin). The structures of these metabolites are shown in Figure 2.

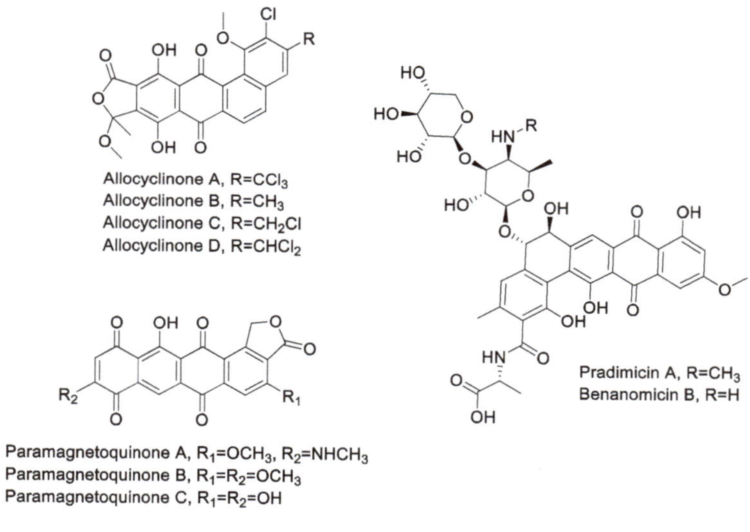

Figure 2. Chemical structures of aromatic polyketides produced by *Actinoallomurus* and described in Section 2.3.

2.4. Polyethers

The extracts from several strains presented large inhibition halos against *S. aureus* but little or no activity against the *E. coli* Δ*tolC* strain. Upon resolution by high performance liquid chromatography (HPLC), the active fractions showed a retention time of 5–11 min and, with one exception, had no ultraviolet (UV) absorption. Mass spectrometry (MS) analysis indicated the presence of *m/z* signals consistent with the formation NH_4^+ and Na^+ adducts but with no detectable H^+ adducts. It should be noted that the extraction procedure and the liquid chromatography (LC)-MS eluent do not contain ammonium or sodium ions. Hence, the observation of NH_4^+ and Na^+ adducts suggests a high cation-binding ability of the active molecules. Moreover, the fragmentation pattern showed losses of 44 amu (free carboxylic acid) and 62 amu (decarboxylation and dehydratation). As demonstrated below, we identified three distinct polyether families within twelve strains: one new compounds, designated α-823, produced by ten independent isolates; an additional new polyether, designated α-770, and the previously reported octacyclomycin, produced by one strain each. Table 2 lists the identified polyether-producing *Actinoallomurus* isolates, along with their origins, accession number of the 16S rRNA gene sequences and the *m/z* value of the most abundant congener.

Table 2. Polyether-producing *Actinoallomurus* strains.

Strain	Origin	Accession Number [a]	m/z [M−Na]$^+$	Compound
ID145265	soil, Nicaragua	MH3933000	937	α-823
ID145554	soil, Mauritius	MH3933001	937	α-823
ID145603	soil, Brazil	MH3933002	937	α-823
ID145770	soil, Niger	MH3933011	857	α-770
ID145802	soil, Nicaragua	MH3933003	937	α-823
ID145804	soil, Cameroon	MH3933004	937	α-823
ID145811	soil, Cameroon	MH3933005	937	α-823
ID145816	soil, Cameroon	MH3933006	937	α-823
ID145817	soil, Cameroon	MH3933010	1039	octacyclomycin
ID145823	soil, Venezuela	MH3933007	937	α-823
ID145828	soil, Nicaragua	MH3933008	937	α-823
ID145830	soil, Nicaragua	MH3933009	937	α-823

[a] On the basis of the 16S rRNA gene sequence.

Several strains produced a likely polyether with major *m/z* signals $[M+NH_4]^+$ 932 and $[M+Na]^+$ 937 (see Figure 4a,b for representative example). The metabolite produced by all these strains appeared identical (Table 2) and those from strain ID145823 were analyzed in detail. The strain produced a complex of related molecules (Figure 4a; Supplementary Materials Table S1) with similar HPLC retention times (they all eluted at ≥90% acetonitrile; see Supplementary Materials Figure S7), appearing as both $[M+NH_4]^+$ and $[M+Na]^+$ adducts, and with similar fragmentation patterns (Supplementary Materials Figure S7). The deduced molecular formulae indicate that the congeners varied in methyl group(s) and oxygen(s) (Supplementary Materials Table S1). The structure of the major congener, designated α-823, was elucidated by a combination of NMR (Supplementary Materials Figures S8–S13) and HR-ESI-MS (Figure S14) and MS/MS analyses. The molecular formula was defined as $C_{48}H_{82}O_{16}Na$ (calculated 937.5495 $[M+Na]^+$, found 937.5510 $[M+Na]^+$). The analysis of ^{1}H-monodimentional spectrum revealed the presence of four singlet and six doublet methyl signals, along with four methoxy groups. Moreover, several diastereotopic methylene signals were observed using 2D-HSQC (bi-dimensional Heteronuclear Single Quantum Coherence) experiments, indicating CH_2 inserted into rigid structures or close to stereocenters. COSY (COrrelated SpectroscopY) and TOCSY (TOtal Correlated SpectroscopY) analyses, along with HMBC (Heteronuclear Multiple Bond Correlation, resulted in the structure shown in Figure 3. α-823 consists of a C_{30} chain with three substituted tetrahydrofuranes and three substituted tetrahydropyranes. Tetrahydrofurane C carries a deoxysugar (Figure 4). Structurally, α-823 closely resembles the polyether SF-2361, produced by an *Actinomadura* sp. [29,30]. Despite an identical molecular formula, α-823 carries a methyl at C-6, while a methyl group in SF-2361 has been assigned to C-2. Indeed, in the α-823 spectrum C-2 is a free methylene with δ_H signal at 2.18–2.54 ppm and δ_C at 44.7 ppm (due to the proximity to a hemiacetal and a carboxylic acid), while C-6 carries no proton (δ_C at 78.6 ppm) and shows HMBC correlations with a methoxy at 3.38 ppm and a singlet methyl at 1.16 ppm.

Strain ID145817 was found to produce a bioactive compound eluting at 9.0 min with no UV adsorption. Upon MS analysis, it showed *m/z* signals $[M+NH_4]^+$ 1034 and $[M+Na]^+$ 1039, corresponding to the NH_4^+ and Na^+ adduct of a molecule of 1016 amu, with major fragments at 990–995 and 972–977 (Supplementary Materials Figure S15). Additionally, a $\Delta m/z$ of 129 suggested the elimination of a deoxysugar. All these properties are consistent with the metabolite produced by strain ID145817 being identical to octacyclomycin, a di-glycosylated polyether previously reported from a *Streptomyces* sp. [31]. NMR analysis of the purified compound confirmed this hypothesis showing signals identical to those reported in literature for octacyclomycin (data not shown). The structure of octacyclomycin is reported in Figure 4. Octacyclomycin, SF-2361 and α-823 derive from a C_{30} chain with identical sequence of tetrahydrofuranes and tetrahydropyranes, but differ for the number and position of methyls, oxygens and glycosyl moieties.

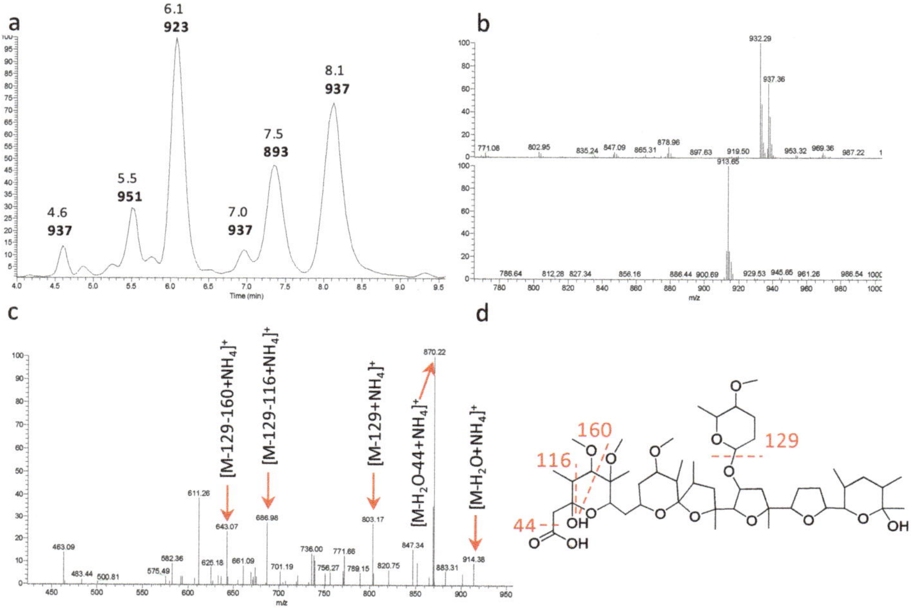

Figure 3. Analysis of α-823. (**a**) Base peak chromatogram of the 4.0–10.0 min portion with retention times and *m/z* [M+MH4]+ values. Data obtained with a partially purified extract of *Actinoallomurus* sp. ID145823 (see Table S1 and Figure S7 for the congeners comparison); (**b**) mass spectrometry (MS) at 8.1 min in positive (above) and negative (below) ionization mode; (**c**) MS^2 analysis of *m/z* $[M+NH_4]^+$ 932; (**d**) putative fragmentation pathway for α-823.

Strain ID145770 was found to produce an active peak eluting at 6.9 min and showing the polyether-diagnostic *m/z* signals $[M+NH_4]^+$ 852 and $[M+Na]^+$ 857, corresponding to a molecule of 834 amu (Figure 5a,b). Unlike the other polyethers, however, this peak showed a UV signal with maximum at 314 nm (Figure 5a) and a $\Delta m/z$ 135 upon MS fragmentation, consistent with presence of a methylsalicylate moiety (Figure 5c), a chromophore previously found in the polyether cationomycin [32]. The active molecule was produced along with a related, $\Delta m/z$ +14 species, consistent with the presence of an extra methyl group atom, as listed in Table S3 and shown in Figure S16. The structure of the major congener α-770 was elucidated by a combination of NMR (Figures S17–S22), HR-ESI-MS (Figure S23) and MS/MS analyses (Figure 5). The molecular formula was defined as $C_{45}H_{70}O_{14}Na$ (calculated 857.4658 $[M+Na]^+$, found 857.4695 $[M+Na]^+$). The analysis of ^1H- and HSQC spectra revealed the presence of 2 methoxy along with 10 methyl groups, with four of them devoid of multiplicity. Moreover, several diastereotopic methylene signals were observed, indicating CH_2 inserted into rigid structures or close to stereocenters. The 2D-NMR-experiments allowed assigning the carbons in a structure consisting four substituted tetrahydrofuranes and one substituted tetrahydropyrane, as shown in Figure 4. The overall structure of α-770 is similar to that of cationomycin, produced by an *Actinomadura* sp. [32]. Both polyethers consists of a C_{27} chain with identical positioning of the methyl groups, suggesting they a common origin from incorporation of the same sequence of propionate and acetate precursors [33]; and both polyethers carry a 6-methylsalicylate moiety linked in an ester bond the C-3 hydroxyl. However, despite these similarities and the small mass difference (16 amu), α-770 and cationomycin differ significantly in the hydroxyl and methoxy decorations. Indeed, 2D-HMBC correlations established that α-770 lacks the hydroxyls at positions 15 and 5′, which are present as methoxy groups in cationomycin, as well as the hydroxyl at position 3. In contrast, α-770 carries methoxys at positions 11 and 21, while cationomycin has no hydroxyls at those positions.

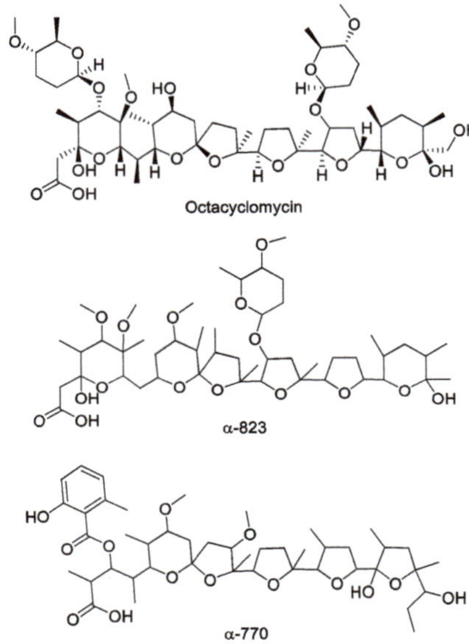

Figure 4. Chemical structure of polyethers produced by *Actinoallomurus* and described in Section 2.4.

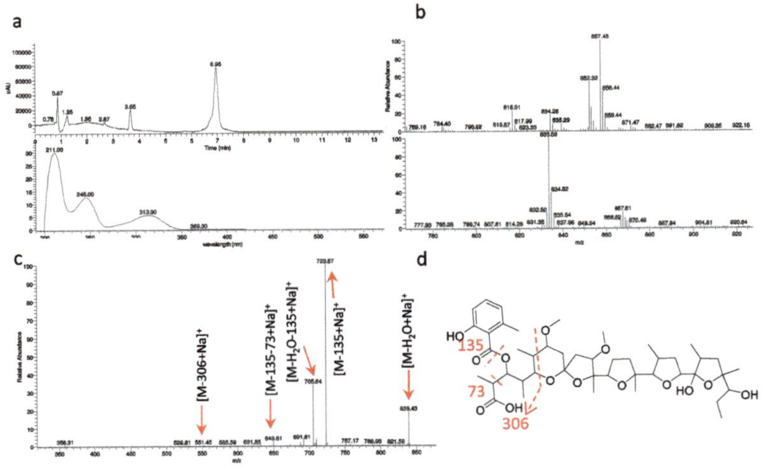

Figure 5. Analysis of α-770. (**a**) UV chromatogram at 230 nm and UV spectrum of 6.9-min peak. Data obtained with a partially purified extract of *Actinoallomurus* sp. ID145770; (**b**) MS at 6.9 min in positive (above) and negative (below) ionization mode; (**c**) MS^2 of *m/z* $[M+NH_4]^+$ 852; (**d**) putative fragmentation pathway for α-770.

The carbon chain of polyethers is assembled by type I PKSs, followed by ring formation by dedicated epoxidases [33]. The carbon chains of α-770 and α-823 are likely to derive from trideca- and pentadeca-ketide precursors, respectively. In addition, the 6-methylsalicylate unit of cationomycin has been show to result from acetate incorporation [34], consistent with the involvement of a type III PKS system.

The antimicrobial activity of polyethers is strictly connected to their ability to insert into cellular membranes and alter transport of metal cations, which leads to changes in the osmotic pressure inside the cytoplasm and cell death [30,35]. However, polyethers generally lack cellular selectivity. The antibacterial activities of α-823 and α-770 are reported in Table 3. They show potent activities against most Gram-positive bacteria, with minimal inhibitory concentrations (MICs) well below 1 μg/mL for α-770, the most active of the three compounds, with 2–4 times lower MICs than salinomycin against most of the tested strains. The polyether α-823 was generally 4–16 times less active than α-770, except for an increased activity against *Mycobacterium smegmatis*. No activities were detected against Gram-negative strains (not shown), except for *Moraxella catarrhalis*.

Table 3. MICs (Minimal Inhibitory Concentrations) of α-770 and α-823 in comparison with salinomycin.

Microorganism [a]	Code	MIC (μg/mL)		
		α-770	α-823	Salinomycin
Staphylococcus aureus (MSSA)	ATCC 6538P	0.125	0.5	0.25
S. aureus (MRSA)	L1400	0.125	2	0.5
S. aureus (GISA)	L3797	0.125	1	0.125
Streptococcus pyogenes	L49	≤0.03	≤0.03	≤0.03
S. pneumoniae	L44	≤0.03	0.125	0.125
S. haemolyticus	L1730	0.125	1	0.5
S. epidermidis	ATCC 12228	0.125	1	0.5
Enterococcus faecalis	L559	≤0.03	0.25	≤0.03
E. faecium	L568	0.25	1	1
Bacillus subtilis	ATCC 6633	≤0.03	0.125	0.125
Micrococcus luteus	ATCC 10240	0.125	0.5	0.5
Mycobacterium smegmatis	mc^2 155	16	2	64
Moraxella catarrhalis	L3292	2	8	16
Clostridium difficile	L4013	0.5	0.125	0.25
Candida albicans	L145	>64	>64	>64

[a] abbreviations: MRSA, methicillin-resistant *Staphylococcus aureus*; MSSA, methicillin-sensitive *Staphylococcus aureus*; GISA, glycopeptide-intermediate *Staphylococcus aureus*.

Some polyethers (e.g., salinomycin, monensin) are commercially used as coccidiostatic agents and salinomycin has also been evaluated as anticancer agents [36]. Recently, salinomycin and other ionophores have been shown to have transmission blocking activity against the etiological agent of malaria [37]. When tested against one chloroquine-sensitive and one chloroquine-resistant strain of *Plasmodium falciparum*, α-770 and α-823 showed inhibitory activity in the 2–10 nM range comparable to those of salinomycin [37]. It should be noted that, although there are over 120 reported polyethers in the literature, their mechanism of action has been studied on a limited number of molecules [30] and we are not aware of studies aimed at making polyethers selective towards a particular cell type.

The 16S rRNA gene sequences of the polyether-producing *Actinoallomurus* strains of Table 2 were determined and compared to those of all described *Actinoallomurus* species. The results are shown in Figure 6. All the ten α-823 producers cluster together in a compact branch that includes the type strains *Actinoallomurus fulvus* and *A. amamiensis*: specifically, the identical 16S rRNA gene sequences from strains ID145265, -145554, -145603 and -145830 are 100% identical to that from *A. fulvus*; the 16S rRNA genes sequences from ID145802 and -145828 are 99.9% and 100% identical, respectively, to that from *A. amamiensis*; while the identical 16S rRNA gene sequences from strains ID145804, -145811 and -145816 and that from strain ID145823 are less related to those of described species (Figure 6). The octacyclomycin producer ID145817 is less closely related (≤99.1% identity) to *A. bryophytorum* and *A. yoronensis*, although all these strains belong to a related phylogenetic branch. The α-770 producer, instead, is distantly related to *A. spadix* (98.7% identity) and belongs to an unrelated branch that includes, among others, strain ID145113, the producer of the aromatic polyketide paramagnetoquinone [18]. Remarkably, all polyether-producing *Actinoallomurus* strains were isolated from soil samples of tropical origin (Table 2), notwithstanding that 64% of the screened strains were of non-tropical origin (mostly from Europe; see Table 1). These observations suggest that the branch including *A. fulvus*, *A. amamiensis* and the α-823 producers might consist of cosmopolitan strains and that polyether production might be mostly associated with *Actinoallomurus* strains from tropical environment. Previous studies have established a correlation between different classes of secondary metabolites and geographic origin [38,39], although we are unaware of previous reports on biogeography of polyether production.

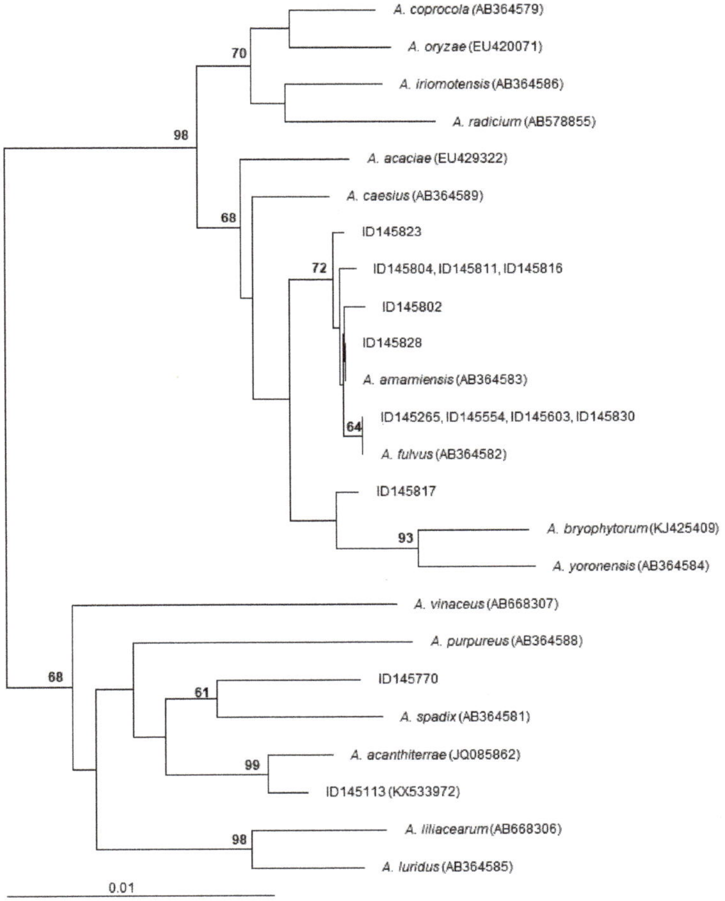

Figure 6. Neighbor-joining tree showing the phylogenetic position of polyether-producing *Actinoallomurus* strains. Type-strains of all described *Actinoallomurus* species are included. The sequence of the paramagnetoquinone producer *Actinoallomurus* sp. ID145113 is also included. The tree is based on 1309 unambiguously aligned positions in the 16S rRNA gene sequences. Numbers at the nodes are bootstrap values based on 100 resamplings; only values higher than 60 are shown. Scale bar represents 1 inferred substitutions per 100 nucleotides. The tree was rooted using *Streptosporangium roseum* 16S rRNA gene sequence (X89947) as outgroup.

3. Materials and Methods

3.1. Bacterial Strains and Media

Actinoallomurus strains are from the NAICONS strain library. Each strain was propagated on S1-5.5 plates (60 g/L oatmeal, 18 g/L agar, 1 mL/L Trace Elements Solution) at 30 °C for two to three weeks. From these plates, the grown mycelium was used to inoculate AF-A medium (10 g/L dextrose monohydrate, 4 g/L soybean meal, 1 g/L yeast extract, 0.5 g/L NaCl, 1.5 g/L 2-(*N*-morpholino) ethanesulfonic acid, pH adjusted to 5.6) in shake-flasks. After 8 days in a rotatory shaker (200 rpm) at 30 °C, cultures were harvested and extracted (see Section 3.2).

PCR amplifications with the eubacterial primers F27 and R1492 and phylogenetic analyses of the 16S rRNA gene sequences were performed as previously described [40]. The 16S rRNA gene sequences have been deposited in GenBank, as listed in Table 2.

3.2. Preparation of Extracts

Three different extracts were prepared from each culture. The culture was centrifuged at 16,000 rcf for 5 min and the resulting pellet was resuspended in 0.4 vol ethanol, while the supernatant was used for ethyl acetate extraction (see below). After shaking 1 h at 55 °C, the suspension was centrifuged once more (16,000 rcf, 5 min) and the supernatant transferred to a new tube, dried under vacuum and resuspended in 10% DMSO at 0.2× the original culture volume. This extract was designated as the mycelium extract.

The supernatant from the first centrifugation step above was extracted with 0.5 vol ethyl acetate. After mixing and phase separation, the organic phase was transferred to a new tube and the aqueous phase extracted again with further 0.5 vol ethyl acetate. The two organic phases were combined, dried and resuspended at 5× the original concentration in 10% DMSO. This extract was designated as EtAc extract. The exhausted aqueous phase was also retained and tested as such.

3.3. Antibacterial Assays

The screening was performed by agar diffusion, using plates containing 15 mL of Müller-Hinton Agar and inoculated with 5×10^5 CFU/mL of the indicator strain. Strains used in this assay were *Staphylococcus aureus* ATCC 6538P and *Escherichia coli* L4242, a Δ*tolC* derivative of MG1061. After spotting 20 µL of the resuspended extract, plates were incubated 18–20 h at 37 °C. After HPLC fractionation, bioactive fractions were identified using the same methodology.

MIC determinations of purified compounds were performed by broth micro dilution in sterile 96-well polystyrene microtiter plates according to CLSI guidelines, using Müller Hinton broth (Difco Laboratories) containing 20 mg/L $CaCl_2$ and 10 mg/L $MgCl_2$ for all strains except for *Streptococcus* spp., which were grown in Todd Hewitt broth. Strains were inoculated at 5×10^5 CFU/mL and incubated at 37 °C for 20−24 h. Strain with an L prefix are from the NAICONS pathogens library.

3.4. Analytical Procedures

For monitoring metabolites production analytical HPLC was performed on Shimadzu Series 10 spectrophotometer (Kyoto, Japan), equipped with a reverse-phase column, LiChrospher RP-18, 5 µm, 4.6 × 125 mm (Merck, Darmstadt, Germany). Phase A was 0.1% trifluoroacetic acid (TFA), phase B acetonitrile, and the flow rate was 1 mL/min. Resolution was achieved with a linear gradient from 10% to 36% phase B in 5 min; from 36% to 50% phase B in 7 min; and from 50% to 80% phase B in 1 min; followed by a 4-min isocratic step at 80% phase B and column re-equilibration. UV detection was at 230 and 270 nm. LC-MS analyses were performed on a Dionex UltiMate 3000 coupled with an LCQ Fleet (Thermo scientific) mass spectrometer equipped with an electrospray interface (ESI) and a tridimensional ion trap. The column was an Atlantis T3 C18 5 µm × 4.6 mm × 50 mm maintained at 40 °C at a flow rate of 0.8 mL/min. Phases A and B were 0.05% TFA in water and acetonitrile, respectively. The elution was with a 14-min multistep program that consisted of 10, 10, 95, 95, 10 and

10% phase B at 0, 1, 7, 12, 12.5 and 14 min, respectively. UV-VIS signals (190–600 nm) were acquired using the diode array detector. The *m/z* range was 110–2000 and the ESI conditions were as follows: spray voltage of 3500 V, capillary temperature of 275 °C, sheath gas flow rate at 35 units and auxiliary gas flow rate at 15 units.

High resolution MS spectra were recorded at Unitech OMICs (University of Milano, Italy) using a Triple TOF® 6600 (Sciex) equipped with an ESI source. The experiments were carried out by direct infusion in positive ionization mode. The ESI parameters were the following: curtain gas 25 units, ion spray voltage floating 5500 v, temperature 50 °C, ion source gas1 10 units, ion source gas2 0 units, declustering potential 80 v, syringe flow rate 10 µL/min, accumulation time 1 s.

Mono- and bi-dimensional NMR spectra were measured in $CDCl_3$ at 298K using an AMX 400 MHz spectrometer. Chemical shifts are reported relative to $CDCl_3$ (δ 7.26 ppm).

3.5. Purification of Polyethers

α-823: Nine parallel 100-mL cultures of *Actinoallomurus* sp. ID145823 in AF-A medium were harvested at seven days and filtered through paper under reduced pressure to separate the mycelium from the clear broth. The latter (860 mL) was extracted three times with 450 mL ethyl acetate while the mycelium was treated with 100 mL acetone, kept on a rotary shaker 1 h and centrifuged. The combined organic phases were dried under reduced pressure and dissolved in 2 mL dichloromethane. The sample was resolved on a 12 g direct-phase Flash column RediSep RF (Teledyne Isco) by using a CombiFlash RF Teledyne Isco medium-pressure chromatography system. The column was previously conditioned at 100% dichloromethane and then eluted at 15 mL/min with a 20-min linear gradient from 0 to 10% methanol. Fractions were analyzed by LC-MS and those with the highest purity were pooled and dried, obtaining 14 mg of purified α-823. Five mg were dissolved in $CDCl_3$ for NMR analysis.

Octacyclomycin: Two parallel 100-mL cultures of *Actinoallomurus* sp. ID145817 in AF-A medium were harvested at seven days and filtered through paper under reduced pressure to separate the mycelium from the clear broth. The latter (170 mL) was extracted three times with 80 mL ethyl acetate while the mycelium was treated with 100 mL ethanol, kept 1 h on a rotary shaker and centrifuged. The combined organic phases were dried under reduced pressure and dissolved in 2 mL dichloromethane. The sample was resolved by medium-pressure chromatography as described above for α-823. Fractions were analyzed by LC-MS and processed as above. Four mg of purified octacyclomycin were obtained.

α-770: Two parallel 100-mL cultures of *Actinoallomurus* sp. ID145770 in medium M8 [41] were harvested at seven days. Mycelium was harvested by centrifugation (10 min at 4000 rpm), treated with 20 mL ethanol, kept 1 h on a rotary shaker and centrifuged. The organic phase was recovered, dried under reduced pressure, dissolved in 2 mL dichloromethane and resolved by medium-pressure chromatography as described for α-823, except that the flow rate was set at 30 mL/min. Fractions 11–14, which showed activity against *S. aureus*, were analyzed by LC-MS and the ones containing similar signals were pooled, dried and dissolved in dichloromethane. A further purification step was performed by preparative thin layer chromatography on silica gel (Analtech Preparative Silica Gel GF with UV254 2000 µm; Sigma-Aldrich, St Louis, MO, USA) in dichloromethane:methanol 9:1. The spot at R_f 0.9 was dried and dissolved in $CDCl_3$ for NMR analysis. Four mg of purified α-770 were obtained.

4. Conclusions

When grown under one routine condition in shake-flasks and only looking at metabolites with growth inhibitory activity towards *S. aureus*, we have been able to show that *Actinoallomurus* strains can express several types of biosynthetic pathways: type I (for making polyethers and spirotetronates), type II (for aromatic polyketides) and type III (for the 6-methylsalicylate moiety of the polyether α-770) polyketide synthases; ribosomally synthesized and post-translationally modified peptides (lantibiotic); aminocoumarins; and short non-ribosomal peptide synthase derived products (diketopiperazines).

Some of the observed metabolites, e.g., the previously reported aromatic polyketide allocyclinones [17] and the polyether α-823, seem to be relatively frequent metabolites in the screened *Actinoallomurus* strains. Other metabolites represent rarer discovery events, with identical or close matches in several *Actinobacteria* genera. Indeed, the genus *Actinoallomurus* resulted from a reclassification of *Actinomadura* spp. within the family *Thermomonosporaceae*, order *Streptosporangiales* [9]. Some of the compounds described here (e.g., pradimicin and benanomicin) and the α-770- and α-823-related polyethers cationomycin and SF-2361, respectively, were previously reported from *Actinomadura* spp. Others of the described metabolites are produced by distantly related taxa, such as NAI-107 by *Microbispora* spp. (family *Streptosporangiaceae*, order *Streptosporangiales*), pyrrolosporin by a *Micromonospora* sp. (order *Micromonosporales*), in addition to the *Streptomyces*-produced coumermycin and octacyclomycin.

Supplementary Materials: The following are available online at http://www.mdpi.com/2079-6382/7/2/47/s1, Figure S1: LC-MS, UV analysis of ethyl acetate extract of *Actinoallomurus* sp. ID145519, Figure S2: LC-MS, UV analysis of ethyl acetate extract of *Actinoallomurus* sp. ID145814, Figure S3: LC-MS, UV analysis of mycelium extract of *Actinoallomurus* sp. ID145640, Figure S4: Comparison between mycelium extract of *Actinoallomurus* sp. ID145640 and NAI-107 standard, Figure S5: LC-MS, UV analysis of ethyl acetate extract of *Actinoallomurus* sp. ID145219, Figure S6: LC-MS, UV analysis of ethyl acetate extract of *Actinoallomurus* sp. ID145114, Figure S7: Fragmentation patterns of the eight different congeners of α-823, Figure S8: 1H-NMR spectrum of α-823, Figure S9: 1H-COSY NMR spectrum of α-823, Figure S10: 1H-TOCSY NMR spectrum of α-823, Figure S11: 1H-13C HSQC NMR spectrum of α-823, Figure S12: 1H-13C HMBC NMR spectrum of α-823, Figure S13: Major COSY and HMBC correlations of α-823, Figure S14: HRESI-MS spectrum of α-823, Figure S15: LC-MS, UV analysis of mycelium extract of *Actinoallomurus* sp. ID145817, Figure S16: Fragmentation patters of the two different congeners of α-770, Figure S17: 1H NMR spectrum of α-770, Figure S18: 1H-COSY NMR spectrum of α-770, Figure S19: 1H-TOCSY NMR spectrum of α-770, Figure S20: 1H-13C HSQC NMR spectrum of α-770, Figure S21: 1H-13C HMBC NMR spectrum of α-770, Figure S22: Major COSY and HMBC correlations of α-770, Figure S23: HRESI-MS spectrum of α-770, Table S1: Different α-823 congeners detected in the extract from *Actinoallomurus* sp. ID145823, Table S2: 1H and 13C NMR data for α-823 in CDCl3, Table S3: α-770 congeners detected in the extract from *Actinoallomurus* sp. ID145770, Table S4: 1H and 13C NMR data for α-770 in CDCl3.

Author Contributions: M.S. and S.D. conceived and designed the experiments; M.I., A.T., J.C.S.C., G.D.G., C.B. and S.I.M. performed the experiments; M.I., A.T., J.C.S.C., G.D.G., C.B., S.I.M., M.S. and S.D. analyzed the data; M.I., A.T. and S.D. wrote the paper.

Acknowledgments: Portion of this work was part of a PhD dissertation of J.C.S.C. at the University of Warwick, UK. This work was partially supported by the European Community's Seventh Framework Programme (FP7/2007-2013) under grant agreement 289285 and by grants from Regione Lombardia. We are grateful to Carlo Mazzetti, Mirko Ornaghi, Roberta Pozzi and Matteo Simone for their early contributions to this project, and to Donatella Taramelli and Silvia Parapini for the anti-plasmodium activity tests. We also thank the Unitech OMICs platform at the University of Milano for HRMS analyses.

Conflicts of Interest: The authors declare no conflict of interest.

References

1. Genilloud, O. Actinomycetes: Still a source of novel antibiotics. *Nat. Prod. Rep.* **2017**, *34*, 1203–1232. [CrossRef] [PubMed]
2. Monciardini, P.; Iorio, M.; Maffioli, S.; Sosio, M.; Donadio, S. Discovering new bioactive molecules from microbial sources. *Microb. Biotechnol.* **2014**, *7*, 209–220. [CrossRef] [PubMed]
3. Wright, G.D. Opportunities for natural products in 21st century antibiotic discovery. *Nat. Prod. Rep.* **2017**, *34*, 694–701. [CrossRef] [PubMed]
4. Jaspars, M.; Challis, G. Microbiology: A talented genus. *Nature* **2014**, *506*, 38–39. [CrossRef] [PubMed]
5. Donadio, S.; Busti, E.; Monciardini, P.; Bamonte, R.; Mazza, P.; Sosio, M.; Cavaletti, L. Sources of polyketides and non-ribosomal peptides. *Ernst Schering Res. Found. Workshop* **2005**, *51*, 19–41.
6. Busti, E.; Cavaletti, L.; Monciardini, P.; Schumann, P.; Rohde, M.; Sosio, M.; Donadio, S. *Catenulispora acidiphila* gen. nov., sp. nov., a novel, mycelium-forming actinomycete, and proposal of *Catenulisporaceae* fam. nov. *Int. J. Syst. Evol. Microbiol.* **2006**, *56*, 1741–1746. [CrossRef] [PubMed]
7. Cavaletti, L.; Monciardini, P.; Schumann, P.; Rohde, M.; Bamonte, R.; Busti, E.; Sosio, M.; Donadio, S. *Actinospica robiniae* gen. nov., sp. nov. and *Actinospica acidiphila* sp. nov.: Proposal for *Actinospicaceae* fam.

nov. and *Catenulisporinae* subord. nov. in the order *Actinomycetales*. *Int. J. Syst. Evol. Microbiol.* **2006**, *56*, 1747–1753. [CrossRef] [PubMed]

8. Monciardini, P.; Cavaletti, L.; Ranghetti, A.; Schumann, P.; Rohde, M.; Bamonte, R.; Sosio, M.; Mezzelani, A.; Donadio, S. Novel members of the family *Micromonosporaceae*, *Rugosimonospora acidiphila* gen. nov., sp. nov. and *Rugosimonospora africana* sp. nov. *Int. J. Syst. Evol. Microbiol.* **2009**, *59*, 2752–2758. [CrossRef] [PubMed]
9. Tamura, T.; Ishida, Y.; Nozawa, Y.; Otoguro, M.; Suzuki, K. Transfer of *Actinomadura spadix* Nonomura and Ohara 1971 to *Actinoallomurus spadix* gen. nov., comb. nov., and description of *Actinoallomurus amamiensis* sp. nov., *Actinoallomurus caesius* sp. nov., *Actinoallomurus coprocola* sp. nov., *Actinoallomurus fulvus* sp. nov., *Actinoallomurus iriomotensis* sp. nov., *Actinoallomurus luridus* sp. nov., *Actinoallomurus purpureus* sp. nov. and *Actinoallomurus yoronensis* sp. nov. *Int. J. Syst. Evol. Microbiol.* **2009**, *59*, 1867–1874. [PubMed]
10. Thamchaipenet, A.; Indananda, C.; Bunyoo, C.; Duangmal, K.; Matsumoto, A.; Takahashi, Y. *Actinoallomurus acaciae* sp. nov., an endophytic actinomycete isolated from *Acacia auriculiformis* A. Cunn. ex Benth. *Int. J. Syst. Evol. Microbiol.* **2010**, *60*, 554–559. [CrossRef] [PubMed]
11. Indananda, C.; Thamchaipenet, A.; Matsumoto, A.; Inahashi, Y.; Duangmal, K.; Takahashi, Y. *Actinoallomurus oryzae* sp. nov., an endophytic actinomycete isolated from roots of a Thai jasmine rice plant. *Int. J. Syst. Evol. Microbiol.* **2011**, *61*, 737–741. [CrossRef] [PubMed]
12. Pozzi, R.; Simone, M.; Mazzetti, C.; Maffioli, S.; Monciardini, P.; Cavaletti, L.; Bamonte, R.; Sosio, M.; Donadio, S. The genus *Actinoallomurus* and some of its metabolites. *J. Antibiot.* **2011**, *64*, 133–139. [CrossRef] [PubMed]
13. Koyama, R.; Matsumoto, A.; Inahashi, Y.; Omura, S.; Takahashi, Y. Isolation of actinomycetes from the root of the plant, *Ophiopogon japonicus*, and proposal of two new species, *Actinoallomurus liliacearum* sp. nov. and *Actinoallomurus vinaceus* sp. nov. *J. Antibiot.* **2012**, *65*, 335–340. [CrossRef] [PubMed]
14. Matsumoto, A.; Fukuda, A.; Inahashi, Y.; Omura, S.; Takahashi, Y. *Actinoallomurus radicium* sp. nov., isolated from the roots of two plant species. *Int. J. Syst. Evol. Microbiol.* **2012**, *62*, 295–298. [CrossRef] [PubMed]
15. Inahashi, Y.; Iwatsuki, M.; Ishiyama, A.; Matsumoto, A.; Hirose, T.; Oshita, J.; Sunazuka, T.; Panbangred, W.; Takahashi, Y.; Kaiser, M.; et al. Actinoallolides A-E, new anti-trypanosomal macrolides, produced by an endophytic actinomycete, *Actinoallomurus fulvus* MK10-036. *Org. Lett.* **2015**, *17*, 864–867. [CrossRef] [PubMed]
16. Mazzetti, C.; Ornaghi, M.; Gaspari, E.; Parapini, S.; Maffioli, S.; Sosio, M.; Donadio, S. Halogenated spirotetronates from *Actinoallomurus*. *J. Nat. Prod.* **2012**, *75*, 1044–1050. [CrossRef] [PubMed]
17. Cruz, J.C.S.; Maffioli, S.; Bernasconi, A.; Brunati, C.; Gaspari, E.; Sosio, M.; Wellington, E.; Donadio, S. Allocyclinones, hyperchlorinated angucyclinones and common metabolites from *Actinoallomurus*. *J. Antibiot.* **2017**, *70*, 73–78. [CrossRef] [PubMed]
18. Iorio, M.; Cruz, J.; Simone, M.; Bernasconi, A.; Brunati, C.; Sosio, M.; Donadio, S.; Maffioli, S.I. Antibacterial paramagnetic quinones from *Actinoallomurus*. *J. Nat. Prod.* **2017**, *80*, 819–827. [CrossRef] [PubMed]
19. Cruz, J.C.S.; Iorio, M.; Monciardini, P.; Simone, M.; Brunati, C.; Gaspari, E.; Maffioli, S.I.; Wellington, E.; Sosio, M.; Donadio, S. Brominated variant of the lantibiotic NAI-107 with enhanced antibacterial potency. *J. Nat. Prod.* **2015**, *78*, 2642–2647. [CrossRef] [PubMed]
20. Kammerer, B.; Kahlich, R.; Laufer, S.; Li, S.M.; Heide, L.; Gleiter, C.H. Mass spectrometric pathway monitoring of secondary metabolites: Systematic analysis of culture extracts of *Streptomyces* species. *Anal. Biochem.* **2004**, *335*, 17–29. [CrossRef] [PubMed]
21. Vieweg, L.; Reichau, S.; Schobert, R.; Leadlay, P.F.; Süssmuth, R.D. Recent advances in the field of bioactive tetronates. *Nat. Prod. Rep.* **2014**, *31*, 1554–1584. [CrossRef] [PubMed]
22. Lam, K.S.; Hesler, G.A.; Gustavson, D.R.; Berry, R.L.; Tomita, K.; MacBeth, J.L.; Ross, J.; Miller, D.; Forenza, S. Pyrrolosporin A, a new antitumor antibiotic from *Micromonospora* sp. C39217-R109-7. I. Taxonomy of producing organism, fermentation and biological activity. *J. Antibiot.* **1996**, *49*, 860–864. [CrossRef] [PubMed]
23. Schroeder, D.R.; Colson, K.L.; Klohr, S.E.; Lee, M.S.; Matson, J.A.; Brinen, L.S.; Clardy, J. Pyrrolosporin A, a new antitumor antibiotic from *Micromonospora* sp. C39217-R109-7. II. Isolation, physic-chemical properties, spectroscopic study and X-ray analysis. *J. Antibiot.* **1996**, *9*, 865–872. [CrossRef]
24. Arnison, P.G.; Bibb, M.J.; Bierbaum, G.; Bowers, A.A.; Bugni, T.S.; Bulaj, G.; Camarero, J.A.; Campopiano, D.J.; Challis, G.L.; Clardy, J.; et al. Ribosomally synthesized and post-translationally modified peptide natural products: Overview and recommendations for a universal nomenclature. *Nat. Prod. Rep.* **2013**, *30*, 108–160. [CrossRef] [PubMed]

25. Lee, M.D. Antibiotics from Microbispora. U.S. Patent 6,551,591, 22 April 2003.
26. Lazzarini, A.; Gastaldo, L.; Candiani, G.; Ciciliato, I.; Losi, D.; Marinelli, F.; Selva, E.; Parenti, F. Antibiotic 107891, Its Factors A1 and A2, Pharmaceutically Acceptable Salts and Compositions, and Use Thereof. U.S. Patent 7,351,687, 1 April 2008.
27. Belin, P.; Moutiez, M.; Lautru, S.; Seguin, J.; Pernodet, J.L.; Gondry, M. The nonribosomal synthesis of diketopiperazines in tRNA-dependent cyclodipeptide synthase pathways. *Nat. Prod. Rep.* **2012**, *29*, 961–979. [CrossRef] [PubMed]
28. Oki, T.; Kinoshi, M.; Tomatsu, K.; Tomita, K.; Saitoh, K.; Tsunakawa, M.; Nishio, M.; Miyaki, T.; Kawaguchi, H. Pradimicin, a novel class of potent antifungal antibiotics. *J. Antibiot.* **1999**, *11*, 1701–1704. [CrossRef]
29. Tadashi, N.; Yukio, T.; Yuko, K.; Mamoru, I.; Taneto, T.; Shinji, M.; Masaji, S.; Michio, K. Novel antibiotic substance SF-2361, production and used thereof. Japan Patent 61260888, 19 November 1986.
30. Kevin, D.A., II; Meujo, D.A.F.; Hamann, M.T. Polyether ionophores: Broad-spectrum and promising biologically active molecules for the control of drug-resistant bacteria and parasites. *Expert Opin. Drug Discov.* **2009**, *4*, 109–146. [CrossRef] [PubMed]
31. Funayama, S.; Nozoe, S.; Tronquet, C.; Anraku, Y.; Komiyama, K.; Omura, S. Isolation and structure of a new polyether antibiotic, octacyclomycin. *J. Antibiot.* **1992**, *45*, 1686–1691. [CrossRef] [PubMed]
32. Nakamura, G.; Kobayashi, K.; Sakurai, T.; Isono, K. Cationomycin, a new polyether ionophore antibiotic produced by *Actinomadura* nov. sp. *J. Antibiot.* **1981**, *34*, 1513–1514. [CrossRef] [PubMed]
33. Minami, A.; Oguri, H.; Watanabe, K.; Oikawa, H. Biosynthetic machinery of ionophore polyether lasalocid: Enzymatic construction of polyether skeleton. *Curr. Opin. Chem. Biol.* **2013**, *17*, 555–561. [CrossRef] [PubMed]
34. Ubukata, M.; Uzawa, J.; Isono, K. Biosynthesis of cationomycin: Direct and indirect incorporation of 13C-acetate and application of homoscalar correlated 2-D carbon-13 NMR and double quantum coherence. *J. Am. Chem. Soc.* **1984**, *106*, 2213–2214. [CrossRef]
35. Rutkowski, J.; Brzezinski, B. Structures and properties of naturally occurring polyether antibiotics. *BioMed Res. Int.* **2013**, *2013*, 162513. [CrossRef] [PubMed]
36. Zhou, S.; Wang, F.; Wong, E.T.; Fonkem, E.; Hsieh, T.C.; Wu, JM.; Wu, E. Salinomycin: A novel anti-cancer agent with known anti-coccidial activities. *Curr. Med. Chem.* **2013**, *20*, 4095–4101. [CrossRef] [PubMed]
37. D'Alessandro, S.; Corbett, Y.; Ilboudo, D.P.; Misiano, P.; Dahiya, N.; Abay, S.M.; Habluetzel, A.; Grande, R.; Gismondo, M.R.; Dechering, K.J.; et al. Salinomycin and other ionophores as a new class of antimalarial drugs with transmission-blocking activity. *Antimicrob. Agents Chemother.* **2015**, *59*, 5135–5144. [CrossRef] [PubMed]
38. Charlop-Powers, Z.; Owen, J.G.; Reddy, B.V.B.; Ternei, M.A.; Brady, S.F. Chemical-biogeographic survey of secondary metabolism in soil. *Proc. Natl. Acad. Sci. USA* **2014**, *111*, 3757–3762. [CrossRef] [PubMed]
39. Charlop-Powers, Z.; Pregitzer, C.C.; Lemetre, C.; Ternei, M.A.; Maniko, J.; Hover, B.M.; Calle, P.Y.; McGuire, K.L.; Garbarino, J.; Forgione, H.M.; et al. Urban park soil microbiomes are a rich reservoir of natural product biosynthetic diversity. *Proc. Natl. Acad. Sci. USA* **2016**, *113*, 14811–14816. [CrossRef] [PubMed]
40. Monciardini, P.; Sosio, M.; Cavaletti, L.; Chiocchini, C.; Donadio, S. New PCR primers for the selective amplifcation of 16S rDNA from different groups of actinomycetes. *FEMS Microbiol. Ecol.* **2002**, *42*, 419–429. [PubMed]
41. Donadio, S.; Monciardini, P.; Sosio, M. Chapter 1. Approaches to discovering novel antibacterial and antifungal agents. *Methods Enzymol.* **2009**, *458*, 3–28. [PubMed]

antibiotics

Article

Specificity of Induction of Glycopeptide Antibiotic Resistance in the Producing Actinomycetes

Elisa Binda [1,2], Pamela Cappelletti [1,2], Flavia Marinelli [1,2] and Giorgia Letizia Marcone [1,2,*]

[1] Department of Biotechnology and Life Sciences, University of Insubria, via J.H. Dunant 3, 21100 Varese, Italy; elisa.binda@uninsubria.it (E.B.); pamela.cappelletti@uninsubria.it (P.C.); flavia.marinelli@uninsubria.it (F.M.)

[2] The Protein Factory Research Center, Politecnico di Milano and University of Insubria, via J.H. Dunant 3, 21100 Varese, Italy

* Correspondence: giorgia.marcone@uninsubria.it; Tel.: +39-0332-421332

Received: 28 February 2018; Accepted: 20 April 2018; Published: 25 April 2018

Abstract: Glycopeptide antibiotics are drugs of last resort for treating severe infections caused by Gram-positive pathogens. It is widely believed that glycopeptide-resistance determinants (*van* genes) are ultimately derived from the producing actinomycetes. We hereby investigated the relationship between the antimicrobial activity of vancomycin and teicoplanins and their differential ability to induce *van* gene expression in *Actinoplanes teichomyceticus*—the producer of teicoplanin—and *Nonomuraea gerenzanensis*—the producer of the teicoplanin-like A40926. As a control, we used the well-characterized resistance model *Streptomyces coelicolor*. The enzyme activities of a cytoplasmic-soluble D,D-dipeptidase and of a membrane-associated D,D-carboxypeptidase (corresponding to VanX and VanY respectively) involved in resistant cell wall remodeling were measured in the actinomycetes grown in the presence or absence of subinhibitory concentrations of vancomycin, teicoplanin, and A40926. Results indicated that actinomycetes possess diverse self-resistance mechanisms, and that each of them responds differently to glycopeptide induction. Gene swapping among teicoplanins-producing actinomycetes indicated that cross-talking is possible and provides useful information for predicting the evolution of future resistance gene combinations emerging in pathogens.

Keywords: actinomycetes; glycopeptide antibiotics; teicoplanin; A40926; *van* resistance genes

---

1. Introduction

Glycopeptide antibiotics (GPAs) are drugs of last resort for treating severe infections caused by Gram-positive pathogens such as *Staphylococcus aureus* (SA), *Enterococcus* spp., and *Clostridioides difficile* [1]. Clinically important GPAs include first-generation vancomycin and teicoplanin—which, although discovered many decades ago, continue to be extensively used in clinical practice—and second-generation telavancin, dalbavancin, and oritavancin, which were recently approved for clinical use for their increased antimicrobial potency and superior pharmacokinetic properties [2–4]. Vancomycin and teicoplanin are natural product GPAs produced by soil-dwelling filamentous actinomycetes. Their common structural motif is a core heptapeptide scaffold containing aromatic amino acids that have undergone extensive oxidative cross-linking and decoration with different moieties, such as sugar residues, chlorine atoms and—in case of teicoplanin—a lipid chain. GPAs inhibit peptidoglycan (PG) synthesis by binding to the D-alanyl-D-alanine (D-Ala-D-Ala) terminus of the peptide stem of PG-precursor lipid II. The binding of GPAs to lipid II by forming five hydrogen bonds locks PG precursors, impeding subsequent cross-linking reactions [5,6]. Second-generation GPAs are semisynthetic derivatives of vancomycin- and teicoplanin-like molecules, where the chemical modifications were introduced outside the D-Ala-D-Ala binding pocket, mainly involving the appendage of hydrophobic aryl or acyl groups that mimic the natural lipid chain of teicoplanin. In fact, the superior antimicrobial potency of teicoplanin-like molecules is due to membrane anchoring of the hydrophobic tail, which strengthens the bond to membrane-localized lipid II [7,8]. Additionally, Dong et al. [9] demonstrated that lipidation is the key functional difference between vancomycin and teicoplanin related to their differing abilities of inducing a GPA resistance response in enterococci. More recently Kwun and Hong [10], using the harmless actinomycete *Streptomyces coelicolor* as a model resistance system, confirmed that teicoplanin-like derivatives are poor inducers of GPA resistance, and a lack of induction accounts for the susceptibility to these molecules.

Many different GPA-resistant phenotypes have been described in enterococci and staphylococci (for an extensive review, see Binda et al., 2014 [2]). In the two most prominent manifestations of resistance (VanA or VanB phenotypes), the GPA-induced expression of *van* genes remodels the bacterial cell wall. The replacement of the dipeptide (D-Ala-D-Ala) terminus of PG precursor with the depsipeptide D-alanyl-D-lactate (D-Ala-D-Lac) reduces by 1000-fold the GPA affinity to their molecular target [11]. Strains displaying this type of resistance are either resistant to both vancomycin and teicoplanin (VanA phenotype), or they are resistant only to vancomycin and susceptible to teicoplanin (VanB phenotype) [12]. Although the proteins directly involved in conferring VanA resistance—i.e., VanH, which converts pyruvate into D-lactate; VanA, a D-Ala-D-Lac ligase; and VanX, a D-Ala-D-Ala dipeptidase—are highly homologous to their counterparts in the VanB phenotype (VanHB, VanB, VanXB), the two-component regulatory systems controlling *van* gene transcription (VanS/VanR in VanA phenotype and VanSB/VanRB in VanB phenotype) are only distantly related [13]. In particular, the membrane-associated sensor domains of VanS and VanSB are unrelated in amino acid sequence, and they respond to GPAs by different mechanisms which account for the difference in induction specificity by vancomycin and teicoplanin and their semi-synthetic derivatives [13–15]. Many efforts [12,14,16,17] have been devoted to identifying the molecular species responsible for differently inducing VanS and VanSB, but their entity and mode of action—i.e., direct binding of the GPAs to the sensor domain or its activation by cell wall intermediates that accumulate as a result of antibiotic action—is still being questioned.

Since it is widely believed that GPA resistance mechanisms are ultimately derived from GPA-producing actinomycetes, which use them to avoid suicide during antibiotic production [2,14], in this paper we investigated the specificity of induction of GPA resistance in teicoplanin- and A40926-producing actinomycetes. Teicoplanin is produced by *Actinoplanes teichomyceticus*, and the teicoplanin-like molecule A40926 [18], which is the natural precursor of the second-generation dalbavancin, is produced by the recently classified *Nonomuraea gerenzanensis* [19,20]. As a control, we used the well-characterized resistance model *S. coelicolor*, which does not synthesize any GPA but does possess a *van* gene cluster conferring inducible resistance to vancomycin but not to teicoplanin, showing the features of the VanB phenotype [10,16,21]. The purpose of this study was to elucidate the relationship between GPA activity and the ability to induce *van* gene expression in the producing actinomycetes, which are considered the evolutionary source of resistance determinants emerging in pathogens, shedding light on the possible evolution of the *van* gene cluster.

2. Results and Discussion

Table 1 reports the minimal inhibitory concentrations (MICs) of vancomycin, teicoplanin, and A40926 against *S. coelicolor*, *A. teichomyceticus*, and *N. gerenzanensis* on solid media. MICs were measured on solid media, since the standard broth dilution method used in unicellular bacteria to determine MICs by turbidity is compromised in mycelial actinomycetes by the formation of multicellular aggregates and by the coexistence of cells in different physiological states (e.g., vegetative mycelium, aerial mycelium, and spores). This is particularly true for difficult-to-cultivate nonstreptomyces actinomycetes such as *Actinoplanes* and *Nonomuraea* strains, which form compact and different-sized clumps when growing in liquid cultures.

Table 1. Minimal inhibitory concentrations (MICs) of glycopeptide antibiotics (GPAs), ramoplanin and bacitracin. The values represent the average of the data from three independent experiments.

Actinomycetes	Vancomycin (μg/mL)	Teicoplanin (μg/mL)	A40926 (μg/mL)	Ramoplanin (μg/mL)	Bacitracin (μg/mL)
S. coelicolor	>100	1.5 ± 0.025	1.5 ± 0.075	0.9 ± 0.045	0.9 ± 0.05
S. coelicolor ΔvanRS	1.25 ± 0.03	1.5 ± 0.02	1.5 ± 0.02	0.9 ± 0.025	0.9 ± 0.035
A. teichomyceticus	90 ± 4.2	20 ± 1.5	32.5 ± 1.5	20 ± 1	20 ± 1.15
N. gerenzanensis	20 ± 1.6	0.9 ± 0.01	4 ± 0.2	20 ± 1.2	20 ± 1.3
N. gerenzanensis pST30	20 ± 1.05	2.75 ± 0.15	17.5 ± 0.875	20 ± 1.1	20 ± 1.2

As expected, *S. coelicolor* was resistant to vancomycin and susceptible to teicoplanin and A40926. Its *ΔvanRS* mutant did not respond to vancomycin, and consequently, it was sensitive to it [10]. For these *Streptomyces* strains, MICs values were slightly higher than those previously reported in a liquid medium [10], which is likely due to the different cultivation method used. Interestingly, *S. coelicolor* and its *ΔvanRS* mutant were equally sensitive to bacitracin and ramoplanin, which are antibiotics structurally unrelated to GPAs that inhibit the late steps of PG synthesis by a diverse mode of action [22,23], evidently not mediated by VanS interaction.

A. teichomyceticus was resistant to all the GPAs tested (Table 1), although its chromosome harbours a canonical *vanHAX* gene cluster including the *vanRS* two component-regulatory system associated with the teicoplanin biosynthetic genes [24]. In this actinomycete, *van* genes are expressed constitutively, even in the absence of the antibiotic, making the cells intrinsically resistant to GPAs [24,25].

In contrast, the A40926-producing strain *N. gerenzanensis* was resistant to vancomycin (less than *A. teichomyceticus* and *S. coelicolor*) but sensitive to teicoplanin and to its own product, albeit at a lower extent. As previously described [26], *N. gerenzanensis* does not possess a *vanHAX* gene cluster, and the only known mechanism of resistance relies on the action of a VanY metallo-D,D-carboxypeptidase (named VanYn) that hydrolyses the C-terminal D-Ala residue of PG pentapeptide precursors. Interestingly, the well-characterized VanA and VanB type enterococci possess, in addition to *vanHAX* genes, an extra *vanY* gene that plays an ancillary unessential role in conferring glycopeptide resistance [27]. Upon vancomycin induction, enterococcal VanX cleaves any residual cytoplasmic D-Ala-D-Ala dipeptide, ensuring that the newly formed PG precursors terminate mostly in D-Ala-D-Lac, whereas VanY just acts on PG precursors that have escaped VanX hydrolysis and converts them into GPA-resistant tetrapeptides [27]. The integration of the pST30 plasmid containing the complete *vanRSHAX* gene cluster from *A. teichomyceticus* [25,28] into the *N. gerenzanensis* chromosome, consistently rendered the host strain more resistant to teicoplanin and A40926 in comparison to the parental strain harbouring only the *vanY* gene (Table 1), confirming the role of *vanHAX* genes in conferring high GPA resistance.

In contrast to *S. coelicolor*, both *A. teichomyceticus* and *N. gerenzanensis* were intrinsically resistant to both bacitracin and ramoplanin. This last result merits further investigation, although it is well-known that PG structure and density, and consequently antibiotic resistance profile, might dramatically vary among different genera of actinomycetes [19,29,30].

To determine the correlation between the antimicrobial activity of the GPAs and their ability to induce the *van* resistance system in actinomycetes, we followed the enzyme activities corresponding to either VanX or VanY peptidases in cells growing in liquid culture in the presence or absence of subinhibitory concentrations of GPAs (Figures 1 and 2). The D,D-dipeptidase activity of VanX (Figure 1) and the D,D-carboxypeptidase activity of VanY (Figure 2) were measured by determining the amount of D-Ala released from the hydrolysis of the D-Ala-D-Ala dipeptide and of the N_ε-acetyl-L-Lys-D-Ala-D-Ala tripeptide, respectively. D-Ala release was measured by using a D-amino acid oxidase coupled to a peroxidase [25,31]. As in the case of enterococci [27], VanX activity was detectable in the cytoplasmic fractions of *S. coelicolor*, *S. coelicolor ΔvanRS*, *A. teichomyceticus*, and *N. gerenzanensis* pST30 (Figure 1), and it was specific for the hydrolysis of D,D-dipeptides, being inactive on the ester D-Ala-D-Lac and on dipeptides substituted at the C or N terminus (data not shown). No VanX activity was detectable in *N. gerenzanensis* cytoplasmic fractions (see Table S1 Supplementary Material). Alternatively, VanY activity was detectable only in the membrane fractions of *N. gerenzanensis* and in its recombinant-derived *N. gerenzanensis* pST30 strain (Figure 2), consistent with the predicted VanY N-terminal structure, which contains a hydrophobic domain [31]. No VanY activity was detectable in *S. coelicolor*, *S. coelicolor ΔvanRS*, and *A. teichomyceticus* (see Table S1 Supplementary Material).

In *S. coelicolor* and *S. coelicolor ΔvanRS*, basal VanX activity (without any GPA addition) reached the maximum value within the 24 h of growth. When *S. coelicolor* parental strain was grown in the presence of subinhibitory concentrations of vancomycin (10 μg/mL), teicoplanin (0.75 μg/mL), and A40926 (0.75 μg/mL), only vancomycin induced an increase in VanX activity, and the level of enzyme activity doubled within the first 24 h from the induction (Figure 1). The addition of ramoplanin and bacitracin (both added at 0.45 μg/mL) did not show any effect on VanX activity (data not shown), thus indicating the specificity of vancomycin induction. Moreover, the addition of subinhibitory concentrations of vancomycin (in this case 0.6 μg /mL), teicoplanin (0.75 μg/mL), A40926 (0.75 μg/mL), and ramoplanin and bacitracin (both added at 0.45 μg/mL) to *S. coelicolor ΔvanRS* did not induce any increase in VanX activity, confirming the role of VanS in responding to vancomycin induction (Figure 1).

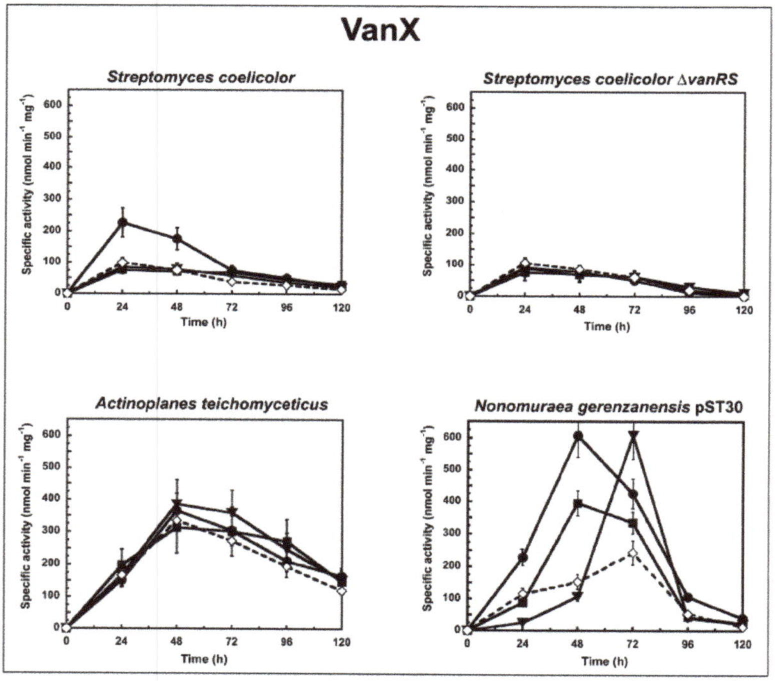

Figure 1. VanX activity in *S. coelicolor*, *S. coelicolor ΔvanRS*, *A. teichomyceticus*, and *N. gerenzanensis* pST30 grown in the absence (dotted line) or in the presence (continuous line) of subinhibitory concentrations of GPAs added at the moment of inoculum (see material and methods). VanX-specific activity is defined as the number of nanomoles of D-Ala released by the hydrolysis of D-Ala-D-Ala dipeptide at 37 °C per minute per milligram of protein contained in cytoplasmic fractions. Symbols represent non-induction (◇), induction with vancomycin (●), or teicoplanin (■), or A40926 (▼). The values represent the averages from three independent experiments, with a standard deviation of <5%.

In *A. teichomyceticus*, basal VanX activity (without any GPA addition) reached the maximum value after 48 h of growth. No significant variation in VanX activity was detectable following the addition of 45 μg/mL of vancomycin, of 10 μg/mL of teicoplanin, or of 15 μg/mL of A40926 (Figure 1). As expected, the MICs of GPAs measured on *A. teichomyceticus* after 48 and 72 h of growth in the presence of teicoplanin were the same as reported in Table 1 (see Table S2 in Supplementary Material). These results are in agreement with the constitutive expression of *vanHAX* genes that, according to Beltrametti et al. [24], is due to an impaired phosphatase function of the mutated VanS, which is consequently locked in the "on" state in this organism [14]. The addition of ramoplanin and bacitracin did not exert any specific effect on the VanX activity in *A. teichomyceticus* (data not shown).

Interesting is what occurred in the recombinant strain *N. gerenzanensis* pST30 that, in addition to *vanYn*, also harbours the heterologous *vanX* from *A. teichomyceticus*. In this strain, the basal D,D-dipeptidase activity of VanX reached the maximum value after 72 h of growth. VanX activity was induced by 10 μg/mL of vancomycin, by 1.75 μg/mL of teicoplanin, and by 10 μg/mL of A40926, albeit with different kinetics and intensity (Figure 1). When the *A. teichomyceticus vanRSHAX* genes were integrated into the host genome, VanX activity was not longer under the control of the parental *vanRS* but became inducible by GPAs, undergoing regulation by a still-unknown circuit present in *N. gerenzanensis*, which responds to subinhibitory concentrations of vancomycin (within the 48 h of growth), A40926 (within the 72 h), and, albeit to less extent, teicoplanin (within the 48 h) (Figure 1).

To shed light on the specificity of induction of GPA resistance in *N. gerenzanensis*, we compared the VanY activity in *N. gerenzanensis* and in its recombinant strain *N. gerenzanensis* pST30 when grown in the absence or presence of GPAs (Figure 2). In both the strains, basal VanY activity reached its maximum activity after 72 h of growth. Upon induction with 10 μg/mL of vancomycin, 0.45 μg/mL of teicoplanin, or 2 μg/mL of A40926, VanY activity in the parental strain increased within the first 24 h (Figure 2), suggesting that VanY-based resistance was induced by the exposure to the three GPAs. As expected, MICs of GPAs measured on *N. gerenzanensis* after 48 and 72 h of growth in the presence of A40926 increased, as reported in Table S2 of Supplementary Material. Again, the addition of ramoplanin and bacitracin did not influence the activity profile (data not shown). Surprisingly, the membrane-associated VanY activity in *N. gerenzanensis* pST30 was significantly induced only by vancomycin, and its detection was delayed in comparison to the parental strain, reaching its maximum value after 96 h of growth (Figure 2). In the presence of A40926 and teicoplanin, the VanY activity was comparable to the basal level previously observed in the noninduced recombinant strain (Figure 2). Intriguingly, these data suggest that swapping heterologous *A. teichomyceticus* genes in *N. gerenzanensis* pST30 did alter the specificity of induction of the homologous VanY activity, although the heterologous VanX activity responded to induction by the three GPAs as the VanY activity did in the parental strain. It seems that in the presence of the heterologous VanX dipeptidase, homologous VanY carboxyesterase tends to play an auxiliary role as it occurs in VanA and VanB enterococci [21]. Irrespective of this finding, further investigations on the still-unknown regulatory genes controlling GPA resistance in *N. gerenzanensis* are needed to better explain these diverse responses to GPA induction. In a previous '*van* genes' swapping experiment conducted by Hutchings et al., 2006, introducing the *Streptomyces toyocaensis* VanRS signal transduction system into *S. coelicolor* Δ*vanRS* switched inducer specificity to that of *S. toyocaensis*, whose resistant genes are induced by A47934 but not by vancomycin [21]. The authors concluded that the inducer specificity was determined by the origin of VanS/VanR [21]. More recently, Kilian et al. reported that the VanRS homologous two-component system VnlRS of *Amycolatopsis balhimycina* (which produces the vancomycin-type balhimycin) activated the transcription of the *vanHAX* genes in *S. coelicolor*, but not in *A. balhimycina* [32]. Surprisingly, the introduction of VnlRS from *A. balhimycina* into *S. coelicolor* induced teicoplanin resistance, most likely by activating further unknown genes required for teicoplanin resistance [32]. Although two members of a putative VanRS-like two-component signal transduction system were previously identified in the A40926 biosynthetic cluster in *N. gerenzanensis* [33], their role in controlling the VanY-based self-resistance mechanism and in responding to GPA induction is still unveiled and merit further investigations.

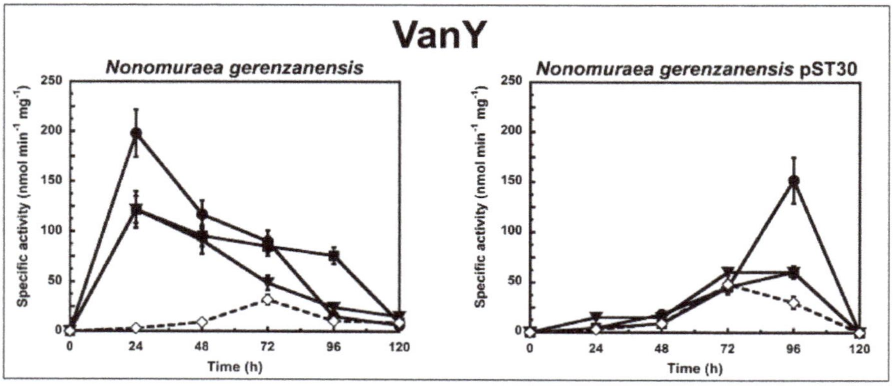

Figure 2. VanY activity in *N. gerenzanensis* and *N. gerenzanensis* pST30 grown in the absence (dotted line) or in the presence (continuous line) of subinhibitory concentrations of GPAs added at the moment of inoculum (see material and methods). VanY-specific activity is defined as the number of nanomoles of D-Ala released by the hydrolysis of N_ε-acetyl-L-Lys-D-Ala-D-Ala tripeptide at 37 °C per minute per milligram of protein contained in membrane extracts. Symbols represent non-induction (◇), induction with vancomycin (●), or teicoplanin (■), or A40926 (▼). The values represent the averages from three independent experiments, with a standard deviation of <5%.

3. Materials and Methods

3.1. Strains and Media

S. coelicolor A3(2) was a gift from Mervyn Bibb, John Innes Institute, Norwich, UK [34]. The Δ*vanRS* mutant of *S. coelicolor* was generously donated by Hee-Jeon Hong, University of Cambridge, UK [21]. *Streptomyces* spp. strains were maintained as spores in 10% (*v/v*) glycerol and propagated in MS agar media [34]. Liquid media for streptomycetes were YEME [34] and BTSB (Difco). Colonies were picked up from agar plates and inoculated into 300 mL Erlenmeyer flasks containing 50 mL of YEME. Flask cultures were incubated on a rotary shaker at 200 rpm and 28 °C.

N. gerenzanensis ATCC 39727 [19], its recombinant strain *N. gerenzanensis* pST30 [25], and *A. teichomyceticus* ATCC 31121 [35] were maintained as lyophilized master cell banks (MCBs). The mycelium from the MCBs was streaked on slants of a salt medium (SM) [34] solidified with agar (15 g/liter). After its growth, the mycelium from a slant was homogenized in 10 mL of 0.9% (*w/v*) NaCl, inoculated into liquid SM, grown for 96 h at 28 °C with aeration, and stored as a working cell bank (WCB) in 1.5 mL cryovials at −80 °C. One vial was used to inoculate each 300 mL Erlenmeyer flask containing 50 mL VSP medium [26], and the flasks were incubated at 28 °C, with shaking at 200 rpm. Previous HPLC data showed that no A40926 or teicoplanin production occurred in this vegetative medium [26,35]. In the case of *N. gerenzanensis* pST30, the VSP medium was added with 50 µg/mL apramycin to maintain plasmid selection. Surface cultures were grown on V0.1 agar [26]. All medium components were from Sigma-Aldrich (St. Louis, MO, USA) unless otherwise stated.

3.2. Induction Experiments

S. coelicolor A3(2), its Δ*vanRS* mutant, *A. teichomyceticus* ATCC 31121, *N. gerenzanensis* ATCC 39727, and *N. gerenzanensis* pST30 were grown as described above. Vancomycin, teicoplanin, A40926, bacitracin, and ramoplanin (all from Sigma-Aldrich) were dissolved in MilliQ water and sterilized by filtration using 0.22 µm filters. Antibiotics were added to the cultures at the moment of inoculation using concentrations calculated as the half-point of MICs (see below and Table 1), except for vancomycin (10 μg/mL) in *S. coelicolor* A3(2), as reported by Kwun et al. [10]. At these concentrations, growth curves of noninduced and induced strains were overlapping.

3.3. MIC Determination

Cryovials of WCBs were thawed at room temperature and used to inoculate a VSP medium for *Nonomuraea* spp. and *Actinoplanes* spp. or YEME for *Streptomyces* spp. in the presence or the absence of GPAs. The strains were grown to exponential phase (approximately 72 h) at 28 °C with shaking. The mycelium was harvested by centrifugation, suspended in 0.9% (*w*/*v*) NaCl, and fragmented by sonication with a Vibracell Albra sonicator 400 W [26]. A suspension of sonicated hyphae (corresponding to 10^7 CFU) was seeded onto V0.1 (*Nonomuraea* spp. and *Actinoplanes* spp.) or MS (*Streptomyces* spp.) agar plates supplemented with increasing concentrations of the following antibiotics: 0 to 100 μg/mL vancomycin in 10-μg/mL increments; 0 to 2 μg/mL teicoplanin in 0.1-μg/mL increments or 0 to 40 μg/mL teicoplanin in 10-μg/mL increments depending on the strain; and 0 to 5 μg/mL or 5 to 50 μg/mL A40926, depending on the strain, in 0.5-μg/mL or 2.5-μg/mL increments. The plates were dried and then incubated at 28 °C. MIC values were determined as the lowest antibiotic concentrations that inhibited visible growth after 10 days of incubation.

3.4. D,D-dipeptidase and D,D-carboxypeptidase Assays

Cells were harvested by centrifugation at 3600× *g* for 20 min at 4 °C, and then they were suspended in 2 mL of 0.9% (*w*/*v*) NaCl per gram of cells (wet weight). All of the following manipulations were conducted at 0 to 4 °C. The mycelium was fragmented by sonication with a Sonics Vibracell VCX 130. Sonication was carried out for 5 min on ice, with cycles of 30 s with an amplitude of 90% (90% of 60 Hz), with breaks of 10 s. The samples were then centrifuged at 39,000× *g* for 15 min, and the supernatants (cytoplasmic fractions) were collected. An alkaline extraction of the pellets (cell debris and membrane fractions) was carried out by adapting a protocol developed previously for extracting membrane-bound proteins [25,36]. The sedimented pellets were resuspended in ice-cold distilled water containing proteinase inhibitors (0.19 mg/mL phenylmethanesulfonyl fluoride and 0.7 μg/mL pepstatin, both purchased from Sigma-Aldrich), and then, immediately before centrifugation (28,000× *g* for 15 min at 4 °C), the pH was adjusted to 12 by adding an appropriate volume of 2.5 N NaOH. Immediately after centrifugation, the supernatants were neutralized to pH 7 by adding 0.5 M sodium acetate (pH 5.4). Enzymatic activities in the supernatant (cytoplasmic fractions) and in the resuspended insoluble fractions (membranes) were assayed as reported previously [36] by measuring the release of D-Ala from commercially available dipeptide (D-Ala-D-Ala, 10 mM; Sigma-Aldrich) and tripeptide (N_ε-acetyl-L-Lys-D-Ala-D-Ala, 10 mM; Sigma-Aldrich). D,D-carboxypeptidase activity was confirmed using 10 mM UDP-MurNAc-L-Ala-D-Glu-*meso*-Dap-D-Ala-D-Ala (UK-BaCWAN, University of Warwick) as a substrate. The release of D-Ala was followed spectrophotometrically at 510 nm with a D-amino acid oxidase–peroxidase coupled reaction. Reaction mixtures contained 5 mM of the peroxidase colorimetric substrate 4-aminoantipyrine (Sigma-Aldrich), 3 U/mL D-amino acid oxidase (Sigma-Aldrich), 7.5 U/mL horseradish peroxidase (Sigma-Aldrich), and 6 mM phenol in 50 mM 1,3-bis[tris(hydroxymethyl)methylamino]propane (pH 7.4) in a final volume of 1 mL, as described in detail in [36]. To compare the D,D-dipeptidase and D,D-carboxypeptidase activities in the cytoplasmic and membrane extracts, the activity was expressed as the number of nanomoles of D-Ala released from the dipeptide or tripeptide per min per mg of protein in the extract.

4. Conclusions

By comparing the specificity of induction of VanX and VanY peptidases in three different soil-dwelling actinomycetes, we confirm that *S. coelicolor* has a VanB-phenotype responding to vancomycin but not to teicoplanin or to teicoplanin-like A40926. *S. coelicolor* does not produce any GPA but possessing resistance genes confers it a selective advantage since it shares the ecological niche (soil) with many other GPAs-producing actinomycetes. Interestingly, by a culture-independent approach using molecular probes, it has been recently estimated that the frequency of encountering a vancomycin-type producer in soil is from 2.5 to 5 times higher than for a teicoplanin-like producing actinomycete [37]. In contrast, the teicoplanin producer *A. teichomyceticus* is highly and constitutively resistant to all the GPAs, including its own product. On the contrary, lower resistance in *N. gerenzanensis* is induced either by vancomycin or teicoplanins, and is based on the action of a VanY carboxypeptidase, which in many enterococci and staphylococci plays an unessential and ancillary role in the presence of VanHAX system [27]. Evidently, although producing structurally similar GPAs, *A. teichomyceticus* and *N. gerenzanensis* do not share the same self-resistant mechanism. Previous comparative analyses of their biosynthetic GPA clusters indicated that many A40926 biosynthetic genes are more related to vancomycin-type genes than to their teicoplanin homologs [38]. The production of A40926 and teicoplanin are most likely the result of a convergent evolution rather than originating from the same common ancestor. Although the identity and the role of the putative VanRS-like two-component signal transduction system in *N. gerenzanensis* needs to be investigated, the results of gene swapping between *A. teichomyceticus* and *N. gerenzanensis* indicate that cross-talking of the two-component systems is possible, as previously demonstrated in other actinomycetes [21,32]. Determining the sequences and the protein structures of the putative VanRS two-component system in *N. gerenzanensis* may help to confirm that the inducer specificity is determined by the origin of VanRS proteins and may provide additional evidence on the role of GPAs as VanS effector ligands. Additionally, since the genes involved in GPAs resistance in pathogens have been recruited from the different antibiotic-producing actinomycetes, gene swapping among different GPA-producing actinomycetes is proving to be useful for unveiling the specificity of regulation and predicting the evolution of future resistance gene combinations emerging in pathogens.

Supplementary Materials: The following are available online at http://www.mdpi.com/2079-6382/7/2/36/s1, Table S1: Maximum VanX and VanY activities during the growth (in the absence of GPAs) of *S. coelicolor*, *S. coelicolor ΔvanRS*, *A. teichomyceticus*, *N. gerenzanensis* and *N. gerenzanensis* pST30. The values represent the average of the data from three independent experiments, Table S2: MICs of GPAs in noninduced and teicoplanins-induced actinomycetes. The values represent the average of the data from three independent experiments.

Author Contributions: E.B., F.M. and G.L.M. conceived and designed the experiments; E.B. and P.C. performed the experiments; E.B., P.C. and G.L.M. analyzed the data; E.B., F.M. and G.L.M. wrote the paper.

Acknowledgments: This work was supported by public grants "Fondo di Ateneo per la Ricerca" 2015, 2016, 2017 to F. Marinelli and G.L. Marcone and by Federation of European Microbiological Societies (FEMS) Research Fellowship 2015 to E. Binda.

Conflicts of Interest: The authors declare that they have no conflict of interest.

References

1. European Centre for Disease Prevention and Control. Available online: https://ecdc.europa.eu/en/about-us/partnerships-and-networks/disease-and-laboratoy-networks/ears-net (accessed on 16 January 2018).
2. Binda, E.; Marinelli, F.; Marcone, G.L. Old and new glycopeptide antibiotics: Action and resistance. *Antibiotics* **2014**, *3*, 572–594. [CrossRef] [PubMed]
3. Marcone, G.L.; Binda, E.; Berini, F.; Marinelli, F. Old and new glycopeptide antibiotics: From product to gene and back in the post-genomic era. *Biotechnol. Adv.* **2018**, *36*, 534–554. [CrossRef] [PubMed]
4. Van Bambeke, F. Lipoglycopeptide antibacterial agents in Gram-positive infections: A comparative review. *Drugs* **2015**, *18*, 2073–2095. [CrossRef] [PubMed]
5. Cooper, M.A.; Williams, D.H. Binding of glycopeptide antibiotics to a model of a vancomycin-resistant bacterium. *Chem. Biol.* **1999**, *6*, 891–899. [CrossRef]
6. Perkins, H.R.; Nieto, M. The chemical basis for the action of the vancomycin group of antibiotics. *Ann. N. Y. Acad. Sci.* **1974**, *235*, 348–363. [CrossRef] [PubMed]
7. Allen, N.E.; Nicas, T.I. Mechanism of action of oritavancin and related glycopeptide antibiotics. *FEMS Microbiol. Rev.* **2003**, *26*, 511–532. [CrossRef] [PubMed]

8. Treviño, J.; Bayón, C.; Ardá, A.; Marinelli, F.; Gandolfi, R.; Molinari, F.; Jimenez-Barbero, J.; Hernáiz, M.J. New insights into glycopeptide antibiotic binding to cell wall precursors using SPR and NMR spectroscopy. *Chemistry* **2014**, *20*, 7363–7372. [CrossRef] [PubMed]
9. Dong, S.D.; Oberthür, M.; Losey, H.C.; Anderson, J.W.; Eggert, U.S.; Peczuh, M.W.; Walsh, C.T.; Kahne, D. The structural basis for induction of VanB resistance. *J. Am. Chem. Soc.* **2002**, *124*, 9064–9065. [CrossRef] [PubMed]
10. Kwun, M.J.; Hong, H.J. The activity of glycopeptide antibiotics against resistant bacteria correlates with their ability to induce the resistance system. *Antimicrob. Agents Chemother.* **2014**, *58*, 6306–6310. [CrossRef] [PubMed]
11. Bugg, T.D.; Wright, G.D.; Dutka-Malen, S.; Arthur, M.; Courvalin, P.; Walsh, C.T. Molecular basis for vancomycin resistance in *Enterococcus faecium* BM4147: Biosynthesis of a depsipeptide peptidoglycan precursor by vancomycin resistance proteins VanH and VanA. *Biochemistry* **1991**, *30*, 10408–10415. [CrossRef] [PubMed]
12. Arthur, M.; Quintiliani, R. Regulation of VanA- and VanB-type glycopeptide resistance in enterococci. *Antimicrob. Agents Chemother.* **2001**, *45*, 375–381. [CrossRef] [PubMed]
13. Evers, S.; Courvalin, P. Regulation of VanB-type vancomycin resistance gene expression by the VanS(B)-VanR(B) two-component regulatory system in *Enterococcus faecalis* V583. *J. Bacteriol.* **1996**, *178*, 1302–1309. [CrossRef] [PubMed]
14. Hong, H.J.; Hutchings, M.I.; Buttner, M.J. Vancomycin resistance VanS/VanR two-component systems. *Adv. Exp. Med. Biol.* **2008**, *631*, 200–213. [CrossRef] [PubMed]
15. Depardieu, F.; Podglajen, I.; Leclercq, R.; Collatz, E.; Courvalin, P. Modes and modulations of antibiotic resistance gene expression. *Clin. Microbiol. Rev.* **2007**, *20*, 79–114. [CrossRef] [PubMed]
16. Kwun, M.J.; Novotna, G.; Hesketh, A.R.; Hill, L.; Hong, H.J. In vivo studies suggest that induction of VanS-dependent vancomycin resistance requires binding of the drug to D-Ala-D-Ala termini in the peptidoglycan cell wall. *Antimicrob. Agents Chemother.* **2013**, *57*, 4470–4480. [CrossRef] [PubMed]
17. Koteva, K.; Hong, H.J.; Wang, X.D.; Nazi, I.; Hughes, D.; Naldrett, M.J.; Buttner, M.J.; Wright, G.D. A vancomycin photoprobe identifies the histidine kinase VanSsc as a vancomycin receptor. *Nat. Chem. Biol.* **2010**, *6*, 327–329. [CrossRef] [PubMed]
18. Goldstein, B.P.; Selva, E.; Gastaldo, L.; Berti, M.; Pallanza, R.; Ripamonti, F.; Ferrari, P.; Denaro, M.; Arioli, V.; Cassani, G. A40926, a new glycopeptide antibiotic with anti-*Neisseria* activity. *Antimicrob. Agents Chemother.* **1987**, *31*, 1961–1966. [CrossRef] [PubMed]
19. Dalmastri, C.; Gastaldo, L.; Marcone, G.L.; Binda, E.; Congiu, T.; Marinelli, F. Classification of *Nonomuraea* sp. ATCC 39727, an actinomycete that produces the glycopeptide antibiotic A40926, as *Nonomuraea gerenzanensis* sp. nov. *Int. J. Syst. Evol. Microbiol.* **2016**, *66*, 912–921. [CrossRef] [PubMed]
20. D'Argenio, V.; Petrillo, M.; Pasanisi, D.; Pagliarulo, C.; Colicchio, R.; Talà, A.; de Biase, M.S.; Zanfardino, M.; Scolamiero, E.; Pagliuca, C.; et al. The complete 12 Mb genome and transcriptome of *Nonomuraea gerenzanensis* with new insights into its duplicated "Magic" RNA polymerase. *Sci. Rep.* **2016**, *6*, 18. [CrossRef] [PubMed]
21. Hutchings, M.I.; Hong, H.J.; Buttner, M.J. The vancomycin resistance VanRS two-component signal transduction system of *Streptomyces coelicolor*. *Mol. Microbiol.* **2006**, *59*, 923–935. [CrossRef] [PubMed]
22. Siewert, G.; Strominger, J.L. Bacitracin: An inhibitor of the dephosphorylation of lipid pyrophosphate, an intermediate in the biosynthesis of the peptidoglycan of bacterial cell walls. *Proc. Natl. Acad. Sci. USA* **1967**, *57*, 767–773. [CrossRef] [PubMed]
23. Hamburger, J.B.; Hoertz, A.J.; Lee, A.; Senturia, R.J.; McCafferty, D.G.; Loll, P.J. A crystal structure of a dimer of the antibiotic ramoplanin illustrates membrane positioning and a potential lipid II docking interface. *Proc. Natl. Acad. Sci. USA* **2009**, *106*, 13759–13764. [CrossRef] [PubMed]
24. Beltrametti, F.; Consolandi, A.; Carrano, L.; Bagatin, F.; Rossi, R.; Leoni, L.; Zennaro, E.; Selva, E.; Marinelli, F. Resistance to glycopeptide antibiotics in the teicoplanin producer is mediated by *van* gene homologue expression directing the synthesis of a modified cell wall peptidoglycan. *Antimicrob. Agents Chemother.* **2007**, *51*, 1135–1141. [CrossRef] [PubMed]
25. Marcone, G.L.; Binda, E.; Carrano, L.; Bibb, M.; Marinelli, F. Relationship between glycopeptide production and resistance in the actinomycete *Nonomuraea* sp. ATCC 39727. *Antimicrob. Agents Chemother.* **2014**, *58*, 5191–5201. [CrossRef] [PubMed]

26. Marcone, G.L.; Beltrametti, F.; Binda, E.; Carrano, L.; Foulston, L.; Hesketh, A.; Bibb, M.; Marinelli, F. Novel mechanism of glycopeptide resistance in the A40926 producer *Nonomuraea* sp. ATCC 39727. *Antimicrob. Agents Chemother.* **2010**, *54*, 2465–2472. [CrossRef] [PubMed]
27. Arthur, M.; Depardieu, F.; Cabanié, L.; Reynolds, P.; Courvalin, P. Requirement of the VanY and VanX D,D-peptidases for glycopeptide resistance in enterococci. *Mol. Microbiol.* **1998**, *30*, 819–830. [CrossRef] [PubMed]
28. Serina, S.; Radice, F.; Maffioli, S.; Donadio, S.; Sosio, M. Glycopeptide resistance determinants from the teicoplanin producer *Actinoplanes teichomyceticus*. *FEMS Microbiol. Lett.* **2004**, *240*, 69–74. [CrossRef] [PubMed]
29. Marcone, G.L.; Carrano, L.; Marinelli, F.; Beltrametti, F. Protoplast preparation and reversion to the normal filamentous growth in antibiotic-producing uncommon actinomycetes. *J. Antibiot.* **2010**, *63*, 83–88. [CrossRef] [PubMed]
30. Marcone, G.L.; Binda, E.; Reguzzoni, M.; Gastaldo, L.; Dalmastri, C.; Marinelli, F. Classification of *Actinoplanes* sp. ATCC 33076, an actinomycete that produces the glycolipodepsipeptide antibiotic ramoplanin, as *Actinoplanes ramoplaninifer* sp. nov. *Int. J. Syst. Evol. Microbiol.* **2017**, *67*, 4181–4188. [CrossRef] [PubMed]
31. Binda, E.; Marcone, G.L.; Pollegioni, L.; Marinelli, F. Characterization of VanYn, a novel D,D-peptidase/D,D-carboxypeptidase involved in glycopeptide antibiotic resistance in *Nonomuraea* sp. ATCC 39727. *FEBS J.* **2012**, *279*, 3203–3213. [CrossRef] [PubMed]
32. Kilian, R.; Frasch, H.J.; Kulik, A.; Wohlleben, W.; Stegmann, E. The VanRS homologous two-component system VnlRSAb of the glycopeptide producer *Amycolatopsis balhimycina* activates transcription of the *vanHAXSc* genes in *Streptomyces coelicolor*, but not in *A. balhimycina*. *Microb. Drug Resist.* **2016**, *22*, 499–509. [CrossRef] [PubMed]
33. Sosio, M.; Stinchi, S.; Beltrametti, F.; Lazzarini, A.; Donadio, S. The gene cluster for the biosynthesis of the glycopeptide antibiotic A40926 by *Nonomuraea* species. *Chem. Biol.* **2003**, *10*, 541–549. [CrossRef]
34. Kieser, T.; Bibb, M.J.; Buttner, M.J.; Chater, K.F.; Hopwood, D.A. *Practical Streptomyces Genetics*; The John Innes Foundation: Norwich, UK, 2000; ISBN 0-7084-0623-8.
35. Taurino, C.; Frattini, L.; Marcone, G.L.; Gastaldo, L.; Marinelli, F. *Actinoplanes teichomyceticus* ATCC 31121 as a cell factory for producing teicoplanin. *Microb. Cell Fact.* **2011**, *10*, 82–94. [CrossRef] [PubMed]
36. Binda, E.; Marcone, G.L.; Berini, F.; Pollegioni, L.; Marinelli, F. *Streptomyces* spp. as efficient expression system for a D,D-peptidase/D,D-carboxypeptidase involved in glycopeptide antibiotic resistance. *BMC Biotechnol.* **2013**, *13*, 24. [CrossRef] [PubMed]
37. Thaker, M.N.; Wang, W.; Spanogiannopoulos, P.; Waglechner, N.; King, A.M.; Medina, R.; Wright, G.D. Identifying producers of antibacterial compounds by screening for antibiotic resistance. *Nat. Biotechnol.* **2013**, *31*, 922–927. [CrossRef] [PubMed]
38. Donadio, S.; Sosio, M.; Stegmann, E.; Weber, T.; Wohlleben, W. Comparative analysis and insights into the evolution of gene clusters for glycopeptide antibiotic biosynthesis. *Mol. Genet. Genom.* **2005**, *274*, 40–50. [CrossRef] [PubMed]

antibiotics

MDPI

Article

Isolation, Characterization, and Antibacterial Activity of Hard-to-Culture Actinobacteria from Cave Moonmilk Deposits

Delphine Adam [1,†], Marta Maciejewska [1,†], Aymeric Naômé [1], Loïc Martinet [1], Wouter Coppieters [2], Latifa Karim [2], Denis Baurain [3] and Sébastien Rigali [1,*]

1 Integrative Biological Sciences (InBioS), Center for Protein Engineering, Liège University, B-4000 Liège, Belgium; delphine.adam@doct.uliege.be (D.A.); maciejewska.m@wp.pl (M.M.); Aymeric.Naome@uliege.be (A.N.); lmartinet@doct.uliege.be (L.M.)

2 Genomics Platform, GIGA (Grappe Interdisciplinaire de Génoprotéomique Appliquée), University of Liège (B34), B-4000 Liège, Belgium; wouter.coppieters@ulg.ac.be (W.C.); lkarim@ulg.ac.be (L.K.)

3 Integrative Biological Sciences (InBioS), PhytoSYSTEMS, Eukaryotic Phylogenomics, University of Liège, B-4000 Liège, Belgium; Denis.Baurain@uliege.be

* Correspondence: srigali@uliege.be; Tel.: +32-4-366-9830

† These authors contributed equally to this work.

Received: 26 February 2018; Accepted: 19 March 2018; Published: 22 March 2018

check for updates

Abstract: Cave moonmilk deposits host an abundant and diverse actinobacterial population that has a great potential for producing novel natural bioactive compounds. In our previous attempt to isolate culturable moonmilk-dwelling Actinobacteria, only *Streptomyces* species were recovered, whereas a metagenetic study of the same deposits revealed a complex actinobacterial community including 46 actinobacterial genera in addition to streptomycetes. In this work, we applied the rehydration-centrifugation method to lessen the occurrence of filamentous species and tested a series of strategies to achieve the isolation of hard-to-culture and rare Actinobacteria from the moonmilk deposits of the cave "Grotte des Collemboles". From the "tips and tricks" that were tested, separate autoclaving of the components of the International Streptomyces Project (ISP) medium number 5 (ISP5) medium, prolonged incubation time, and dilution of the moonmilk suspension were found to most effectively improve colony forming units. Taxonomic analyses of the 40 isolates revealed new representatives of the *Agromyces*, *Amycolatopsis*, *Kocuria*, *Micrococcus*, *Micromonospora*, *Nocardia*, and *Rhodococcus* species, as well as additional new streptomycetes. The applied methodologies allowed the isolation of strains associated with both the least and most abundant moonmilk-dwelling actinobacterial operational taxonomic units. Finally, bioactivity screenings revealed that some isolates displayed high antibacterial activities, and genome mining uncovered a strong potential for the production of natural compounds.

Keywords: Cave microbiology; secondary metabolite; antibiotics; rare Actinobacteria; Streptomyces; Amycolatopsis; unculturability; siderophore

1. Introduction

Bioprospecting for natural compounds from microorganisms dwelling in poorly explored and extreme environments has gained renewed interest [1,2], boosted by the need to fight resistance to compounds currently used as antimicrobials, herbicides, antivirals, and anticancer agents [3]. Caves, despite being highly oligotrophic environmental niches, support rich and diverse microbial life, a phenomenon called "the Paradox of the Plankton", which suggests that, in spite of limited nutrient resources, an unexpectedly wide range of species coexist [4]. The extremely starved nature of subsurface habitats presumably stimulates unique strategies of indigenous microbiomes, among which the fine-tuning of the secondary metabolism (in terms of quantity and diversity of molecules) might be one of the key features enabling life in such challenging environments [5,6]. The observation that Actinobacteria, which are the most prolific producers of specialized (secondary) metabolites [7], are also abundant in limestone caves [8–10] is not likely a coincidence. Therefore, Actinobacteria isolated from oligotrophic environments, particularly the rare representatives of the phylum, are expected to own a unique metabolome that could potentially be an important source of novel drugs.

In our first attempt, aimed at isolating antibiotic-producing Actinobacteria from cave carbonate deposits called moonmilk originating from the cave "Grotte des Collemboles", we successfully isolated representatives of a single genus, *Streptomyces* [11,12]. However, high-throughput sequencing (HTS) of DNA extracted from the same moonmilk deposits demonstrated that we failed to isolate representatives of at least 46 additional actinobacterial genera [13]. Such an outcome might be related not only to the fact that *Streptomyces* are adapted to use a wide range of nutrients and therefore grow effectively on many selective media, but also to the fact that their growth is faster in comparison to that of other actinobacterial genera. For this reason, non-streptomycetes, which are more challenging to isolate in pure cultures, are regarded as rare Actinobacteria. Nonetheless, the 78 culturable *Streptomyces* strains isolated in our previous screening [11] constitute a minor fraction of those dwelling in the studied moonmilk deposits, because we only recovered relatives of 5 out of the 19 operational taxonomic units (OTUs) affiliated with the *Streptomyces* genus [13], suggesting that this environment hosts hard-to-culture *Streptomyces* species that are potentially highly diverged from their soil-dwelling counterparts.

Bacterial unculturability is one of the major problems in microbiology, as many species are present in the environment in a so-called viable but non-cultivable state (VBNC) [14]. The phenomenon known as the "great plate count anomaly" clearly depicts this issue—no matter what kind of approach will be applied to the isolation of microorganisms, it is possible to cultivate only about 1% of what is present in a sample [15]. As demonstrated in the recently published new "Tree of Life", the large majority of "known" microorganisms have been identified exclusively via genome-based approaches and are not represented by any culturable strain [16]. For this reason, a series of innovations in cultivation techniques are being introduced to increase the recovery yield of microorganisms from their natural habitats, including: (i) dilution of the sample to minimize the competition of fast-growing organisms and reduce the effect of growth inhibitors [17]; (ii) addition of signaling compounds/growth factors [18,19]; (iii) extension of the incubation time for the recovery of slow-growing bacteria [20]; (iv) in situ cultivation [21]; (v) use of polymers as growth substrates to reduce osmotic shock [20]; and (vi) limitation of the production of reactive oxygen species, which prevent bacterial growth [22,23]. In addition, various pretreatment methods are being applied to selective isolation, which targets (to favor or instead to exclude) a bacterial group of interest [24]. In the case of Actinobacteria, a series of chemical and physical treatments are used [20], which take advantage of the high resistance of actinobacterial spores to many factors such as ultraviolet (UV), ultrasonic waves, desiccation, and many chemicals [25–27].

In light of the above findings, and taking into account the fact that moonmilk deposits host a broad diversity of Actinobacteria (243 OTUs) [13], we have made an attempt to improve their recovery, particularly of rare taxa, in pure cultures by combining several approaches. Here, we present the "tips and tricks" used to isolate rare moonmilk-dwelling Actinobacteria, evaluate their potential at producing antibacterial compounds, and describe their genetic capacity to produce other specialized metabolites.

2. Results and Discussion

2.1. Assessment of Various Strategies for Isolating Moonmilk-Dwelling Rare Actinobacteria

Different strategies were implemented in order to increase the odds of isolating hard-to-cultivate Actinobacteria from the moonmilk samples of the cave "Grotte des Collemboles". For this purpose, we applied the rehydration-centrifugation (RC) method to the preparation of the moonmilk suspension. Applying the RC methodology has been shown previously to lessen the occurrence of filamentous and long spore-chain forming species (such as most *Streptomyces*) and non-motile bacteria, thereby enriching the isolation of rare and zoosporic Actinobacteria [27]. The moonmilk suspension was inoculated in the two media that showed the most contrasting isolation efficiency in our previous screening [12]—(i) the International Streptomyces Project (ISP) medium number 5 (ISP5) medium in which no colonies could be observed; and (ii) the starch nitrate (SN) medium, which gave the highest number of colony forming units (CFUs). As the existence of unculturable bacteria in certain environments has been proposed to depend on siderophore-producing neighboring microorganisms [18,28], we facilitated iron acquisition in these two media by (i) increasing the concentration of iron (20 μM $FeCl_3$); (ii) including 1 μM of the siderophore desferrioxamine B (DFB), which has been shown to be essential for growth in certain media [28]; and (iii) adding both $FeCl_3$ and DFB. Moreover, in order to reduce the formation of highly toxic reactive oxygen species (ROS) that might be generated during autoclaving through the interaction of components of the media, particularly phosphate and agar [22,23], ISP5 and SN were prepared in two different manners—(i) either with all of the components autoclaved together or (ii) with agar and phosphate autoclaved separately from all of the other components. Further evaluation of the possible toxic effect of ROS generated during autoclaving was tested by adding a solution of catalase to the center of the plates, as described previously [23]. All media were additionally supplemented with selective agents (antibiotic and antifungal compounds), inoculated with serially diluted moonmilk suspension (up to 10^{-4}), and incubated for up to two months.

The combination of strategies described above increased the magnitude of isolated bacteria by about 15-fold in comparison to the colony counts obtained in our previous screening (~5×10^5 CFUs per gram of inoculum in this study (Figure 1), in contrast to ~3×10^4 CFUs in [12]). Dilution of the inoculum (from 100- to 1000-fold) increased CFUs (compare bar plots in blue and those in red in Figure 1). Considering that bacteria living in extremely oligotrophic environments are thought to have developed mutualistic interactions, this result is rather unexpected. Indeed, dilution increases the distance between colonies—which limits the exchange of growth factors between 'helpers' and unculturable organisms—and should eventually yield a lower number of colonies. The fact that diluting the inoculum instead resulted in higher CFUs suggests either that one or more of the strains inoculated at the lower dilution secreted highly active antibacterial agents or that the collected moonmilk deposits could possibly contain growth inhibitors that were attenuated by the sample dilution.

Furthermore, the putative effect of ROS on bacterial cultivability was clearly recorded on the ISP5 medium (Figures 1 and 2). Sterilization of phosphate and agar separate from the other components of the ISP5 medium resulted in abundant growth, whereas no colonies could be observed when the same medium was prepared according to the standard procedure (Figures 1 and 2). The probable effect of ROS-preventing bacterial growth on the traditionally prepared ISP5 medium was supported by the observation that adding a solution containing a catalase (decomposing hydrogen peroxide to water and oxygen) restored colony formation (Figure 2b). In contrast, supplementation of the media by iron (at final concentration of 20 μM), DFOB (at a final concentration of 1 μM), or both did not significantly improve the bacterial count (Figure 1). However, regardless of the lack of improved CFUs, the growing/isolated bacteria still might be different from those grown in the same media deprived from additional iron/siderophore supply. Finally, increased incubation time (from 16 to 44 days) increased CFUs in only a limited number of conditions (see for instance Fe + DFB in starch nitrate medium, Figure 1).

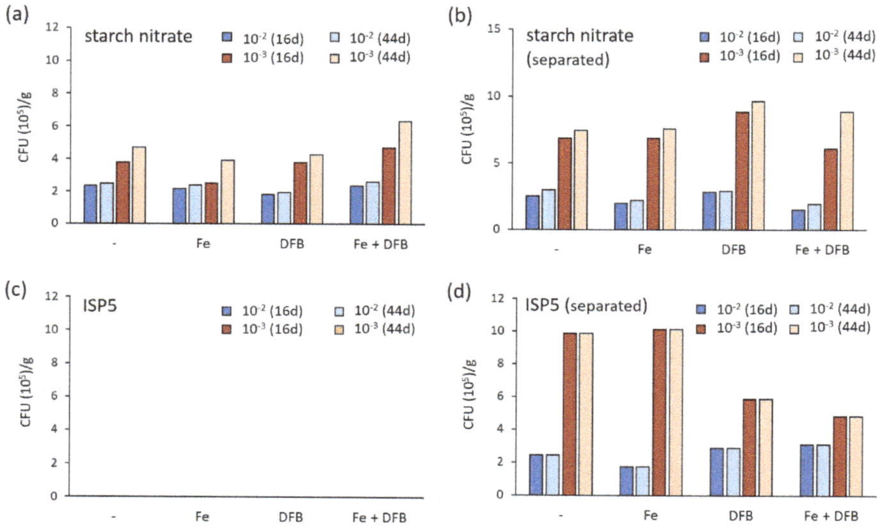

Figure 1. Total number of colony forming units (CFUs) according to the different media and methods used. The different media were inoculated with a diluted moonmilk suspension (10^{-2} in blue and 10^{-3} in red) prepared according to the rehydration-centrifugation method and incubated for up to two months. Data obtained after 16 days (16 d) and 44 days (44 d) of incubation are displayed. The "separated" label indicates that the autoclaving of agar and phosphate was performed separately from the other components of the media. Abbreviations: Fe, 20 μM of $FeCl_3$; DFB, 1 μM of desferrioxamine B.

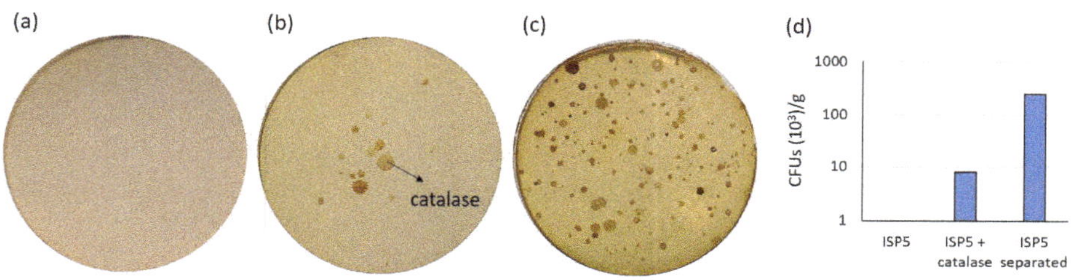

Figure 2. Effect of reactive oxygen species on bacterial growth in International Streptomyces Project (ISP) medium number 5 (ISP5) medium. Growth in (**a**) ISP5 medium prepared according to the standard protocol, in (**b**) the same medium but with catalase injected on the paper disc in the center of the plate, and in (**c**) the ISP5 medium prepared with autoclaving of agar and phosphate separate from the other media components. (**d**) CFUs on (**a**), (**b**) and (**c**).

2.2. Characterization of Culturable Moonmilk-Derived Actinobacterial Isolates

100 colonies were initially selected for further characterization, but only 42 survived after two rounds of inoculation/cultivation to obtain pure isolates (Table 1). This significant loss (58%) was likely caused by the lack of growth factors emanating from neighboring colonies in pure cultures. Hence, the strains that passed the purification steps might represent bacteria able to feed only on nutrients present in the synthetic medium, whereas species most adapted to cooperative growth and life in oligotrophic and inorganic environments would constitute a significant part of the lost isolates. The remaining 42 isolates were cultured in liquid LB, ISP1, or ISP2 media for DNA isolation and

subsequent sequencing of the 16S (SSU) rRNA gene, either at full length (extracted from the genome sequence) or nearly full length (through PCR amplification).

Table 1 lists the closest hits for members of our collection deduced from BLAST searches, and Figure 3 shows their phenotype. Two of the isolates (MMun130 and MMun142), closely related to *Paenibacillus* sp. DSL09-3 and *Paracoccus* sp. clone 54, respectively (Table 1), are not Actinobacteria and were not investigated further in this study. The remaining 40 isolates represented eight actinobacterial genera, namely *Agromyces, Amycolatopsis, Kocuria, Micrococcus, Micromonospora, Nocardia, Streptomyces*, and *Rhodococcus* (Table 1 and Figure 3). Finding members of the *Rhodococcus* genus, which was the most abundant in terms of the absolute number of sequences (17% of actinobacterial 16S rRNA sequences), and of the *Streptomyces* genus, which displayed the largest diversity (19 OTUs) based on DNA extracted from the investigated moonmilk deposits [13], was rather expected. However, finding representatives of genera *Nocardia* (0.5%), *Agromyces* (0.2%), *Kocuria* (0.1%), and *Amycolatopsis* (0.03%), which correspond to less abundantly represented taxa, demonstrated the efficiency of the tested cultivation approaches to also recover rare moonmilk-dwelling Actinobacteria. Additionally, our work reports the first isolation of strains belonging to the *Micrococcus, Kocuria* (Micrococcaceae), and *Micromonospora* (Micromonosporaceae) genera from moonmilk deposits. While other members of Micrococcaceae (*Arthrobacter*) had been previously reported in moonmilk [10], representatives of the Micromonosporaceae family (MMun172) had not been identified so far in this type of speleothems.

The overrepresentation of *Streptomyces* isolates (27 out of 40 strains) despite using the RC protocol might be explained by specific morphological features of the isolated moonmilk-dwelling species. While *Streptomyces* filamentous pellets would be excluded from the final suspension by the centrifugation steps in the RC protocol, the inoculum could still contain single streptomycetes spores. *Streptomyces lunaelactis* [29], the most frequently observed *Streptomyces* species in the studied deposits, was found to form collapsed/fragmented packs of spores instead of long aerial spore chains. This morphological feature would explain why we isolated two strains of the *S. lunaelactis* species (MMun143 and MMun152) using the RC method and could be a common characteristic of the other *Streptomyces* isolates obtained in this study. In addition to these two novel *S. lunaelactis* strains, another 14 out of 27 *Streptomyces* strains obtained in this study have an identical 16S rRNA sequence compared to strains previously isolated from the same moonmilk deposits (MM18, MM44, MM63, MM82, MM110, MM126 [12], Figure 4 and Table 1). These 14 "MMun" strains were therefore assigned to phylotypes defined in our first screening (phylotypes marked with an asterisk in Table 1) [12], and the other 11 *Streptomyces* strains were distributed into 9 novel phylotypes (from XXXII to XL) based on their 16S rRNA sequence (Figure 4).

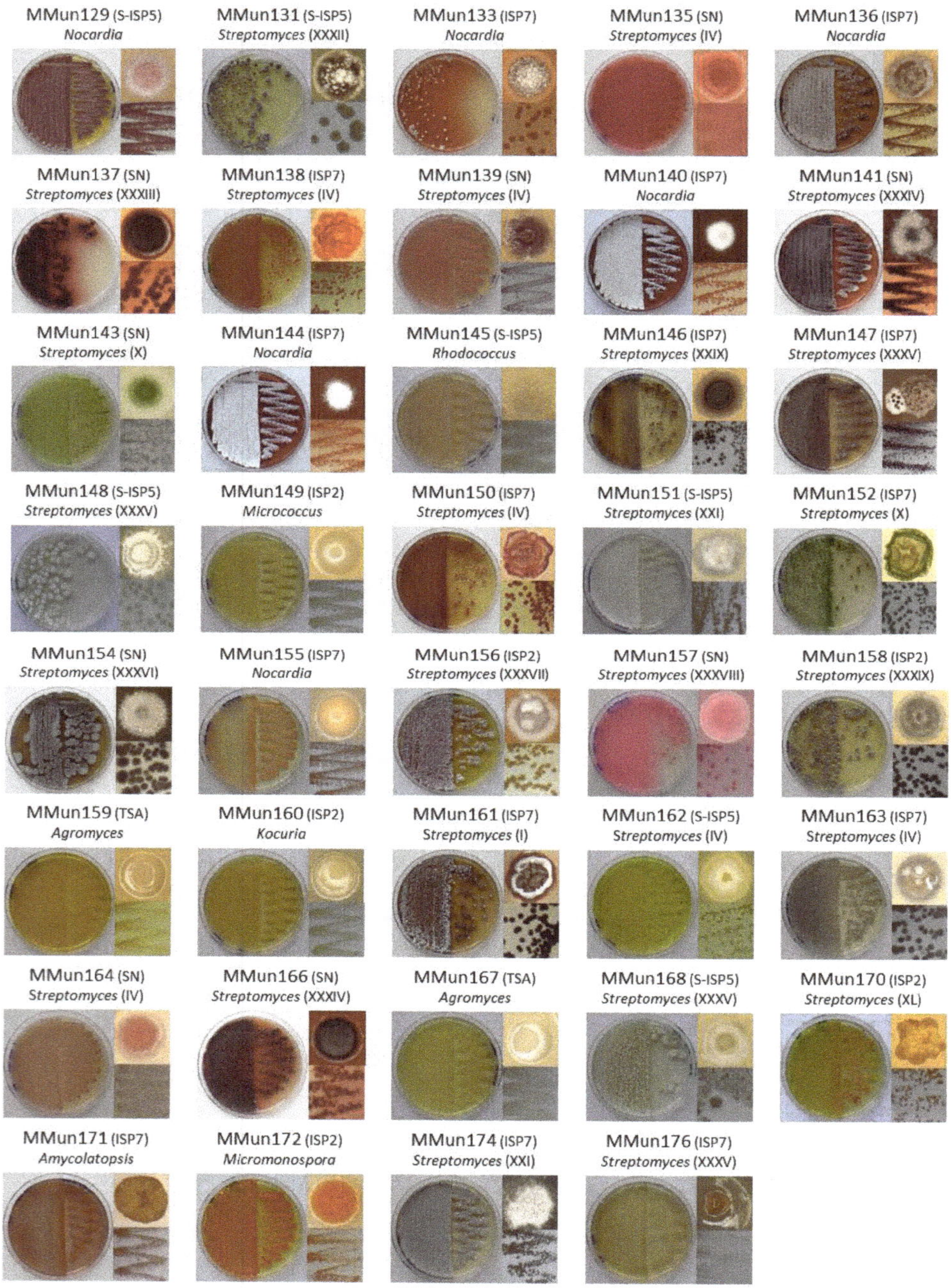

Figure 3. Phenotypes of actinobacterial strains isolated in this study. For each isolate, the front and back of the Petri dish-grown bacteria are presented together with the phenotype of a single colony. The phylotype number for *Streptomyces* isolates is indicated in parentheses.

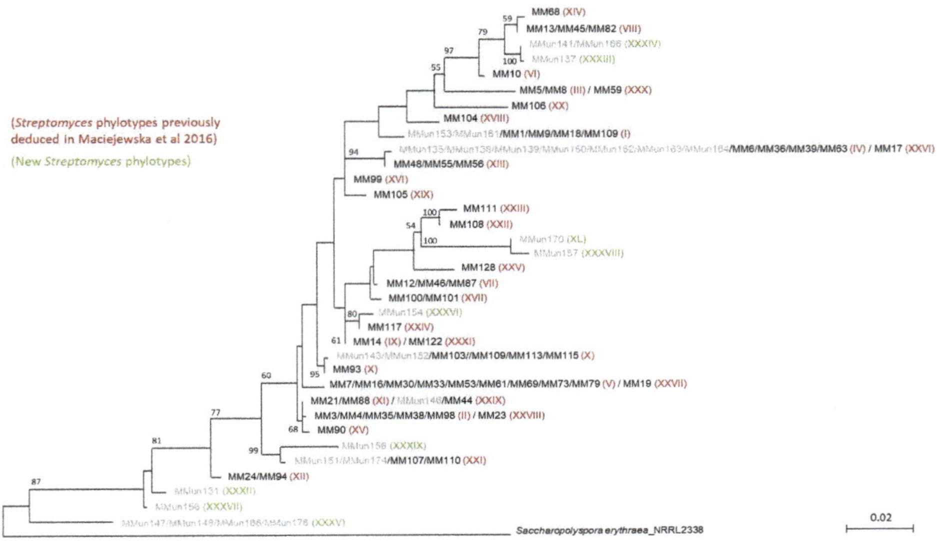

Figure 4. 16S rRNA-based phylogenetic tree of moonmilk-dwelling *Streptomyces* strains. Strains isolated using the rehydration-centrifugation (RC) method (MMun) are shown in grey, and strains isolated in our previous study (MM) are in black. Phylotype affiliations are indicated in parentheses. The alignment of nearly complete 16S rRNA sequences had 1538 unambiguously aligned positions for 92 strains, but only the 1342 positions without missing nucleotides were used to infer the tree. The evolutionary model was GTR + Γ_4 and bootstrap values are based on 100 pseudoreplicates (bootstrap values <50% are not displayed). *Saccharopolyspora erythraea* was used as outgroup. The scale bar represents 0.02 substitutions per site.

An important question is "to which of the 243 actinobacterial OTUs (Table S1) identified by HTS [13] in the studied moonmilk deposits are the 40 new actinobacterial isolates associated". In other words, did the isolation protocols used in this study enable the recovery of abundant and widespread strains or Actinobacteria with much more limited abundance and distribution in these speleothems (or both). For this purpose, we compared the 16S rRNA sequences (trimmed to the V6-V7 variable region) of the new isolates with the 16S rRNA amplicons (V6–V7 region) of the 243 actinobacterial OTUs obtained in our metagenetic study on DNA extracted from the same moonmilk deposits [13]. The phylogenetic tree positioning the 40 MMun strains (Table 1) with the 243 OTUs [13] is displayed in Supplementary Figures S1 and S2 and the results are summarized in Figure 5.

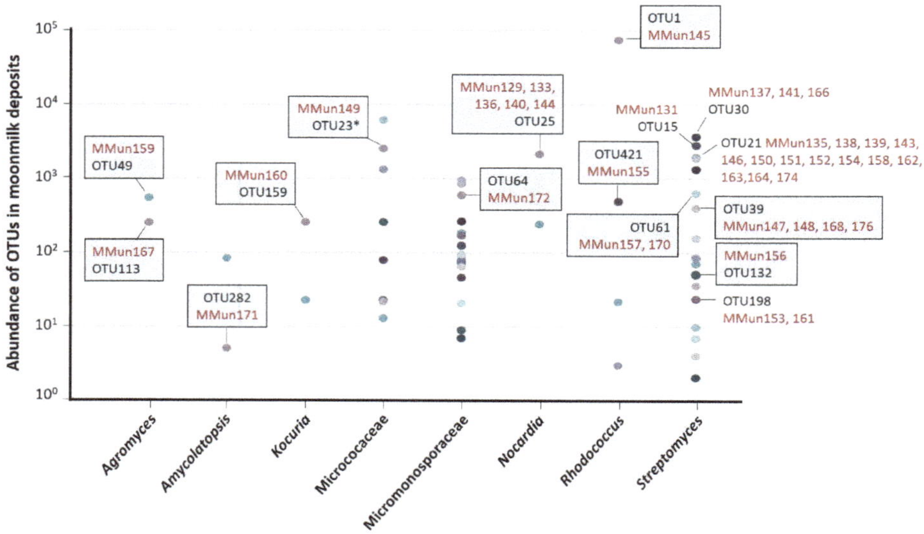

Figure 5. Membership of MMun strains isolated in this study to the actinobacterial operational taxonomic units (OTUs) identified by the high-throughput sequencing (HTS) metagenetic study. OTUs in blue boxes are those for which we isolated a first representative strain in this study. Note that for MMun149 and MMun172, the search of their associated OTU was performed at the family level (Micrococcaceae and Micromonosporaceae) and not at the genus level, as the HTS study [13] did not allow us to unambiguously assign the sequences of the 16S rRNA V6–V7 amplicons to the *Micrococcus* and *Micromonospora* genera. Symbol: *, due to the phylogenetic divergence of its V6–V7 region and based on the full sequence of its 16S rRNA gene (extracted from the genome sequence), MMun149 was manually affiliated to OTU23 even though its "closest" OTU in the tree was OTU141 (Figure S2).

Figure 5 reveals that our study enabled the isolation of a strain (MMun145) associated with the most abundant OTU of the studied carbonate deposits—OTU1, which was assigned to the *Rhodococcus* genus and accounted for 16.45% (75,143 sequences) of the total number of 16S rRNA amplicon sequences obtained by our HTS analysis [13]. BLAST analysis of the full-length 16S rRNA sequence of MMun145 found *Rhodococcus* sp. strain MAK1 [30] as the closest hit, a strain isolated from diesel oil-contaminated soil and therefore possibly adapted to life in extreme environments. In another (and opposite) extreme case, MMun171 (*Amycolatopsis*) was associated with one of the less abundant OTUs—OTU282, which corresponded to only five 16S rRNA amplicon sequences identified (0.001%) in our metagenetic study [13]. This extremely rare Actinobacteria, only detected in collection point number 4 of the Grotte des Collemboles [13], has been able to grow in the ISP5 medium with separate autoclaving of the phosphate and agar components, as well as in media supplemented by iron and DFB (Table 1).

The other isolated MMun strains belong to OTUs that stand between these two extreme ends, but generally, they belong to some of the most abundant OTUs of their respective genus/family (Figure 5). It must be noted that eight of the newly identified *Streptomyces* strains are associated with OTUs for which culturable representatives had not been isolated in our previous screening [12] (Figure 5). However, no culturable *Streptomyces* strains have been isolated from twelve of the 19 *Streptomyces* OTUs, which highlights how many moonmilk-dwelling species remain to be discovered.

2.3. Evaluation of the Antibacterial Activity of the New Moonmilk Isolates

The potential to produce compounds with antimicrobial activity against Gram-positive and Gram-negative bacteria was evaluated for each strain via cross-streak assays (see the heatmap presented in Figure 6). Strains were streaked on ISP2, S-ISP5, ISP7, and TSA agar plates, and incubated for 14 days at 28 °C. Two types of antibacterial activities were recorded: (i) those that fully inhibited the growth (GI, growth inhibition) of the tested reference strains (see Figure 7 MMun156 against *B. subtilis* as a GI example) and (ii) those that allowed only partial growth (IG, impaired growth, see Figure 7 MMun160 against *E. coli* as an IG example). On several occasions, both GI and IG were observed (see Figure 7 MMun171 against *B. subtilis* as an example).

As expected, much stronger overall inhibitory activities (GI and IG) were recorded against Gram-positive bacteria (~87% of the MMun strains) compared to those against Gram-negative bacteria (~59% of the MMun strains) (Figure 8). Two isolates were particularly active against the tested Gram-negative bacteria—MMun160 (*Kocuria*) and MMun171 (*Amycolatopsis*) (Figures 6 and 7). However, these two strains exhibited very different activity patterns depending on the culture conditions: while MMun171 displayed strong inhibitory activity under all culture conditions, MM160 only secreted its antibiotic(s) when cultured in the ISP2 medium (Figure 6).

Regarding the antibacterial activities against the Gram-positive strains, their growth was affected by more actinobacterial isolates and was overall more strongly inhibited (Figures 6 and 7). Three moonmilk *Streptomyces* strains (MMun141, MMun146, and MMun156) were found to be particularly active, as they exhibited strong inhibitory activities against all tested Gram-positive species and under each culture condition tested (Figure 6 and examples shown in Figure 7). Interestingly, the same isolates showed virtually no inhibitory effect against Gram-negative bacteria. In addition, the moonmilk isolates most active against Gram-negative pathogens (MMun160 and MMun171) also displayed quite strong inhibitory profiles against Gram-positive strains.

In contrast, a few strains did not display any (or extremely weak) antibacterial activity, such as isolates MMun145 and MMun155 that belong to the *Rhodococcus* genus, the two *Agromyces* strains (MMun159 and MMun167), and the *Micromonospora* isolate MMun172 (Figures 6–8). Similarly, the five *Nocardia* isolates did not secrete compounds with antibacterial activity under most conditions, except isolate MMun136, which displayed weak antibacterial activity against *M. luteus*. Therefore, while we managed to uncover conditions to isolate and grow these rare Actinobacteria, the conditions that enable production of their putative antibiotics remain undefined.

Gram -

MMun strains	*K. pneumoniae* ISP7	*K. pneumoniae* S-ISP5	*K. pneumoniae* ISP2	*K. pneumoniae* TSA	*E. coli* ISP7	*E. coli* S-ISP5	*E. coli* ISP2	*E. coli* TSA	*C. freundii* ISP7	*C. freundii* S-ISP5	*C. freundii* ISP2	*C. freundii* TSA	*P. aeruginosa* ISP7	*P. aeruginosa* S-ISP5	*P. aeruginosa* ISP2	*P. aeruginosa* TSA
129	2.9	0	0	0	0	0	0	0	0	0	0	0	0	0	0.7	0
131	ng	ng	ng	ng	ng	ng	ng	ng	ng	ng	ng	ng	ng	ng	ng	ng
133	3	0	0	0	0	0	0	0	2	0	0	0	0	0	0.9	0
135	0.1	0	0	0	0	0	0	0	0.5	0	0	0	0	0	0.5	0
136	2.8	0	0	0	0	0	0	0	2	0	0.4	0	0	0	0	0
137	0	0	0	0	0	0	0	0	0	0	0	0	0	0	0.6	0
138	1.5	0	0	0	0	0	0	0	0	0	0	0	0	0	0.7	0
139	1.7	0	0	0	0	0	0	0	0.1	0.1	0	0	0	0	0.5	0
140	4.4	0	0	0	0	0	0	0	0	0	0	0	0	0	0.8	0
141	0	0	0	0	0	0.6	0	0	0	0	0	0	0	0	0	0
143	0.5	0	0	0	0	0	0	0	0	0	0.5	0	0	0	0.8	0
144	4.4	0	0	0	0	0	0	0	3.4	0	0	0	0	0	0.7	0
145	0	0	0	0	0	0	0	0	0	0	0	0	0	0	0	0
146	2	0	0	0	0	0.8	0	0	0.1	1	0	0	0	0	1	0
147	2.3	0	0	0	0	0	0	0	0	0	0	0	0	0	0	0
148	0	0	0	0	0	0	0	0	0	0	0	0	0	0	0	0
149	2.2	0	0	0	0	0	0	0	0	0	0	0	0	0	0	0
150	0.1	0	0	0	0	0	0.1	0	0	0	0	0	0	0	0	0
151	1.6	0	0	0	0	0	0	0	0	0	0	0	0	0	0	0
152	0	0	0	0	0	0	0	0	0	0	0	0	0	0	0	0
154	1	0.7	0	0	0	0	0	0	0	0	0	0	0	0	0	0
155	0	0	0	0	0	0	0	0	0	0	0	0	0	0	0	0
156	0	0	0	0	0	0	0	0	0	0	0	0	0	0	0	0
157	1	0	0	0	0	0	0	0	0	0	0	0	0	0	0	0
158	0	0	0	0	0	0	0	1.5	0	0	0	0	0	0	0	0
159	0	0	0	0	0	0	0	0	0	0	0	0	0	0	0	0
160	0	0	4.5	0	0	0	4.5	0	0	0	4.5	0	0	0	4.5	0
161	0.5	0	0	0	0	0	0	0	0	0	0	0	0	0	0	0
162	0.1	0	0	0	0	0	0	0	0	0	0	0	0	0	1	0
163	2.5	0	0	0	0	0	0	0	0	0	0	0	0	0	1	0
164	0.5	0	0	0	0	0	0	0	0	0	0.1	0	0	0	0	0
166	0	0	0	0	0	0	0	0	0	0	0	0	0	0	1.4	0
167	0	0	0	0	0	0	0	0	0	0	0	0	0	0	0	0
168	0	0	0	0	0	0	0	0	0	0.3	0	0	0	0	0	0
170	0	0	0	0	0	0	0	0	0	0	0	0	0	0	0	0
171	2.8	1.2	4.5	4.5	1.8	1.2	3.7	4.5	2.2	0.8	2	4.5	0	0	1.5	0
172	1.6	0	0	0	0	0	0	0	0	0	0	0	0	0	0	0
174	0	0	0	0	0	0	0	0	0	0	0	0	0	0	0	0
176	0	0	0	0	0	0	0	0	0	0	0	0	0	0	0	0

Gram +

MMun strains	*S. aureus* ISP7	*S. aureus* S-ISP5	*S. aureus* ISP2	*S. aureus* TSA	*B. subtilis* ISP7	*B. subtilis* S-ISP5	*B. subtilis* ISP2	*B. subtilis* TSA	*M. luteus* ISP7	*M. luteus* S-ISP5	*M. luteus* ISP2	*M. luteus* TSA
129	x	x	0	0	1.1	0	0	0	0	0	0	0
131	x	x	ng	ng	ng	ng	ng	ng	ng	ng	ng	ng
133	x	x	0	0	0	0	0	0	1.5	0	0	0
135	x	x	0	0	0.1	0	0.3	0.1	1	1	0	0
136	x	x	0	0	0	0	0	0	2.5	0.1	0.3	0
137	x	x	0.7	0.6	0.6	0.2	0.7	0.5	0.2	0.3	0.8	0
138	x	x	1	0	0.5	0.1	1.1	0.2	0.5	0	0.9	0.1
139	x	x	0	0	0	0	0	0	0.5	0.7	0	0
140	x	x	0	0	0	0	0	0	0	0	0	0
141	x	x	2.8	0.7	4.3	3	2.3	0.7	3.1	2.7	1.9	0.9
143	x	x	0.3	0	0	0.3	0.5	0	0	0	0	0
144	x	x	0	0	0	0	0	0	0	0.1	0	0
145	x	x	0	0	0	0	0	0	0	0	0	0
146	x	x	0.7	0	3.6	3.4	1	0	4.5	4.5	2.7	0
147	x	x	0	0	0	0	0	0	0.5	3.3	0	0
148	x	x	0	0	0	0	0	0	0	1.8	0	0
149	x	x	0	0	1	0	0	4.5	0	0	0	0
150	x	x	0	0.1	1	1	0.1	1	1	1.6	0.1	0.2
151	x	x	0	0	1.7	0.6	0	0	2.5	1.3	0	0
152	x	x	0	0	0	0.1	0	0	0	0.1	0	0.1
154	x	x	0	0	0	0	0.3	0	0.3	0	0.1	0
155	x	x	0	0	0	0	0	0	0	0	0	0.5
156	x	x	3.5	2.5	3.6	4.5	4.5	4.5	3.4	3.5	1.6	3
157	x	x	0.6	0	0	0.5	1	0	1.6	1.5	0.4	0
158	x	x	0.4	1.2	0	1.1	1.7	1.6	2	0.1	0.8	1.5
159	x	x	0	0	0	0	0	0	0	0	0	0
160	x	x	0	0.5	0	0	4.5	0	0	0	4.5	1
161	x	x	0	0.1	2.4	2.5	0	0.1	2.5	3.3	0	0.1
162	x	x	0.1	0	0	0.1	0	0	0.1	0.1	0.1	0
163	x	x	0	0.1	3.9	0	0	0.1	3.5	0	0	0.1
164	x	x	0	0	1.3	0.1	0.1	0.7	0.9	0.1	0	0.1
166	x	x	2	0.5	0.6	0.5	1.1	0.5	0.5	0.3	1.4	0.3
167	x	x	0	0	0	0	0	1.7	0	0	0	0
168	x	x	0	0	0	0	0	0	0	3	0	0
170	x	x	0.1	0	0	4.5	0	1.5	0	4.5	0	0.1
171	x	x	2.5	0	1.6	0.8	1.5	3.2	0.5	0.7	2.8	1.5
172	x	x	0	0	0	0	0	0	0	0	0	0.2
174	x	x	0.2	0	1.6	0.9	1	0	2.3	1.8	0.2	0
176	x	x	0	0	0	0	0	0	0.5	3	0	0

Legend: Growth inhibition; Impaired growth; Impaired growth + growth inhibition; Eagle effect; x No growth of *S. aureus*; ng Poor or no growth of MMun131

Figure 6. Heatmap illustrating antibacterial activities of the moonmilk actinobacteria isolated in this study. The size of the inhibition zone (in cm) is given for each strain.

Table 1. Moonmilk strains isolated using the RC method and according to selective media.

MMun Strains	Family	Closest Match (BLAST)	16S rRNA Ident. %; gaps	Isolation Medium	Phylotype	OTU
129 ♣	Nocardiaceae	*Nocardia* sp. strain 2M-SSA4	99.40; 1	S-SN + Fe + DFB	-	25
130 ♣	Paenibacillaceae	*Paenibacillus* sp. DSL09-3	98.96; 1	S-SN + Fe + DFB	-	na
131 ♣	Streptomycetaceae	*Streptomyces* sp. 2323.1	99.61; 0	S-SN + Fe + DFB	XXXII	15
133 ♣	Nocardiaceae	*Nocardia* sp. strain 2M-SSA4	99.14; 3	S-SN + Fe + DFB	-	25
135 ♣	Streptomycetaceae	*Streptomyces* sp. MM63	100; 0	SN + Fe	IV *	21
136 ♣	Nocardiaceae	*Nocardia soli* strain DSM 44488	100; 0	SN + Fe	-	25
137 ♣	Streptomycetaceae	*Streptomyces* sp. MM82	99.08; 2	SN + DFB	XXXIII	30
138 ♣	Streptomycetaceae	*Streptomyces* sp. MM63	100; 0	SN	IV *	21
139 ♣	Streptomycetaceae	*Streptomyces* sp. MM63	100; 0	SN	IV *	21
140 ♣	Nocardiaceae	*Nocardia* sp. strain 2M-SSA4	99.54; 1	SN	-	25
141 ♣	Streptomycetaceae	*Streptomyces* sp. MM82	99.15; 2	SN	XXXIV	30
142 ♣	Rhodobacteraceae	Uncult. *Paracoccus* sp. clone 54	99.93; 0	SN + catalase	-	na
143 ♣	Streptomycetaceae	*Streptomyces lunaelactis* strain MM126	100; 0	SN + catalase	X *	21
144 ♣	Nocardiaceae	*Nocardia* soli strain Y48	99.34; 0	SN + catalase	-	25
145 ♣	Nocardiaceae	*Rhodococcus* sp. MTM3W5.2	99.80; 1	SN + catalase	-	1
146 ♣	Streptomycetaceae	*Streptomyces* sp. MM44	100; 0	S-SN + DFB	XXIX *	21
147 ♣	Streptomycetaceae	*Streptomyces xiamenensis* strain 318	98.69; 8	S-SN + DFB	XXXV	39
148 ♣	Streptomycetaceae	*Streptomyces xiamenensis* strain 318	98.69; 8	S-SN + DFB	XXXV	39
149	Micrococcaceae	*Micrococcus* sp. 3455	100; 0	SN	-	23
150 ♣	Streptomycetaceae	*Streptomyces* sp. MM63	100; 0	SN	IV *	21
151 ♣	Streptomycetaceae	*Streptomyces* sp. MM110	100; 0	SN	XXI *	21
152	Streptomycetaceae	*Streptomyces lunaelactis* strain MM126	100; 0	SN + Fe	X *	21
153	Streptomycetaceae	*Streptomyces* sp. MM18	99.93; 1	SN + DFB	I * (lost)	198
154 ♣	Streptomycetaceae	*Streptomyces* sp. ND04-1H	99.93; 0	SN	XXXVI	21
155	Nocardiaceae	*Rhodococcus* sp. strain MAK1	99.86; 2	SN	-	421
156	Streptomycetaceae	*Streptomyces* sp. 1C-HV8	99.86; 2	S-SN + Fe + DFB	XXXVII	132
157 ♣	Streptomycetaceae	*Streptomyces* sp. 2M-TWYE1	99.47; 0	SN + catalase	XXXVIII	61
158	Streptomycetaceae	*Streptomyces* sp. SpC090624KE_06	99.51; 0	S-SN + Fe + DFB	XXXIX	21
159	Microbacteriaceae	*Agromyces* sp. strain 4K403B	97.88; 14	S-SN + DFB	-	49
160	Micrococcaceae	*Kocuria rhizophila* strain R-42745	99.93; 0	S-SN	-	159
161	Streptomycetaceae	*Streptomyces* sp. MM18	100; 0	S-SN + Fe + DFB	I *	198
162	Streptomycetaceae	*Streptomyces* sp. MM63	99.93; 1	S-SN + Fe + DFB	IV *	21
163	Streptomycetaceae	*Streptomyces* sp. MM63	100; 0	S-SN	IV *	21
164	Streptomycetaceae	*Streptomyces* sp. MM63	100; 0	S-SN	IV *	21

Table 1. *Cont.*

MMun Strains	Family	Closest Match (BLAST)	16S rRNA Ident. %; gaps	Isolation Medium	Phylotype	OTU
166	Streptomycetaceae	*Streptomyces* sp. strain A301	99.23; 3	S-ISP5 + Fe + DFB	XXXIV	30
167	Microbacteriaceae	*Agromyces cerinus* strain DSM 8595	99.58; 0	ISP5 + catalase	-	113
168	Streptomycetaceae	*Streptomyces* sp. 28a-5-1	98.60; 8	ISP5 + catalase	XXXV	39
170	Streptomycetaceae	*Streptomyces* sp. 2C-SSA16-1	99.64; 0	S-ISP5 + Fe + DFB	XL	61
171	Pseudonocardiaceae	*Amycolatopsis* sp. CA11	98.25; 16	S-ISP5 + Fe + DFB	-	282
172	Micromonosporaceae	*Micromonospora maoerensis* strain NEAU-MES19	99.93; 0	S-ISP5 + Fe + DFB	-	64
174	Streptomycetaceae	*Streptomyces* sp. MM110	100; 0	ISP5 + catalase	XXI *	21
176	Streptomycetaceae	*Streptomyces* sp. 28a-5-1	98.67; 7	S-ISP5	XXXV	39

Numbers in the phylotype column have been deduced from the phylogenetic tree presented in Figure 4 and based on our previous screening [11]. The last column indicates to which OTU identified in our metagenetic study [13] the MMun strains are affiliated (according to the topology and patristic distances of the tree in Supplementary Figures S1 and S2). Abbreviations: S- before SN and ISP5, indicates that media were prepared by autoclaving of agar and phosphate separate from other media components; DFB, desferrioxamine B; na, not Actinobacteria; *, phylotypes described in our previous study [12]; lost, culture from mycelium stocks of MMun153 are no longer possible, and this strain is now considered as lost; and ♣, MMun isolates for which the genome has been sequenced.

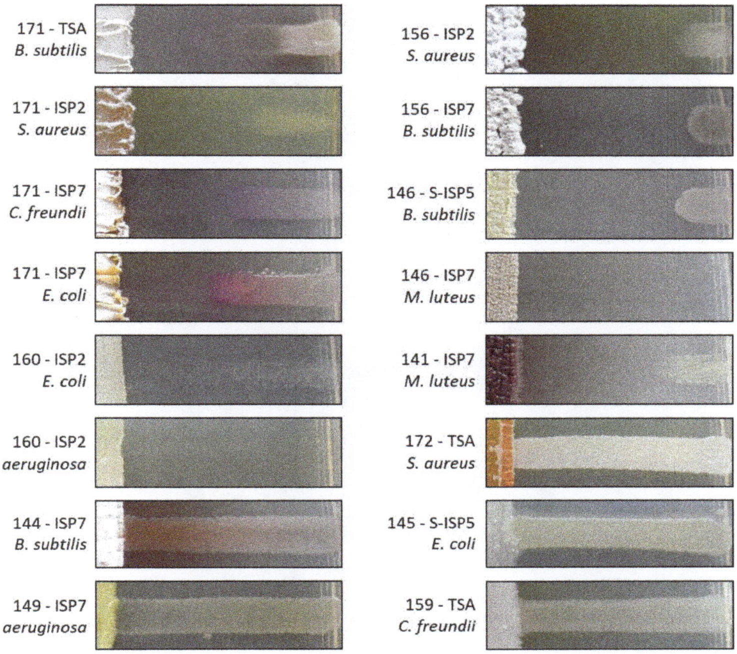

Figure 7. Selected representative examples of the cross-streak results. MMun strain numbers, media, and tested cultures are mentioned.

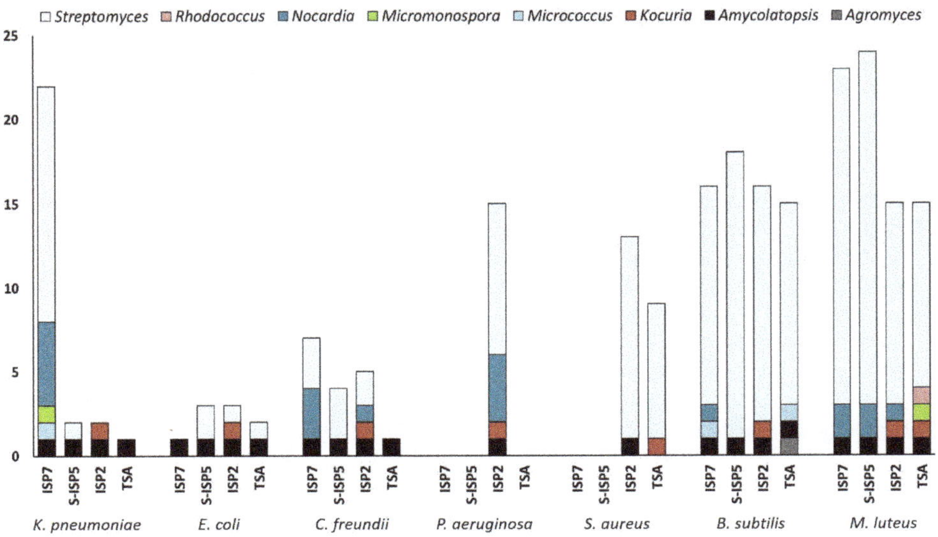

Figure 8. Number of strains with antibacterial activity, broken down by actinobacterial genus and test conditions.

Interestingly, unusual growth/susceptibility of *Klebsiella pneumonia* was frequently observed in the ISP7 medium, as reported in our first antibiotic activity screening assays [11]. Examples of different atypical growth responses are presented in Figure 9. In the depicted cases, we observed a non-linear response to the diffusion gradient of the secreted antibiotics, which has been described as the 'Eagle effect' (or paradoxical zone phenomenon) [31]. *K. pneumoniae* is indeed often able to grow near the actinobacterial strain, while its susceptibility to the secreted compounds is increased at a higher distance. Occasionally, the 'Eagle effect' response was also observed with *Citrobacter freundii*.

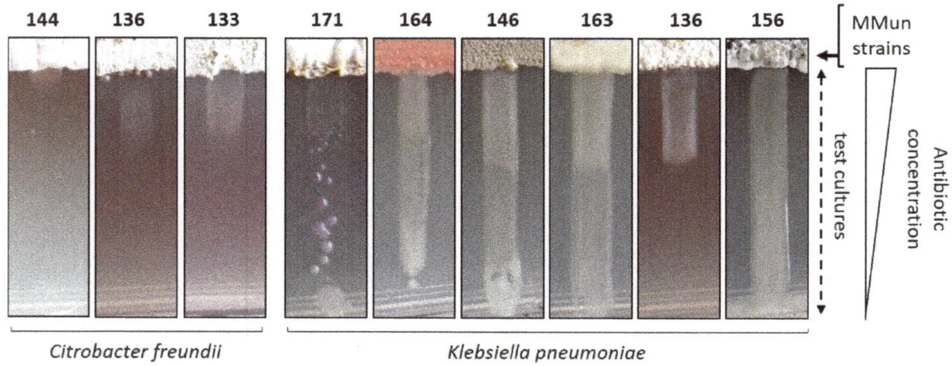

Figure 9. Illustration of the paradoxical zone phenomenon (or Eagle effect) observed with *Klebsiella pneumonia* and *Citrobacter freundii*. MMun156 was chosen as a control to show non-affected growth of *K. pneumoniae* on the ISP7 medium. All conditions where an Eagle effect was observed are marked with a "star" symbol in Figure 6.

Finally, draft genome assemblies of 20 MMun isolates were mined for biosynthetic gene clusters (BGCs) in order to estimate their potential at producing natural compounds. More precisely, we examined the presence of BGCs containing polyketide synthases (PKSs, including PKS-I, PKS-II, and PKS-III) and NRPS domains, as described previously [12]. The evaluation of the genetic predisposition to produce natural compounds of isolates for which we obtained genome assemblies is summarized in Table 2.

All investigated isolates encode in their genomes at least four and up to 34 NRPS domains. All strains (except MMun154 and MMun145) also possess multiple PKS-I (up to 21, MMun146) and PKS-II (up to 6, MMun131, 137, and 141) domains. PKS-IIIs were expectedly found less frequently (maximum 2), albeit 70% contained at least one BGC including a PKS-III domain. Interestingly, the highest number of BGCs was found in all isolated *Nocardia* strains, with MMun133 harboring the largest amount of biosynthetic cluster domains (45 in total). The fact that the strains that harbor the highest number of BGCs (Table 2) did not display any—or extremely weak or only Eagle-effect—antibacterial activity (Figure 6) suggests that if we indeed found conditions to isolate and grow them in synthetic media, we still have to discover the culture conditions that will trigger their secondary metabolism.

Table 2. Number of biosynthetic gene clusters (BGCs) based on genome mining of 22 MMun strains.

MMun	PKS-I	PKS-II	PKS-III	NRPS	Total	Genus
129	2	2	1	34	39	*Nocardia*
131	2	6	1	27	36	*Streptomyces*
133	7	4	1	33	45	*Nocardia*
135	10	4	1	16	31	*Streptomyces*
136	6	2	1	31	40	*Nocardia*
137	9	6	0	6	21	*Streptomyces*
138	11	4	1	16	32	*Streptomyces*
139	4	4	0	13	21	*Streptomyces*
140	2	2	1	32	37	*Nocardia*
141	9	6	0	6	21	*Streptomyces*
143	6	4	1	9	20	*Streptomyces*
144	4	2	2	31	39	*Nocardia*
145	5	0	0	12	17	*Rhodococcus*
146	21	4	0	9	34	*Streptomyces*
147	10	2	1	4	17	*Streptomyces*
148	10	2	1	4	17	*Streptomyces*
150	12	4	1	14	31	*Streptomyces*
151	4	4	0	22	30	*Streptomyces*
154	0	4	1	10	15	*Streptomyces*
157	12	4	0	9	25	*Streptomyces*

3. Materials and Methods

3.1. Preparation of Media and Moonmilk Suspension for Isolation of Rare Actinobacteria

Starch nitrate (SN) [32] and International Streptomyces Project (ISP) medium number 5 (ISP5) [33] were prepared in two different manners: either according to the standard protocols with all of the components autoclaved together (SN/ISP5) or with the agar and phosphate solution autoclaved separately and mixed before media pouring (S-SN/S-ISP5) in order to avoid hydrogen peroxide (H_2O_2) formation [23]. Media were supplemented with nalidixic acid (75 µg/mL) and nystatin (50 µg/mL) to suppress the growth of Gram-negative bacteria and fungi, respectively. Each medium was additionally supplemented with factors known or predicted to increase microbial growth, including: (i) 20 µM $FeCl_3$; (ii) 1 µM of desferrioxamine B (Sigma-Aldrich, St. Louis, MO, USA); and (iii) 10 µl of 10 mg/mL bovine liver catalase (Sigma-Aldrich, St. Louis, MO, USA) (on the Whatman 3 MM paper disc (Sigma-Aldrich, St. Louis, MO, USA) in the center of the Petri dish) [23].

Isolation of Actinobacteria was performed by preparing a moonmilk suspension via the rehydration-centrifugation method according to Hop et al. (2011) [25]. Briefly, 0.45 g of freeze-dried moonmilk (0.15 g of each collection site—COL1, COL3, and COL4) was suspended in 50 mL of 0.01 M phosphate buffer (pH 7.0), vortexed for 5 min, and incubated for 90 min at 30 °C. Then, 8 mL of the upper part of the supernatant was transferred to a 50 mL conical centrifuge tube and centrifuged at room temperature for 10 min at 3000 rpm. After a 30-min incubation at 30 °C, 100 µL of serial dilutions (up to 10^{-4}) were inoculated on the different solid media. All plates were incubated at 15 °C for about 2 months. The number of CFUs was evaluated after 16, 44, and 69 days of incubation. Among the 100 colonies selected from the culturable population, 42 isolates (named MMun) were successfully subcultured twice to obtain pure strains and stored both as 25% glycerol mycelium stock at −20 °C and in solid media at 4 °C.

3.2. Isolation and Sequencing of Genomic DNA

Genomic DNA from MMun strains was extracted with the GenElute Bacterial Genomic DNA Kit (Sigma-Aldrich, St. Louis, MO, USA) according to the manufacturer's instructions from the mycelium grown in liquid LB, ISP1, or ISP2 media at 28 °C. Genomic libraries of moonmilk isolates were constructed using the Nextera XT kit (Illumina, Inc., San Diego, CA, USA). Sequencing was carried out on the Illumina MiSeq platform with 2 × 300-bp read configuration. Complete genomes were assembled de novo from raw sequence data with SPAdes v3.6.2 [34] using the "careful" option, and the quality of the assemblies was subsequently assessed with QUAST v2.3 [35].

3.3. Phylogenetic Analyses of Actinobacterial Strains

The phylogenetic analysis of the moonmilk-derived strains shown in Figure 4 was based on either the full-length or the nearly full-length sequences of their 16S rRNA gene, which were either (i) recovered from the sequenced genomes (Table 2) or (ii) amplified using bacterial universal primers 8F and 1541R [36], as reported previously [27]. A multiple sequence alignment (1538 sites) was built with MUSCLE v3.8.31 [37] (default parameters) and filtered using the script ali2phylip.pl (from the Bio-MUST-Core software package; D. Baurain; https://metacpan.org/release/Bio-MUST-Core), so as to only keep sites with no missing character states. The filtered alignment (92 sequences × 1342 sites) was then submitted to phylogenetic inference using the rapid bootstrap analysis of RAxML v8.1.17 ([38]; 100 pseudoreplicates) under the model GTR + Γ_4.

The V6–V7 regions of the 40 MMun strains were extracted from their 16S rRNA full-length sequences and combined to the corresponding regions of the 243 metagenetic OTUs to yield the tree shown in Supplementary Figure S1. Briefly, the final alignment was carefully crafted using a combination of automatic and manual approaches, including MAFFT v7.273 [39], Clustal Omega v1.1.0 [40], and the alignment editor (ed) of the MUST software package [41]. OTU403 was discarded because of its identity to OTU1, except for an aberrant five-terminal stretch. The optimized alignment (296 sites) was then filtered using ali2phylip.pl to remove sites present in <5% of the sequences, resulting in a final dataset of 282 sequences × 261 sites (0.12% of missing character states). Phylogenetic inference was carried out as for the full-length 16S rRNA genes above. The tree was first formatted in FigTree v1.4.21 (http://tree.bio.ed.ac.uk/software/figtree/) and then further arranged using Inkscape v0.91 (https://inkscape.org/). Patristic distances were then obtained using SeaView v4.5.4 [42] and the OTU closest to each MMun V6-V7 region was automatically extracted using a custom Perl script. Both the tree topology and the patristic distances were considered to affiliate MMun strains to metagenetic OTUs (Supplementary Figures S1 and S2).

The accession numbers of the 16S rRNA deposited in the GenBank database are listed in the Supplementary Table S2.

3.4. Antimicrobial Activities of Rare Moonmilk Actinobacterial Strains

Antibacterial activities of "MMun" strains of our collection were evaluated via cross-streak assays as described previously [11]. Each strain was inoculated from a mycelium stock as a single line in the center of the square plate and incubated for 14 days at 28 °C before cross-streaked with bacterial reference strains: *Escherichia coli* (ATCC 25922), *Pseudomonas aeruginosa* (ATCC 27853), *Citrobacter freundii* (ATCC 43864), *Klebsiella pneumoniae* (ATCC 13883), *Bacillus subtilis* (ATCC 19659), *Staphylococcus aureus* (ATCC 25923), and *Micrococcus luteus* (ATCC 9341).

3.5. Genome Mining for Gene Clusters Involved in Secondary Metabolite Production.

Genomes of 20 MMun strains were screened for NRPS and PKS genes, as described previously [11]. Only contigs $\geq$10 kb were considered and genes were identified as NRPS and PKS-I only when they displayed adenylation and acyltransferase domains, respectively.

4. Conclusions

Motivated by the results of our metagenetic study that had revealed a high diversity of the actinobacterial microbiome dwelling in the moonmilk deposits of the "Grotte des Collemboles" [13], we set up a series of protocols and culture conditions to isolate representatives of some of the most promising actinobacterial genera for the production of natural compounds. Remarkably, we isolated for the first time rare Actinobacteria belonging to the *Agromyces, Amycolatopsis, Kocuria, Micrococcus, Micromonospora, Nocardia,* and *Rhodococcus* genera, as well as additional new members of the *Streptomyces* genus. Remarkably, we succeeded to isolate for the first time from moonmilk deposits actinobacterial strains that belong to the *Micrococcus, Micromonospora,* and *Kocuria* genera. Another remarkable outcome of our research is the isolation of an *Amycolatopsis* strain that belongs to one of the less abundant OTUs identified in the studied moonmilk, accounting for less than 0.001% of the actinobacterial microbiome. This strain has also been revealed to display the highest propensity for producing antibacterial agents, both in terms of intensity and spectrum of activity.

Despite our success at isolating strains belonging to different actinobacterial genera, we however encountered a significant loss of the isolated colonies (58%) during our steps to obtain pure cultures. This most likely highlights the necessity of these strains to grow in the presence of specific growth factors of their environmental niche or an obligation to evolve in a mutualistic population with other moonmilk-dwelling bacteria. This is particularly problematic in bioprospecting, as the lost strains have probably been unable to grow as pure isolates because they are particularly (too) well-adapted to their unique natural habitat, and this adaptation might be a consequence of a unique specialized (secondary) metabolism. Therefore, we might have lost some of the most interesting strains in terms of potential producers of novel bioactive compounds.

In line with the conclusion mentioned above, though in this work we managed to isolate novel strains (and most likely novel species) thanks to our adapted protocols; the comparison of the 16S rRNA sequences of the new isolates with the 16S rRNA amplicons obtained in our metagenetic study [13] revealed the weak representativeness of our collection compared to Actinobacteria actually dwelling in the studied moonmilk deposits. Indeed, the identified 132 culturable strains are affiliated with 16 (6.5%) of the 243 actinobacterial OTUs identified from the DNA extracted from the moonmilk deposits. Considering that one single OTU can include lots of different species and strains (see OTU21 as example, Figure 4), this further emphasizes how much our collection is not exhaustively representative of the actinobacterial community inhabiting these carbonate deposits.

Finally, finding novel species/strains is only the first step towards the discovery of novel natural compounds. The next step is to make them trigger their secondary metabolism, which is an equally challenging task. Indeed, even if some strains already display high antibacterial activity, many others remain metabolically silent under the tested culture conditions. This is the case for all *Nocardia* strains isolated in this study that were unable to display antibacterial activity but were revealed to possess the largest number of BGCs (up to 45). As most of our new moonmilk strains have been isolated using a series of "tips and tricks" to cultivate the unculturable, they are unlikely to grow in many different media, which limits the culture conditions to be tested to awaken their secondary metabolism. Mining their genomes to unveil the triggers and cues of their secondary metabolism is a priority of our ongoing research [43].

Supplementary Materials: The following are available online at http://www.mdpi.com/2079-6382/7/2/28/s1: Figure S1: Phylogenetic tree of newly isolated moonmilk-dwelling Actinobacteria strains based on the V6–V7 region of the 16S (SSU) rRNA gene; Figure S2: Closest OTUs (V6–V7 regions) for the 40 MMun strains; Table S1: OTUs and their abundance in moonmilk deposits of the "Grotte des Collemboles"; Table S2: Accession number of 16S rRNA sequences of MMun strains. Please provide the Table S2 that can be edited and put Figure S1 and Tables S1 and S2 into Word file.

Acknowledgments: D.A. and M.M.'s work was supported by a Research Foundation for Industry and Agriculture (FRIA) grant. Computational resources ("durandal" grid computer) were funded by three grants from the University of Liège, "Fonds spéciaux pour la recherche", "Crédit de démarrage 2012" (SFRD-12/03 and SFRD-12/04), and "Crédit classique 2014" (C-14/73) and by a grant from the F.R.S.-FNRS "Crédit de recherche 2014" (CDR J.0080.15). This work is supported in part by the Belgian program of Interuniversity Attraction Poles initiated by the Federal Office for Scientific Technical and Cultural Affairs (PAI No. P7/44). SR is a Research Associate at Belgian Fund for Scientific Research (F.R.S.-FNRS).

Author Contributions: D.A., M.M., A.N., L.M. and S.R. designed and performed experiments. Bioinformatic analyses were performed by A.N., D.B. and S.R. Data were analyzed by all authors. The manuscript was written and/or corrected by all authors.

Conflicts of Interest: The authors declare no conflict of interest.

References

1. Hill, P.; Heberlig, G.W.; Boddy, C.N. Sampling Terrestrial Environments for Bacterial Polyketides. *Molecules* **2017**, *22*. [CrossRef] [PubMed]
2. Goodfellow, M.; Fiedler, H.-P. A guide to successful bioprospecting: Informed by actinobacterial systematics. *Antonie Leeuwenhoek* **2010**, *98*, 119–142. [CrossRef] [PubMed]
3. World Health Organization (WHO). *Antimicrobial Resistance: Global Report on Surveillance 2014*; AMR Report; WHO: Geneva, Switzerland, 2014.
4. Hutchinson, G.E. The Paradox of the Plankton. *Am. Nat.* **1961**, *95*, 137–145. [CrossRef]
5. Bhullar, K.; Waglechner, N.; Pawlowski, A.; Koteva, K.; Banks, E.D.; Johnston, M.D.; Barton, H.A.; Wright, G.D. Antibiotic resistance is prevalent in an isolated cave microbiome. *PLoS ONE* **2012**, *7*. [CrossRef] [PubMed]
6. Lavoie, K.; Ruhumbika, T.; Bawa, A.; Whitney, A.; de Ondarza, J. High levels of antibiotic resistance but no antibiotic production detected along a gypsum gradient in great Onyx Cave, KY, USA. *Diversity* **2017**, *9*. [CrossRef]
7. Hopwood, D.A. *Streptomyces in Nature and Medicine: The Antibiotic Makers*; Oxford University Press: New York, NY, USA, 2007; ISBN 978-0-19-515066-7.
8. Dhami, N.K.; Mukherjee, A.; Watkin, E.L.J. Microbial Diversity and Mineralogical-Mechanical Properties of Calcitic Cave Speleothems in Natural and in Vitro Biomineralization Conditions. *Front. Microbiol.* **2018**, *9*. [CrossRef] [PubMed]
9. Portillo, M.C.; Gonzalez, J.M. Moonmilk Deposits Originate from Specific Bacterial Communities in Altamira Cave (Spain). *Microb. Ecol.* **2011**, *61*, 182–189. [CrossRef] [PubMed]
10. Rooney, D.C.; Hutchens, E.; Clipson, N.; Baldini, J.; McDermott, F. Microbial Community Diversity of Moonmilk Deposits at Ballynamintra Cave, Co. *Waterford, Ireland. Microb. Ecol.* **2010**, *60*, 753–761. [CrossRef] [PubMed]
11. Maciejewska, M.; Adam, D.; Naômé, A.; Martinet, L.; Tenconi, E.; Całusińska, M.; Delfosse, P.; Hanikenne, M.; Baurain, D.; Compère, P.; et al. Assessment of the Potential Role of Streptomyces in Cave Moonmilk Formation. *Front. Microbiol.* **2017**, *8*, 1181. [CrossRef] [PubMed]
12. Maciejewska, M.; Adam, D.; Martinet, L.; Naômé, A.; Całusińska, M.; Delfosse, P.; Carnol, M.; Barton, H.A.; Hayette, M.-P.; Smargiasso, N.; et al. A Phenotypic and Genotypic Analysis of the Antimicrobial Potential of Cultivable Streptomyces Isolated from Cave Moonmilk Deposits. *Front. Microbiol.* **2016**, *7*, 1455. [CrossRef] [PubMed]
13. Maciejewska, M.; Calusinska, M.; Cornet, L.; Adam, D.; Pessi, I.S.; Malchair, S.; Delfosse, P.; Baurain, D.; Barton, H.A.; Carnol, M.; et al. High-throughput sequencing analysis of the actinobacterial spatial diversity in moonmilk deposits. *Antibiotics* **2018**. [CrossRef]
14. Oliver, J.D. The viable but nonculturable state in bacteria. *J. Microbiol.* **2005**, *43*, 93–100. [PubMed]

15. Staley, J.T.; Konopka, A. Measurement of in situ activities of nonphotosynthetic microorganisms in aquatic and terrestrial habitats. *Annu. Rev. Microbiol.* **1985**, *39*, 321–346. [CrossRef] [PubMed]
16. Hug, L.; Baker, B.; Anantharaman, K.; Brown, C.; Probst, A.; Castelle, C.; Butterfield, C.; Hernsdorf, A.; Amano, Y.; Ise, K.; et al. A new view of the tree of life. *Nat. Microbiol.* **2016**, *1*, 16048. [CrossRef] [PubMed]
17. Button, D.K.; Schut, F.; Quang, P.; Martin, R.; Robertson, B.R. Viability and isolation of marine bacteria by dilution culture: Theory, procedures, and initial results. *Appl. Environ. Microbiol.* **1993**, *59*, 881–891. [PubMed]
18. D'Onofrio, A.; Crawford, J.M.; Stewart, E.J.; Witt, K.; Gavrish, E.; Epstein, S.; Clardy, J.; Lewis, K. Siderophores from Neighboring Organisms Promote the Growth of Uncultured Bacteria. *Chem. Biol.* **2010**, *17*, 254–264. [CrossRef] [PubMed]
19. Bruns, A.; Cypionka, H.; Overmann, J. Cyclic AMP and acyl homoserine lactones increase the cultivation efficiency of heterotrophic bacteria from the central Baltic Sea. *Appl. Environ. Microbiol.* **2002**, *68*, 3978–3987. [CrossRef] [PubMed]
20. Sait, M.; Hugenholtz, P.; Janssen, P.H. Cultivation of globally distributed soil bacteria from phylogenetic lineages previously only detected in cultivation-independent surveys. *Environ. Microbiol.* **2002**, *4*, 654–666. [CrossRef] [PubMed]
21. Nichols, D.; Cahoon, N.; Trakhtenberg, E.M.; Pham, L.; Mehta, A.; Belanger, A.; Kanigan, T.; Lewis, K.; Epstein, S.S. Use of Ichip for High-Throughput In Situ Cultivation of "Uncultivable" Microbial Species. *Appl. Environ. Microbiol.* **2010**, *76*, 2445–2450. [CrossRef] [PubMed]
22. Tanaka, T.; Kawasaki, K.; Daimon, S.; Kitagawa, W.; Yamamoto, K.; Tamaki, H.; Tanaka, M.; Nakatsu, C.H.; Kamagata, Y. A hidden pitfall in the preparation of agar media undermines microorganism cultivability. *Appl. Environ. Microbiol.* **2014**, *80*, 7659–7666. [CrossRef] [PubMed]
23. Kawasaki, K.; Kamagata, Y. Phosphate-Catalyzed Hydrogen Peroxide Formation from Agar, Gellan, and κ-Carrageenan and Recovery of Microbial Cultivability via Catalase and Pyruvate. *Appl. Environ. Microbiol.* **2017**, *83*. [CrossRef] [PubMed]
24. Jiang, C.-Y.; Dong, L.; Zhao, J.-K.; Hu, X.; Shen, C.; Qiao, Y.; Zhang, X.; Wang, Y.; Ismagilov, R.F.; Liu, S.-J.; et al. High-Throughput Single-Cell Cultivation on Microfluidic Streak Plates. *Appl. Environ. Microbiol.* **2016**, *82*, 2210–2218. [CrossRef] [PubMed]
25. Hop, D.V.; Sakiyama, Y.; Binh, C.T.T.; Otoguro, M.; Hang, D.T.; Miyadoh, S.; Luong, D.T.; Ando, K. Taxonomic and ecological studies of actinomycetes from Vietnam: Isolation and genus-level diversity. *J. Antibiot.* **2011**, *64*, 599–606. [CrossRef] [PubMed]
26. Otoguro, M.; Hayakawa, M.; Yamazaki, T.; Iimura, Y. An integrated method for the enrichment and selective isolation of Actinokineospora spp. in soil and plant litter. *J. Appl. Microbiol.* **2001**, *91*, 118–130. [PubMed]
27. Hayakawa, M.; Otoguro, M.; Takeuchi, T.; Yamazaki, T.; Iimura, Y. Application of a method incorporating differential centrifugation for selective isolation of motile actinomycetes in soil and plant litter. *Antonie Leeuwenhoek* **2000**, *78*, 171–185. [CrossRef] [PubMed]
28. Arias, A.A.; Lambert, S.; Martinet, L.; Adam, D.; Tenconi, E.; Hayette, M.-P.; Ongena, M.; Rigali, S. Growth of desferrioxamine-deficient Streptomyces mutants through xenosiderophore piracy of airborne fungal contaminations. *FEMS Microbiol. Ecol.* **2015**, *91*. [CrossRef] [PubMed]
29. Maciejewska, M.; Pessi, I.S.; Arguelles-Arias, A.; Noirfalise, P.; Luis, G.; Ongena, M.; Barton, H.; Carnol, M.; Rigali, S. Streptomyces lunaelactis sp. nov., a novel ferroverdin A-producing Streptomyces species isolated from a moonmilk speleothem. *Antonie Leeuwenhoek* **2015**, *107*, 519–531. [CrossRef] [PubMed]
30. Kis, Á.E.; Laczi, K.; Zsíros, S.; Kós, P.; Tengölics, R.; Bounedjoum, N.; Kovács, T.; Rákhely, G.; Perei, K. Characterization of the Rhodococcus sp. MK1 strain and its pilot application for bioremediation of diesel oil-contaminated soil. *Acta Microbiol. Immunol. Hung.* **2017**, *64*, 463–482. [CrossRef] [PubMed]
31. Eagle, H.; Musselman, A.D. The rate of bactericidal action of penicillin in vitro as a function of its concentrations, and its paradoxically reduced activity at high concentrations against certain organisms. *J. Exp. Med.* **1948**, *88*, 99–131. [CrossRef] [PubMed]
32. Waksman, S.; Lechevalier, H. *The Actinomycetales, Classification, Identification and Description of Genera and Species*; Williams & Wilkins Co.: Baltimore, MD, USA, 1961.
33. Shirling, E.B.; Gottlieb, D. Methods for characterization of Streptomyces species. *Int. J. Syst. Bacteriol.* **1966**, *16*, 313–340. [CrossRef]

34. Bankevich, A.; Nurk, S.; Antipov, D.; Gurevich, A.A.; Dvorkin, M.; Kulikov, A.S.; Lesin, V.M.; Nikolenko, S.I.; Pham, S.; Prjibelski, A.D.; et al. SPAdes: A new genome assembly algorithm and its applications to single-cell sequencing. *J. Comput. Biol.* **2012**, *19*, 455–477. [CrossRef] [PubMed]
35. Gurevich, A.; Saveliev, V.; Vyahhi, N.; Tesler, G. QUAST: Quality assessment tool for genome assemblies. *Bioinformatics* **2013**, *29*, 1072–1075. [CrossRef] [PubMed]
36. Onstott, T.C.; McGown, D.J.; Bakermans, C.; Ruskeeniemi, T.; Ahonen, L.; Telling, J.; Soffientino, B.; Pfiffner, S.M.; Sherwood-Lollar, B.; Frape, S.; et al. Microbial communities in subpermafrost saline fracture water at the Lupin Au mine, Nunavut, Canada. *Microb. Ecol.* **2009**, *58*, 786–807. [CrossRef] [PubMed]
37. Edgar, R.C. MUSCLE: Multiple sequence alignment with high accuracy and high throughput. *Nucleic Acids Res.* **2004**, *32*, 1792–1797. [CrossRef] [PubMed]
38. Stamatakis, A. RAxML version 8: A tool for phylogenetic analysis and post-analysis of large phylogenies. *Bioinformatics* **2014**, *30*, 1312–1313. [CrossRef] [PubMed]
39. Katoh, K.; Standley, D.M. MAFFT multiple sequence alignment software version 7: Improvements in performance and usability. *Mol. Biol. Evol.* **2013**, *30*, 772–780. [CrossRef] [PubMed]
40. Sievers, F.; Wilm, A.; Dineen, D.; Gibson, T.J.; Karplus, K.; Li, W.; Lopez, R.; McWilliam, H.; Remmert, M.; Söding, J.; et al. Fast, scalable generation of high-quality protein multiple sequence alignments using Clustal Omega. *Mol. Syst. Biol.* **2011**, *7*, 539. [CrossRef] [PubMed]
41. Philippe, H. MUST, a computer package of Management Utilities for Sequences and Trees. *Nucleic Acids Res.* **1993**, *21*, 5264–5272. [CrossRef] [PubMed]
42. Gouy, M.; Guindon, S.; Gascuel, O. SeaView version 4: A multiplatform graphical user interface for sequence alignment and phylogenetic tree building. *Mol. Biol. Evol.* **2010**, *27*, 221–224. [CrossRef] [PubMed]
43. Rigali, S.; Anderssen, S.; Naômé, A.; van Wezel, G.P. Cracking the Regulatory Code of Biosynthetic Gene Clusters as a Strategy for Natural Product Discovery. *Biochem. Pharmacol.* **2018**. [CrossRef] [PubMed]

antibiotics

MDPI

Article

High-Throughput Sequencing Analysis of the Actinobacterial Spatial Diversity in Moonmilk Deposits

Marta Maciejewska [1], Magdalena Całusińska [2], Luc Cornet [3], Delphine Adam [1], Igor S. Pessi [1], Sandrine Malchair [4], Philippe Delfosse [2], Denis Baurain [3], Hazel A. Barton [5], Monique Carnol [4] and Sébastien Rigali [1,*]

1 InBioS—Centre for Protein Engineering, Institut de Chimie B6a, University of Liège, B-4000 Liège, Belgium; maciejewska.m@wp.pl (M.M.); delphine.adam@doct.ulg.ac.be (D.A.); ispessi@alumni.ulg.ac.be (I.S.P.)

2 Environmental Research and Innovation Department, Luxembourg Institute of Science and Technology, Belvaux, Luxembourg; magdalena.calusinska@list.lu (M.C.); philippe.delfosse@list.lu (P.D.)

3 InBioS—PhytoSYSTEMS, Eukaryotic Phylogenomics, University of Liège, B-4000 Liège, Belgium; luc.cornet@uliege.be (L.C.); Denis.Baurain@uliege.be (D.B.)

4 InBioS—Plant and Microbial Ecology, Botany B22, University of Liège, B-4000 Liège, Belgium; S.Malchair@uliege.be (S.M.); S.Malchair@uliege.be (M.C.)

5 Department of Biology, University of Akron, Akron, OH 44325, USA; bartonh@uakron.edu

* Correspondence: srigali@uliege.be; Tel.: +32-4-366-9830

Received: 12 February 2018; Accepted: 16 March 2018; Published: 21 March 2018

check for updates

Abstract: Moonmilk are cave carbonate deposits that host a rich microbiome, including antibiotic-producing Actinobacteria, making these speleothems appealing for bioprospecting. Here, we investigated the taxonomic profile of the actinobacterial community of three moonmilk deposits of the cave "Grotte des Collemboles" via high-throughput sequencing of 16S rRNA amplicons. Actinobacteria was the most common phylum after Proteobacteria, ranging from 9% to 23% of the total bacterial population. Next to actinobacterial operational taxonomic units (OTUs) attributed to uncultured organisms at the genus level (~44%), we identified 47 actinobacterial genera with *Rhodoccocus* (4 OTUs, 17%) and *Pseudonocardia* (9 OTUs, ~16%) as the most abundant in terms of the absolute number of sequences. Streptomycetes presented the highest diversity (19 OTUs, 3%), with most of the OTUs unlinked to the culturable *Streptomyces* strains that were previously isolated from the same deposits. Furthermore, 43% of the OTUs were shared between the three studied collection points, while 34% were exclusive to one deposit, indicating that distinct speleothems host their own population, despite their nearby localization. This important spatial diversity suggests that prospecting within different moonmilk deposits should result in the isolation of unique and novel Actinobacteria. These speleothems also host a wide range of non-streptomycetes antibiotic-producing genera, and should therefore be subjected to methodologies for isolating rare Actinobacteria.

Keywords: antibiotics; geomicrobiology; Illumina sequencing; microbiome diversity; *Streptomyces*; Actinobacteria

1. Introduction

Molecular approaches evaluating microbial communities in caves have revealed a level of diversity greater than initially expected [1]. Microorganisms have been found to inhabit virtually all subterranean niches, including cave walls, ceilings, speleothems, soils, sediments, pools, and aquifers [2]. Cave bacteria often represent novel taxonomic groups [3–7], which are frequently more closely related to other cave-derived bacterial lineages than to the microbiota of other environments [8–10].

Among cave speleothems, moonmilk draws a particular scientific attention due to its distinctive crystal morphology. The origins of various moonmilk crystalline habits, including monocrystalline rods, polycrystalline chains, and nanofibers, are tentatively attributed to the moonmilk indigenous microbial population [11]. Among a moonmilk microbiome comprising Archaea, Fungi, and Bacteria [9,10,12–19], the indigenous filamentous Fungi [20] and Actinobacteria [11,21] are believed to mediate moonmilk genesis with cell surfaces promoting $CaCO_3$ deposition [11,20,21]. Actinobacteria were additionally reported to be metabolically capable of inducing favorable conditions for $CaCO_3$ precipitation, or even directly precipitating carbonate minerals [12,21]. Members of the phylum Actinobacteria are routinely found in this speleothem [9,10,12–14,18,19], as well as in the other subterranean deposits within limestone caves [3,8,22,23], volcanic caves [24–26], and ice caves [27]. The broad distribution of Actinobacteria in the subsurface systems stimulates investigation in order to understand the factors driving their existence in mainly inorganic and highly oligotrophic environments, and the processes that enable them to mediate speleogenesis. The successful adaptation of Actinobacteria to a wide range of environments could probably be a consequence of their broad-spectrum metabolism, which includes prolific secreted hydrolytic systems that are capable of generating nutrient sources from various substrates, along with their extraordinary faculty to produce specialized metabolites (metal chelators, antimicrobials, hormones, etc.) [28].

As recently reported, moonmilk Actinobacteria represent novel microorganisms, which is a discovery that opens great avenues for the bioprospecting of novel drugs [6,10,18]. Rooney et al. (2010) [13] showed that spatially separated moonmilk speleothems in Ballynamintra Cave are inhabited by taxonomically distinct fungal and bacterial communities. Instead, in our attempt to isolate moonmilk-dwelling Actinobacteria for assessing their potential for participating in the genesis of these speleothems [21] and producing antimicrobial compounds [10], we only recovered members of the genus *Streptomyces*. Such a dominance of streptomycetes was rather unexpected, according to other moonmilk microbial diversity studies performed through culture-dependent [10,12,13,18] and culture-independent approaches using clone libraries [9], denaturing gradient gel electrophoresis (DGGE) fingerprinting [14,16,17], automated ribosomal intergenic spacer analysis (ARISA) [13], and, more recently, high-throughput sequencing (HTS) [19]. The actinobacterial genera identified in those studies included *Rhodococcus, Pseudonocardia, Propionibacterium, Nocardia, Amycolatopsis, Saccharothrix, Geodermatophilus, Mycobacterium, Aeromicrobium, Kribella, Nocardioides, Actinomycetospora, Nonomuraea, Euzebya, Rubrobacter*, and *Arthrobacter*, in addition to *Streptomyces*. Nonetheless, the diversity of the moonmilk actinobacterial microbiome still remains largely unknown and, beyond evaluating *"what and how much have we missed in our culture-dependent bioprospecting approach"* [10], a major important question that arises is: *to what extent are moonmilk-dwelling Actinobacteria different between the moonmilk deposits within a single cave, or in different caves?*

In this work, we carried out a comparative (HTS) of 16S small subunit (SSU) rRNA gene from DNA extracted from spatially separated moonmilk deposits within the same cave, "Grotte des Collemboles" (Springtails' Cave) in Comblain-au-Pont, Belgium (Figure S1), in order to draw a detailed taxonomic picture of the intra-phylum diversity. Identifying the presence of rare Actinobacteria and unveiling to which degree they exhibit a spatial variability would help determining whether it is worth prospecting from different moonmilk deposits to isolate unique and novel natural compound producers.

2. Results

2.1. Actinobacterial Abundance within the Whole Moonmilk Bacterial Microbiome

Libraries spanning the V4–V6 variable regions of the 16S rRNA gene using universal bacterial primers were used to assess the proportion of Actinobacteria in comparison to the whole bacterial community of three moonmilk deposits of the cave "Grotte des Collemboles" (Table S1a). The observed bacterial communities differed in species richness, evenness, and diversity between the three sampling points (Table 1). Phylotype richness (total number of operational taxonomic units (OTUs) per site) was the highest in COL4 (1863 OTUs), followed by COL1 and COL3, with 1332 and 1161 OTUs, respectively (Table 1, Figure 1a). Across the three sampling points, we found a total of 2301 different OTUs, amongst which 710 (31%) were common to all of the deposits (Figure 1a). Interestingly, pairwise comparison revealed highly similar percentages (~31.7 ± 0.53%) of shared bacterial OTUs between moonmilk deposits (Table 2, Figure 1a). A total of 956 OTUs (42%) were found to be exclusive to one sampling site, with COL4 having the highest number of unique bacterial phylotypes (584 OTUs), along with the most diverse bacterial population, as reflected by the highest diversity indices (Table 1, Figure 1a).

Table 1. Richness, specificity, diversity, and evenness of the bacterial and actinobacterial communities in the three moonmilk deposits of the "Grotte des Collemboles". OTUs: operational taxonomic units.

Target Group	Site	Total OTUs (Richness)	Unique OTUs (Specificity)	Inverse Simpson Index (Diversity)	Simpson Index (Evenness)
Bacteria	COL1	1332	238 (17.9%)	13.23	0.01
	COL3	1161	134 (11.6%)	58.94	0.05
	COL4	1863	584 (31.3%)	155.31	0.08
Actinobacteria	COL1	150	14 (9.3%)	6.21	0.04
	COL3	147	15 (10.2%)	7.74	0.05
	COL4	211	54 (25.6%)	24.13	0.11

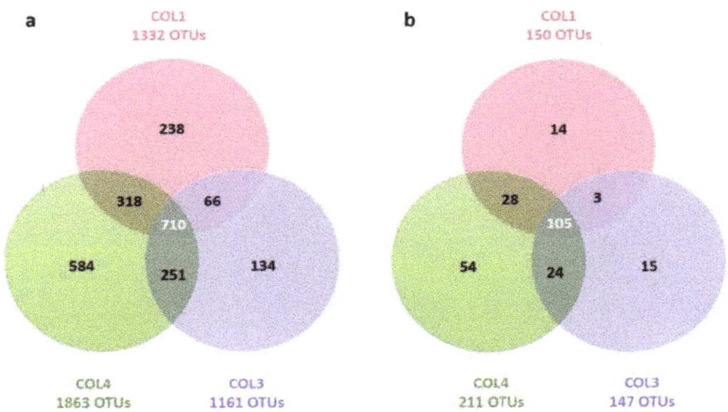

Figure 1. Venn diagrams showing the numbers of shared and unique bacterial (**a**) and actinobacterial (**b**) OTUs between the three moonmilk sampling points (COL1, COL3, COL4).

Table 2. Pairwise comparisons of shared OTUs between the moonmilk deposits.

Target Group	COL1 and COL3	COL1 and COL4	COL4 and COL3
Bacteria	776/2493 (31.1%)	1028/3195 (32.2%)	961/3024 (31.8%)
Actinobacteria	108/297 (36.4%)	133/361 (36.9%)	129/358 (36.0%)

Bacterial OTUs were grouped into 21 phyla and 18 candidate phyla (Table S2, Figure 2). Actinobacteria represented 9%, 23%, and 10% of the total bacterial population in COL1, COL3, and COL4, respectively (Table S2, Figure 2a). In terms of abundance, they were the most common phylum after Proteobacteria, which accounted for 52%, 34%, and 30% of the total community in COL1, COL3, and COL4, respectively (Table S2, Figure 2a). The other major phyla of the moonmilk microbiome included Acidobacteria, Nitrospirae, Chloroflexi, Gemmatimonadetes, Planctomycetes, Latescibacteria, Verrucomicrobia, Zixibacteria, Armatimonadetes, Bacteroidetes, and Parcubacteria (Table S2, Figure 2a). Together, these phyla constituted 93.4%, 94.7%, and 91.5% of the total community in COL1, COL3, and COL4, respectively (Table S2, Figure 2a). The remaining phyla (with a relative abundance of <1%) were pooled as 'other' (Figure 2a), and included most of the candidate divisions identified in this study (Figure 2b). Sequences that could not be affiliated to any bacterial phylum accounted for 4%, 3%, and 5% of the sequences in COL1, COL3, and COL4, respectively (Table S2). Some fraction of the moonmilk microbial diversity still remains to be discovered for all of the three sampling sites, as the rarefaction curves did not reach a plateau (Figure S2a).

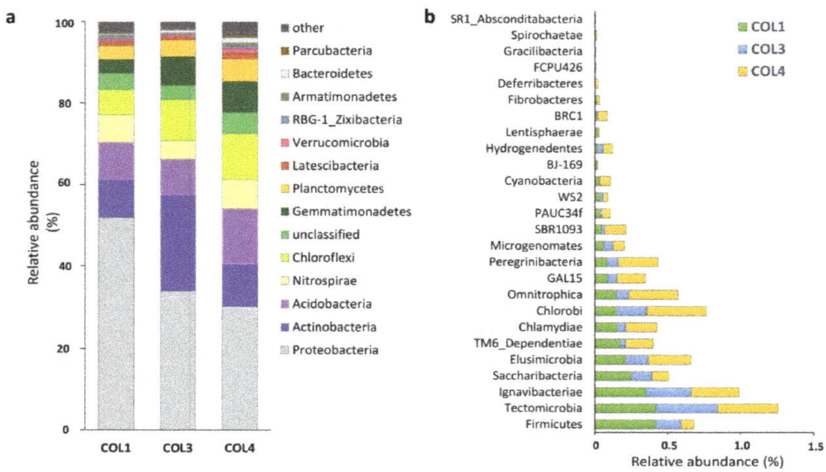

Figure 2. Taxonomic profiles of the moonmilk-associated microbiome at the phylum level across the three moonmilk sampling points (COL1, COL3, COL4). The main phyla of the microbiome are presented on the left (**a**), while the pattern of low-abundance taxa, named as 'other' (with a relative abundance of <1%) is displayed on the right (**b**).

2.2. Actinobacterial Diversity in Moonmilk Deposits

Evaluation of the actinobacterial profile was performed with libraries spanning the V6–V7 variable regions of 16S rRNA gene, and using modified Actinobacteria-specific primers (Table S1a). The specificity of the primers was confirmed by the detection of only 1%, 0.2%, and 2% of non-actinobacterial sequences in COL1, COL3, and COL4, respectively (Figure S3). In contrast to the bacterial dataset, the diversity of Actinobacteria appeared to be exhaustively sampled with the phylum-specific primers (Figure S2b).

The diversity indices for Actinobacteria showed the same trends as the ones observed for the whole Bacteria domain, i.e., evenness and diversity were the highest in COL4, followed by COL3, and COL1 (Table 1). Phylotype richness was the highest in COL4 with 211 OTUs, followed by COL1 and COL3 with 150 OTUs and 147 OTUs, respectively (Figure 1b and Table 1). Among the 243 different OTUs, 105 OTUs (43%) were found in all three of the studied moonmilk deposits (Figure 1b). Hence, the moonmilk-associated actinobacterial community appeared to be more conservative than the moonmilk-associated bacterial population (31%, Figure 1a). If we also include OTUs shared between at least two sampling points, the level of conservation rises to 66% of OTUs for Actinobacteria, and 58% for Bacteria. Still, 34% of the 243 OTUs (14, 15, and 54 OTUs in COL1, COL3, and COL4, respectively) remained specific to a moonmilk deposit, despite the close localization of collection points within the studied cave (Figure 1b). COL4 was characterized not only with the highest number of unique phylotypes (54 OTUs) (Figure 1b), but also with the most diverse population, as revealed by diversity indices (Table 1). As observed for the bacterial dataset, pairwise comparisons showed highly similar percentages (~36.4% ± 0.41%) of shared actinobacterial OTUs between moonmilk deposits (Table 2).

A taxonomic assignment of actinobacterial OTUs revealed the presence of two major classes—Acidimicrobiia and Actinobacteria, next to the low-abundant Thermoleophilia class (Table S3, Figure 3a). Acidimicrobiia was represented by one single order, the Acidimicrobiales, which dominated sample COL4, constituting 55.3% of the population (Table S3, Figure 3a). The Acidimicrobiales order consisted of two families, i.e., Acidimicrobiaceae and Iamiaceae (Table S3). The Actinobacteria class was represented by 15 orders, with Corynebacteriales dominating in COL1, and Pseudonocardiales in COL3 and COL4 (Table S3, Figure 3b). The most abundant families among the Actinobacteria class were Pseudonocardiaceae, Nocardiaceae, and Streptomycetaceae (Table S3). The proportion of unclassified and uncultured sequences at the family level ranged from 9% in COL3, to 25% in COL1, and 53% in COL4 (Table S3).

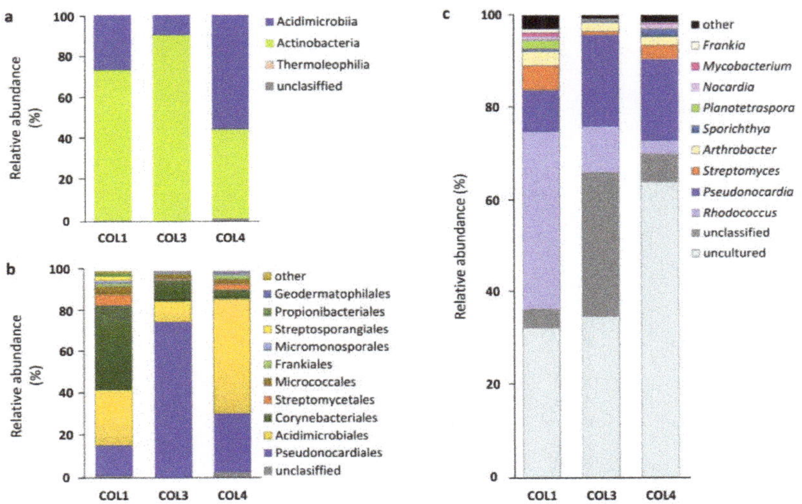

Figure 3. Taxonomic profiles of moonmilk-associated Actinobacteria at different taxonomic levels—(**a**) class; (**b**) order; (**c**) family—observed across the three moonmilk-sampling points (COL1, COL3, COL4). 'Other' includes orders and families with a relative abundance of <1%.

Among 28 families, a total of 47 genera were identified across the investigated samples (Table S3 and Table 3), with 35 genera identified for the first time in moonmilk (Table 3). COL1 was dominated by *Rhodococcus* (38.37%), while uncultured and unclassified Actinobacteria were the most abundant in COL3 and COL4 (Table 3). When only known genera were taken into account, *Pseudonocardia* prevailed in those samples, accounting for 20% and 18% of the population in COL3 and COL4, respectively (Table 3). Other genera, which constituted more than 1% of the population in at least one moonmilk deposit, included *Streptomyces, Arthrobacter, Sporichthya, Planotetraspora, Nocardia, Mycobacterium,* and *Frankia* (Table 3). While accounting in average for only 3% of the actinobacterial community, streptomycetes displayed the highest diversity, with 19 OTUs identified across the three moonmilk deposits (Table 3).

Some taxa showed important differences in their relative abundance between investigated samples, particularly *Rhodococcus*, which was approximately four and 14 times more abundant in COL1 than in COL3 and COL4, respectively (Table 3). The *Streptomyces* genus represented only 0.8% of the population in COL3, while it was detected at the level of 5.3% and 3% in the COL1 and COL4, respectively (Table 3). An important discrepancy in the relative abundance between speleothems was also observed for the genera *Planotetraspora, Mycobacterium,* and *Frankia*, whereas some taxa (e.g., *Pseudoclavibacter, Lentzea, Propionibacterium*) were exclusively found in a single sampling site (Table 3).

Table 3. Actinobacterial genera pattern in moonmilk deposits of the "Grotte des Collemboles" based on 16S rRNA amplicon libraries.

Genus	COL1			COL3			COL4			TOTAL		
	Seq.	%	OTUs	Seq.	%	OTUs	Seq.	%	OTUs	Seq.	Av. %	Diff. OTUs
Uncultured	46,346	32.38	65	58,903	34.80	63	92,444	63.98	90	197,693	43.7	96
Rhodococcus	54,920	38.37	3	16,636	9.83	3	4102	2.84	3	75,658	17.0	4
Pseudonocardia **†**	12,913	9.02	7	33,640	19.87	8	25,545	17.68	9	72,098	15.5	9
Unclassified	5762	4.03	18	52,908	31.26	20	8668	6.00	26	67,338	13.8	32
Streptomyces **†**	7628	5.33	10	1292	0.76	11	4347	3.01	18	13,267	3.0	19
Arthrobacter	4323	3.02	4	3214	1.90	4	2649	1.83	5	10,186	2.3	5
Sporichthya *****	771	0.54	1	556	0.33	2	2515	1.74	2	3842	0.9	2
Planotetraspora *****	2769	1.93	1	294	0.17	1	113	0.08	1	3176	0.7	1
Nocardia	1023	0.71	2	157	0.09	2	1212	0.84	2	2392	0.5	2
Mycobacterium **†**	1299	0.91	2	127	0.08	2	440	0.30	3	1866	0.4	3
Frankia *****	1067	0.75	2	77	0.05	1	126	0.09	3	1270	0.3	3
Luedemannella *****	555	0.39	2	212	0.13	2	276	0.19	2	1043	0.2	2
Longispora *****	371	0.26	1	374	0.22	1	110	0.08	1	855	0.2	1
Agromyces *****	310	0.22	2	136	0.08	2	355	0.25	2	801	0.2	2
Actinoplanes *****	591	0.41	2	85	0.05	2	111	0.08	2	787	0.2	2
Nakamurella *****	416	0.29	1	114	0.07	1	101	0.07	1	631	0.1	1
Nocardioides **†**	360	0.25	4	52	0.03	3	158	0.11	8	570	0.1	9
Geodermatophilus **†**	78	0.05	1	70	0.04	2	374	0.26	2	522	0.1	2
Catellatospora *****	95	0.07	3	70	0.04	2	141	0.10	4	306	0.07	4
Kribbella **†**	120	0.08	1	36	0.02	1	135	0.09	2	291	0.07	2
Kocuria *****	261	0.18	1	23	0.01	1	-	-	-	284	0.1	2
Actinomyces *****	247	0.17	5	1	0.001	1	6	0.004	1	254	0.06	5
Corynebacterium *****	151	0.11	1	32	0.02	3	10	0.01	2	193	0.04	5
Rhizocola *****	-	-	-	27	0.02	1	145	0.10	1	172	0.06	1
Microbacterium *****	108	0.08	1	48	0.03	1	5	0.003	1	161	0.04	1
Iamia *****	107	0.07	1	-	-	-	31	0.02	1	138	0.05	1
Pseudoclavibacter *****	138	0.10	1	-	-	-	-	-	-	138	0.1	1
Lentzea *****	-	-	-	-	-	-	111	0.08	1	111	0.08	1
Aeromicrobium **†**	73	0.05	1	-	-	-	28	0.02	2	101	0.04	2
Amycolatopsis	86	0.06	1	-	-	-	5	0.003	1	91	0.03	2
Cryptosporangium *****	-	-	-	71	0.04	1	10	0.01	1	81	0.02	1
Glycomyces *****	35	0.02	1	-	-	-	45	0.03	1	80	0.03	1
Streptosporangium *****	61	0.04	1	-	-	-	19	0.01	1	80	0.03	1
Smaragdicoccus *****	43	0.03	1	-	-	-	34	0.02	1	77	0.03	1
Propionibacterium	57	0.04	1	-	-	-	-	-	0	57	0.04	1
Kineosporia *****	44	0.03	1	-	-	-	8	0.01	1	52	0.02	1
Jatrophihabitans *****	-	-	-	21	0.01	1	24	0.02	1	45	0.01	1
Promicromonospora *****	-	-	-	22	0.01	1	21	0.01	1	43	0.01	2
Millisia *****	-	-	-	22	0.01	1	6	0.004	1	28	0.01	1
Rothia *****	2	0.001	1	13	0.01	1	7	0.005	1	22	0.005	1
Tessaracoccus *****	-	-	-	17	0.01	1	-	-	-	17	0.01	1
Acidothermus *****	-	-	-	-	-	-	16	0.01	1	16	0.01	1
Marmoricola *****	-	-	-	-	-	-	14	0.01	2	14	0.01	2
Dermacoccus *****	-	-	-	-	-	-	11	0.008	1	11	0.01	1
Ponticoccus *****	-	-	-	8	0.005	1	-	-	-	8	0.005	1
Stackebrandtia *****	-	-	-	-	-	-	8	0.006	1	8	0.01	1
Umezawaea *****	-	-	-	-	-	-	2	0.001	1	2	0.001	1
Actinospica *****	-	-	-	-	-	-	1	0.001	1	1	0.001	1
Propionimicrobium *****	-	-	-	-	-	-	1	0.001	1	1	0.001	1

For each taxon, the number of obtained sequences (Seq.) and their relative abundance (%), together with the number of OTUs, are given. The total number of sequences, average relative abundance, and total number of different OTUs obtained per genus are shown in the last three columns. Taxa marked with an asterisk (*) were reported for the first time in moonmilk deposits in this studyTaxa marked with a cross (†) were detected in moonmilk deposits in this work, and in the high-throughput sequencing (HTS)-based study of Dhami et al. [19]. Taxa underlined represent the ones that were also detected in other moonmilk microbial diversity studies [12–15,18]. Cases filled in grey highlight the most abundant genera in each studied sampling point. Abbreviations: Seq.—number of sequences identified.

2.3. Analysis of the Most Abundant Actinobacterial OTUs

In order to obtain more information about the most dominant moonmilk-dwelling Actinobacteria, a detailed analysis was conducted for the 41 most abundant OTUs (~17% of all of the OTUs) accounting together for 90% (413,739 out of 456,878) of the sequences obtained via our HTS approach (Table 4). Out of the subset of 41 OTUs, 16 phylotypes belonged to the class Acidimicrobiia, with most of them being uncultured at the family level, and the remaining 25 OTUs belonged to the class Actinobacteria (Table 4). In the latter case, all of the OTUs were associated with major families previously identified in moonmilk deposits, including Pseudonocardiaceae, Propionibacteriaceae, Micrococcaceae, Nocardiaceae, Streptomycetaceae, and Streptosporangiaceae (Table 4). Only 16

OTUs could be classified at the genus level and were affiliated to genera *Rhodococcus, Pseudonocardia, Arthrobacter, Sporichthya, Streptomyces, Planotetraspora,* and *Nocardia* (Table 4).

Table 4. The relative abundance (%) and taxonomy assignment of the most abundant actinobacterial OTUs found across moonmilk samples within the "Grotte des Collemboles".

OTU	COL1	COL3	COL4	Av. %	Class	Family	Genus
OTU1	38.28	9.68	2.74	16.90	Actinobacteria	Nocardiaceae	*Rhodococcus*
OTU2	0.31	28.89	2.13	10.44	Actinobacteria	Pseudonocardiaceae	unclassified
OTU8	0.75	13.96	0.58	5.10	Actinobacteria	Pseudonocardiaceae	uncultured
OTU4	3.35	1.13	11.47	5.32	Acidimicrobiia	uncultured	uncultured
OTU3	0.52	7.80	4.87	4.40	Actinobacteria	Pseudonocardiaceae	*Pseudonocardia*
OTU262	3.45	4.12	5.89	4.49	Actinobacteria	Pseudonocardiaceae	*Pseudonocardia*
OTU6	7.53	1.00	3.98	4.17	Acidimicrobiia	uncultured	uncultured
OTU12	2.97	6.61	1.82	3.80	Actinobacteria	Pseudonocardiaceae	uncultured
OTU5	3.53	1.70	5.87	3.70	Acidimicrobiia	uncultured	uncultured
OTU13	0.46	4.91	3.65	3.01	Actinobacteria	Pseudonocardiaceae	*Pseudonocardia*
OTU98	0.99	1.07	6.94	3.00	Acidimicrobiia	uncultured	uncultured
OTU203	1.73	0.31	5.92	2.66	Acidimicrobiia	uncultured	uncultured
OTU432	3.68	1.40	1.66	2.25	Actinobacteria	Pseudonocardiaceae	*Pseudonocardia*
OTU142	0.72	1.91	2.35	1.66	Actinobacteria	Pseudonocardiaceae	unclassified
OTU7	0	1.70	3.31	1.67	Actinobacteria	Pseudonocardiaceae	uncultured
OTU19	2.06	0.52	1.65	1.41	Actinobacteria	Micrococcaceae	*Arthrobacter*
OTU190	0.46	0.31	2.84	1.20	Acidimicrobiia	uncultured	uncultured
OTU10	0.81	0.29	1.68	0.93	Acidimicrobiia	Acidimicrobiaceae	uncultured
OTU9	0.24	0.15	2.39	0.93	Acidimicrobiia	uncultured	uncultured
OTU251	0.77	0.65	1.12	0.85	Actinobacteria	Pseudonocardiaceae	*Pseudonocardia*
OTU14	0.54	0.31	1.73	0.86	Actinobacteria	Sporichthyaceae	*Sporichthya*
OTU30	2.11	0.13	0.31	0.85	Actinobacteria	Streptomycetaceae	*Streptomyces*
OTU360	0.40	0.17	1.68	0.75	Acidimicrobiia	uncultured	uncultured
OTU24	1.93	0.17	0.08	0.73	Actinobacteria	Streptosporangiaceae	*Planotetraspora*
OTU11	0.38	0.03	1.72	0.71	Acidimicrobiia	uncultured	uncultured
OTU20	0.71	0.47	0.81	0.67	Acidimicrobiia	Iamiaceae	uncultured
OTU47	0.75	0.26	0.98	0.66	Acidimicrobiia	uncultured	uncultured
OTU15	0.30	0.17	1.48	0.65	Actinobacteria	Streptomycetaceae	*Streptomyces*
OTU192	0.01	1.65	0.01	0.56	Actinobacteria	Pseudonocardiaceae	uncultured
OTU16	0.24	0.51	1.11	0.62	Acidimicrobiia	uncultured	uncultured
OTU50	0.88	0.54	0.37	0.60	Actinobacteria	Pseudonocardiaceae	uncultured
OTU22	1.61	0.16	0.06	0.61	Actinobacteria	Propionibacteriaceae	unclassified
OTU23	0.27	1.19	0.10	0.52	Actinobacteria	Micrococcaceae	*Arthrobacter*
OTU54	0.13	0.92	0.46	0.50	Actinobacteria	Pseudonocardiaceae	*Pseudonocardia*
OTU18	0.48	0.12	1.04	0.54	Acidimicrobiia	uncultured	uncultured
OTU25	0.56	0.09	0.83	0.49	Actinobacteria	Nocardiaceae	*Nocardia*
OTU21	0.80	0.19	0.34	0.45	Actinobacteria	Streptomycetaceae	*Streptomyces*
OTU99	1.05	0.09	0.17	0.44	Actinobacteria	Streptomycetaceae	*Streptomyces*
OTU36	0.22	0.04	0.96	0.41	Acidimicrobiia	uncultured	uncultured
OTU44	0.80	0.14	0.26	0.40	Acidimicrobiia	uncultured	uncultured
OTU32	0.22	0.03	0.92	0.39	Actinobacteria	unclassified	unclassified

Taking into account the spatial differences in terms of the most abundant taxa across the cave, COL1 was highly dominated by OTU1, affiliated to the genus *Rhodococcus*, and accounting for 38% of the total population in this speleothem (Table 4). This phylotype highly outnumbered other two *Rhodococcus* OTUs detected in COL1 (Table 3), which together constituted only 0.1% (data not shown). The predominant phylotypes identified in speleothems COL3 and COL4 were OTU2, representing an unclassified Pseudonocardiaceae in COL3 (29%), and OTU4, representing uncultured bacterium from Acidimicrobiia class in COL4 (11%) (Table 4). Among the known genera, *Rhodococcus* (OTU1, 10%) was also prevailing in COL3, while *Pseudonocardia* (OTU262, 6%) was found to be the most abundant in COL4 (Table 4).

In total, 40 out of 41 OTUs were present in all the three studied moonmilk deposits, often with an extreme variation in terms of their relative abundance across the different collection points. This is well demonstrated by OTU2 (Pseudonocardiaceae, unclassified at the genus level), which largely dominated the actinobacterial community in COL3 (29%), while only representing 0.3% of the actinobacterial microbiome in COL1 (Table 4).

2.4. Comparison of Moonmilk Streptomyces OTUs and Streptomyces Strains Isolated via the Culture-Dependent Approach

The true diversity of microbial communities is known to be strongly biased by cultivation-based methods in comparison to molecular techniques; therefore, we wanted to assess how much of the *Streptomyces* moonmilk-dwelling community we managed to isolate in our previous bioprospection work [10]. For this purpose, we compared the 16S rRNA sequences of the 19 *Streptomyces* OTUs retrieved from the HTS approach with the sequences of the 31 previously isolated *Streptomyces* phylotypes (MM strains), which were trimmed to the corresponding V6–V7 variable regions of HTS amplicons. Figure 4 presents the phylogenetic tree generated by maximum likelihood with all the 252 nt 16S rRNA sequences from the *Streptomyces* phylotypes (MM strains) and OTUs. The identity threshold for clustering sequences in the same branch of the tree was fixed to 97%, i.e., the same threshold as the one used to define OTUs in our HTS approach (see methods for details). As deduced from the generated phylogenetic tree, the 31 isolated *Streptomyces* strains matched with only five of the 19 *Streptomyces* OTUs, suggesting that the isolated strains represent a minor fraction of the *Streptomyces* species dwelling in the moonmilk deposits of the studied cave. Expectedly, Figure 4 further shows that we isolated *Streptomyces* species that are associated with the most abundant *Streptomyces* OTUs, e.g., OTU15, OTU21, OTU30, and OTU99 (Table 4), which together represent 79% of the *Streptomyces* sequences retrieved by our HTS approach. Moreover, 21 out of the 31 phylotype strains (68%) clustered together with OTU21 (Figure 4). Finally, two *Streptomyces* isolates, i.e., MM24 and MM106, did not cluster with any of the identified *Streptomyces* OTUs (Figure 4).

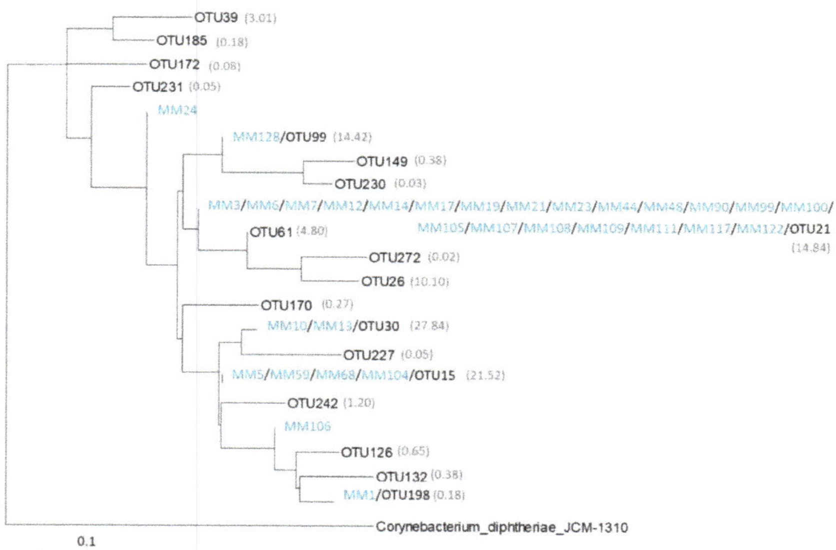

Figure 4. Phylogenetic relationships between culturable and non-culturable *Streptomyces* originating from moonmilk of "Grotte des Collemboles". The tree was inferred by maximum likelihood. Scale bar is in substitution per site. Numbers between brackets reflect the predicted mean abundance of *Streptomyces* OTUs in the studied deposits based on the percentage of sequences retrieved from the HTS analysis. *Streptomyces* phylotypes isolated in our previous bioprospection study (MM strains) are marked in blue.

3. Discussion

3.1. New Insights into Moonmilk Bacterial Diversity Revealed by High-Throughput Sequencing

Previous investigations on the moonmilk microbiome revealed a very diverse microbial community in these deposits [9,13–17,19]. The high-throughput sequencing approach used in this work complemented previous findings by providing an in-depth picture of the bacterial population, together with a detailed taxonomic fingerprint of the phylum Actinobacteria.

Comparison of the bacterial diversity in moonmilk between earlier investigations and the present work is limited to some extent by the differences in experimental procedures, such as DNA isolation and PCR-based approaches, and the sensitivity of the sequencing techniques. Nonetheless, the profile of the major taxonomic groups found in this work is consistent with that observed for the moonmilk communities in the caves "Grotta della Foos" and "Bus della Genziana" in Italy, which were obtained from 16S rRNA clone libraries [9]. All of the phyla detected in the above-mentioned caves, including Bacteroidetes, Acidobacteria, Chloroflexi, Planctomycetes, Verrucomicrobia, Actinobacteria, Firmicutes, Nitrospirae, Chlorobi, Proteobacteria, and WS3 (now Latescibacteria), were also identified in the cave "Grotte des Collemboles", although their relative abundance varied between the studies. While Proteobacteria were found to be the most abundant phylum in both cases, the second most abundant population identified in Italian caves was the phylum Bacteroidetes, which constituted a minor part of the bacterial community in the present study. The Actinobacteria population was found to be an important part of the moonmilk microbiome in the "Grotte des Collemboles" (from 9% to 23%), but instead represented only a minor fraction (<2%) of the bacterial population in the two Italian caves investigated by Engel et al. (2013). Very recently, a study by Dhami et al. has reported the moonmilk microbiome profile in the Australian "Lake Cave" using an HTS approach [19], as in this work. The presence of Proteobacteria, Actinobacteria, Acidobacteria, Chloroflexi, Nitrospirae, Gemmatimonadetes, Firmicutes, and Bacteroidetes were detected in the moonmilk deposit of the "Lake Cave", similarly to the "Grotte des Collemboles". However, many of the low-abundance taxa identified in the Belgian cave were not reported, possibly because their phylogenetic profiles were based on different regions of 16S rRNA gene—V3/V4 for the "Lake Cave", and V6–V7 for the "Grotte des Collemboles". Interestingly, unlike in Italian and Belgian caves, the "Lake Cave" moonmilk deposit was strongly dominated by Actinobacteria, which were more than twice as abundant as the Proteobacteria [19]. The highly sensitive HTS amplicon sequencing approach employed in this work revealed the presence of 26 phyla within the moonmilk microbiome that had not been previously described in this speleothem. These included Zixibacteria (formerly RBG-1), Armatimonadetes (formerly OP10), and Parcubacteria (formerly OD1) among the main phyla of moonmilk microbiome (Figure 2a), which have been previously reported from other subterranean environments [24,29–33], and 23 low-abundant taxa that were found below the level of 1%, and included many candidate divisions (Figure 2b).

This new study uncovered a surprisingly diverse Actinobacteria taxonomic profile that demonstrates the limitations of our previous cultivation-based screening, in which only the *Streptomyces* species could be isolated from the three moonmilk deposits [10]. Here, a total of 47 actinobacterial genera from 28 families were identified across the investigated samples. Beyond the previously reported members of the Actinomycetales family—including *Nocardia* and *Rhodococcus* (Nocardiaceae) [15,18], *Pseudonocardia*, *Amycolatopsis* and *Saccharothrix* (not identified in our study) (Pseudonocardiaceae) [12,14,19], *Propionibacterium* (Propionibacteriaceae) [14], *Streptomyces* (Streptomycetaceae) [10,12,18,19], *Arthrobacter* (Micrococcaceae) [13], *Mycobacterium* (Mycobacteriaceae) [19], *Nocardioides*, *Aeromicrobium*, and *Kribbella* (Nocardioidaceae) [19], and *Geodermatophilus* (Geodermatophilaceae) [19]—35 other genera were identified in the moonmilk deposits of the "Grotte des Collemboles". The population of each investigated sample was also found to include representatives of the Acidimicrobiia class, which were not previously reported in moonmilk. Their presence in all the three sampling sites, with an abundance up to 55% in COL4, and the dominance of the unclassified Acidimicrobiia phylotype (OTU 4) within the community of COL4, suggest that the chemical composition of the investigated moonmilk would be particularly suitable for the development of the representatives of this class of Actinobacteria, of which the ecology and metabolism are still largely unknown.

3.2. Moonmilk Deposits as Appealing Source of Novel Producers of Bioactive Compounds

Extreme environmental niches have recently become the main targets for intense bioprospecting, as they are expected to host diverse yet-unknown microorganisms, which could offer unexplored chemical diversity. While *Streptomyces* are reported as the most prolific "antibiotic makers", advances in the cultivation and characterization of rare Actinobacteria revealed similarly promising capabilities for the production of bioactive natural compounds [34–36]. The results obtained in this work suggest a significant biodiversity of the moonmilk-dwelling actinobacterial population, with a wide spectrum of rare genera. Next to *Streptomyces*, other members of Actinobacteria with valuable secondary metabolism were detected at a high proportion, such as *Pseudonocardia*, *Amycolatopsis*, *Streptosporangium*, *Nocardia*, *Nocardioides*, and *Rhodococcus*. Such findings clearly prompt to apply appropriate selective cultivation methods to isolate rare Actinobacteria from moonmilk deposits.

Moreover, particular importance should be also focused on Acidimicrobiia, which constituted an important part of the community in the studied deposits. Members of this class are a recently identified taxonomic unit [37] that is considered to represent an early-branching lineage within the phylum [38]. Due to their phylogenetic isolation and novelty, they are likely to hide a yet-uncovered valuable bioactive arsenal.

The great potential of moonmilk as a source of diverse and metabolically beneficial Actinobacteria is illustrated by the comparison of *Streptomyces* isolated in our previous study and the *Streptomyces* OTUs identified in this work (Figure 4). Most *Streptomyces* OTUs are phylogenetically distinct from culturable representatives (Figure 4), indicating that a great number of species still remain to be isolated. On the other hand, our culture-dependent study identified *Streptomyces* strains (MM24 and MM106, Figure 4) that were not associated with OTUs deduced from the HTS approach, confirming that both strategies are complementary, and should be used in parallel for microbial diversity assessment [39,40]. In addition, next to the identification methods themselves, our data suggests that the diversity level can be also biased by the identity threshold that is used for OTU definition. The tree revealed that a single OTU (OTU21, Figure 4) clustered together with most of the phylotypes deduced from MLSA (multilocus sequence analysis), each most likely representing a distinct species [10]. This indicates that the 97% sequence homology threshold applied to the comparative analysis of the V6–V7 regions of the 16S RNA gene largely underestimated the number of *Streptomyces* species dwelling in a studied environmental niche.

4. Materials and Methods

4.1. Site description and Sampling

The cave "Grotte des Collemboles" (Springtails' Cave), located in Comblain-au-Pont (GPS coordinates 50°28′41″ N, 5°36′35″ E), Belgium (Figure S1, Maciejewska et al. 2017 for full description), was formed in Visean limestone and has the shape of a 70-m long meander. White to brown–orange moonmilk deposits are found on the walls in the first narrow chamber located at the entrance of the cave, as well as in the narrow passages leading deeper into the cave (Figure S1). Moonmilk samples used for total DNA extractions were aseptically collected in January 2012 from three spatially separated locations along about a 20-m transect in the cave. Soft moonmilk speleothems were scratched with sterile scalpels into sterile Falcon tubes from the wall in the first chamber, adjacent to the cave entrance (COL4), and from the walls in a narrow passage after the first chamber (COL1, COL3) (Figure S1). COL4 was located approximately 6 m from COL1, and 20 m from COL3 (Figure S1).

Samples were immediately transferred to the laboratory, freeze-dried on a VirTis Benchtop SLC Lyophilizer (SP Scientific, Warminster, PA, USA), and stored at −20 °C.

4.2. Total DNA Extraction and 16S rRNA Gene Amplicon High-Throughput Sequencing

The metagenetic approach applied in this work was performed on DNA extracted from three moonmilk deposits (COL1, COL3, and COL4) originating from the "Grotte des Collemboles". Environmental genomic DNA isolation was carried out from 200 mg of the freeze-dried moonmilk samples COL1, COL3, and COL4 (Figure S1), using the PowerClean Soil DNA kit (MoBio, Carlsbad, CA, USA), according to manufacturer's instructions. The integrity of purified DNA was assessed by agarose gel electrophoresis (1% *w*/*v*), and the dsDNA concentration was evaluated by Qubit fluorometer (Invitrogen, Carlsbad, CA, USA).

The 16S rRNA gene amplicon libraries were generated using bacterial (S-D-Bact-0517-a-S-17/S-D-Bact-1061-a-A-17 spanning V4–V6 region [41]) and actinobacterial (Com2xf/Ac1186r, spanning V6–V7 region [42]) specific primer pairs. The Illumina platform-compatible dual index paired-end approach was designed as previously described [43] (detailed description provided in the Table S1a). Each forward and reverse primer consisted of an Illumina-compatible forward/reverse primer overhang attached to the 5′ end. Additionally, a heterogeneity spacer of four degenerate nucleotides (Ns) was added to the forward primer, between the primer overhang and the locus-specific sequence. The Illumina barcodes and sequencing adapters were added during the subsequent cycle-limited amplification step using Nextera XT Index kit (Illumina, San Diego, CA, USA). Triplicated PCR reactions were performed for each sample in 25 µL of volume containing 2.5 µL of total DNA, 5 µL of each primer (1 µM), and 12.5 µL of 2× Q5 High-Fidelity Master Mix (New England Biolabs, Ipswich, MA, USA). Amplification conditions for each set of primers are listed in Table S1b. The triplicated amplicons were visualized on 3% agarose gel, pulled, purified with Agencourt AMPure XP beads (Beckman Coulter, Brea, CA, USA), and quantified with the Qubit HS dsDNA assay kit (Invitrogen, Carlsbad, CA,) before being processed for index ligation, using the Nextera XT Index kit (Illumina, San Diego, CA, USA). The PCR amplifications were performed with the same enzyme and cycling conditions as described above [43], with the total number of cycles reduced to eight, and an annealing temperature of 55 °C. The resulting amplicons were purified with the Agencourt AMPure XP magnetic beads (Beckman Coulter, CA, USA), quantified, and pooled in equimolar concentrations. The library concentration was quantified by qPCR using a Kappa SYBR FAST kit (Kapa Biosystems, Wilmington, MA, USA), and subsequently, the library was normalized to 4 nM, denaturated, and diluted to the final concentration of 8 p.m. The resulting pool was mixed with the PhiX control and subjected to 2 × 300 bp paired-end sequencing on Illumina MiSeq platform (Illumina, San Diego, CA, USA). Raw sequences were deposited in the NCBI Sequence Read Archive (SRA) database under the Bioproject PRJNA428798 with accession numbers SRX3540524–SRX3540529.

4.3. 16S rRNA Amplicon Analysis

16S rRNA amplicon analysis was based for both Bacteria and Actinobacteria on forward reads only, owing to the poor quality of reverse reads. Quality trimming (prohibiting mismatches and ambiguities, ensuring a minimum quality score of 20 and removing the four degenerate nucleotides from the 5′ end) was carried out using CLC Genomic Workbench (Qiagen, Hilden, Germany). USEARCH [44] was applied for length trimming (minimum length = 240 nt) and dereplication. Operational taxonomic units (OTUs) for both bacterial and actinobacterial datasets were defined using a 97% identity threshold on 16S rRNA sequences. OTUs were clustered using the UPARSE algorithm [45], and their taxonomic position was assigned by MOTHUR [46] with SILVA v128 database [47]. OTUs were further classified using BLASTN [48] analyses against a local mirror of NCBI nt database (downloaded on 9 August 2017), through manual and automatic analyses. For the automatic approach, a last common ancestor (LCA) classification was performed with a custom parser mimicking the MEGAN algorithm [49], which we developed for analyses of genome contamination (Cornet et al., 2017, under review). A maximum number of 100 hits per OTU were taken into account. To consider a BLASTN hit, the E-value threshold was set at 1e-15, the minimum identity threshold was set at 95.5%, the minimum bit score was set at 200, and the bit score percentage threshold was set at 99% of the best hit. These thresholds were defined through preliminary analyses (data not shown). When the BLASTN hits are too numerous, the MEGAN-like algorithm frequently yields high-ranking LCAs (e.g., Bacteria) that are not informative in practice. In order to minimize this effect, we decided to skip uncultured/unclassified hits whenever other, more informative, hits also passed the thresholds. Moreover, when computing LCAs, we only considered the most frequent taxa, provided that they represented ≥95% of the (up to 100) accumulated BLASTN hits, so as to avoid uninformative classifications due to a few (possibly aberrant) outliers.

Normalized OTU abundance data was used to calculate α-diversity and β-diversity estimators using MOTHUR [46]. Community richness, evenness, diversity, and differential OTU abundance between samples were calculated using sobs, the Simpson index, the inverse Simpson index and Venn diagrams, respectively.

The 19 OTUs identified as *Streptomyces* were combined to 31 sequences (16S rRNA region V6–V7) from previously isolated *Streptomyces* phylotypes (MM strains) and dereplicated with the UCLUST algorithm [44] using an identity threshold of 97%. This yielded 21 clusters, to which we added the homologous region of *Corynebacterium diphtheriae* JCM-1310 as an outgroup. A multiple sequence alignment was built with MUSCLE [50] (default parameters), and then analyzed with PhyML [51] under a K80 + Γ_4 model. Due to the limited amount of phylogenetic signal (short sequences from very related organisms), the resolution of the tree was low (bootstrap proportions <50 for nearly all nodes; data not shown).

5. Conclusions

Before the advent of metagenomics, bioprospecting was carried out blindly, with poor knowledge on the real potential of an ecological niche mined for novel organisms, enzymes, or bioactive compounds. The results of the metagenetic study presented here confirmed that different moonmilk deposits host their own indigenous microbial population, and thus each individual speleothem can be a source of a great biodiversity. Consequently, the observed important differences in the spatial diversity of Actinobacteria imply that bioprospecting within different moonmilk deposits—from different caves or within the same cave—could result in the isolation of unique and novel natural compound producers. Our study also revealed how many and which actinobacterial genera have been missed in our first attempt to isolate antibiotic producers. We now know that the *Streptomyces* strains of our collection isolated from the moonmilk deposits of the cave 'Grotte des Collemboles' [10] are just the tip of the iceberg. These results prompted us to apply a series of '*tips and tricks*' to isolate other *Streptomyces* and representatives of other antibiotic-producing Actinobacteria that are present in different proportions in each moonmilk deposit. The results of our adapted protocols for the isolation of rare Actinobacteria are presented in the article '*Isolation, Characterization, and Antibacterial Activity of Hard-to-Culture Actinobacteria from Cave Moonmilk Deposits*', which is published in the same special issue [52].

Supplementary Materials: The following are available online at http://www.mdpi.com/2079-6382/7/2/27/s1, Figure S1: Localization of the "Grotte des Collemboles" (Springtails' Cave) together with the cave map and visualization of the moonmilk deposit sampling points, Figure S2: Rarefaction curves of OTUs clustered at 97% sequence identity across the three moonmilk-sampling points for Bacteria (a) and Actinobacteria (b), Figure S3: Taxonomic profile of bacterial phyla generated with Actinobacteria-specific primers. Note the high specificity of Actinobacteria primers, Table S1: Details of the PCR primers used for community profiling of moonmilk samples (a) and—PCR conditions used for 16S rRNA amplification from moonmilk samples (b), Table S2: Relative abundance (%) of bacterial phyla identified in the three moonmilk deposits in "Grotte des Collemboles". Low-abundant taxa with relative abundance <1% are marked in red, Table S3: Relative abundance (%) of the phylum Actinobacteria at different taxonomic levels identified in the three moonmilk deposits in the "Grotte des Collemboles".

Acknowledgments: The authors are grateful to Luc Willems for the introduction to the subject and help with sampling. MM, DA, and LC work was supported by a Research Foundation for Industry and Agriculture (FRIA) grant. MC and PD were supported by the Luxembourg National Research Fund (FNR CORE 2011 project

GASPOP; C11/SR/1280949: Influence of the Reactor Design and the Operational Parameters on the Dynamics of the Microbial Consortia Involved in the Biomethanation Process). Computational resources ("durandal" grid computer) were funded by three grants from the University of Liège, "Fonds spéciaux pour la recherche", "Crédit de démarrage 2012" (SFRD-12/03 and SFRD-12/04) and "Crédit classique 2014" (C-14/73) and by a grant from the F.R.S.-FNRS "Crédit de recherche 2014" (CDR J.0080.15). This work is supported in part by the Belgian program of Interuniversity Attraction Poles initiated by the Federal Office for Scientific Technical and Cultural Affairs (PAI No. P7/44). SR is a Research Associate at Belgian Fund for Scientific Research (F.R.S-FNRS). The authors declare no conflict of interest. We dedicate the work to the memory of Leonard Maculewicz (1936–2017) who always supported our work with great enthusiasm.

Author Contributions: M.M., Ma.C., S.M., Mo.C., P.D., and S.R. designed and performed experiments. Bioinformatic analyses were performed by M.M., Mo.C., Ma.C., L.C., D.B., and S.R. Data were analyzed by all authors. The manuscript was written and/or corrected by all authors.

Conflicts of Interest: The authors declare no conflict of interest.

References

1. Engel, A.S. Microbial diversity of cave ecosystems. In *Geomicrobiology: Molecular and Environmental Perspective*; Springer: Berlin, Germany, 2010, ISBN 9789048192038.
2. Barton, H.A.; Northup, D.E. Geomicrobiology in cave environments: Past, current and future perspectives. *J. Cave Karst Stud.* **2007**, *69*, 163–178.
3. Groth, I.; Vettermann, R.; Schuetze, B.; Schumann, P.; Saiz-Jimenez, C. Actinomycetes in Karstic caves of northern Spain (Altamira and Tito Bustillo). *J. Microbiol. Methods* **1999**, *36*, 115–122. [CrossRef]
4. Jurado, V.; Kroppenstedt, R.M.; Saiz-Jimenez, C.; Klenk, H.P.; Mouniée, D.; Laiz, L.; Couble, A.; Pötter, G.; Boiron, P.; Rodríguez-Nava, V. Hoyosella altamirensis gen. nov., sp. nov., a new member of the order Actinomycetales isolated from a cave biofilm. *Int. J. Syst. Evol. Microbiol.* **2009**, *59*, 3105–3110. [CrossRef] [PubMed]
5. Jurado, V.; Groth, I.; Gonzalez, J.M.; Laiz, L.; Saiz-Jimenez, C. Agromyces salentinus sp. nov. and Agromyces neolithicus sp. nov. *Int. J. Syst. Evol. Microbiol.* **2005**, *55*, 153–157. [CrossRef] [PubMed]
6. Maciejewska, M.; Pessi, I.S.; Arguelles-Arias, A.; Noirfalise, P.; Luis, G.; Ongena, M.; Barton, H.; Carnol, M.; Rigali, S. *Streptomyces lunaelactis* sp. nov., a novel ferroverdin A-producing *Streptomyces* species isolated from a moonmilk speleothem. *Antonie van Leeuwenhoek Int. J. Gen. Mol. Microbiol.* **2015**, *107*, 519–531. [CrossRef] [PubMed]
7. Lee, S.D.; Kang, S.O.; Hah, Y.C. Hongia gen, nov., a new genus of the order Actinomycetales. *Int. J. Syst. Evol. Microbiol.* **2000**, *50*, 191–199. [CrossRef] [PubMed]
8. Ortiz, M.; Neilson, J.W.; Nelson, W.M.; Legatzki, A.; Byrne, A.; Yu, Y.; Wing, R.A.; Soderlund, C.A.; Pryor, B.M.; Pierson, L.S.; et al. Profiling Bacterial Diversity and Taxonomic Composition on Speleothem Surfaces in Kartchner Caverns, AZ. *Microb. Ecol.* **2013**, *65*, 371–383. [CrossRef] [PubMed]
9. Engel, A.S.; Paoletti, M.G.; Beggio, M.; Dorigo, L.; Pamio, A.; Gomiero, T.; Furlan, C.; Brilli, M.; Dreon, A.L.; Bertoni, R.; et al. Comparative microbial community composition from secondary carbonate (moonmilk) deposits: Implications for the Cansiliella servadeii cave hygropetric food web. *Int. J. Speleol.* **2013**, *42*, 181–192. [CrossRef]
10. Maciejewska, M.; Adam, D.; Martinet, L.; Naômé, A.; Calusinska, M.; Smargiasso, N.; De Pauw, E.; Barton, H.; Carnol, M.; Hanikenne, M.; et al. A Phenotypic and Genotypic Analysis of the Antimicrobial Potential of Cultivable *Streptomyces* isolated from Cave Moonmilk Deposits. *Front. Microbiol.* **2016**. [CrossRef] [PubMed]
11. Cañaveras, J.C.; Cuezva, S.; Sanchez-Moral, S.; Lario, J.; Laiz, L.; Gonzalez, J.M.; Saiz-Jimenez, C. On the origin of fiber calcite crystals in moonmilk deposits. *Naturwissenschaften* **2006**, *93*, 27–32. [CrossRef] [PubMed]
12. Cañaveras, J.C.; Hoyos Gómez, M.; Sánchez-Moral, S.; Sanz Rubio, E.; Bedoya, J.; Hoyos, V.; Groth, I.; Schumann, P.; Laiz Trobajo, L. Microbial Communities Associated with Hydromagnesite and Needle-Fiber Aragonite Deposits in a Karstic Cave (Altamira, Northern Spain). *Geomicrobiol. J.* **1999**, *16*, 9–25.
13. Rooney, D.C.; Hutchens, E.; Clipson, N.; Baldini, J.; McDermott, F. Microbial Community Diversity of Moonmilk Deposits at Ballynamintra Cave, Co. Waterford, Ireland. *Microb. Ecol.* **2010**, *60*, 753–761. [CrossRef] [PubMed]
14. Portillo, M.C.; Gonzalez, J.M. Moonmilk Deposits Originate from Specific Bacterial Communities in Altamira Cave (Spain). *Microb. Ecol.* **2011**, *61*, 182–189. [CrossRef] [PubMed]

15. Reitschuler, C.; Lins, P.; Schwarzenauer, T.; Spotl, C.; Wagner, A.O.; Illmer, P. New Undescribed Lineages of Non-extremophilic Archaea Form a Homogeneous and Dominant Element within Alpine Moonmilk Microbiomes. *Geomicrobiol. J.* **2015**, *32*. [CrossRef]
16. Reitschuler, C.; Spötl, C.; Hofmann, K.; Wagner, A.O.; Illmer, P. Archaeal Distribution in Moonmilk Deposits from Alpine Caves and Their Ecophysiological Potential. *Microb. Ecol.* **2016**, *71*, 686–699. [CrossRef] [PubMed]
17. Reitschuler, C.; Lins, P.; Wagner, A.O.; Illmer, P. Cultivation of moonmilk-born non-extremophilic Thaum and Euryarchaeota in mixed culture. *Anaerobe* **2014**, *29*, 73–79. [CrossRef] [PubMed]
18. Axenov-Gibanov, D.V.; Voytsekhovskaya, I.V.; Tokovenko, B.T.; Protasov, E.S.; Gamaiunov, S.V.; Rebets, Y.V.; Luzhetskyy, A.N.; Timofeyev, M.A. Actinobacteria isolated from an underground lake and moonmilk speleothem from the biggest conglomeratic karstic cave in Siberia as sources of novel biologically active compounds. *PLoS ONE* **2016**, *11*. [CrossRef] [PubMed]
19. Dhami, N.K.; Mukherjee, A.; Watkin, E. Characterisation of Mineralogical-Mechanical-Microbial properties of calcitic speleothems and the in vitro biomineralization potential of associated microbial communities. *Front. Microbiol.* **2018**, *9*, 40. [CrossRef] [PubMed]
20. Bindschedler, S.; Milliere, L.; Cailleau, G.; Job, D.; Verrecchia, E.P. Calcitic nanofibres in soils and caves: A putative fungal contribution to carbonatogenesis. *Geol. Soc. Lond. Spec. Publ.* **2010**, *336*, 225–238. [CrossRef]
21. Maciejewska, M.; Adam, D.; Naômé, A.; Martinet, L.; Tenconi, E.; Calusinska, M.; Delfosse, P.; Hanikenne, M.; Baurain, D.; Compère, P.; et al. Assessment of the potential role of *Streptomyces* in cave moonmilk formation. *Front. Microbiol.* **2017**, *8*. [CrossRef] [PubMed]
22. Nimaichand, S.; Devi, A.M.; Tamreihao, K.; Ningthoujam, D.S.; Li, W.J. Actinobacterial diversity in limestone deposit sites in Hundung, Manipur (India) and their antimicrobial activities. *Front. Microbiol.* **2015**, *6*. [CrossRef] [PubMed]
23. Wu, Y.; Tan, L.; Liu, W.; Wang, B.; Wang, J.; Cai, Y.; Lin, X. Profiling bacterial diversity in a limestone cave of the western Loess Plateau of China. *Front. Microbiol.* **2015**, *6*. [CrossRef] [PubMed]
24. Northup, D.E.; Melim, L.A.; Spilde, M.N.; Hathaway, J.J.M.; Garcia, M.G.; Moya, M.; Stone, F.D.; Boston, P.J.; Dapkevicius, M.L.N.E.; Riquelme, C. Lava Cave Microbial Communities within Mats and Secondary Mineral Deposits: Implications for Life Detection on Other Planets. *Astrobiology* **2011**, *11*, 601–618. [CrossRef] [PubMed]
25. Riquelme, C.; Hathaway, J.J.M.; Dapkevicius, M.d.L.N.E.; Miller, A.Z.; Kooser, A.; Northup, D.E.; Jurado, V.; Fernandez, O.; Saiz-Jimenez, C.; Cheeptham, N. Actinobacterial diversity in volcanic caves and associated geomicrobiological interactions. *Front. Microbiol.* **2015**, *6*. [CrossRef] [PubMed]
26. Cheeptham, N.; Sadoway, T.; Rule, D.; Watson, K.; Moote, P.; Soliman, L.C.; Azad, N.; Donkor, K.K.; Horne, D. Cure from the cave: Volcanic cave actinomycetes and their potential in drug discovery. *Int. J. Speleol.* **2013**, *42*, 35–47. [CrossRef]
27. Tebo, B.M.; Davis, R.E.; Anitori, R.P.; Connell, L.B.; Schiffman, P.; Staudigel, H. Microbial communities in dark oligotrophic volcanic ice cave ecosystems of Mt. Erebus, Antarctica. *Front. Microbiol.* **2015**, *6*. [CrossRef] [PubMed]
28. Chater, K.F.; Biró, S.; Lee, K.J.; Palmer, T.; Schrempf, H. The complex extracellular biology of *Streptomyces*. *FEMS Microbiol. Rev.* **2010**, *34*, 171–198. [CrossRef] [PubMed]
29. Castelle, C.J.; Hug, L.A.; Wrighton, K.C.; Thomas, B.C.; Williams, K.H.; Wu, D.; Tringe, S.G.; Singer, S.W.; Eisen, J.A.; Banfield, J.F. Extraordinary phylogenetic diversity and metabolic versatility in aquifer sediment. *Nat. Commun.* **2013**, *4*. [CrossRef] [PubMed]
30. Schabereiter-Gurtner, C.; Saiz-Jimenez, C.; Pinar, G.; Lubitz, W.; Rolleke, S. Phylogenetic diversity of bacteria associated with Paleolithic paintings and surrounding rock walls in two Spanish caves (Llonin and La Garma). *FEMS Microbiol. Ecol.* **2004**, *47*, 235–247. [CrossRef]
31. Pedersen, K.; Arlinger, J.; Ekendahl, S.; Hallbeck, L. 16S rRNA gene diversity of attached and unattached bacteria in boreholes along the access tunnel to the Äspö hard rock laboratory, Sweden. *FEMS Microbiol. Ecol.* **1996**, *19*, 249–262. [CrossRef]
32. Zhou, J.; Gu, Y.; Zou, C.; Mo, M. Phylogenetic diversity of bacteria in an earth-cave in Guizhou Province, Southwest of China. *J. Microbiol.* **2007**, *45*, 105–112. [PubMed]
33. De Mandal, S.; Chatterjee, R.; Kumar, N.S. Dominant bacterial phyla in caves and their predicted functional roles in C and N cycle. *BMC Microbiol.* **2017**, *17*, 90. [CrossRef] [PubMed]

34. Tiwari, K.; Gupta, R.K. Rare actinomycetes: A potential storehouse for novel antibiotics. *Crit. Rev. Biotechnol.* **2012**, *32*, 108–132. [CrossRef] [PubMed]
35. Choi, S.S.; Kim, H.J.; Lee, H.S.; Kim, P.; Kim, E.S. Genome mining of rare actinomycetes and cryptic pathway awakening. *Process Biochem.* **2015**, *50*, 1184–1193. [CrossRef]
36. Rigali, S.; Anderssen, S.; Naômé, A.; van Wezel, G.P. Cracking the regulatory code of biosynthetic gene clusters as a strategy for natural product discovery. *Biochem. Pharmacol.* **2018**. [CrossRef] [PubMed]
37. Norris, P. Class II. Acidimicrobiia class. nov. In *Bergey's Manual of Systematic Bacteriology*; Whitman, W., Goodfellow, M., Kämpfer, P., Busse, H.-J., Trujillo, M., Ludwig, W., Suzuki, K.-I., Parte, A., Eds.; Springer: New York, NY, USA, 2012; Volume 5, p. 1968.
38. Bull, A.T. Actinobacteria of the extremobiosphere. In *Extremophiles Handbook*; Springer: Tokyo, Japan, 2011; pp. 1203–1240, ISBN 9784431538981.
39. Stefani, F.O.P.; Bell, T.H.; Marchand, C.; de la Providencia, I.E.; El Yassimi, A.; St-Arnaud, M.; Hijri, M. Culture-Dependent and -Independent Methods Capture Different Microbial Community Fractions in Hydrocarbon-Contaminated Soils. *PLoS ONE* **2015**, *10*, e0128272. [CrossRef] [PubMed]
40. Vaz-Moreira, I.; Egas, C.; Nunes, O.C.; Manaia, C.M. Culture-dependent and culture-independent diversity surveys target different bacteria: A case study in a freshwater sample. *Antonie van Leeuwenhoek, Int. J. Gen. Mol. Microbiol.* **2011**, *100*, 245–257. [CrossRef] [PubMed]
41. Klindworth, A.; Pruesse, E.; Schweer, T.; Peplies, J.; Quast, C.; Horn, M.; Glöckner, F.O. Evaluation of general 16S ribosomal RNA gene PCR primers for classical and next-generation sequencing-based diversity studies. *Nucleic Acids Res.* **2013**, *41*. [CrossRef] [PubMed]
42. Schäfer, J.; Jäckel, U.; Kämpfer, P. Development of a new PCR primer system for selective amplification of Actinobacteria. *FEMS Microbiol. Lett.* **2010**, *311*, 103–112. [CrossRef] [PubMed]
43. Goux, X.; Calusinska, M.; Fossépré, M.; Benizri, E.; Delfosse, P. Start-up phase of an anaerobic full-scale farm reactor—Appearance of mesophilic anaerobic conditions and establishment of the methanogenic microbial community. *Bioresour. Technol.* **2016**, *212*, 217–226. [CrossRef] [PubMed]
44. Edgar, R.C. Search and clustering orders of magnitude faster than BLAST. *Bioinformatics* **2010**, *26*, 2460–2461. [CrossRef] [PubMed]
45. Edgar, R.C. UPARSE: Highly accurate OTU sequences from microbial amplicon reads. *Nat. Methods* **2013**, *10*, 996–998. [CrossRef] [PubMed]
46. Schloss, P.D.; Westcott, S.L.; Ryabin, T.; Hall, J.R.; Hartmann, M.; Hollister, E.B.; Lesniewski, R.A.; Oakley, B.B.; Parks, D.H.; Robinson, C.J.; et al. Introducing mothur: Open-source, platform-independent, community-supported software for describing and comparing microbial communities. *Appl. Environ. Microbiol.* **2009**, *75*, 7537–7541. [CrossRef] [PubMed]
47. Pruesse, E.; Peplies, J.; Glöckner, F.O. SINA: Accurate high-throughput multiple sequence alignment of ribosomal RNA genes. *Bioinformatics* **2012**, *28*, 1823–1829. [CrossRef] [PubMed]
48. Boratyn, G.M.; Schäffer, A.A.; Agarwala, R.; Altschul, S.F.; Lipman, D.J.; Madden, T.L. Domain enhanced lookup time accelerated BLAST. *Biol. Direct* **2012**, *7*. [CrossRef] [PubMed]
49. Huson, D.H.; Auch, A.F.; Qi, J.; Schuster, S.C. MEGAN analysis of metagenomic data. *Genome Res.* **2007**, *17*, 377–386. [CrossRef] [PubMed]
50. Edgar, R.C. MUSCLE: Multiple sequence alignment with high accuracy and high throughput. *Nucleic Acids Res.* **2004**, *32*, 1792–1797. [CrossRef] [PubMed]
51. Guindon, S.; Dufayard, J.F.; Lefort, V.; Anisimova, M.; Hordijk, W.; Gascuel, O. New algorithms and methods to estimate maximum-likelihood phylogenies: Assessing the performance of PhyML 3.0. *Syst. Biol.* **2010**, *59*, 307–321. [CrossRef] [PubMed]
52. Adam, D.; Maciejewska, M.; Naômé, A.; Martinet, L.; Coppieters, W.; Karim, L.; Baurain, D.; Rigali, S. Isolation, Characterization, and Antibacterial Activity of Hard-to-Culture Actinobacteria from Cave Moonmilk Deposits. *Antibiotics* **2018**, in press. [CrossRef]

antibiotics

MDPI

Article

Biosynthesis of Rishirilide B

Philipp Schwarzer [1,†], Julia Wunsch-Palasis [1,†], Andreas Bechthold [1,*] and Thomas Paululat [2,*]

1 Pharmaceutical Biology and Biotechnology, University of Freiburg, Stefan-Meier-Str. 19, 79104 Freiburg im Breisgau, Germany; philipp.schwarzer@pharmazie.uni-freiburg.de (P.S.); Wujul@gmx.de (J.W.-P.)

2 Organic Chemistry, University of Siegen, Adolf-Reichwein-Str. 2, 57068 Siegen, Germany

* Correspondence: andreas.bechthold@pharmazie.uni-freiburg.de (A.B.); thomas.paululat@uni-siegen.de (T.P.); Tel.: +49-761-203-8371 (A.B.); +49-271-740-4241 (T.P.)

† These authors contributed equally to this work.

Received: 14 February 2018; Accepted: 1 March 2018; Published: 7 March 2018

Abstract: Rishirilide B was isolated from *Streptomyces rishiriensis* and *Streptomyces bottropensis* on the basis of its inhibitory activity towards alpha-2-macroglobulin. The biosynthesis of rishirilide B was investigated by feeding experiments with different ^{13}C labelled precursors using the heterologous host *Streptomyces albus* J1074::cos4 containing a cosmid encoding of the gene cluster responsible for rishirilide B production. NMR spectroscopic analysis of labelled compounds demonstrate that the tricyclic backbone of rishirilide B is a polyketide synthesized from nine acetate units. One of the acetate units is decarboxylated to give a methyl group. The origin of the starter unit was determined to be isobutyrate.

Keywords: streptomyces; rishirilide; biosynthesis; polyketides

---

1. Introduction

Polyketides represent a large and diverse group of natural products from different biological sources such as bacteria, plants and fungi. Numerous polyketides are pharmacologically useful and in clinical use in many areas of application [1].

Rishirilide B (Figure 1), a product of a type II polyketide synthase (PKS), was first isolated from *Streptomyces rishiriensis* OFR-1056 in 1984 [2]. Later, *Streptomyces bottropensis* (formerly *Streptomyces* sp. Gö C4/4) was described as another producer of rishirilide B [3]. Rishirilide B inhibits alpha-2-macroglobulin, a plasma protein that effects the blood coagulation system by inhibiting a large variety of proteinases. Thus alpha-2-macroglobulin inhibition is an effective mechanism to prevent and treat fibrinolytic accentuated thrombosis [3]. Structurally, rishirilide B is a tricyclic compound with an isopentyl sidechain, and further substitutions including one methyl, one carboxlic acid, and three hydroxyl groups. The isopentyl side chain is uncommon for aromatic polyketides. This branched chain may originate from activated isobutyrate and could be the starter unit of polyketide synthesis [4]. The origin of this starter unit might be valine-derived isobutyl-CoA as described in literature for other natural compounds, such as tautomycin and virginiamycin M [5,6]. Isobutyryl-CoA derives from L-valine via transamination and decarboxylation [7].

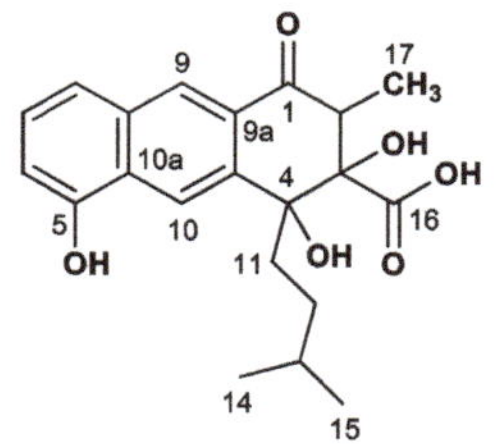

Figure 1. Structure of rishirilide B.

Aromatic polyketide formation undergoes a complex building process under the control of the PKS. CoA activated carboxylic acids are assembled via decarboxylative Claisen condensation by an iteratively used set of proteins leading to the formation of a highly reactive polyketide chain. Within this process, many variables determine the unique order of events and therewith the final structure of the emerging polyketide. The variables include the choice of starter units, the determination of chain length as well as cyclisation patterns. Post PKS events, induced by tailoring enzymes, increase structural variation [4,8]. A review by Staunton and Weissman provides a complete overview about polyketides [9]. The gene cluster responsible for Rishirilide B production contains 28 ORF's that encode a type II PKS, tailoring enzymes, regulatory proteins, and transporters [3].

Insights in the early formation process of polyketides can be obtained by incorporation studies with labelled precursors. These experiments lead to substantial information about the starter unit and the number of incorporated extender units and also indicate the order of events during polyketide formation and give valuable hints about possible rearrangement steps that might occur throughout the biosynthesis. In this way, the carbon backbone of many known natural products like malonomicin, polyketomycin and clavulanic acid could be examined [10–12]. Recent incorporation studies with labelled [1-^{13}C]acetate, [2-^{13}C]acetate, and [1,2-$^{13}C_2$]acetate and [$^{13}c_6$]-L-isoleucine on trixocarcin, a natural compound produced by *Streptomyces bottropensis* DO-45 led to the identification of eight extender units and to 2-methylbutyryl-CoA, a previously unknown starter unit which derives from isoleucine [13]. Here we describe incorporation studies with [1-^{13}C]acetate, [2-^{13}C]acetate, and [1,2-$^{13}C_2$]acetate, as well as [$^{13}c_5$,$^{15}n_1$]-L-valine that clarify the origin of all carbon atoms in rishirilide B. These experiments also give insight into the rearrangement process that occur during rishirilide biosynthesis.

2. Results

A cosmid encoding the entire rishirilide B gene cluster was transformed into *Streptomyces albus* J1074, yielding *S. albus* J1074::cos4. Stable heterologous production of rishirilide B was observed with *S. albus* J1074::cos4 and therefore all feeding experiments have been carried out with this strain. The origin of each carbon atom in rishirilide B was determined by supplementing the production medium with [1-^{13}C]-, [1,2-$^{13}C_2$]-, [2-^{13}C]acetate, and [$^{13}c_5$,$^{15}n_1$]-L-valine (Figure 2 and Table 1, assignment in Table S1).

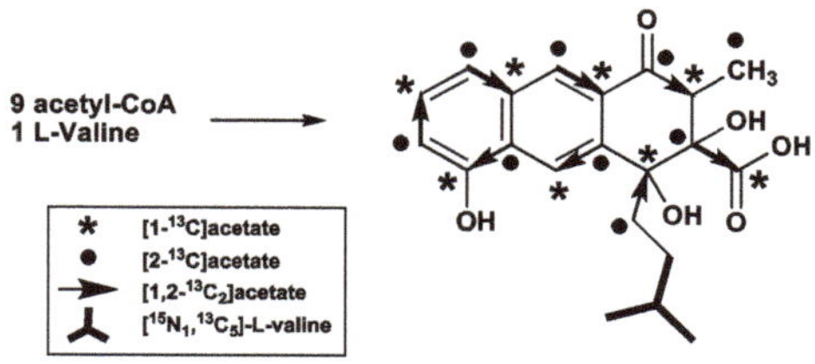

Figure 2. Labelling positions from feeding experiments using different labelled acetates and L-valine.

Table 1. ^{13}C-NMR signals of rishirilide B together with specific incorporations and coupling constants after feeding with [1-^{13}C]acetate (I), [2-^{13}C]acetate (II), [1,2-$^{13}C_2$]acetate (III) and [$^{13}c_5$,^{15}n]-L-valine (IV).

Pos.	δ_C (ppm)	I$^{(b)}$ (%)	II$^{(a)}$ (%)	III J_{CC} (Hz)	III$^{(c)}$ J Partner	IV J_{CC} (Hz)	IV J Partner
1	197.3	−0.4	3.2	41	C-2		(d)
2	47.9	3.8	−0.6	41	C-1		(d)
3	83.0	0.8	4.4	51	C-16		(d)
4	76.6	4.9	−0.3	39	C-11		(d)
4a	140.8	–	6.3	64	C-10		(d)
5	152.9	4.6	−0.5	64	C-10a		(d)
6	109.9	−0.4	2.6	55	C-7		(d)
7	126.0	7.9	−0.4	55	C-6		(d)
8	119.6	−0.3	1.3	55	C-8a		(d)
8a	132.2	5.1	−0.6	55	C-8		(d)
9	125.3	0.4	3.8	64	C-9a		(d)
9a	130.3	4.2	0.4	64	C-9		(d)
10	119.3	6.7	−0.3	64	C-10a		(d)
10a	126.3	−0.8	2.0	64	C-10		(d)
11	35.0	−0.3	2.4	39	C-4		(d)
12	31.1	−0.03	−0.2	-	-	35	C-13$^{(c)}$
13	27.9	−0.2	−0.1	-	-	35,35	C-12$^{(c)}$, C-14$^{(c)}$, C-15$^{(c)}$
14	22.4	(b)	(b)	-	-	35	C-13$^{(c)}$
15	22.6	−0.1	−0.4	-	-	35	C-13$^{(c)}$
16	173.8	5.0	−0.1	51	C-3		(d)
17	10.1	−0.3	1.4	-	-		(d)

(**a**) Relative enrichments were normalized to peak intensity of the C-14 signal; (**b**) Reference signal; (**c**) Supported from Inadequate NMR spectrum; (**d**) Weak enrichment, pattern similar to [1,2-$^{13}C_2$]acetate incorporation through biosynthetic transformation of valine.

Incorporation studies with [1-^{13}C]acetate show ^{13}C isotopic labelling in positions C-2, C-4, C-5, C-7, C-8a, C-9a, C-10 and C-16 as detected from ^{13}C NMR signal enhancements (Table S2, Figures S1 and S2). Feeding of uniformly labelled [1,2-$^{13}C_2$]acetate reveals eight intact acetate units (Table S3, Figures S3–S5). When denoted according to the direction of the proposed biosynthetic pathway, these acetate units correspond to C11/C4, C4a/C-10, C-10a/C-5, C-6/C-7, C-8/C-8a, C-9/C-9a, C-1/C-2 and C-3/C-16. C-17 was proposed to come from methionine, but a feeding experiment with L-[methyl-^{13}c]methionine did not produce a labelled position in rishirilide B (Table S4, Figure S6). After feeding of [2-^{13}C]acetate, enrichment in positions C-1, C-3, C-4a, C-6, C-8, C-9, C-10a, C-11 and C-17 was observed (Table S5, Figures S7 and S8). This implies that an acetate unit affords the C-17 methyl group after decarboxylation. This result also demonstrates that a nonaketide is the precursor to the tricyclic backbone of rishirilide B.

The C-12–C-15 moiety, which represents the starter unit of the polyketide, was not affected by any feeding experiment with labelled acetate or methionine, indicating that it derives from a different precursor. However, feeding experiments with [$^{13}c_5$,$^{15}n_1$]-L-valine led to a mass increase of +4 (m/z) compared to unlabeled rishirilide B (Figure S9). ^{13}C-NMR spectroscopic analysis likewise showed enrichments at positions C-12, C-13, C-14, and C-15 (Table S6, Figures S10–S13). Both MS and NMR data are consistent with a starter unit that is a valine derived isobutyryl-CoA.

3. Discussion

The biosynthetic origin of all carbon atoms of rishirilide B has been determined through labelled precursor feeding experiments. The tricyclic skeleton shows the expected isotopic labelling pattern that arises from a nonaketide that adopts an S-mode folding prior to condensation and aromatization [14]. Decarboxylation of one acetate shortens the polyketide, resulting in a residual methyl group. The starter unit arises from isobutyryl-CoA, which in turn is derived from valine. Interestingly, when feeding with [^{13}c$_5$,^{15}n]-L-valine was performed, we observed strong enrichment of C12-C15 but also weak enrichment pattern in other carbons similar to [1,2-$^{13}C_2$]acetate incorporation. Despite feeding the labelled precursors in several pulses, some L-valine degraded to acetate.

The labelling pattern arising from acetate feeding indicates that a rearrangement occurs during rishirilide B biosynthesis, otherwise C-17 and C-3/C-16 could not show the detected isotopic enrichments. We propose that a Bayer-Villiger type oxidation followed by hydrolytic ring opening and an aldol condensation would produce the labelling pattern detected in our feeding experiments (Figure 3). A similar sequence of Bayer-Villiger oxidation and hydrolysis is proposed for the conversion of questin to desmethylsulochrin [15]. Likewise, Bayer-Villiger oxidations play a role in the biosynthesis of premithramycin B-lactone or murayaquinone, respectively [16,17]. Enterocin, produced by *Streptomyces maritimus*, also has an unprecedented carbon skeleton that is derived from an aromatic polyketide biosynthetic pathway. Its caged tricyclic, nonaromatic core is derived from a linear poly-beta-ketide precursor that undergoes a Favorskii-like oxidative rearrangement [18,19].

Further examples on molecular level show that Bayer-Villiger rearrangements can be catalyzed by oxygenases, like the monooxygenases GilOII in Gilovarcin biosynthesis and BexE in BE-7585A biosynthesis [20,21]. Several enzymes like the luciferase like monooxygenase RslO1 and the putative monooxygenase of the antibiotic biosynthesis monooxygenase (ABM) superfamily RslO4, are potential candidates involved in the rearrangement. An experimental confirmation is still pending.

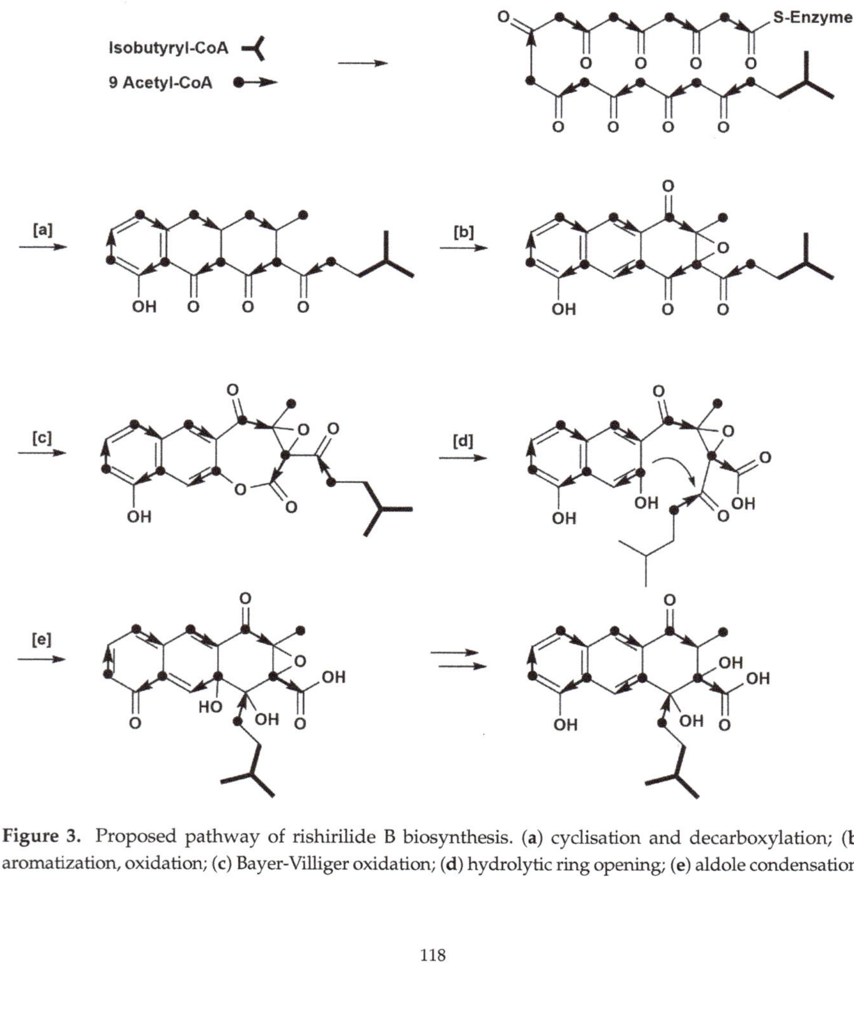

Figure 3. Proposed pathway of rishirilide B biosynthesis. (**a**) cyclisation and decarboxylation; (**b**) aromatization, oxidation; (**c**) Bayer-Villiger oxidation; (**d**) hydrolytic ring opening; (**e**) aldole condensation.

4. Materials and Methods

4.1. Bacterial Strains and Cultivation

Experiments were performed in *Streptomyces albus* J1074::cos4, a transformant, which harbors a cosmid clone of the entire gene cluster of rishirilide B. *Streptomyces albus* J1074::cos4 was cultivated in TSB media (CASO bouillon 30 g·L^{-1}, Carl Roth, Karlsruhe, Germany) supplemented with appropriate antibiotics and incubated in shake flasks at 28 °C for 24 h.

4.2. Feeding Experiments

Sodium[1-^{13}C]acetate, sodium[1,2-$^{13}C_2$]acetate and sodium[2-^{13}C]acetate were purchased from Sigma Aldrich and have 99% ^{13}C atom purity, whereas [$^{13}c_5$,$^{15}n_1$]-L-valine was purchased from cortecnet (98% ^{13}C, 95% ^{15}N enriched). Feeding experiments with labelled sodium[1-^{13}C]acetate, sodium[1,2-$^{13}C_2$]acetate and sodium[2-^{13}C] acetate were performed in HA media (glucose 4 g·L^{-1}, yeast extract 4 g·L^{-1}, malt extract 10 g·L^{-1}, pH 7.4), whereas experiments with [$^{13}c_5$,$^{15}n_1$]-L-valine were carried out in DNPM media (Bacto Soytone 7.5 g·L^{-1}, dry yeast 5 g·L^{-1}, MOPS 21 g·L^{-1}, pH 6.8) both supplemented with appropriate antibiotics. The media was inoculated (1% *v*/*v*) with a 24 h old TSB culture of *Streptomyces albus* J1074::cos4. After 20 h of cultivation, determined as the starting point of rishirilide B production, the isotope labelled precursors were added aseptically into the production media. The feeding was done in 10 pulses—every six hours with a 12 h overnight break, hence 3 feedings within 24 h. The final concentration for [1-^{13}C]acetate, [1,2-$^{13}C_2$]acetate and [2-^{13}C]acetate was 6.1 mM and for feeding with [$^{13}c_5$,$^{15}n_1$]-L-valine, 2 mM. Rishirilide B was isolated after 5–6 days when production reaches its maximum.

4.3. Analysis of Rishirilide B by HPLC/MS

Rishirilide B production was monitored on a HPLC/MS system equipped with a XBridge C18 precolumn (3.5 µm; 20 mm × 4.6 mm) and an XBridge C18 main column (3.5 µm, 100 mm × 4.6 mm) coupled to a DAD UV detector and MS (Agilent, 1100 series, Agilent Technologies, Waldbronn, Germany). The column was run at a flow rate of 0.5 mL·min^{-1} beginning with 80% of buffer A (H_2O + 0.5% acetic acid) and 20% of buffer B (MeCN + 0.5% acetic acid). After 6 min, buffer B was raised to 30% within 1 min followed by an 18 min linear gradient from 30% to 95%. After a delay of 3 min at 95%, the conditions returned to the starting values within 2 min followed by 5 min equilibration. Rishirilide B was detected at 254 nm.

4.4. Isolation and Purification of Rishirilide B

Rishirilide B was isolated from 2 L of culture broth for all feeding experiments. After adjustment to pH 4 the culture broth was centrifuged and extracted with EtOAc. The crude extract was fractionated with increasing methanol content (10% increments) by solid phase extraction (Oasis HLB 35cc (6g) LP Extraction Cartridge, Waters GmbH, Eschborn, Germany). The 70% and 80% methanol fractions containing rishirilide B were combined and further purified by semi preparative HPLC using a Zorbax SB-C18 precolumn (5 µm, 9.4 mm × 20 mm) and a Zorbax B-C18 main column (5 µm, 9.4 mm × 150 mm) coupled to a PDA detector. The column was eluted with buffer A and buffer B at a flow rate of 0.5 mL·min^{-1}.

The gradients for purification of rishirilide B from acetate feeding and L-valine feeding experiments were slightly different. The starting conditions for purification of rishirilide B from acetate feeding experiments were 50% buffer A and 50% buffer B held for 4 min followed by a 7 min linear gradient to 95% buffer B, where rishirilide B was collected. With a delay of 1 min at 95% buffer B, the conditions were changed to starting conditions and maintained for 15 min.

The starting conditions for purification of rishirilide B from L-valine feeding experiment were 55% buffer A and 45% buffer B. After 3 min the concentrations were changed to 80% buffer B over a 6 min linear gradient. After 9 min, buffer B was quickly increased to 95%. With a delay of 3 min under these conditions, buffer B was lowered to starting conditions at minute 12, and the column was washed for a further 3 min.

Finally, the samples were further purified on a column (39.5 cm × 1.4 cm) packed with Sephadex® LH20 (GE Healthcare GmbH, Solingen, Germany) with MeOH as solvent. Flow rate was determined by gravity. The fractions were analyzed by HPLC-MS and rishirilide B containing fractions were combined and evaporated to dryness.

4.5. NMR Analysis of Rishirilide B

NMR spectra were recorded using a Varian VNMR-S 600 equipped with a Nalorac 3 mm broadband probe. Spectra were recorded in 150 μL DMSO-d_6 at 35 °C. Solvent signals were used as internal standard (DMSO-d_6: δ_H = 2.50 ppm, δ_C = 39.5 ppm). Calculation of enrichment and specific enrichment was done according to Scott et al. [22].

5. Conclusions

Investigations of rishirilide B biosynthesis have been carried out from feeding experiments with different isotopic labelled acetates and L-valine. The carbon skeleton of rishirilide B is assembled from nine acetate units and one isobutyrate (derived from valine). One acetate undergoes decarboxylation to afford a methyl group. The pattern of the labelling positions from feeding experiments with labelled acetate reveals that a rearrangement of the carbon skeleton occurs during rishirilide B biosynthesis. Future studies involving gene knock out experiments and isolation of reaction intermediates will provide insight into the enzymes and mechanisms involved in this rearrangement.

Supplementary Materials: The following are available online at http://www.mdpi.com/2079-6382/7/1/20/s1; Table S1: NMR data of rishirilide B (600/150 MHz, DMSO-d_6, 35 °C); Table S2: Calculation of enrichment and specific enrichment for rishirilide B from feeding experiment with [1-^{13}C]acetate; Figure S1: Labelling positions of rishirilide B after feeding experiment with [1-^{13}C]acetate; Figure S2: ^{13}C NMR spectra (150 MHz, DMSO-d_6, 35 °C) of rishirilide B after feeding of [1-^{13}C]acetate in comparison to rishirilide B at natural abundance; Table S3: Table of NMR data of rishirilide B from feeding experiment with [1,2-$^{13}C_2$]acetate; Figure S3: Labelling positions of rishirilide B after feeding experiment with [1,2-$^{13}C_2$]acetate; Figure S4: ^{13}C NMR spectra (150 MHz, DMSO-d_6, 35 °C) of rishirilide B after feeding of [1,2-$^{13}C_2$]acetate in comparison to rishirilide B at natural abundance; Figure S5: Inadequate (150 MHz, DMSO-d_6, 35 °C) of rishirilide B after feeding of [1,2-$^{13}C_2$]acetate; Table S4: Results of feeding experiment with L-[methyl-^{13}c]methionine; Figure S6: ^{13}C NMR spectra (150 MHz, DMSO-d_6, 35 °C) of rishirilide B from feeding experiment with L-[methyl-^{13}c]methionine in comparison to rishirilide B at natural abundance; Table S5: Calculation of enrichment and specific enrichment for rishirilide B from feeding experiment with [2-^{13}C]acetate; Figure S7: Labelling positions of rishirilide B after feeding experiment with [2-^{13}C]acetate; Figure S8: ^{13}C NMR spectra (150 MHz, DMSO-d_6, 35 °C) of rishirilide B after feeding of [2-^{13}C]acetate in comparison to rishirilide B at natural abundance; Figure S9: Mass spectrum of rishirilide B from feeding experiment with [$^{13}c_5$, $^{15}n_1$]-L-valine; Table S6: Table of NMR data rishirilide B from feeding experiment with [$^{13}c_5$, $^{15}n_1$]-L-valine; Figure S10: Labelling positions of rishirilide B after feeding experiment with [$^{13}c_5$, $^{15}n_1$]-L-valine; Figure S11: ^{13}C NMR spectra (150 MHz, DMSO-d_6, 35 °C) of rishirilide B from feeding experiment with [$^{13}c_5$, $^{15}n_1$]-L-valine in comparison to rishirilide B at natural abundance; Figure S12: Inadequate (150 MHz, DMSO-d_6, 35 °C) of rishirilide B from feeding experiment with [$^{13}c_5$, $^{15}n_1$]-L-valine; Figure S13: Expansion of Inadequate (150 MHz, DMSO-d_6, 35 °C) of rishirilide B from feeding experiment with [$^{13}c_5$, $^{15}n_1$]-L-valine.

Acknowledgments: This work was supported by the Deutsche Forschungsgemeinschaft (RTG 2202), grand to Andreas Bechthold. We thank David L. Zechel, Queen's University Kingston, Ontario for manuscript reading.

Author Contributions: Andreas Bechthold and Thomas Paululat conceived and designed the experiments. Philipp Schwarzer performed the feeding experiment with labelled valine and isolation of the product. Julia Wunsch-Palasis performed the feeding experiment and product isolation with the other precursors. Thomas Paululat measured the NMR spectra and analysed the data. Philipp Schwarzer, Andreas Bechthold and Thomas Paululat wrote the paper.

Conflicts of Interest: The authors declare no conflict of interest.

References

1. O'Hagan, D. Biosynthesis of polyketide metabolites. *Nat. Prod. Rep.* **1992**, *9*, 447–479. [CrossRef] [PubMed]
2. Iwaki, H.; Nakayama, Y.; Takahashi, M.; Uetsuki, S.; Kido, M. Structures of rishirilides A and B, α_2-macroglobulin inhibitors produced by *Streptomyces rishiriensis* OFR-1056. *J. Antibiot.* **1984**, *XXXVII*, 9–11.
3. Yan, X.; Probst, K.; Linnenbrink, A.; Arnold, M.; Paululat, T.; Zeeck, A.; Bechthold, A. Cloning and heterologous expression of three type II PKS gene clusters from *Streptomyces bottropensis*. *Chembiochem* **2012**, *13*, 224–230. [CrossRef] [PubMed]
4. Moore, B.S.; Hertweck, C. Biosynthesis and attachment of novel bacterial polyketide synthase starter units. *Nat. Prod. Rep.* **2002**, *19*, 70–99. [PubMed]
5. Ubukata, M.; Cheng, X.C.; Uzawa, J.; Isono, K. Biosynthesis of the dialkylmaleic anhydride-containing antibiotics, tautomycin and tautomycetin. *J. Chem. Soc. Perkin Trans. 1* **1995**, 2399–2404. [CrossRef]
6. Kingston, D.G.; Kolpak, M.X.; LeFevre, J.W.; Borup-Grochtmann, I. Biosynthesis of antibiotics of the virginiamycin family. 3. Biosynthesis of virginiamycin M1. *J. Am. Chem. Soc.* **1983**, *105*, 5106–5110. [CrossRef]
7. Omura, S.; Tsuzuki, K.; Tanaka, Y.; Sakakibara, H.; Aizawa, M.; Lukacs, G. Valine as a precursor of *n*-butyrate unit in the biosynthesis of macrolide aglycone. *J. Antibiot.* **1983**, 614–616. [CrossRef]
8. Chan, Y.A.; Podevels, A.M.; Kevany, B.M.; Thomas, M.G. Biosynthesis of polyketide synthase extender units. *Nat. Prod. Rep.* **2009**, *26*, 90–114. [CrossRef] [PubMed]
9. Staunton, J.; Weissman, K.J. Polyketide biosynthesis: A millennium review. *Nat. Prod. Rep.* **2001**, *18*, 380–416. [CrossRef] [PubMed]
10. Schipper, D.; Van Der Baan, J.L.; Bickelhaupt, F. Biosynthesis of malonomicin. Part 1. ^{13}C nuclear magnetic resonance spectrum and feeding experiments with ^{13}C-labelled precursors. *J. Chem. Soc. Perkin Trans. 1* **1979**, *0*, 2017–2022. [CrossRef]
11. Paululat, T.; Gutterer, J.M.; Fiedler, H.-P.; Zeeck, A. Biosynthesis of polyketomycin produced by *Streptomyces diastatochromogenes* Tü 6028. *J. Antibiot. (Tokyo)* **1999**, *52*, 96–101. [CrossRef] [PubMed]
12. Elson, S.W.; Oliver, R.S. Studies on the biosynthesis of clavulanic acid. Incorporation of ^{13}C-labelled precursors. *J. Antibiot. (Tokyo)* **1978**, *31*, 586–592. [CrossRef] [PubMed]
13. Zhang, M.; Hou, X.-F.; Qi, L.-H.; Yin, Y.; Li, Q.; Pan, H.-X.; Chen, X.-Y.; Tang, G.-L. Biosynthesis of trioxacarcin revealing a different starter unit and complex tailoring steps for type II polyketide synthase. *Chem. Sci.* **2015**, *6*, 3440–3447. [CrossRef]
14. Thomas, R. A biosynthetic classification of fungal and streptomycete fused-ring aromatic polyketides. *ChemBioChem* **2001**, *2*, 612–627. [CrossRef]
15. Fujii, I.; Ebizuka, Y.; Sankawa, U. A novel anthraquinone ring cleavage enzyme from *Aspergillus terreus*. *J. Biochem.* **1988**, *103*, 878–883. [CrossRef] [PubMed]
16. Beam, M.P.; Bossermann, M.A.; Noinaj, N.; Wehenkel, M.; Rohr, J. Crystal structure of Baeyer-Villiger monooxygenase MtmOIV, the key enzyme of the mithramycin biosynthetic pathway. *J. Biochem.* **2009**, *48*, 4476–4487. [CrossRef] [PubMed]
17. Gao, G.; Liu, X.; Xu, M.; Wang, Y.; Zhang, F.; Xu, L.; Lv, J.; Long, Q.; Kang, Q.; Ou, H.Y.; Wang, Y.; Rohr, J.; Deng, Z.; Jiang, M.; Lin, S.; Tao, M. Formation of an angular aromatic polyketide from a linear anthrene precursor via oxidative rearrangement. *Cell Chem. Biol.* **2017**, *24*, 881–891. [CrossRef] [PubMed]
18. Xiang, L.; Kalaitzis, J.A.; Moore, B.S. EncM, a versatile enterocin biosynthetic enzyme involved in Favorskii oxidative rearrangement, aldol condensation, and heterocycle-forming reactions. *Proc. Natl. Acad. Sci. USA* **2009**, *101*, 15609–15614. [CrossRef] [PubMed]
19. Teufel, R.; Miyanaga, A.; Michaudel, Q.; Stull, F.; Louie, G.; Noel, J.P.; Baran, P.S.; Palfey, B.; Moore, B.S. Flavin-mediated dual oxidation controls an enzymatic Favorskii-type rearrangement. *Nature* **2013**, *503*, 552–556. [CrossRef] [PubMed]
20. Tibrewal, N.; Pahari, P.; Wang, G.; Kharel, M.K.; Morris, C.; Downey, T.; Hou, Y.; Bugni, T.S.; Rohr, J. Baeyer-Villiger C-C bond cleavage reaction in gilvocarcin and jadomycin biosynthesis. *J. Am. Chem. Soc.* **2012**, *134*, 18181–18184. [CrossRef] [PubMed]

21. Jackson, D.R.; Yu, X.; Patel, A.B.; Calveras, J.; Barajas, J.F.; Sakaki, E.; Metsä-Ketelä, M.; Liu, H.-W.; Rohr, J.; Tsai, S.-C. Insights into complex oxidation during BE-7585A biosynthesis: Structural determination and analysis of the polyketide monooxygenase BexE. *ACS Chem. Biol.* **2016**, *11*, 1137–1147. [CrossRef] [PubMed]
22. Scott, A.I.; Townsend, C.A.; Okada, K.; Kajiwara, M.; Cushley, R.J.; Whitman, P.J. Biosynthesis of corrins. II. Incorporation of ^{13}C-labeled substrates in vitamins B12. *J. Am. Chem. Soc.* **1974**, *96*, 8069–8080. [CrossRef] [PubMed]

antibiotics

MDPI

Article

Diversification of Secondary Metabolite Biosynthetic Gene Clusters Coincides with Lineage Divergence in *Streptomyces*

Mallory J. Choudoir, Charles Pepe-Ranney and Daniel H. Buckley *

School of Integrative Plant Science, Bradfield Hall 705, Cornell University, Ithaca, NY 14853, USA; mjchoudoir@gmail.com (M.J.C.); chuck.peperanney@gmail.com (C.P.-R.)

* Correspondence: dbuckley@cornell.edu; Tel.: +1-607-255-1716

Received: 11 January 2018; Accepted: 7 February 2018; Published: 13 February 2018

Abstract: We have identified *Streptomyces* sister-taxa which share a recent common ancestor and nearly identical small subunit (SSU) rRNA gene sequences, but inhabit distinct geographic ranges demarcated by latitude and have sufficient genomic divergence to represent distinct species. Here, we explore the evolutionary dynamics of secondary metabolite biosynthetic gene clusters (SMGCs) following lineage divergence of these sister-taxa. These sister-taxa strains contained 310 distinct SMGCs belonging to 22 different gene cluster classes. While there was broad conservation of these 22 gene cluster classes among the genomes analyzed, each individual genome harbored a different number of gene clusters within each class. A total of nine SMGCs were conserved across nearly all strains, but the majority (57%) of SMGCs were strain-specific. We show that while each individual genome has a unique combination of SMGCs, this diversity displays lineage-level modularity. Overall, the northern-derived (NDR) clade had more SMGCs than the southern-derived (SDR) clade (40.7 ± 3.9 and 33.8 ± 3.9, mean and S.D., respectively). This difference in SMGC content corresponded with differences in the number of predicted open reading frames (ORFs) per genome (7775 ± 196 and 7093 ± 205, mean and S.D., respectively) such that the ratio of SMGC:ORF did not differ between sister-taxa genomes. We show that changes in SMGC diversity between the sister-taxa were driven primarily by gene acquisition and deletion events, and these changes were associated with an overall change in genome size which accompanied lineage divergence.

Keywords: *Streptomyces*; biogeography; comparative genomics; diversification; secondary metabolite biosynthetic gene clusters; SMGC; natural products

---

1. Introduction

Microbial secondary metabolism encapsulates a remarkable diversity of natural products with an extensive range of biological activities. Secondary metabolites differ from primary metabolites in that they are not involved in essential catabolic and anabolic activities required for normal growth and reproduction, but may contribute significantly to an individual's fitness [1]. While primary metabolic pathways are often conserved deeply within a phylogeny, secondary metabolic pathways are more divergent, often being species or strain-specific, with conservation sometimes observed among closely related species and genera [2]. This phylogenetic pattern suggests an adaptive role for secondary metabolites, and if secondary metabolism pathways provide adaptive benefits, their evolution might drive or reinforce evolutionary processes that result in microbial diversification and speciation [3].

The values of natural products to humanity are widely recognized, yet because most research has focused on their discovery and human-centric relevance, we are still far from understanding their biological role in natural systems. The discovery and application of antibiotics revolutionized medicine in the 1940's, sparking the "golden age" of antibiotics between 1950 and 1960, during which time approximately half of the microbial-derived drugs we use today were discovered [4]. Presently, thousands of bioactive compounds with antibacterial, antifungal, and antitumor activities are cataloged [5,6], and yet these represent only a fraction of actual natural product diversity [7]. In addition, microbial populations in situ are exposed to natural products at concentrations far below the lethal clinical dose, and hence these compounds may serve different functions in the environment from those observed during therapeutic application. We know that secondary metabolites can mediate diverse biotic interactions including mutualistic interactions, competition for nutrients, metal scavenging, and plant-microbe and insect-microbe symbioses [8–10], which can all have profound impacts on microbial fitness. It is clear that natural products must have considerable impacts on microbial ecology and evolution and that understanding the biology and evolutionary history of natural products will enhance our ability to use these agents therapeutically.

Soil-dwelling actinomycetes are the predominant source of microbial-derived therapeutic natural products, and the majority of described bioactive compounds originate from the genus *Streptomyces* [6,7]. The *Streptomyces* life cycle resembles that of many fungi, consisting of filamentous growth, formation of mycelia, and production of aerial hyphae and spores. Indeed, *Streptomyces* were thought to be an intermediary between bacteria and fungi until as recently as the 1950's [11]. However, *Streptomyces* are Gram-positive *Actinobacteria* with long linear chromosomes that have a high G+C content [12]. Traditionally, *Streptomyces* species are often known to produce several secondary metabolites when grown in culture. Genome sequencing, however, reveals that *Streptomyces* contain an enormous reservoir of "cryptic secondary metabolites" which are not expressed under standard laboratory conditions [13]. For instance, while *Streptomyces coelicolor* A3(2) was known to produce several well characterized secondary metabolites, genome sequencing discovered that it actually contained >20 biosynthetic gene clusters not expressed when grown in culture [14]. Genes within secondary metabolite biosynthetic gene clusters (SMGCs) are co-localized as operons within discrete genomic regions. SMGCs have recognizable functional domains, so SMGCs are readily predicted using bioinformatics [15]. Phylogenetic conservation of SMGCs between closely related microbes suggests that these secondary metabolites may have ecological roles which facilitate microbial diversification [16–18].

The evolutionary and ecological processes that govern SMGC diversity remain largely unexplored. The richness of SMGCs within soils is linked to both edaphic and biotic factors [19,20]. For example, the production of antibiotics by *Streptomyces* isolated from prairie soils is highly variable between strains and correlates poorly with 16S rRNA gene phylogeny, suggesting a role of selection acting at small spatial scales [21,22]. Conversely, at larger spatial scales, *Streptomyces* SMGC composition varies in relation to both spatial distance and environmental dissimilarity [23]. Furthermore, evidence within *Streptomyces* for endemism at inter-continental and regional geographic scales [16,24,25] suggests limits to dispersal at large spatial scales. These data indicate that both adaptive and neutral processes contribute to patterns of SMGC biogeography.

Microbial biogeography is readily explored with geographically explicit microbial culture collections (reviewed in [26]), and the genus *Streptomyces* is an ideal model system to evaluate the influence of SMGC dynamics on patterns of diversification. We previously assembled a culture collection of *Streptomyces* from sites spanning the United States, and we observed evidence for dispersal limitation, as well as a latitudinal gradient of species riches and intraspecific nucleotide diversity [27,28]. From this culture collection, we have identified *Streptomyces* sister-taxa that have geographic ranges delimited by latitude and have patterns of gene flow and genomic diversity consistent with their diversification from a recent common ancestor [28,29]. Here, we evaluate changes in SMGC diversity between these *Streptomyces* sister-taxa to explore SMGC evolutionary dynamics during the divergence of *Streptomyces* species.

2. Results and Discussion

2.1. Genomic Divergence between Streptomyces Sister-Taxa

We used comparative genomics to analyze patterns of genomic diversity and SMGC content in 24 *Streptomyces* representing sister-taxa and related strains. These strains were identified through a phylogenetic analysis of a *Streptomyces* culture collection [28] generated from soils of ecologically similar grassland sites spanning 6000 km across the continental United States (Figure 1, Table S1). The sister-taxa, which we have designated the northern-derived (NDR) and southern-derived (SDR) clades, were defined by their geographic range and genomic similarity (Figure 1). Each clade contains ten isolates, and an additional four genomes represent intermediate (INT) taxa.

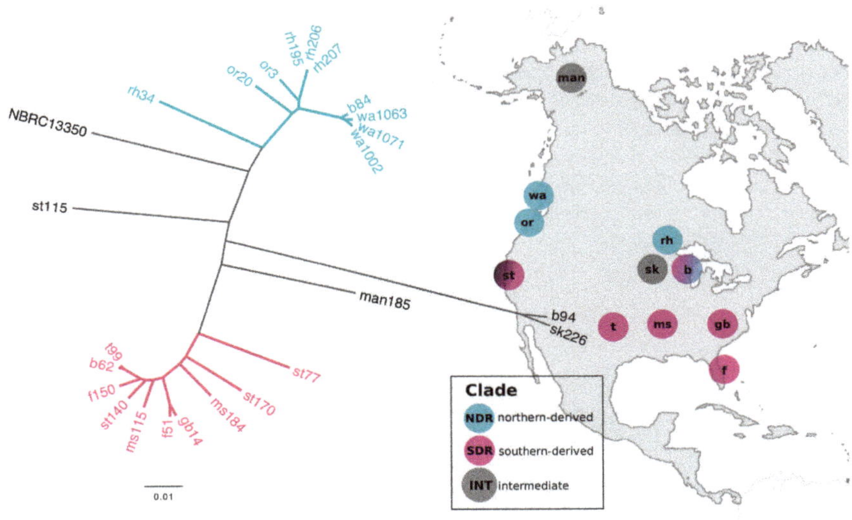

Figure 1. The northern-derived (NDR) and southern-derived (SDR) clades are closely related sister-taxa and yet were isolated from soils of different latitude. The un-rooted tree was constructed from multiple whole genome alignments with maximum likelihood and a GTRGAMMA model of evolution. Scale bar represents nucleotide substitutions per site. Colored branches depict the northern-derived (NDR) and southern-derived (SDR) clades. Strain names reflect the sample site they were isolated from (Table S1). Genome NBRC 13350 is the publically available type strain *Streptomyces griseus* subsp. *griseus* NBRC 13350. Sample locations are shown in the right panel and labeled with the site code. Circles are colored to reflect the geographic distribution of clades. (Figure modified from [29]).

Assembled genomes are 7.5–9.1 Mb with a G+C content of 71.4–72.5% and 6776–8078 predicted open reading frames (ORFs) (Table S2). The core gene content across all 24 strains is comprised of 3234 orthologous genes (representing 2778 single-copy genes), with a total of 22,054 genes in the overall pan-genome. All isolates affiliate taxonomically with the *Streptomyces griseus* species cluster [30] and share >90% average nucleotide identity (ANI) with the type strain *Streptomyces griseus* subsp. *griseus* NBRC 13350 (Figure 1).

The NDR core genome is comprised of 4234 genes, and the SDR core genome is comprised of 4400 genes. The NDR and SDR clades share a recent phylogenetic ancestor and have nearly identical 16S rRNA genes (inter-lineage nucleotide dissimilarity of 0–0.21% between strains). Strains within each clade have a whole genome ANI value ranging from 95.6% to 99.9%, while the ANI between strains of NDR and SDR range from 92.6% to 93.3% (Figure 1). Distinct microbial species are typically distinguished by ANI in the range of 95–96% [31]. Comparative population genomics reveals signatures of genomic differentiation and gene flow limitation between NDR and SDR consistent with expectations of allopatric diversification [29]. Collectively, these results indicate that NDR and SDR clades represent distinct microbial species which have recently diverged from a common ancestor.

2.2. Secondary Metabolite Biosynthetic Gene Cluster (SMGC) Identification and Classification

We used antiSMASH [32] to identify SMGCs in the genomes of our *Streptomyces* sister-taxa. To assess the novelty of these SMGCs, we utilized antiSMASH's downstream annotation pipeline, which annotates SMGCs based on similarity to genes and pathways present within the Minimum Information about a Biosynthetic Gene cluster (MIBiG) database. The antiSMASH pipeline annotated 120 SMGCs across the 24 strains (Table S3). Each genome had between 28 and 47 SMGCs which ranged in size from 1 to 137 Kb (20.9 ± 15.7 Kb, mean ± S.D., respectively) (Figure 2). This range in SMGC content is consistent with the results obtained from previous genomic surveys of *Streptomyces* [14,33–36]. The NDR clade has a greater number of SMGCs per genome than the SDR clade (40.7 ± 3.9, 33.8 ± 3.9, mean ± S.D., respectively; *t*-test, $p < 0.001$; Figure 2a). The NDR clade also has a greater number of ORFs per genome than the SDR clade (7775 ± 196 and 7093 ± 205, mean and S.D., respectively; *t*-test, $p < 0.001$; Table S2). Correspondingly, NDR strains also have larger genomes than SDR strains (8.7 ± 0.25 Mb and 7.9 ± 0.21 Mb, mean ± S.D., respectively; *t*-test, $p < 0.001$; Table S2). We observed a strong positive correlation between genome size and number of SMGCs across all genomes examined (Pearson's $r = 0.66$, $p < 0.001$).

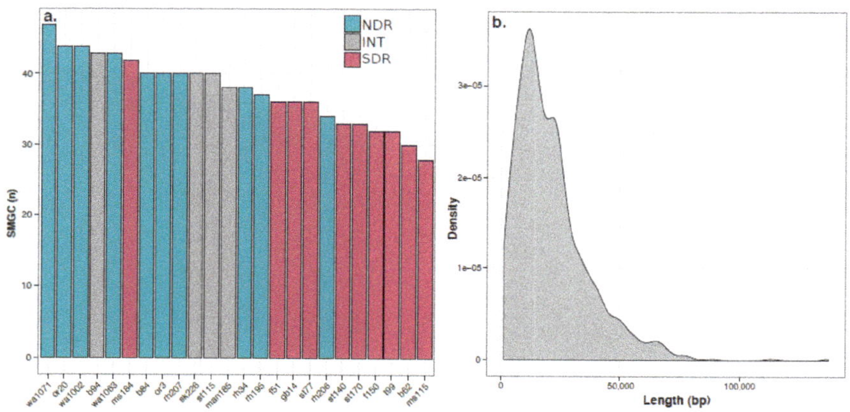

Figure 2. NDR strains have more secondary metabolite biosynthetic gene clusters (SMGCs) than SDR strains (*t*-test, $p < 0.001$). (**a**). Bars indicate the number of SMGCs identified in each genome and are colored according to clade affiliation, and genome names reflect the site of isolation as identified in Table S1; (**b**). Kernal density plot shows the distribution of SMGC length (bp).

Only 21% ($n = 25$) of the MIBiG-annotated SMGCs represent well-characterized biosynthetic gene clusters (in which ≥70% of the genes in a SMGC show similarity to genes within the most similar known cluster from the MIBiG database) (Table S3). In addition, each genome harbors five to 25 potentially novel SMGCs with low similarity to biosynthetic pathways within the MIBiG database. These findings indicate that the diversity of *Streptomyces* SMGCs found within public databases remains low and that a vast reservoir of *Streptomyces* SMGC diversity remains to be characterized within natural populations.

The SMGCs predicted by antiSMASH within our *Streptomyces* sister-taxa encompass 22 classes of natural products. Most of these classes, including bacteriocin, butyrolactones, ectoine, lantipeptide, melanin, non-ribosomal peptide synthases (NRPS), siderophore, polyketide synthases (PKS), and terpene gene clusters, are widely conserved at the genus level [2]. The most abundant SMGC classes in our genomes are NRPS and terpene clusters (Figure 3, Table S3). Many of the predicted gene clusters are NRPS-PKS hybrids (Table S3). Given the similar structure and activity between NRPS and PKS [37], it is unsurprising that hybrid NRPS-PKS clusters are commonly detected in *Streptomyces* genomes [38,39]. Most SMGC classes are present in both NDR and SDR clades, but the relative abundance of each class differs between genomes, as well as between clades (Figure 3). We observe the significant enrichment of melanin and ladderane gene clusters in NDR compared to SDR (*t*-test with Bonferrori correction, $p < 0.002$). Additionally, NDR genomes harbor linaridin gene clusters, which are entirely absent from SDR genomes (Figure 3) but are found in the type strain *Streptomyces griseus* NBRC subsp. *griseus* 13350 [40]. Interestingly, antiSMASH did not identify aminoglycoside biosynthetic clusters in our *Streptomyces* isolates, and all of these genomes presumably lack genes for streptomycin biosynthesis (Figure 3). Schatz and Waksman reported the isolation of streptomycin from *Streptomyces griseus* in 1944, and this was the first antibiotic used to successfully combat tuberculosis [41]. However, not all *Streptomyces griseus* isolates produce streptomycin [42,43].

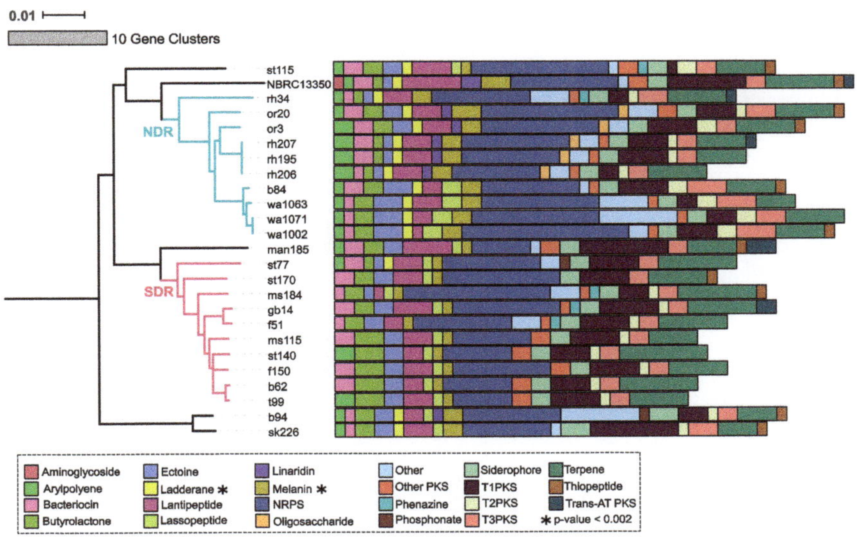

Figure 3. A total of 22 SMGC classes were observed in NDR and SDR genomes by antiSMASH [32]. The tree reflects phylogenetic relationships between *Streptomyces* sister-taxa genomes and was constructed from multiple whole genome alignments (see Figure 1). Scale bar represents nucleotide substitutions per site. Tree branches are colored according to clade affiliation. Bars depict the number of gene clusters belonging to each class for each genome. Colors illustrate gene cluster class as provided by the legend. Asterisks note gene cluster classes that are significantly enriched between clades (*t*-test and Bonferonni correction for multiple comparisons, $p < 0.002$).

2.3. Core and Accessory SMGCs of Streptomyces Sister-Taxa

Comparative population genomics and pan-genome analyses can offer powerful insights into the processes underlying species divergence [44,45]. Given that many of our SMGCs have low similarity to biosynthetic pathways in public databases, we determined shared orthologous SMGCs within our genomes using an annotation-independent approach that compares SMGCs based on similarity in nucleotide composition and gene content (see Materials and Methods). This approach identified 310 non-redundant SMGCs within the pan-genome of all 24 strains (Figures 4 and 5); this number is greater than the number of MIBiG-annotated SMGCs because it classified both known and unknown pathways into distinct non-redundant gene clusters. Only two SMGCs are conserved in all 24 genomes, an ectoine gene cluster and the siderophore desferrioxamine B (Figure 6). Desferrioxamine siderophores are commonly observed in other species of *Streptomyces* and acintomycetes [46,47].

We observed that core SMGC content increased with phylogenetic similarity, but that more than half of the SMGCs were strain-specific (Figures 4 and 5). NDR and SDR shared nine core SMGCs (present in ≥80% of genomes), while NDR strains shared 11 core SMGCs (nine in the conserved core and two in the NDR-specific core), and SDR strains shared 15 core SMGCs (nine in the conserved core and six in the SDR-specific core) (Figure 6). In addition, there were 158 accessory SMGCs (present in <80% genomes) in NDR and 114 accessory SMGCs in SDR (Figure 4). Most SMGCs were observed at low to intermediate frequencies (Figures 4 and 5), and 177 SMGCs were strain-specific, with each *Streptomyces* genome harboring one to 19 exclusive SMGCs. These estimates are generally consistent with previous observations that indicate each different *Streptomyces* species will harbor a distinct repertoire of natural product pathways [17]. For example, Seipke [36] estimated 18 core SMGCs for six *Streptomyces albus* isolates. However, despite the phylogenetic conservation of core SMGC content, even *Streptomyces* with identical 16S rRNA gene sequences can have distinct secondary metabolite profiles [48], indicating that SMGC content exhibits significant strain to strain variability within a species. Thus, we propose that core SMGCs reflect the shared evolutionary history of *Streptomyces* genomes, while patterns of the accessory SMGC carriage suggest lineage and strain-specific processes across more recent evolutionary time scales.

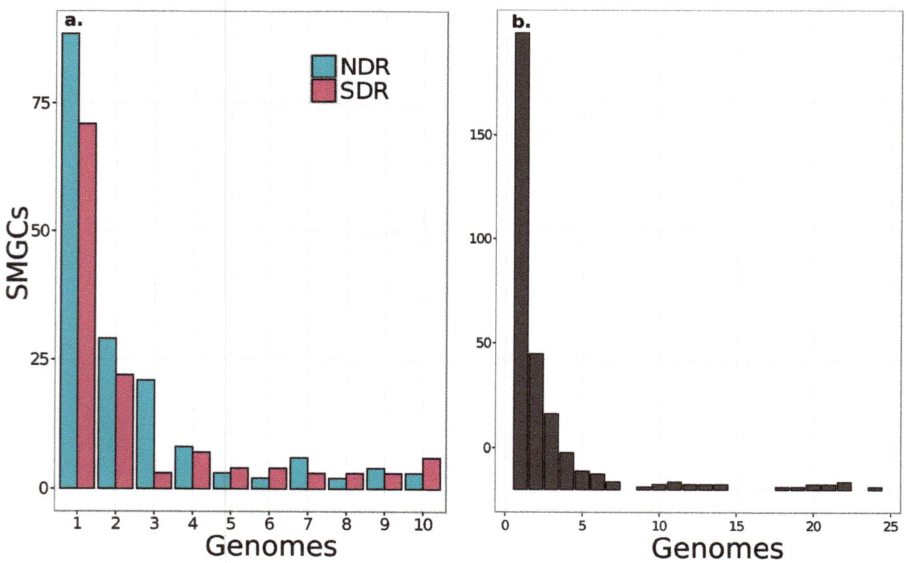

Figure 4. The frequency distribution of SMGCs across strains shows that most SMGCs are strain-specific and fewer are species-specific. Results are shown both for NDR and SDR. (**a**) and for all 24 genomes; (**b**). Non-redundant orthologous SMGCs were defined using our annotation-independent approach (see Materials and Methods).

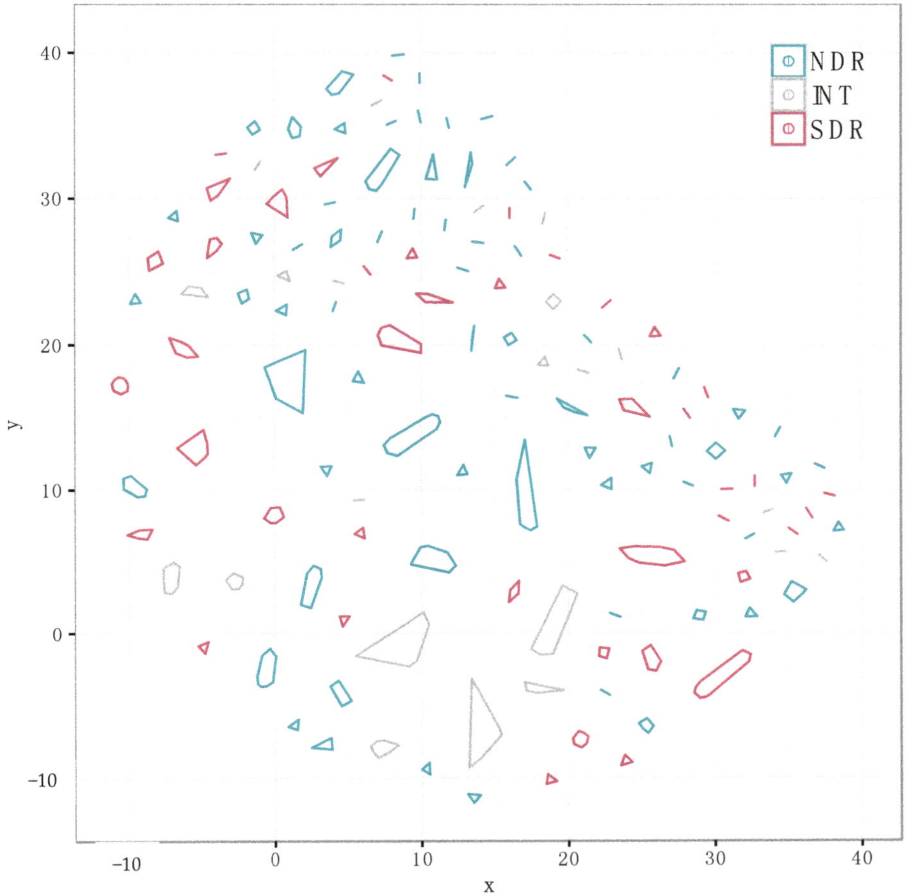

Figure 5. We identified 310 non-redundant distinct SMGCs using our annotation-independent gene clustering approach (see Materials and Methods). Each point represents a unique SMGC from a single genome, and colors correspond to clade affiliation. SMGCs with a similar gene composition are clustered spatially, and cluster membership is depicted with polygons. The same data is presented in a different network diagram in Figure S1.

2.4. Evolutionary Dynamics of Core and Accessory SMGCs

To address potential lineage-specific mechanisms of divergence, we next evaluated the evolutionary dynamics of SMGCs. Most shared SMGCs occur within rather than between clades (Figures 5 and S1). A total of 78 SMGCs are shared among two or more NDR genomes, and 55 are shared among SDR genomes, but only 37 SMGCs are shared across clade boundaries (i.e., found in both NDR and SDR genomes). Furthermore, network analysis reveals unique patterns of SMGC sharing that manifests as nodes of connectivity within clades (Figure S1). This network indicates that there is a core set of SMGC content which links NDR and SRD and which must be ancestral, that there is a clade-specific core set of SMGCs which link the strains of each clade together based on shared SMGC content, and that there are a large number of strain-specific SMGCs (Figures 5 and S1).

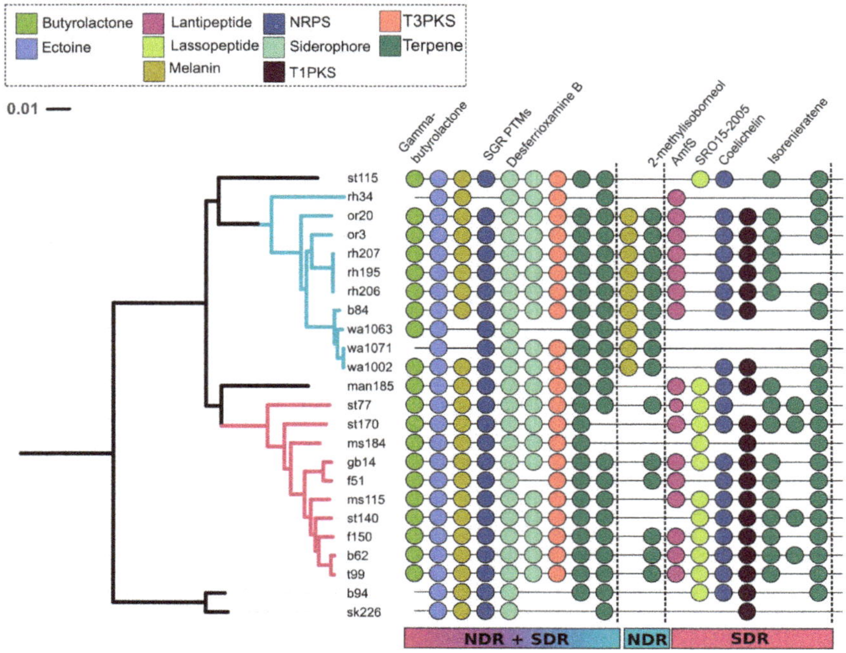

Figure 6. A total of nine core (i.e., conserved in ≥80% of genomes) SMGCs were found in both NDR and SDR. The NDR clade had 11 core SMGCs and the SDR clade had 15 core SMGCs. The tree reflects phylogenetic relationships between *Streptomyces* sister-taxa genomes and was constructed from multiple whole genome alignments (see Figure 1). Scale bar represents nucleotide substitutions per site. Tree branches are colored according to clade affiliation. Core orthologous SMGCs (depicted by colored circles) were determined using the antiSMASH [32] MIBiG annotation pipeline or were defined using our annotation-independent approach (see Materials and Method). Colors correspond to SMGC class (see legend), and natural product annotations are labeled if available.

Differences in gene content between closely related microbes ultimately result from gene gain and loss events [49–51]. Although deletion bias is strong in bacterial genomes [52], gene acquisitions can drive rapid genome innovation and evolution [53]. Gene clusters are often acquired through horizontal gene transfer leading to the formation of new operons in bacterial genomes [54], and many SMGCs in actinomycetes are believed to be the result of horizontal gene transfer [16,18,33,55]. Parsimony predicts that low frequency and strain-specific genes are likely the result of a recent acquisition, while high frequency "near core" genes are the likely result of recent deletion events [56]. Hence, we are able to infer SMGC gain and loss dynamics in our *Streptomyces* sister-taxa from SMGC frequency distributions (Figure 4).

The majority of SMGCs observed within the sister-clades occurred in only one or a few strains, and this suggests that gene acquisition is a major force that drives the diversity of SMGC pathways in *Streptomyces*. However, each clade has a distinct set of core and accessory SMGCs (Figures 3, 5, 6 and S1), and this suggests that SMGC composition (Figure 7) may underlie ecological traits that promote or reinforce lineage divergence. For example, nearly all genomes within the NDR clade (with the exception of rh34) harbor a melanin gene cluster which is absent from both the intermediate (INT) and SDR genomes, suggesting that horizontal gene transfer of the melanin gene cluster into the immediate ancestor of NDR accompanied lineage divergence (Figures 6 and 7). Overall, NDR has more low frequency SMGCs (present in one to three strains) than SDR (139 and 96, respectively) (Figure 4). This result suggests a greater rate of gene acquisition in NDR than in SDR and is consistent with the observation that NDR has more SMGCs (Figure 2) and larger genomes overall than SDR. While this difference in gene content is potentially adaptive, it could also be explained as a consequence of neutral demographic processes such as genome surfing (reviewed in [57]). However, the distribution of SMGC frequencies does not differ significantly between clades (Kolmogorov-Smirnov test, p = 0.4). Hence, while it seems clear that gene acquisition is a major driver of SMGC biodiversity, the role of gene acquisition in driving lineage divergence remains unclear.

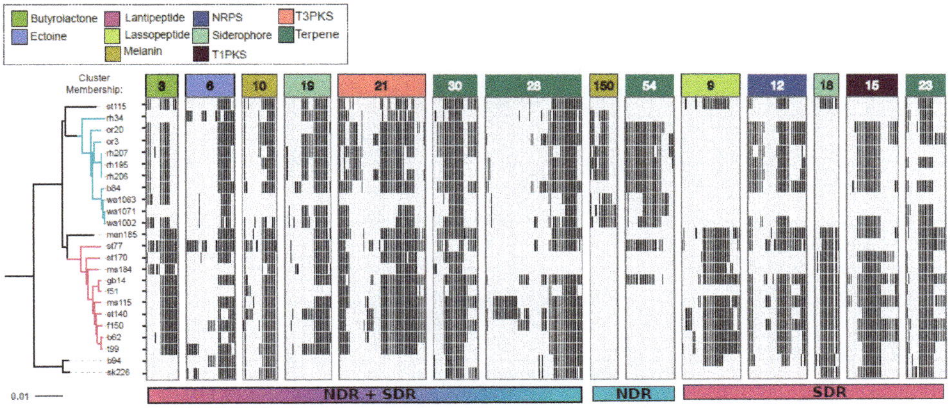

Figure 7. Gene content of core SMGCs vary within and between clades as a result of gene acquisition and deletion events. Panels depict the gene content (i.e., genetic architecture) of core SMGCs (i.e., conserved in ≥80% of genomes), the NDR-specific SMGC core, and the SDR-specific SMGC core. Black bars within the panels represent orthologous genes. The tree reflects phylogenetic relationships between *Streptomyces* sister-taxa genomes and was constructed from multiple whole genome alignments (see Figure 1). Scale bar represents nucleotide substitutions per site. Panel colors correspond to SMGC class (see legend). Panels are labeled with the SMGC cluster membership (see Table S3) defined using our annotation-independent approach (see Materials and Methods).

We also see evidence that NDR has undergone the deletion of SMGC-associated genes inherited from the common ancestor of NRD and SDR. For example, the SRO15-2005 lassopeptide gene cluster is conserved in SDR and found in INT but absent from NDR, suggesting that deletion of this lassopeptide accompanied NDR divergence (Figures 6 and 7). We also find that core SMGC gene loss is more common in NDR than SDR (strain-level deletions occur in six out of nine core gene clusters within NDR and two out of nine core gene clusters in SDR) (Figure 6). Similarly, we can observe SDR species-specific core gene clusters (AmfS, coelichelin, a T1PKS, and a terpene) that are found in only 70% (i.e., near core) of NDR strains (Figure 6). This pattern suggests that these SMGCs were present in the common ancestor of the two clades and subsequently deleted from NDR isolates. In addition, the butyrolactone operon (cluster 3) is comprised of more genes in SDR than in NDR, and this likely indicates active gene loss within this pathway for NDR strains (Figure 7).

Taken together, these results suggest that the sister-clades are under different evolutionary pressures which drive dissimilarity in SMGC composition. NDR genomes have increased in size relative to their ancestors suggesting an overall increase in the rate of gene acquisition via horizontal gene exchange, and this increase in gene acquisition has resulted in an increase in strain-specific SMGC content in NDR. In addition, the presence of NDR-specific core SMGCs (e.g., melanin gene cluster) indicates that some horizontally acquired SMGC have gone to fixation within NDR. At the same time, deletion events in NDR have pruned away SMGCs inherited from ancestral lineages (i.e., those clusters present in SDR and INT). We hypothesize that these changes in SMGC content are likely to have effects on fitness which should act to reinforce lineage divergence either as a result of antagonism or niche differentiation.

3. Materials and Methods

3.1. *Streptomyces* Isolation and DNA Extraction

We built a culture collection of >1000 *Streptomyces* isolated from grassland soils (pH 3.9–7.3) sampled at 0–5 cm from sites across the United States [27]. Pure *Streptomyces* cultures were obtained from air-dried soils on glycerol-arginine agar (pH 8.7) containing antifungals as previously described [58]. Genomic DNA was extracted using a standard phenol/chloroform/isoamyl alcohol protocol from liquid cultures grown in yeast extract-malt extract medium (YEME) with 0.5% glycine [5] for 72 h shaking at 30 °C.

3.2. Whole Genome Sequencing, Assembly, and Annotation

Streptomyces genomic sequencing libraries were prepped with the Nextera DNA Library Preparation Kit (Illumina, San Diego, CA, USA), and draft genomes were generated using the Illumina HiSeq2500 platform (Illumina, San Diego, CA, USA) and paired-end 2 × 100 bp reads at the Cornell University Biotechnology Resource Center (BRC). Quality control and assembly was performed with the A5 pipeline [59], and genomes were annotated using the online RAST Server [60]. Multiple whole genome alignments were obtained with Mugsy [61], and trimAL v1.2 removed poorly aligned regions [62]. Orthologous genes were identified using ITEP [63] with MCL clustering parameters as follows: inflation value = 2.0, cutoff = 0.04, maxbit score. Average nucleotide identity (ANI) was determined using mother [64]. Genome sequences are available at NCBI under BioProject ID PRJNA401484 accession numbers SAMN07606143–SAMN07606166.

3.3. Phylogenetic Reconstruction

The phylogenetic relationship between genomes was reconstructed from DNA sequences of multiple whole genome alignments using maximum likelihood (ML) with the generalized time reversible nucleotide substitution model [65] with gamma distributed rate heterogeneity among sites (GTRGAMMA) in RAxML v7.3.0 [66]. Bootstrap support was determined using the RAxML rapid bootstrapping algorithm [67].

3.4. Secondary Metabolite Biosynthetic Gene Cluster (SMGC) Identification

Secondary metabolite biosynthetic gene clusters (SMGC) were predicted and annotated using the online server antiSMASH 3.0 [32]. We also used an annotation-independent approach to identify SMGCs shared between genomes. For each SMGC identified by antiSMASH, we used Prodigal [68] to call open reading frames (ORFs) and Parasail with default parameters to identify orthologous genes and orthologous gene groups [69]. We used the R package igraph [70] to cluster similar SMGCs, define cluster membership, and thus determine which SMGCs are shared between genomes. Cluster membership was determined based on gene content using a binary (i.e., Jaccard) dissimilarity distance of ≤4.0 generated from an orthologous group presence/absence table. Dissimilarity distances of >4.0 did not result in an appreciable gain in the number of total clusters. The SMGC network was visualized and analyzed with Cytoscape 3.3.0 [71].

4. Conclusions

We used comparative genomics to examine SMGC diversity within strains of two closely related *Streptomyces* species that recently diverged from a common ancestor. Our objective was to observe and explore the evolutionary dynamics of SMGCs that accompany evolutionary diversification and to assess SMGC conservation within and between closely related species. It is clear that gene gain and loss events drive major differences in SMGC composition, both within and between species. While both species share conserved core SMGCs, each clade has its own species-specific SMGC core, and the majority of SMGCs were strain-specific. This pattern indicates that these SMGCs, not present in shared ancestors, were acquired recently due to horizontal gene exchange.

In addition, we observe that SMGCs that have been inherited from a shared ancestor can vary considerably in gene content, both due to the acquisition and deletion of individual genes within each gene cluster. We observe SMGC gain and loss dynamics that differ between clades and identify SMGC acquisition and deletion events that correspond to ancestral diversification events. These findings show that SMGC modification is associated with lineage divergence, though whether these changes cause or reinforce divergence directly or are an indirect product of evolutionary divergence remains to be seen. A limitation of the comparative genomics approach is that we cannot assess the ecological activity of a pathway from genome sequence data. It is possible that some (or all) of the strain-specific pathways, if acquired by recent horizontal exchange, may be non-functional. It is also possible that changes in SMGC architecture and gene content could alter pathway functionality and that pathways deemed orthologous on the basis of genetic similarity may have different functions in different strains.

Finally, we can conclude that, while strains within a species will share a core set of SMGCs, the number of accessory SMGC within a given species can be quite large, with each strain having its own repertoire of strain-specific SMGCs. Furthermore, the majority of these strain-specific SMGCs remain uncharacterized and lack similarity to SMGCs documented in public databases.

Supplementary Materials: The following are available online at http://www.mdpi.com/2079-6382/7/1/12/s1, Figure S1: Each clade has a distinct SMGC network. The network illustrates inter- and intra-clade sharing of SMGC content. Large circles represent the genomes of Streptomyces strains and are labeled with isolate names and colored according to clade affiliation. Smaller circles represent non-redundant distinct SMGCs identified using our annotation-independent approach (see Materials and Methods). Lines connect each SMGC to the strains in which they are found. Network nodes and edges are scaled in proportion to the number of connections and colored according to gene cluster class (see legend). Network is arranged in the organic layout using Cytoscape 3.3.0 [71]. Core SMGCs can be observed as larger central nodes while strain specific and low frequency SMGCs occur around the edges of the graph; Table S1: The 24 Streptomyces genomes were isolated from 11 sites. Isolate names begin with the site code from which they were isolated from followed by strain number; Table S2: Genome and assembly characteristics for 24 Streptomyces genomes. The clade affiliations include the northern-derived (NDR), southern-derive (SDR), and intermediate (INT). Sample site of each isolate can be found in Table S1. Values report assembled draft genome size, genome-wide G+C content, the number of predicted open reading frames (ORFs), and the number of predicted secondary metabolite biosynthetic gene clusters (SMGCs) per genome; Table S3: SMGCs are predicted by antiSMASH [32] in our 24 Streptomyces genomes. For each SMGC, columns report the affiliated genome, clade, gene cluster class (hybrids are indicated by hyphens), gene cluster length (bp), natural product annotation provided by antiSMASH, cluster membership (Clust Memb), MIBiG database identification, the portion of genes with similarity to genes within the most similar known cluster from the MIBiG database (% Genes w/Similarity). Cluster membership was determined using our annotation-independent approach (see Materials and Methods). NA indicates information is not available.

Acknowledgments: This material is based upon work supported by the National Science Foundation under Grant No. DEB-1456821 awarded to Daniel H. Buckley.

Author Contributions: Mallory J. Choudoir and Daniel H. Buckley conceived and designed the study; Mallory J. Choudoir performed the analyses and analyzed the data; Charles Pepe-Ranney contributed to the analyses; Mallory J. Choudoir and Daniel H. Buckley wrote the paper.

Conflicts of Interest: The authors declare no conflict of interest. The founding sponsors had no role in the design of the study; in the collection, analyses, or interpretation of data; in the writing of the manuscript, and in the decision to publish the results.

References

1. Kossel, A. Ueber die chemische zusammensetzung der zelle. *Arch. Physiol.* **1891**, *4*, 181–186.
2. Doroghazi, J.R.; Metcalf, W.W. Comparative genomics of actinomycetes with a focus on natural product biosynthetic genes. *BMC Genom.* **2013**, *14*, 611. [CrossRef] [PubMed]
3. Karlovsky, P. *Secondary Metabolites in Soil Ecology*; Springer: New York, NY, USA, 2008; pp. 1–19.
4. Davies, J. Where have all the antibiotics gone? *Can. J. Infect. Dis. Med. Microbiol.* **2006**, *17*, 287–290. [CrossRef] [PubMed]

5. Kieser, T.; Bibb, M.J.; Buttner, M.J.; Charter, K.F.; Hopwood, D.A. *Practical Streptomyces Genetics*; John Innes Foundation: Norwich, UK, 2000.
6. Bérdy, J. Bioactive microbial metabolites. *J. Antibiot.* **2005**, *58*, 1–26. [CrossRef] [PubMed]
7. Watve, M.; Tickoo, R.; Jog, M.; Bhole, B. How many antibiotics are produced by the genus *Streptomyces*? *Arch. Microbiol.* **2001**, *176*, 386–390. [CrossRef] [PubMed]
8. O'Brien, J.; Wright, G.D. An ecological perspective of microbial secondary metabolism. *Curr. Opin. Biotechnol.* **2011**, *22*, 552–558. [CrossRef] [PubMed]
9. Vaz Jauri, P.; Bakker, M.G.; Salomon, C.E.; Kinkel, L.L. Subinhibitory antibiotic concentrations mediate nutrient use and competition among soil *Streptomyces*. *PLoS ONE* **2013**, *8*, e81064. [CrossRef] [PubMed]
10. Traxler, M.F.; Kolter, R. Natural products in soil microbe interactions and evolution. *Nat. Prod. Rep.* **2015**, *32*, 956–970. [CrossRef] [PubMed]
11. Hopwood, D.A. Soil to genomics: The *Streptomyces* chromosome. *Annu. Rev. Genet.* **2006**, *40*, 1–23. [CrossRef] [PubMed]
12. Kämpfer, P. The family Streptomycetaceae Part I: Taxonomy. In *The Prokaryotes*; Springer: New York, NY, USA, 2006; pp. 538–604.
13. Van der Meij, A.; Worsley, S.F.; Hutchings, M.I.; van Wezel, G.P. Chemical ecology of antibiotic production by actinomycetes. *FEMS Microbiol. Rev.* **2017**, *41*, 392–416. [CrossRef] [PubMed]
14. Bentley, S.D.; Chater, K.F.; Cerdeño-Tárraga, A.M.; Challis, G.L.; Thomson, N.R.; James, K.D.; Harris, D.E.; Quail, M.A.; Kieser, H.; Harper, D.; et al. Complete genome sequence of the model actinomycete *Streptomyces coelicolor* A3(2). *Nature* **2002**, *417*, 141–147. [CrossRef] [PubMed]
15. Blin, K.; Medema, M.H.; Kazempour, D.; Fischbach, M.A.; Breitling, R.; Takano, E.; Weber, T. AntiSMASH 2.0—A versatile platform for genome mining of secondary metabolite producers. *Nucleic Acids Res.* **2013**, *41*, W204–W212. [CrossRef] [PubMed]
16. Jensen, P.R.; Williams, P.G.; Oh, D.-C.; Zeigler, L.; Fenical, W. Species-specific secondary metabolite production in marine actinomycetes of the genus *Salinispora*. *Appl. Environ. Microbiol.* **2007**, *73*, 1146–1152. [CrossRef] [PubMed]
17. Doroghazi, J.R.; Albright, J.C.; Goering, A.W.; Ju, K.S.; Haines, R.R.; Tchalukov, K.A.; Labeda, D.P.; Kelleher, N.L.; Metcalf, W.W. A roadmap for natural product discovery based on large-scale genomics and metabolomics. *Nat. Chem. Biol.* **2014**, *10*, 963–968. [CrossRef] [PubMed]
18. Ziemert, N.; Lechner, A.; Wietz, M.; Millán-Aguiñaga, N.; Chavarria, K.L.; Jensen, P.R. Diversity and evolution of secondary metabolism in the marine actinomycete genus *Salinispora*. *Proc. Natl. Acad. Sci. USA* **2014**, *111*, E1130–E1139. [CrossRef] [PubMed]
19. Wawrik, B.; Kerkhof, L.; Zylstra, G.J.; Kukor, J.J. Identification of unique type II polyketide synthase genes in soil. *Appl. Environ. Microbiol.* **2005**, *71*, 2232–2238. [CrossRef] [PubMed]
20. Charlop-Powers, Z.; Owen, J.G.; Reddy, B.V.B.; Ternei, M.A.; Brady, S.F. Chemical-biogeographic survey of secondary metabolism in soil. *Proc. Natl. Acad. Sci. USA* **2014**, *111*, 3757–3762. [CrossRef] [PubMed]
21. Davelos, A.L.; Kinkel, L.L.; Samac, D.A. Spatial variation in frequency and intensity of antibiotic interactions among *Streptomycetes* from prairie soil. *Appl. Environ. Microbiol.* **2004**, *70*, 1051–1058. [CrossRef] [PubMed]
22. Davelos Baines, A.L.; Xiao, K.; Kinkel, L.L. Lack of correspondence between genetic and phenotypic groups amongst soil-borne *Streptomycetes*. *FEMS Microbiol. Ecol.* **2007**, *59*, 564–575. [CrossRef] [PubMed]
23. Charlop-Powers, Z.; Owen, J.G.; Reddy, B.V.B.; Ternei, M.A.; Guimarães, D.O.; de Frias, U.A.; Pupo, M.T.; Seepe, P.; Feng, Z.; Brady, S.F. Global biogeographic sampling of bacterial secondary metabolism. *Elife* **2015**, *4*, e05048. [CrossRef] [PubMed]
24. Wawrik, B.; Kutliev, D.; Abdivasievna, U.A.; Kukor, J.J.; Zylstra, G.J.; Kerkhof, L. Biogeography of actinomycete communities and type II polyketide synthase genes in soils collected in New Jersey and Central Asia. *Appl. Environ. Microbiol.* **2007**, *73*, 2982–2989. [CrossRef] [PubMed]
25. Reddy, B.V.B.; Kallifidas, D.; Kim, J.H.; Charlop-Powers, Z.; Feng, Z.; Brady, S.F. Natural product biosynthetic gene diversity in geographically distinct soil microbiomes. *Appl. Environ. Microbiol.* **2012**, *78*, 3744–3752. [CrossRef] [PubMed]
26. Choudoir, M.J.; Campbell, A.N.; Buckley, D.H. Grappling with Proteus: Population level approaches to understanding microbial diversity. *Front. Microbiol.* **2012**, *3*. [CrossRef] [PubMed]

27. Andam, C.P.; Doroghazi, J.R.; Campbell, A.N.; Kelly, P.J.; Choudoir, M.J.; Buckley, D.H. A Latitudinal diversity gradient in terrestrial bacteria of the genus *Streptomyces*. *MBio* **2016**, *7*. [CrossRef] [PubMed]
28. Choudoir, M.J.; Doroghazi, J.R.; Buckley, D.H. Latitude delineates patterns of biogeography in terrestrial *Streptomyces*. *Environ. Microbiol.* **2016**, *18*, 4931–4945. [CrossRef] [PubMed]
29. Choudoir, M.J.; Buckley, D.H. Phylogenetic conservatism of thermal traits explains dispersal limitation and genomic differentiation of *Streptomyces* sister-taxa. *ISME J.* **2018**, under review.
30. Rong, X.; Huang, Y. Taxonomic evaluation of the *Streptomyces griseus* clade using multilocus sequence analysis and DNA-DNA hybridization, with proposal to combine 29 species and three subspecies as 11 genomic species. *Int. J. Syst. Evol. Microbiol.* **2010**, *60*, 696–703. [CrossRef] [PubMed]
31. Konstantinidis, K.T.; Ramette, A.; Tiedje, J.M. The bacterial species definition in the genomic era. *Philos. Trans. R. Soc. Lond. B Biol. Sci.* **2006**, *361*, 1929–1940. [CrossRef] [PubMed]
32. Weber, T.; Blin, K.; Duddela, S.; Krug, D.; Kim, H.U.; Bruccoleri, R.; Lee, S.Y.; Fischbach, M.A.; Müller, R.; Wohlleben, W.; et al. AntiSMASH 3.0—A comprehensive resource for the genome mining of biosynthetic gene clusters. *Nucleic Acids Res.* **2015**, *43*, W237–W243. [CrossRef] [PubMed]
33. Ōmura, S.; Ikeda, H.; Ishikawa, J.; Hanamoto, A.; Takahashi, C.; Shinose, M.; Takahashi, Y.; Horikawa, H.; Nakazawa, H.; Osonoe, T.; et al. Genome sequence of an industrial microorganism *Streptomyces avermitilis*: Deducing the ability of producing secondary metabolites. *Proc. Natl. Acad. Sci. USA* **2001**, *98*, 12215–12220. [CrossRef] [PubMed]
34. Ohnishi, Y.; Ishikawa, J.; Hara, H.; Suzuki, H.; Ikenoya, M.; Ikeda, H.; Yamashita, A.; Hattori, M.; Horinouchi, S. Genome sequence of the streptomycin-producing microorganism *Streptomyces griseus* IFO 13350. *J. Bacteriol.* **2008**, *190*, 4050–4060. [CrossRef] [PubMed]
35. Aigle, B.; Lautru, S.; Spiteller, D.; Dickschat, J.S.; Challis, G.L.; Leblond, P.; Pernodet, J.L. Genome mining of *Streptomyces ambofaciens*. *J. Ind. Microbiol. Biotechnol.* **2014**, *41*, 251–263. [CrossRef] [PubMed]
36. Seipke, R.F. Strain-level diversity of secondary metabolism in *Streptomyces albus*. *PLoS ONE* **2015**, *10*, e0116457. [CrossRef] [PubMed]
37. Du, L.; Sánchez, C.; Shen, B. Hybrid peptide–polyketide natural products: Biosynthesis and prospects toward engineering novel molecules. *Metab. Eng.* **2001**, *3*, 78–95. [CrossRef] [PubMed]
38. Medema, M.H.; Blin, K.; Cimermancic, P.; de Jager, V.; Zakrzewski, P.; Fischbach, M.A.; Weber, T.; Takano, E.; Breitling, R. AntiSMASH: Rapid identification, annotation and analysis of secondary metabolite biosynthesis gene clusters in bacterial and fungal genome sequences. *Nucleic Acids Res.* **2011**, *39*, W339–W346. [CrossRef] [PubMed]
39. Olano, C.; García, I.; González, A.; Rodriguez, M.; Rozas, D.; Rubio, J.; Sánchez-Hidalgo, M.; Braña, A.F.; Méndez, C.; Salas, J.A. Activation and identification of five clusters for secondary metabolites in *Streptomyces albus* J1074. *Microb. Biotechnol.* **2014**, *7*, 242–256. [CrossRef] [PubMed]
40. Claesen, J.; Bibb, M.J. Biosynthesis and regulation of grisemycin, a new member of the linaridin family of ribosomally synthesized peptides produced by *Streptomyces griseus* IFO 13350. *J. Bacteriol.* **2011**, *193*, 2510–2516. [CrossRef] [PubMed]
41. Zetterström, R. Selman, A. Waksman (1888–1973) Nobel Prize in 1952 for the discovery of streptomycin, the first antibiotic effective against tuberculosis. *Acta Paediatr.* **2007**, *96*, 317–319. [CrossRef] [PubMed]
42. Waksman, S.A.; Reilly, H.C.; Johnstone, D.B. Isolation of streptomycin-producing strains of *Streptomyces griseus*. *J. Bacteriol.* **1946**, *52*, 393–397. [PubMed]
43. Hotta, K.; Ishikawa, J. Strain- and species-specific distribution of the streptomycin gene cluster and kan-related sequences in *Streptomyces griseus*. *J. Antibiot.* **1988**, *41*, 1116–1123. [CrossRef] [PubMed]
44. Tettelin, H.; Riley, D.; Cattuto, C.; Medini, D. Comparative genomics: The bacterial pan-genome. *Curr. Opin. Microbiol.* **2008**, *11*, 472–477. [CrossRef] [PubMed]
45. Lefébure, T.; Bitar, P.D.P.; Suzuki, H.; Stanhope, M.J. Evolutionary dynamics of complete *Campylobacter* pan-genomes and the bacterial species concept. *Genome Biol. Evol.* **2010**, *2*, 646–655. [CrossRef] [PubMed]
46. Imbert, M.; Béchet, M.; Blondeau, R. Comparison of the main siderophores produced by some species of *Streptomyces*. *Curr. Microbiol.* **1995**, *31*, 129–133. [CrossRef]
47. Roberts, A.A.; Schultz, A.W.; Kersten, R.D.; Dorrestein, P.C.; Moore, B.S. Iron acquisition in the marine actinomycete genus *Salinispora* is controlled by the desferrioxamine family of siderophores. *FEMS Microbiol. Lett.* **2012**, *335*, 95–103. [CrossRef] [PubMed]

48. Antony-Babu, S.; Stien, D.; Eparvier, V.; Parrot, D.; Tomasi, S.; Suzuki, M.T. Multiple *Streptomyces* species with distinct secondary metabolomes have identical 16S rRNA gene sequences. *Sci. Rep.* **2017**, *7*, 11089. [CrossRef] [PubMed]
49. Gogarten, J.P.; Townsend, J.P. Horizontal gene transfer, genome innovation and evolution. *Nat. Rev. Microbiol.* **2005**, *3*, 679–687. [CrossRef] [PubMed]
50. Marri, P.R.; Hao, W.; Golding, G.B. Gene gain and gene loss in *Streptococcus*: Is it driven by habitat? *Mol. Biol. Evol.* **2006**, *23*, 2379–2391. [CrossRef] [PubMed]
51. Reno, M.L.; Held, N.L.; Fields, C.J.; Burke, P.V.; Whitaker, R.J. Biogeography of the *Sulfolobus islandicus* pan-genome. *Proc. Natl. Acad. Sci. USA* **2009**, *106*, 8605–8610. [CrossRef] [PubMed]
52. Mira, A.; Ochman, H.; Moran, N.A. Deletional bias and the evolution of bacterial genomes. *Trends Genet.* **2001**, *17*, 589–596. [CrossRef]
53. Ochman, H.; Lawrence, J.G.; Groisman, E.A. Lateral gene transfer and the nature of bacterial innovation. *Nature* **2000**, *405*, 299–304. [CrossRef] [PubMed]
54. Homma, K.; Fukuchi, S.; Nakamura, Y.; Gojobori, T.; Nishikawa, K. Gene cluster analysis method identifies horizontally transferred genes with high reliability and indicates that they provide the main mechanism of operon gain in 8 species of γ-Proteobacteria. *Mol. Biol. Evol.* **2006**, *24*, 805–813. [CrossRef] [PubMed]
55. Penn, K.; Jenkins, C.; Nett, M.; Udwary, D.W.; Gontang, E.A.; McGlinchey, R.P.; Foster, B.; Lapidus, A.; Podell, S.; Allen, E.E.; et al. Genomic islands link secondary metabolism to functional adaptation in marine Actinobacteria. *ISME J.* **2009**, *3*, 1193–1203. [CrossRef] [PubMed]
56. Bolotin, E.; Hershberg, R. Gene loss dominates as a source of genetic variation within clonal pathogenic bacterial species. *Genome Biol. Evol.* **2015**, *7*, 2173–2187. [CrossRef] [PubMed]
57. Choudoir, M.J.; Panke-Buisse, K.; Andam, C.P.; Buckley, D.H. Genome surfing as driver of microbial genomic diversity. *Trends Microbiol.* **2017**, *25*, 624–636. [CrossRef] [PubMed]
58. Doroghazi, J.R.; Buckley, D.H. Widespread homologous recombination within and between *Streptomyces* species. *ISME J.* **2010**, *4*, 1136–1143. [CrossRef] [PubMed]
59. Tritt, A.; Eisen, J.A.; Facciotti, M.T.; Darling, A.E. An integrated pipeline for de novo assembly of microbial genomes. *PLoS ONE* **2012**, *7*, e42304. [CrossRef] [PubMed]
60. Aziz, R.K.; Bartels, D.; Best, A.A.; DeJongh, M.; Disz, T.; Edwards, R.A.; Formsma, K.; Gerdes, S.; Glass, E.M.; Kubal, M.; et al. The RAST server: Rapid annotations using subsystems technology. *BMC Genom.* **2008**, *9*, 75. [CrossRef] [PubMed]
61. Angiuoli, S.V.; Salzberg, S.L. Mugsy: Fast multiple alignment of closely related whole genomes. *Bioinformatics* **2011**, *27*, 334–342. [CrossRef] [PubMed]
62. Capella-Gutierrez, S.; Silla-Martinez, J.M.; Gabaldon, T. TrimAl: A tool for automated alignment trimming in large-scale phylogenetic analyses. *Bioinformatics* **2009**, *25*, 1972–1973. [CrossRef] [PubMed]
63. Benedict, M.N.; Henriksen, J.R.; Metcalf, W.W.; Whitaker, R.J.; Price, N.D. ITEP: An integrated toolkit for exploration of microbial pan-genomes. *BMC Genom.* **2014**, *15*, 8. [CrossRef] [PubMed]
64. Schloss, P.D.; Westcott, S.L.; Ryabin, T.; Hall, J.R.; Hartmann, M.; Hollister, E.B.; Lesniewski, R.A.; Oakley, B.B.; Parks, D.H.; Robinson, C.J.; et al. Introducing mothur: Open-source, platform-independent, community-supported software for describing and comparing microbial communities. *Appl. Environ. Microbiol.* **2009**, *75*, 7537–7541. [CrossRef] [PubMed]
65. Tavaré, S. Some probabilistic and statistical problems in the analysis of DNA sequences. In *Lectures on Mathematics in the Life Sciences*; American Mathematical Society: Providence, RI, USA, 1986; Volume 17, pp. 57–86.
66. Stamatakis, A. RAxML-VI-HPC: Maximum likelihood-based phylogenetic analyses with thousands of taxa and mixed models. *Bioinformatics* **2006**, *22*, 2688–2690. [CrossRef] [PubMed]
67. Stamatakis, A.; Hoover, P.; Rougemont, J.; Renner, S. A rapid bootstrap algorithm for the RAxML Web Servers. *Syst. Biol.* **2008**, *57*, 758–771. [CrossRef] [PubMed]
68. Hyatt, D.; LoCascio, P.F.; Hauser, L.J.; Uberbacher, E.C. Gene and translation initiation site prediction in metagenomic sequences. *Bioinformatics* **2012**, *28*, 2223–2230. [CrossRef] [PubMed]
69. Daily, J. Parasail: SIMD C library for global, semi-global, and local pairwise sequence alignments. *BMC Bioinform.* **2016**, *17*, 81. [CrossRef] [PubMed]

70. Csardi, G.; Nepusz, T. The Igraph Software Package for Complex Network Research. 2006. Available online: http://www.interjournal.org/manuscript_abstract.php?361100992 (accessed on 6 February 2018).
71. Shannon, P.; Markiel, A.; Ozier, O.; Baliga, N.S.; Wang, J.T.; Ramage, D.; Amin, N.; Schwikowski, B.; Ideker, T. Cytoscape: A software environment for integrated models of biomolecular interaction networks. *Genome Res.* **2003**, *13*, 2498–2504. [CrossRef] [PubMed]

antibiotics

MDPI

Review

Mining Actinomycetes for Novel Antibiotics in the Omics Era: Are We Ready to Exploit This New Paradigm?

Olga Genilloud

Fundación MEDINA, Avda Conocimiento 34, 18016 Granada, Spain; olga.genilloud@medinaandalucia.es

Received: 6 August 2018; Accepted: 21 September 2018; Published: 25 September 2018

check for updates

Abstract: The current spread of multi-drug resistance in a number of key pathogens and the lack of therapeutic solutions in development to address most of the emerging infections in the clinic that are difficult to treat have become major concerns. Microbial natural products represent one of the most important sources for the discovery of potential new antibiotics and actinomycetes have been one of the most relevant groups that are prolific producers of these bioactive compounds. Advances in genome sequencing and bioinformatic tools have collected a wealth of knowledge on the biosynthesis of these molecules. This has revealed the broad untapped biosynthetic diversity of actinomycetes, with large genomes and the capacity to produce more molecules than previously estimated, opening new opportunities to identify the novel classes of compounds that are awaiting to be discovered. Comparative genomics, metabolomics and proteomics and the development of new analysis and genetic engineering tools provide access to the integration of new knowledge and better understanding of the physiology of actinomycetes and their tight regulation of the production of natural products antibiotics. This new paradigm is fostering the development of new genomic-driven and culture-based strategies, which aims to deliver new chemical classes of antibiotics to be developed to the clinic and replenish the exhausted pipeline of drugs for fighting the progression of infection diseases in the near future.

Keywords: actinomycetes; antibiotics; secondary metabolism; culture-based approaches; omics

1. Introduction

Actinomycetes are accepted as one of the most relevant bacterial groups as prolific producers of secondary metabolites (SM). For decades, they were intensively exploited by industrial discovery programs that deliveredmost of the natural products, which were subsequently used as scaffolds to derive a large number of the antibiotics that are currently in the clinic [1,2]. Despite this past success, the current spread of multi-drug resistance in a number of key pathogens (*Enterococcus faecium*, *Staphylococcus aureus*, *Klebsiella pneumoniae*, *Acinetobacter baumannii*, *Pseudomonas aeruginosa* and *Enterobacter* spp.) and the lack of therapeutic solutions in development to address most of the emerging infections in the clinic that are difficult to treat have become major health concerns and require a new and sustained research effort to respond to the need for new antibiotics [3–5]. The progressive abandonment of the field by large pharmaceutical companies, moving away from the high costs of development and poor incentives as part of a broken economic model, has left academic groups and small biotech companies as the only players in the discovery field. These research teams are confronted with the major challenge of identifying novel antibiotic classes that they may not have the capacity to develop alone to reach the market [6,7]. The international initiatives launched in the last few years are supporting the development of new antibiotics and are revitalizing new preclinical and clinical developments. In contrast, basic research and early discovery efforts still remain poorly addressed by these programs and the existing research gaps are not enabling the establishment of a sustainable early discovery pipeline for future new compounds from the innovative approaches that have emerged in the field in last few decades [8,9]. This limited number of clinical development programs is in direct contrast to the dynamic and prolific activity that has developed in the field of microbial natural products during the last decade [10,11]. Research in natural products has evolved to represent one of the most important sources for the discovery of potential new antibiotics, compared to the lack of success of synthetic molecules. The poor suitability of synthetic libraries that were used to select molecules from target-based in vitro screens, or to develop rational drug designs based on ligand binding devoid of suitable physiochemical properties for penetrating bacterial membranes has been extensively discussed in previous reports [12–15]. Despite the rediscovery problem of known compounds, which has been represented as one of the major burdens for the continued investment in traditional NP discovery programs in the past, the advances in genome sequencing and bioinformatic tools have permitted us to collect a wealth of knowledge on the biosynthesis of natural products, which has revealed the broad untapped biosynthetic diversity of microorganisms, especially actinomycetes [16]. These talented filamenting bacteria have been shown to encode a previously unexpected diversity of biosynthetic gene clusters, opening new opportunities to identify novel classes of compounds that are awaiting to be discovered [16–18]. New advances in cultivation techniques from unexploited microbial niches have provided new insights into the diversity of actinobacteria and the possibility to access a broader chemical space of bioactive compounds. Comparative genomics, metabolomics and proteomics and the development of new analysis tools are generating a wealth of integrated information, which is fostering the emergence of new strategies that are aimed at obtaining a better understanding of the physiology of actinomycetes and their production of natural products. The new methods developed to mine genomes and predict chemical structures, to activate silent clusters and enable pathway engineering and to directly express metagenome gene clusters in heterologous hosts are providing the necessary tools to set up the basis for a new antibiotic discovery paradigm. Many recent excellent reviews cover each of the individual aspects of these new developments in the field and their impact on the identification of new antibiotics [19–22]. This short review is mostly focused on revisiting the most recent contributions in the field of actinomycetes in terms of their natural products, which have an impact on the discovery of novel scaffolds. These new approaches could bring new opportunities for the production of novel classes of compounds to address the current therapeutic needs in the multi-resistance era.

2. Exploiting Diversity of Cultured Actinomycetes

The continued interest in exploring the environment to untap novel sources of microbial diversity has resulted in extensive prospective studies on a broad variety of sources, which ranges from specific terrestrial extreme environments to unique microbial assemblages in plant–host associations and marine ecosystems. The distribution of some microbial species presents biogeographic patterns that are mostly determined by micro-environmental conditions and the most recent efforts have been focused at exploiting these still poorly explored habitats to discover new chemical diversity. The exploration of the extreme conditions of desert and arid habitats, such as the Atacama desert with high salinity and high levels of UV radiation, has permitted the isolation of new species of actinomycetes that are well adapted to surviving under these conditions [23,24]. These actinomycetes communities have shown to produce a wide diversity of novel compounds from different natural product classes, such as the new ansamycins chaxamycins or the β-diketone polyketides asenjonamides A–C. This only names a few compounds with antibacterial activities that have been described recently [25,26] (Table 1) (Figure 1). Cave microbiomes are another pristine eco-system that have frequently been studied in the search for novel producing strains and many reports have highlighted the isolation of novel species of actinomycetes that produce bioactive compounds [27]. The actinobacterial diversity in calcite moonmilk deposits are one of the most recent examples showing the potential of these new isolates to create products with antibacterial activities [28]. Alternatively, the use of diffusion chamber methods to grow previously uncultured soil bacteria has permitted to isolate new antibiotics, such as lassomycin [29].

The marine environment has been traditionally another source of new actinomycetes [42,43]. The broad diversity of marine ecosystems, ranging from mangroves, shallow waters, deep sea sediments and associated invertebrates, have continued to attract the interest of microbiologists. Their isolation programs have ensured a continued discovery of new strains that produce new compounds or new analogs with biological activity (Table 1 and Figure 1) [30–34,43–48]. Despite this prolific description of marine-derived strains and new antibiotic producers, this environment remains poorly studied in terms of microbial diversity and functional diversity. Marine sediments have been the focus of recent studies, which has shown that actinobacteria are in fact only a minor component of this microbial community. More interestingly, these findings have suggested that the production of many of the secondary metabolites have a deep impact on the microbial community composition [49]. There is no doubt that the parallel advances in the metagenomic assessment of microbial diversity have allowed us to explore the dynamics of the microbial populations of interest in the communities currently being studied. These technologies are opening new avenues to investigate the potential roles of these members and the effects of the antibacterial compounds on the microbiome composition. The results derived from these studies will guide the isolation and selection of the most promising members of these microbial communities and their screening for the production of novel bioactive compounds.

Table 1. New antibiotics and analogs described from Actinomycetes since 2013, following different mining approaches.

Antibiotic	Structural Class	Producing Species	Antibiotic Spectrum	Discovery Approach	Reference
Asenjonamides A–C	di-ketone polyketides	*S. asenjonii*	Gram positive/ negative	Extreme environment	[26]
Lassomycin	cyclic peptide	*Lentzea kentukyensis*	*M. tuberculosis*	Diffusion chambers	[29]
Anthracimycin	tricarboxilic	*Streptomyces* sp.	Gram positive	Marine source	[30]
Salinamide F	depsipeptide	*Streptomyces* sp.	Gram positive/ negative	Marine source	[31]
Kocurin	thiazolylpeptide	*Kocuria lacustris* /*Micrococcus* sp.	Gram positive	Marine source	[32]
Micromonohalimanes A and B	diterpenoids	*Micromonospora* sp.	Gram positive	Marine source	[33]
Phocoenamicins B and C	spirotetronates	*Micromonospora* sp.	Gram positive	Marine source	[34]
Argolaphos A/B	phosphonopeptide	*Streptomyces monomycini*	Gram positive/negative	Genome-driven	[35]
Thiolactomycin and analogs	thiotetronic acids	*Salinispora/S. afghaniensis*	Gram positive	Genome-driven	[36]
taromycin A	lipopetide	*Saccharomonospora* sp.	Gram positive	Genome-driven	[37]
Enterocin	polyketide	*Salinispora pacifica*	Gram positive	Genome-driven	[38]
Difluostatin A	angucycline	*Micromonospora rosaria*	Gram positive	Genome-driven	[39]
Alchivemycin A and B	heterocyclic	*S. endus* + *Tsukamurella pulmonis*	Gram positive	Co-cultivation	[40]
Ciromicins	polyketide	*Nocardiopsis* sp. + *Rhodococcus wratislaviensis*	Not determined	Co-cultivation	[41]

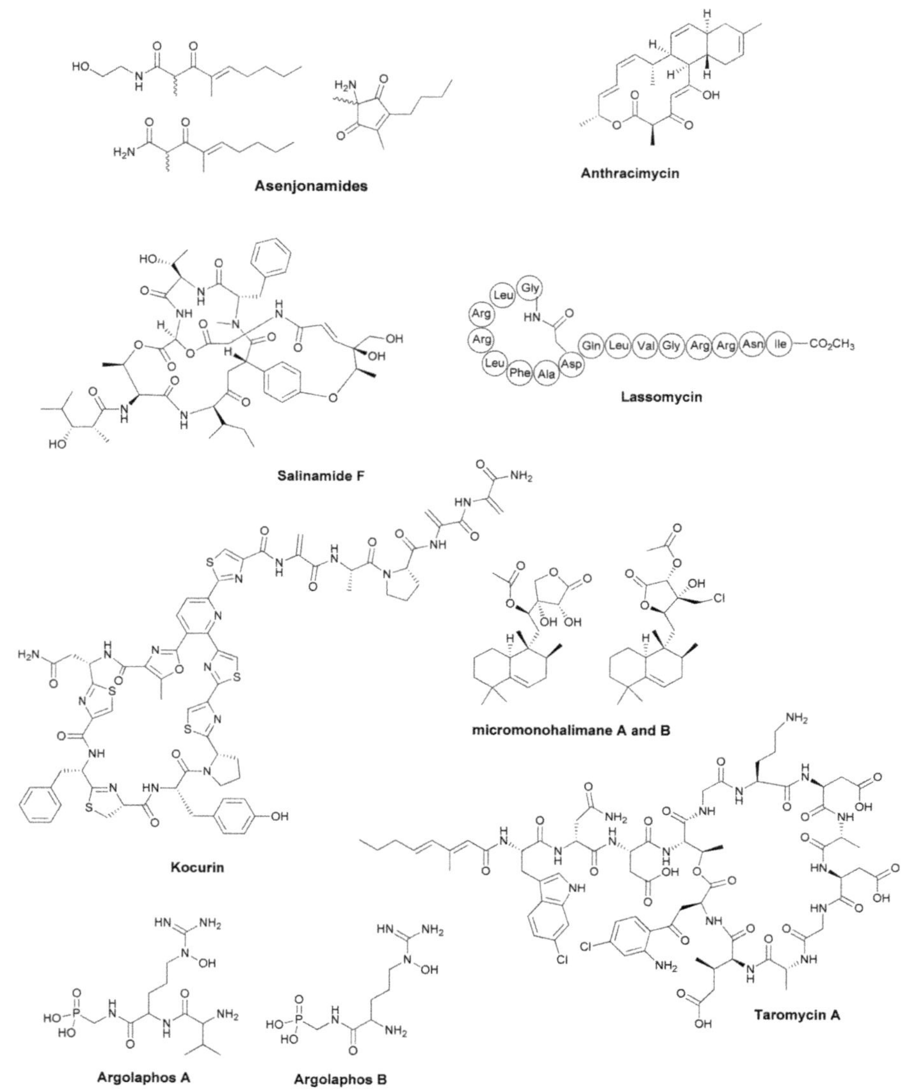

Figure 1. *Cont.*

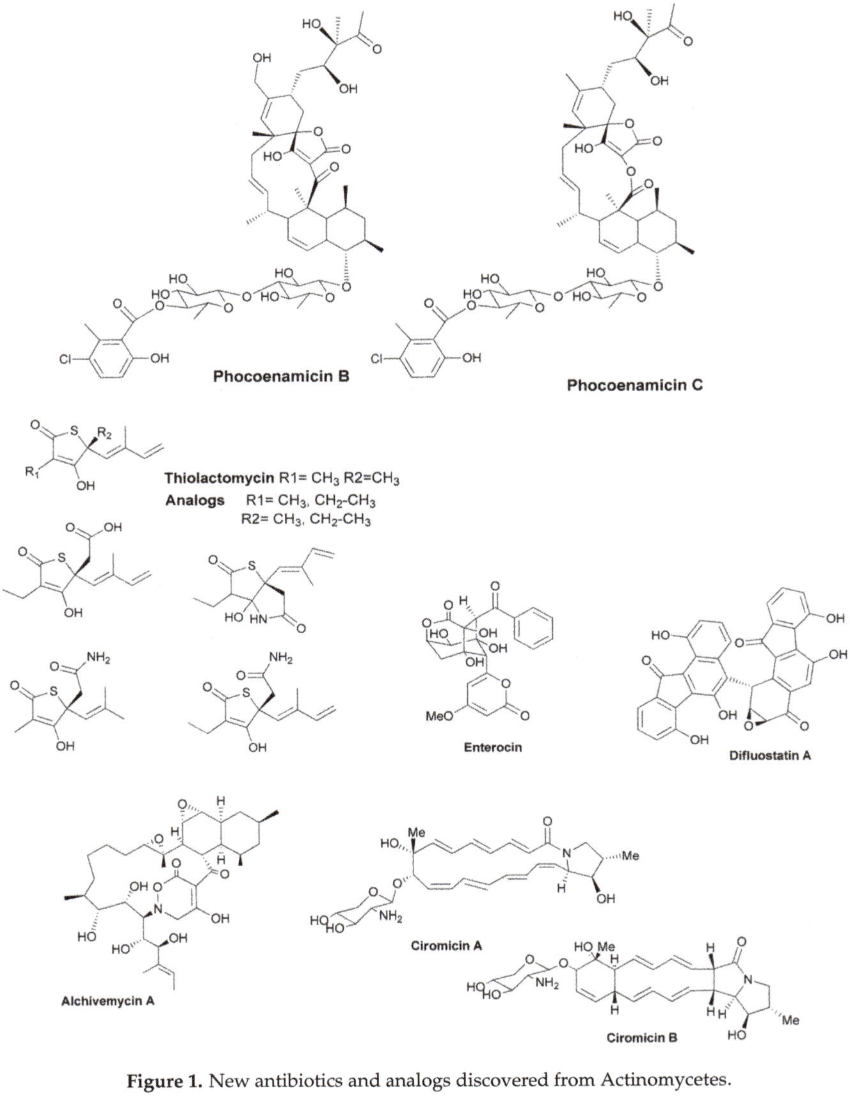

Figure 1. New antibiotics and analogs discovered from Actinomycetes.

3. Genomics Driven Discovery

In parallel, the exponential increase in the number of partial and complete genome sequencing projects on the actinomycetes species available in the public databases not only have confirmed their broad biosynthetic diversity across the different lineages, but also have enabled intensive genome mining approaches to untap new natural products scaffolds. The identification of new biosynthetic pathways of potentially interesting new molecules has fostered the search for new biosynthetic clusters (BGCs) on draft genomes, which has taken advantage of improved bioinformatic tools for cluster identification and gene annotation, such as AntiSMASH [50–52]. Furthermore, the increasing numbers of almost complete genomes has permitted the extensive comparative genomic analyses of members of this bacterial group, and the identification of the genomic components and the evolutionary history of many different species [53–55]. These comparative studies are revealing the existence of a core genome within some members of actinomycetes as well as a divergence of BGCs among the different lineages. These results are providing a basis for understanding the functional evolution of species as shown for *Streptomyces* [56,57]. Another relevant aspect of the impact of the increasing number of BGCs sequence information on antibiotic discovery is the possibility of developing specific targeted genome mining searches in genomic libraries based on specific genomic signatures related to the biosynthesis of privileged scaffolds or functionalizations that could drive the discovery of novel compounds and chemical spaces (Table 1 and Figure 1) [35,36]. The integration of genomics with transcriptomics, proteomics and metabolomics is providing unique information for assessing the functional evolution of actinomycetes species. These data are encouraging the development of new approaches for the expression and engineering of these biosynthetic pathways and the identification of new bioactive compounds from cultured actinomycetes not only from underexplored habitats but also from large microbial collections that are still the untapped treasures of biosynthetic diversity (Figure 2).

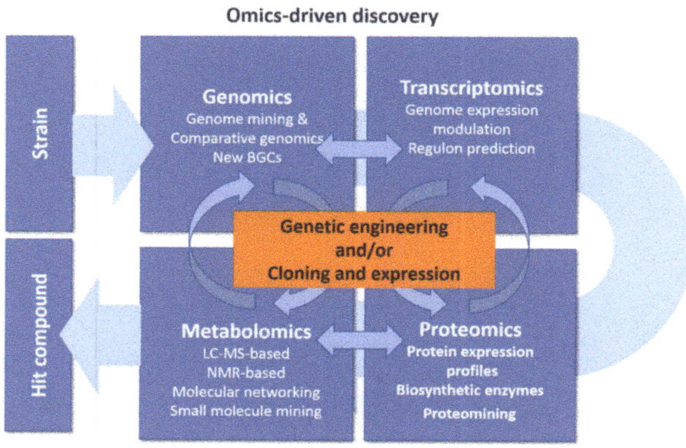

Figure 2. Omics-driven discovery.

One of the major challenges that still remains in this context is the efficient cloning and expression of BGCs that are originally silent or poorly expressed in their natural host after the use of refactoring by the replacement of the regulatory elements and further detection of the synthetized compounds [37,58–60]. Thus far, many different in vitro DNA assembly or direct capture methods have been described to clone BGCs in heterologous hosts [38,39,61–63]. Many BGCs cannot be detected by the rule-based bioinformatic tools due to the absence of signature genes, but the application of prediction tools based on the frequencies of Pfam domains occurring in BGCs have improved the identification of additional clusters [51]. New genomic bacterial artificial chromosome (BAC) libraries built from large 100-Kb fragments of *Streptomyces* spp. genomic DNA are also being used in the high throughput functional screening approaches to identify non-predicted BGCs by heterologous expression [64].

4. Eliciting Production from Silent Pathways

The activation of silent BGCs in the well-characterized producing strains by direct gene manipulation is an alternative approach when the strains are amenable to genetic engineering. Many recent successful examples have been reported, which have described the use of the newest genetic engineering tools in modifying metabolic pathways, altering metabolic fluxes that are blocking unrelated metabolic pathways, inactivating transcriptional repressors, over-expressing pathway-specific activator genes or even multiplying the BGC copies in the original producing strains to increase titers [63,65–67]. The recent development of optimal ribosomal binding sequences and strong terminators to be applied in the control of the metabolic pathways of actinomycetes has opened new possibilities for gene expression modulation and represent promising tools for metabolic engineering (Table 2) [68].

Table 2. Strategies to elicit antibiotic production in actinomycetes.

Eliciting Production Approaches	Methods and Targets	References
Genetic engineering:	Genome expression modulation: Transcriptional repressors inactivation; Transcriptional activators overexpression Optimized ribosomal binding sequences Strong terminators Increase BGCs copies Alter metabolic fluxes	[63,65–68]
Culture-based approaches:	Small molecule signaling: Sub-inhibitory small molecule elicitors Co-cultivation: cell–cell signaling	[69,70] [71]
Analytical mining:	Comparative metabolomics: LC-MS-based metabolomics NMR-based metabolomics Proteomining	[72–75] [75] [76,77]
Regulation of primary and secondary metabolism	Pathway specific regulatory elements: Two-component systems Sigma factors Pathways specific regulators	[78–80]
	Global regulatory metabolic networks Master regulators	[81]
	Comparative genomics: Identification of Transcription factor orthologs	[82]
	Regulon prediction: identification of regulatory networks Primary metabolism transcription factors Signaling cascades Pleiotropic regulators	[83,84] [85–87]
	Primary metabolism gene expansion	[88]

Despite this success, a large proportion of wild-type and industrial strains remain recalcitrant to being manipulated and many silent BGCs cannot be directly activated using genetic tools. Culture-based approaches based on multiple nutritional conditions have been one of the most common methods employed to explore the media components required for the production of new compounds. The concept of using small molecules as elicitors by perturbing biological systems and signaling pathways dates back several decades and has been extensively used to activate silent or poorly expressed pathways with a broad range of small molecules, including the sub-inhibitory concentrations of antibiotics [69,70]. The first large scale elicitor screening reported was performed with more than 30,000 compounds on *S. coelicolor* and it identified a small number of compounds that are able to stimulate the production of some of the secondary metabolites by several times [89]. Many examples of hormesis have been described with sub-inhibitory concentrations of antibiotics and other well-known bioactive natural products, which has elicited a response that is associated with the major activation of secondary metabolism, induction of cryptic gene clusters and production of novel compounds [70,90]. The activation effect cannot be predicted from the antibiotic mode of actions and the lack of universal effectors to awaken all silent BGCs has added another level of complexity in the identification of new effector molecules [70,71]. The same effect is pursued by co-cultivation, which is an approach that has been used extensively with many different types of cultivation formats and strain combinations. The approach takes advantage of the effect that small concentrations of signaling or antibiotic molecules from one of the strains can have on another strain. The difficulty to scale-up this approach as a general method is related to the impossibility of predicting which combinations will result in an effective response, which normally does not account for more than 15–20% of the cases studied [71]. Mycolic acids have been shown to play a role in the physical interaction and the activation of some silent pathways. New antibacterial compounds, such as alchivemycin A and B, arcyriaflavin E or ciromicins, were described after co-culturing different *Rhodoccoci* with the species of *Streptomyces*, *Tsukamurella* and *Nocardiopsis* (Table 1) (Figure 1) [40,41]. In other situations, the induction does not require cell-to-cell interaction and is only mediated by diffusible small effector molecules. One of the most recent examples is the production of the cryptic natural product keyicin from the co-cultivation of the producer *Micromonospora* strain with *Rhodococcus* [91] (Table 2).

Most recent studies on the differential metabolomic analysis of metabolites in response to cultivation conditions have shown that the chemical potential of actinomycetes is far from being fully characterized [92]. From a methodological perspective, the modern LC-MS and NMR analytical tools and differential metabolomic analysis have been the determinant for detecting the production of novel compounds in complex mixtures and mapping the response to external chemical signals. Comparative metabolomics have been efficiently used to identify the induced expression of secondary metabolites from *S. coelicolor* cryptic genes resulting from the exposure to multiplexed perturbations and to identify the subsets of primary and secondary metabolites that respond similarly across a large variety of stimuli [72]. The major challenge for these methods is related to the identification of novel bioactive molecules within the complexity of the metabolomic profiles. New dereplication and identification approaches are continuously being developed, which are based on the similarity of MS/MS patterns in natural product databases and NMR-based metabolomics [73–75]. Proteomining is another method developed to support this identification. The analysis links natural products to biosynthetic enzymes as it correlates protein expression profiles of biosynthetic enzymes to the metabolome of the producing strains based on the statistical analysis of strains cultured under different conditions [76,77].

5. Harnessing Regulation of Primary and Secondary Metabolisms

The production of secondary metabolites in actinomycetes is tightly regulated and responds to external stimuli from the environment. This regulation is the result of different classes of pathway-specific regulatory elements involving two-component systems, extra-cytoplasmic sigma factors or pathway specific regulators, such as the most recently described LmbU family [78–80]. Triggering the expression of BGCs frequently involves an additional transcriptional response through the master regulators involved in global regulatory metabolic networks that are not always pathway specific [81]. Understanding the right combinations of regulatory elements and transcription factors that regulate a BGC has been proposed as the "cracking the code" approach to be followed in identifying the key regulatory signals and the eliciting signals needed to set up culture conditions to activate a specific BGC [83]. The regulon predictor program PREDetector was developed to identify signaling cascades deduced from the in silico searches of regulatory elements, which revealed that primary metabolism transcription factors were also involved in controlling pathway-specific regulators [84]. One of the best examples are the pleiotropic regulators, DasR and CebR, that control the uptake of chitin and cellulose, for which responsive elements were identified upstream of many pathway-specific regulators [85–87]. The increasing number of available new genomes are enabling the new in silico approaches derived from comparative genomics to detect specific binding motifs of transcription factor orthologues [82]. Regulon prediction has been proposed as a successful strategy to identify the regulatory networks involved in the control of BGC expression and the most promising strains to be explored for the induction of silent pathways. This finely tuned participation of regulatory requirements for the specialized metabolite production also requires the precise provision of chemical precursors from primary metabolism. Recent reports highlight the role of primary metabolism gene expansions in secondary metabolite producing strains with impact on metabolic adaptation and strain fitness and how they may represent another target for future genetic engineering interventions to improve production and activate silent pathways [88] (Table 2).

6. Conclusions and Future Prospects

Microbial prospection studies have continued to reveal that there is a huge and still poorly explored diversity of actinomycetes in the environment that is waiting to be mined for the isolation of new bioactive compounds. In addition, thousands of selected strains that are preserved in microbial collections and distributed across laboratories worldwide also should be revisited as they represent a unique reservoir of silent biosynthetic diversity that traditional approaches did not manage to unlock in its full extension. The continued increase in new genome sequences and the development of new genome-mining and genome-directed engineering tools to trigger the production of new natural products are paralleling the development of integrated analytical tools and open access databases. The application of comparative genomics, metabolomics and proteomics to culture-based studies is providing a wealth of knowledge on the physiology and the regulation systems, opening new avenues to approach the activation of silent or poorly expressed pathways. There is no doubt that all these advances are setting the new foundations for a new paradigm in natural product discovery, especially from actinomycetes. A sustained and integrated multidisciplinary research effort should respond to the major challenge of discovering new chemical classes of antibiotics that will be required to replenish the preclinical development pipeline in the near future.

Funding: This research received no external funding.

Conflicts of Interest: The author declares no conflict of interest.

References

1. Berdy, J.J. Thoughts and facts about antibiotics: Where we are now and where we are heading. *J. Antibiot.* **2012**, *65*, 385–395. [CrossRef] [PubMed]
2. Landwehr, W.; Wolf, C.; Wink, J. Actinobacteria and myxobacteria-two of the most important bacterial resources for novel antibiotics. *Curr. Top. Microbiol. Immunol.* **2016**, *398*, 273–302. [PubMed]
3. Fischbach, M.A.; Walsh, C.T. Antibiotics for emerging pathogens. *Science* **2009**, *325*, 1089–1093. [CrossRef] [PubMed]
4. Silver, L.L. Challenges of antibacterial discovery. *Clin. Microbiol. Rev.* **2011**, *24*, 71–109. [CrossRef] [PubMed]
5. Walsh, C.T.; Wencewicz, T.A. Prospects for new antibiotics: A molecule-centered perspective. *J. Antibiot.* **2014**, *67*, 7–22. [CrossRef] [PubMed]
6. Rex, J.H.; Outterson, K. Antibiotic reimbursement in a sales-delinked model: Context and a benchmark-based global approach. *Lancet Infect. Dis.* **2016**, *16*, 500–505. [CrossRef]
7. Editorial. Wanted: A reward for antibiotic development. *Nat. Biotechnol.* **2018**, *36*, 555. [CrossRef] [PubMed]
8. Rex, J.H.; Goldberger, M.; Eisenstein, B.; Harney, C. The evolution of the regulatory framework for antibacterial agents. *Ann. N. Y. Acad. Sci.* **2014**, *1323*, 11–21. [CrossRef] [PubMed]
9. Simpkin, V.L.; Renwick, M.J.; Kelly, R.; Mossialos, E. Incentivising innovation in antibiotic drug discovery and development: Progress, challenges and next steps. *J. Antibiot.* **2017**, *70*, 1087–1096. [CrossRef] [PubMed]
10. Antibiotics Currently in Global Clinical Development. Available online: http://www.pewtrusts.org/-/media/assets/2018/03/antibiotics_clinical_dev_table_february2018.pdf (accessed on 1 February 2018).
11. Brown, E.D.; Wright, G.D. Antibacterial drug discovery in the resistance era. *Nature* **2016**, *529*, 336–343. [CrossRef] [PubMed]
12. Payne, D.J.; Gwynn, M.N.; Holmes, D.J.; Pompliano, D.L. Drugs for bad bugs: Confronting the challenges of antibacterial discovery. *Nat. Rev. Drug Discov.* **2007**, *6*, 29–40. [CrossRef] [PubMed]
13. Silver, L.L. Are natural products still the best source for antibacterial discovery? The bacterial entry factor. *Expert Opin. Drug Discov.* **2008**, *3*, 487–500. [CrossRef] [PubMed]
14. Wright, G.D. Something old, something new: Revisiting natural products in antibiotic drug discovery. *Can. J. Microbiol.* **2014**, *60*, 147–154. [CrossRef] [PubMed]
15. Lewis, K. New approaches to antimicrobial discovery. *Biochem. Pharmacol.* **2017**, *134*, 87–98. [CrossRef] [PubMed]

16. Baltz, R.H. Gifted microbes for genome mining and natural product discovery. *J. Ind. Microbiol. Biotechnol.* **2017**, *44*, 573–588. [CrossRef] [PubMed]
17. Nett, M.; Ikeda, H.; Moore, B.S. Genomic basis for natural product biosynthetic diversity in the actinomycetes. *Nat. Prod. Rep.* **2009**, *26*, 1362–1384. [CrossRef] [PubMed]
18. Rutledge, P.J.; Challis, G.L. Discovery of microbial natural products by activation of silent biosynthetic gene clusters. *Nat. Rev. Microbiol.* **2015**, *13*, 509–523. [CrossRef] [PubMed]
19. Genilloud, O. Actinomycetes: Still a source of novel antibiotics. *Nat. Prod. Rep.* **2017**, *34*, 1203–1232. [CrossRef] [PubMed]
20. Katz, L.; Baltz, R.H. Natural product discovery: Past, present, and future. *J. Ind. Microbiol. Biotechnol.* **2016**, *43*, 155–176. [CrossRef] [PubMed]
21. Kealey, C.; Creaven, C.A.; Murphy, C.D.; Brady, C.B. New approaches to antibiotic discovery. *Biotechnol. Lett.* **2017**, *39*, 805–817. [CrossRef] [PubMed]
22. Wencewicz, T.A. New antibiotics from nature's chemical inventory. *Bioorg. Med. Chem.* **2016**, *24*, 6227–6252. [CrossRef] [PubMed]
23. Bull, A.T.; Asenjo, J.A.; Goodfellow, M.; Gomez-Silva, B. The atacama desert: Technical resources and the growing importance of novel microbial diversity. *Ann. Rev. Microbiol.* **2016**, *70*, 215–234. [CrossRef] [PubMed]
24. Mohammadipanah, F.; Wink, J. Actinobacteria from arid and desert habitats: Diversity and biological activity. *Front. Microbiol.* **2016**, *6*, 1541. [CrossRef] [PubMed]
25. Rateb, M.E.; Ebel, R.; Jaspars, M. Natural product diversity of actinobacteria in the atacama desert. *Antonie Van Leeuwenhoek* **2018**, *111*, 1467–1477. [CrossRef] [PubMed]
26. Abdelkader, M.S.A.; Philippon, T.; Asenjo, J.A.; Bull, A.T.; Goodfellow, M.; Ebel, R.; Jaspars, M.; Rateb, M.E. Asenjonamides A-C, antibacterial metabolites isolated from *Streptomyces* asenjonii strain KNN 42.f from an extreme-hyper arid Atacama Desert soil. *J. Antibiot.* **2018**, *71*, 425–431. [CrossRef] [PubMed]
27. Ghosh, S.; Kuisiene, N.; Cheeptham, N. The cave microbiome as a source for drug discovery: Reality or pipe dream? *Biochem. Pharmacol.* **2017**, *134*, 18–34. [CrossRef] [PubMed]
28. Adam, D.; Maciejewska, M.; Naômé, A.; Martinet, L.; Coppieters, W.; Karim, L.; Baurain, D.; Rigali, S. Isolation, characterization, and antibacterial activity of hard-to-culture actinobacteria from cave moonmilk deposits. *Antibiotics* **2018**, *7*, 28–48. [CrossRef] [PubMed]
29. Gavrish, E.; Sit, C.S.; Cao, S.; Kandror, O.; Spoering, A.; Peoples, A.; Ling, L.A.; Hughes, D.; Bissell, A.; Torrey, H.; et al. Lassomycin, a ribosomally synthesized cyclic peptide, kills mycobacterium tuberculosis by targeting the ATP-dependent protease ClpC1P1P2. *Chem. Biol.* **2014**, *21*, 509–518. [CrossRef] [PubMed]
30. Jang, K.H.; Nam, S.-J.; Locke, J.B.; Kauffman, C.A.; Beatty, D.S.; Paul, L.A.; Fenical, W. Anthracimycin, a potent anthrax antibiotic from a marine-derived actinomycete. *Chem. Int. Ed.* **2013**, *52*, 7822–7824. [CrossRef] [PubMed]
31. Hassan, H.M.; Degen, D.; Jang, K.H.; Ebright, R.H.; Fenical, W.; Salinamide, F. New depsipeptide antibiotic and inhibitor of bacterial RNA polymerase from a marine-derived *Streptomyces* sp. *J. Antibiot.* **2015**, *68*, 206–209. [CrossRef] [PubMed]
32. Palomo, S.; González, I.; Martín, J.; de la Cruz, M.; Vicente, F.; Reyes, F.; Tormo, J.R.; Anderson, M.; Hill, R.T.O.; Genilloud, O. Sponge-derived *Kocuria* and *Micrococcus* spp. as sources of the new thiazolyl peptide antibiotic kocurin. *Mar. Drugs* **2013**, *11*, 1071–1086. [CrossRef] [PubMed]
33. Zhang, Y.; Adnani, N.; Braun, D.R.; Ellis, G.A.; Barns, K.J.; Parker-Nance, S.; Guzei, I.A.; Bugni, T.S. Micromonohalimanes A and B: Antibacterial halimane-type diterpenoids from a marine *Micromonospora* species. *J. Nat. Prod.* **2016**, *79*, 2968–2972. [CrossRef] [PubMed]
34. Pérez-Bonilla, M.; Oves-Costales, D.; de la Cruz, M.; Kokkini, M.; Martín, J.; Vicente, F.; Genilloud, O.; Reyes, F. Phocoenamicins B and C, New antibacterial spirotetronates isolated from a marine *Micromonospora* sp. *Mar. Drugs* **2018**, *16*. [CrossRef] [PubMed]
35. Ju, K.-S.; Gao, J.; Doroghazi, J.R.; Wang, K.-K.A.; Thibodeaux, C.J.; Lia, S.; Metzger, E.; Fudala, J.; Su, J.; Zhang, J.K.; et al. Discovery of phosphonic acid natural products by mining the genomes of 10,000 actinomycetes. *Proc. Natl. Acad. Sci. USA* **2015**, *112*, 12175–12180. [CrossRef] [PubMed]
36. Tang, X.; Li, J.; Millán-Aguiñaga, N.; Zhang, J.J.; O'Neill, E.C.; Ugalde, J.A.; Jensen, P.R.; Mantovani, S.M.; Moore, B.S. Identification of thiotetronic acid antibiotic biosynthetic pathways by target-directed genome mining. *ACS Chem. Biol.* **2015**, *10*, 2841–2849. [CrossRef] [PubMed]

37. Yamanaka, K.; Reynolds, K.A.; Kersten, R.D.; Ryan, K.S.; Gonzalez, D.J.; Nizet, V.; Dorrestein, P.C.; Moore, B.S. Direct cloning and refactoring of a silent lipopeptide biosynthetic gene cluster yields the antibiotic taromycin A. *Proc. Natl. Acad. Sci. USA* **2014**, *111*, 1957–1962. [CrossRef] [PubMed]
38. Bonet, B.; Teufel, R.; Cruüsemann, M.; Ziemert, N.; Moore, B.S. Direct capture and heterologous expression of Salinispora natural product genes for the biosynthesis of enterocin. *J. Nat. Prod.* **2014**, *78*, 539–542. [CrossRef] [PubMed]
39. Yang, C.; Huang, C.; Zhang, W.; Zhu, Y.; Zhang, C. Heterologous expression of fluostatin gene cluster leads to a bioactive heterodimer. *Org. Lett.* **2015**, *17*, 5324–5327. [CrossRef] [PubMed]
40. Onaka, H. Novel antibiotic screening methods to awaken silent or cryptic secondary metabolic pathways in actinomycetes. *J. Antibiot.* **2017**, *70*, 865–870. [CrossRef] [PubMed]
41. Derewacz, D.K.; Covington, B.C.; McLean, J.A.; Bachmann, B.O. Mapping microbial response metabolomes for induced natural product discovery. *ACS Chem. Biol.* **2015**, *10*, 1998–2006. [CrossRef] [PubMed]
42. Fenical, W.; Jensen, P.R. Developing a new resource for drug discovery: Marine actinomycete bacteria. *Nat. Chem. Biol.* **2006**, *2*, 666–673. [CrossRef] [PubMed]
43. Bull, A.T.; Stach, J.E.M. Marine actinobacteria: New opportunities for natural product search and discovery. *Trends Microbiol.* **2007**, *15*, 491–499. [CrossRef] [PubMed]
44. Choudhary, A.; Naughton, L.M.; Montánchez, I.; Dobson, A.D.W.; Rai, D.K. Current status and future prospects of marine natural products (MNPs) as antimicrobials. *Mar. Drugs* **2017**, *15*, 272. [CrossRef] [PubMed]
45. Dhakal, D.; Pokhrel, A.R.; Shrestha, B.; Sohng, J.K. Marine rare actinobacteria: Isolation, characterization, and strategies for harnessing bioactive compounds. *Front. Microbiol.* **2017**, *8*, 1106. [CrossRef] [PubMed]
46. Xu, D.; Han, L.; Li, C.; Cao, Q.; Zhu, D.; Barrett, N.H.; Harmody, D.; Chen, J.; Zhu, H.; McCarthy, P.J.; et al. Bioprospecting deep-sea actinobacteria for novel anti-infective natural products. *Front. Microbiol.* **2018**, *9*, 787. [CrossRef] [PubMed]
47. Jose, P.A.; Jha, B. Intertidal marine sediment harbours actinobacteria with promising bioactive and biosynthetic potential. *Sci. Rep.* **2017**, *7*, 10041. [CrossRef] [PubMed]
48. Sarmiento-Vizcaíno, A.; González, V.; Braña, A.F.; Palacios, J.J.; Otero, L.; Fernández, J.; Molina, A.; Kulik, A.; Vázquez, F.; Acuña, J.L.; et al. Pharmacological potential of phylogenetically diverse actinobacteria isolated from deep-sea coral ecosystems of the submarine Avilés canyon in the cantabrian sea. *Microb. Ecol.* **2017**, *73*, 338–352. [CrossRef] [PubMed]
49. Patin, N.V.; Schorn, M.; Aguinaldo, K.; Lincecum, T.; Moore, B.S.; Jensen, P.R. Effects of actinomycete secondary metabolites on sediment microbial communities. *Appl. Environ. Microbiol.* **2017**, *83*, e02676-16. [CrossRef] [PubMed]
50. Medema, M.H.; Fischbach, M.A. Computational approaches to natural product discovery. *Nat. Chem. Biol.* **2015**, *11*, 639–648. [CrossRef] [PubMed]
51. Weber, T.; Kim, H.U. The secondary metabolite bioinformatics portal: Computational tools to facilitate synthetic biology of secondary metabolite production. *Synth. Syst. Biotechnol.* **2016**, *1*, 69–79. [CrossRef] [PubMed]
52. Blin, K.; Medema, M.H.; Kottmann, R.; Lee, S.Y.; Weber, T. The antiSMASH database, a comprehensive database of microbial secondary metabolite biosynthetic gene clusters. *Nucleic Acids Res.* **2017**, *45*, D555–D559. [CrossRef] [PubMed]
53. Sánchez-Hidalgo, M.; González, I.; Díaz-Muñoz, C.; Martínez, G.; Genilloud, O. Comparative Genomics and Biosynthetic Potential Analysis of Two Lichen-Isolated *Amycolatopsis* Strains. *Front. Microbiol.* **2018**, *9*, 369. [CrossRef] [PubMed]
54. Adamek, M.; Alanjary, M.; Sales-Ortells, H.; Goodfellow, M.; Bull, A.T.; Winkler, A.; Wibberg, D.; Kalinowski, J.; Ziemert, N. Comparative genomics reveals phylogenetic distribution patterns of secondary metabolites in *Amycolatopsis* species. *BMC Genomics* **2018**, *19*, 426. [CrossRef] [PubMed]
55. Zhou, Z.; Gu, J.; Li, Y.-Q.; Wang, Y. Genome plasticity and systems evolution in *Streptomyces*. *BMC Bioinform.* **2012**, *13*, S8. [CrossRef] [PubMed]
56. Kim, J.-N.; Kim, Y.; Jeong, Y.; Roe, J.-H.; Kim, B.-G.; Cho, B.-K. Comparative Genomics Reveals the Core and Accessory Genomes of *Streptomyces* Species. *J. Microbiol. Biotechnol.* **2015**, *25*, 1599–1605. [CrossRef] [PubMed]
57. Choudoir, M.J.; Pepe-Ranney, C.; Buckley, D.H. Diversification of Secondary Metabolite Biosynthetic Gene Clusters Coincides with Lineage Divergence in *Streptomyces*. *Antibiotics* **2018**, *7*, 12. [CrossRef] [PubMed]
58. Zhang, M.M.; Wang, Y.; Ang, E.L.; Zhao, H. Engineering microbial hosts for production of bacterial natural products. *Nat. Prod. Rep.* **2016**, *33*, 963–987. [CrossRef] [PubMed]

59. Myronovskyia, M.; Luzhetskyy, A. Native and engineered promoters in natural product discovery. *Nat. Prod. Rep.* **2016**, *33*, 1006–1020. [CrossRef] [PubMed]
60. Weber, T.; Charusanti, P.; Musiol-Kroll, E.M.; Jiang, X.; Tong, Y.; Hyun Uk Kim, H.U.; Lee, S.Y. Metabolic engineering of antibiotic factories: New tools for antibiotic production in actinomycetes. *Trends Biotechnol.* **2015**, *33*, 15–26. [CrossRef] [PubMed]
61. Ren, H.; Wang, B.; Zhao, H. Breaking the silence: New strategies for discovering novel natural products. *Curr. Opin. Biotechnol.* **2017**, *48*, 21–27. [CrossRef] [PubMed]
62. Li, L.; Zhenga, G.; Chena, J.; Gec, M.; Jianga, W.; Lu, Y. New strategies and approaches for engineering biosynthetic gene clusters of microbial natural products. *Metab. Eng.* **2017**, *40*, 80–92. [CrossRef] [PubMed]
63. Tong, Y.; Robertsen, H.L.; Blin, K.; Weber, T.; Lee, S.Y. CRISPR-Cas9 Toolkit for Actinomycete Genome Editing. In *Synthetic Metabolic Pathways*; Humana Press: New York, NY, USA, 2018; pp. 163–184.
64. Xu, M.; Wang, Y.; Zhao, Z.; Gao, G.; Huang, S.-X.; Kang, Q.; He, X.; Lin, S.; Pang, X.; Deng, Z.; et al. Functional genome mining for metabolites encoded by large gene clusters through heterologous expression of a whole-genome bacterial artificial chromosome library in *Streptomyces* spp. *Appl. Environ. Microbiol.* **2016**, *82*, 5795–5805. [CrossRef] [PubMed]
65. Arakawa, K. Manipulation of metabolic pathways controlled by signaling molecules, inducers of antibiotic production, for genome mining in *Streptomyces* spp. *Antonie van Leeuwenhoek* **2018**, *111*, 743–751. [CrossRef] [PubMed]
66. Baltz, R.H. Genetic manipulation of secondary metabolite biosynthesis for improved production in *Streptomyces* and other actinomycetes. *J. Ind. Microbiol. Biotechnol.* **2016**, *43*, 343–370. [CrossRef] [PubMed]
67. Hwang, K.-S.; Kim, H.U.; Charusanti, P.; Palsson, B.O.; Lee, S.Y. Systems biology and biotechnology of *Streptomyces* species for the production of secondary metabolites. *Biotechnol. Adv.* **2014**, *32*, 255–268. [CrossRef] [PubMed]
68. Horbal, L.; Siegl, T.; Luzhetskyy, A. A set of synthetic versatile genetic control elements for the efficient expression of genes in Actinobacteria. *Sci. Rep.* **2018**, *8*, 491. [CrossRef] [PubMed]
69. Goh, E.B.; Yim, G.; Tsui, W.; McClure, J.; Surette, M.G.; Davies, J. Transcriptional modulation of bacterial gene expression by subinhibitory concentrations of antibiotics *Proc. Natl. Acad. Sci. USA* **2002**, *99*, 17025–17030. [CrossRef] [PubMed]
70. Rosen, P.C.; Seyedsayamdost, M.R. Though much is taken, much abides: Finding new antibiotics using old ones. *Biochemistry* **2017**, *56*, 4925–4926. [CrossRef] [PubMed]
71. Okada, B.K.; Seyedsayamdost, M.R. Antibiotic dialogues: Induction of silent biosynthetic gene clusters by exogenous small molecules. *FEMS Microbiol. Rev.* **2017**, *41*, 19–33. [CrossRef] [PubMed]
72. Goodwin, C.R.; Covington, B.C.; Derewacz, D.K.; McNees, C.R.; Wikswo, J.P.; McLean, J.A.; Bachmann, B.O. Structuring microbial metabolic responses to multiplexed stimuli via self-organizing metabolomics maps. *Chem. Biol.* **2015**, *22*, 661–670. [CrossRef] [PubMed]
73. Wang, M.; Carver, J.J.; Phelan, V.V.; Sanchez, L.M.; Garg, N.; Peng, Y.; Nguyen, D.D.; Watrous, J.; Kapono, C.A.; Luzzatto-Knaan, T.; et al. Sharing and community curation of mass spectrometry data with GNPS. *Nat. Biotechnol.* **2016**, *34*, 828–837. [CrossRef] [PubMed]
74. Pérez-Victoria, I.; Martín, J.; Reyes, F. Combined LC/UV/MS and NMR strategies for the dereplication of marine natural products. *Planta Med.* **2016**, *82*, 857–871. [CrossRef] [PubMed]
75. Wu, C.; Du, C.; Ichinose, K.; Choi, Y.H.; van Wezel, G.P. Discovery of C-Glycosylpyranonaphthoquinones in *Streptomyces* sp. MBT76 by a Combined NMR-Based Metabolomics and Bioinformatics Workflow. *J. Nat. Prod.* **2017**, *80*, 269. [CrossRef] [PubMed]
76. Gubbens, J.; Zhu, H.; Girard, G.; Song, L.; Florea, B.I.; Aston, P.; Ichinose, K.; Filippov, D.V.; Choi, Y.H.; Overkleeft, H.S.; et al. Natural product proteomining, a quantitative proteomics platform, allows rapid discovery of biosynthetic gene clusters for different classes of natural products. *Chem. Biol.* **2014**, *21*, 707. [CrossRef] [PubMed]
77. Chao Du, C.; van Wezel, G.P. Mining for microbial gems: Integrating proteomics in the postgenomic natural product discovery pipeline. *Proteomics* **2018**. [CrossRef]
78. Hou, B.; Lin, Y.; Wu, H.; Guo, M.; Petkovic, H.; Tao, L.; Zhu, X.; Ye, J.; Zhang, H. The novel transcriptional regulator LmbU promotes lincomycin biosynthesis through regulating expression of its target genes in *Streptomyces lincolnensis*. *J. Bacteriol.* **2018**, *200*, e00447-17. [CrossRef] [PubMed]
79. Liu, G.; Chater, K.F.; Chandra, G.; Niu, G.; Tan, H. Molecular regulation of antibiotic biosynthesis in *Streptomyces*. *Microbiol. Mol. Biol. Rev.* **2013**, *77*, 112–143. [CrossRef] [PubMed]

80. Van Wezel, G.P.; Mcdowall, K.J. The regulation of the secondary metabolism of *Streptomyces*: New links and experimental advances. *Nat. Prod. Rep.* **2011**, *28*, 1311–1333. [CrossRef] [PubMed]
81. Hoskisson, P.A.; Fernández-Martínez, L.T. Regulation of specialised metabolites in actinobacteria–expanding the paradigms. *Environ. Microbiol. Rep.* **2018**, *10*, 231–238. [CrossRef] [PubMed]
82. Bailey, T.L.J.; Grant, C.E.; Noble, W.S. The MEME suite. *Nucleic Acids Res.* **2015**, *43*, W39–W49. [CrossRef] [PubMed]
83. Rigali, S.; Anderssen, S.; Naômé, A.; van Wezel, G.P. Cracking the regulatory code of biosynthetic gene clusters as a strategy for natural product discovery. *Biochem. Pharmacol.* **2018**, *153*, 24–34. [CrossRef] [PubMed]
84. Hiard, S.; Marée, R.; Colson, S.; Hoskisson, P.A.; Titgemeyer, F.; van Wezel, G.P.B.; Wehenkel, J.L.; Rigali, S. PREDetector: A new tool to identify regulatory elements in bacterial genomes. *Biochem. Biophys. Res. Commun.* **2007**, *357*, 861–864. [CrossRef] [PubMed]
85. Rigali, S.; Titgemeyer, F.; Barends, S.; Mulder, S.; Thomae, A.W.; Hopwood, D.A.; van Wezel, G.P. Feast or famine: The global regulator DasR links nutrient stress to antibiotic production by *Streptomyces*. *EMBO Rep.* **2008**, *9*, 670–675. [CrossRef] [PubMed]
86. Craig, M.; Lambert, S.; Jourdan, S.; Tenconi, E.; Colson, S.; Maciejewska, M.; Ongena, M.; Martin, J.F.; van Wezel, G.; Rigali, S. Unsuspected control of siderophore production by *N*-acetylglucosamine in streptomycetes. *Environ. Microbiol. Rep.* **2012**, *4*, 512–521. [CrossRef] [PubMed]
87. Jourdan, S.; Francis, I.M.; Kim, M.J.; Salazar, J.J.C.; Planckaert, S.; Frère, J.-M.; Matagne, A.; Kerff, F.; Devreese, B.; Loria, R.; et al. The CebE/MsiK transporter is a doorway to the cello-oligosaccharide-mediated Induction of *Streptomyces scabies* pathogenicity. *Sci. Rep.* **2016**, *6*, 27144. [CrossRef] [PubMed]
88. Schniete, J.K.; Cruz-Morales, P.; Selem-Mojica, N.; Fernández-Martínez, L.T.; Hunter, I.S.; Barona-Gómez, F.; Hoskisson, P.A. Expanding primary metabolism helps generatethe metabolic robustness to facilitate antibioticbiosynthesis in *Streptomyces*. *mBio* **2018**, *9*, e02283-17. [CrossRef] [PubMed]
89. Craney, A.; Ozimok, C.; Pimentel-Elardo, S.M.; Capretta, A.; Nodwell, J.R. Chemical perturbation of secondary metabolism demonstrates important links to primary metabolism. *Chem. Biol.* **2012**, *19*, 1020–1027. [CrossRef] [PubMed]
90. Seyedsayamdost, M.R. High-throughput platform for the discovery of elicitors of silent bacterial gene clusters. *Proc. Natl. Acad. Sci. USA* **2014**, *111*, 7266–7271. [CrossRef] [PubMed]
91. Adnani, N.; Chevrette, M.G.; Adibhatla, S.N.; Zhang, F.; Yu, Q.; Braun, D.R.; Nelson, J.; Simpkins, S.W.; McDonald, B.R.; Myers, C.L.; et al. Coculture of marine invertebrate-associated bacteria and interdisciplinary technologies enable biosynthesis and discovery of a new antibiotic, keyicin. *ACS Chem. Biol.* **2017**, *12*, 3093–3102. [CrossRef] [PubMed]
92. Sengesa, C.H.R.; Al-Dilaimib, A.; Marchbankc, D.H.; Wibbergb, D.; Winklerb, A.; Haltlic, B.; Nowrousiand, M.; Kalinowskib, J.; Kerrc, R.G.; Bandow, J.E. The secreted metabolome of *Streptomyces chartreusis* and implications for bacterial chemistry. *Proc. Natl. Acad. Sci. USA* **2018**, *115*, 2490–2495. [CrossRef] [PubMed]

antibiotics

MDPI

Review

Acyltransferases as Tools for Polyketide Synthase Engineering

Ewa Maria Musiol-Kroll * and Wolfgang Wohlleben

Interfakultäres Institut für Mikrobiologie und Infektionsmedizin, Eberhard Karls Universität Tübingen, Auf der Morgenstelle 28, 72076 Tübingen, Germany; wolfgang.wohlleben@biotech.uni-tuebingen.de

* Correspondence: ewa.musiol@biotech.uni-tuebingen.de; Tel.: +49-7071-29-78839

Received: 2 June 2018; Accepted: 16 July 2018; Published: 18 July 2018

Abstract: Polyketides belong to the most valuable natural products, including diverse bioactive compounds, such as antibiotics, anticancer drugs, antifungal agents, immunosuppressants and others. Their structures are assembled by polyketide synthases (PKSs). Modular PKSs are composed of modules, which involve sets of domains catalysing the stepwise polyketide biosynthesis. The acyltransferase (AT) domains and their "partners", the acyl carrier proteins (ACPs), thereby play an essential role. The AT loads the building blocks onto the "substrate acceptor", the ACP. Thus, the AT dictates which building blocks are incorporated into the polyketide structure. The precursor- and occasionally the ACP-specificity of the ATs differ across the polyketide pathways and therefore, the ATs contribute to the structural diversity within this group of complex natural products. Those features make the AT enzymes one of the most promising tools for manipulation of polyketide assembly lines and generation of new polyketide compounds. However, the AT-based PKS engineering is still not straightforward and thus, rational design of functional PKSs requires detailed understanding of the complex machineries. This review summarizes the attempts of PKS engineering by exploiting the AT attributes for the modification of polyketide structures. The article includes 253 references and covers the most relevant literature published until May 2018.

Keywords: natural products; polyketides; polyketide synthases; acyltransferases; engineering; new bioactive compounds

1. Introduction

Polyketides are a large class of structurally complex compounds with interesting and valuable activities. Those agents are widely used as antibiotics, antifungals, and drugs for other clinical applications [1–3]. Erythromycin [4,5], mupirocin [6,7], rapamycin [8,9], FK506 [10,11], and epothilone B [12,13] are examples of antimicrobial, immunosuppressant, and anticancer drugs, respectively (Figure 1). In general, polyketides can be obtained from biological sources (e.g., actinomycetes) or chemically synthesized (semi- or total chemical synthesis). While the isolation of compounds from biological material is rather easy to implement, the chemical synthesis is quite challenging, often limited to certain type of reactions and resulting in low quantities [14]. Therefore, bioactive product discovery methods, which are based on traditional, bioactivity-guided screening and compound isolation are still relevant. There is even a renewed interest in this approach as a strategy for the identification of new drugs [15–19]. In particular, the developments in molecular biology, synthetic biology, sequencing technology, chemistry, and bioinformatics provide new opportunities for the classic screening programs and compound recovery from the natural environment. For example, the combination of the bioactivity-guided screening with sequencing and genome mining approaches supports and accelerates the downstream process of early drug development after a compound has been identified. Furthermore, the recently established methods provide new opportunities, not only for the optimization of the production of low-yield natural products, but they also enable the in vivo/in vitro modification of the molecules, leading to advances in structural diversity. The new perspectives are particularly important for diversification of polyketides, as their structures are complex and thus difficult to access by chemical synthesis. It is the polyketide biosynthesis that offers many opportunities for implementation of the innovative technologies and tools to generate novel analogs. One of such "hotspots" enabling structural modifications of polyketides, are the core assembly lines and their components (Figure 2).

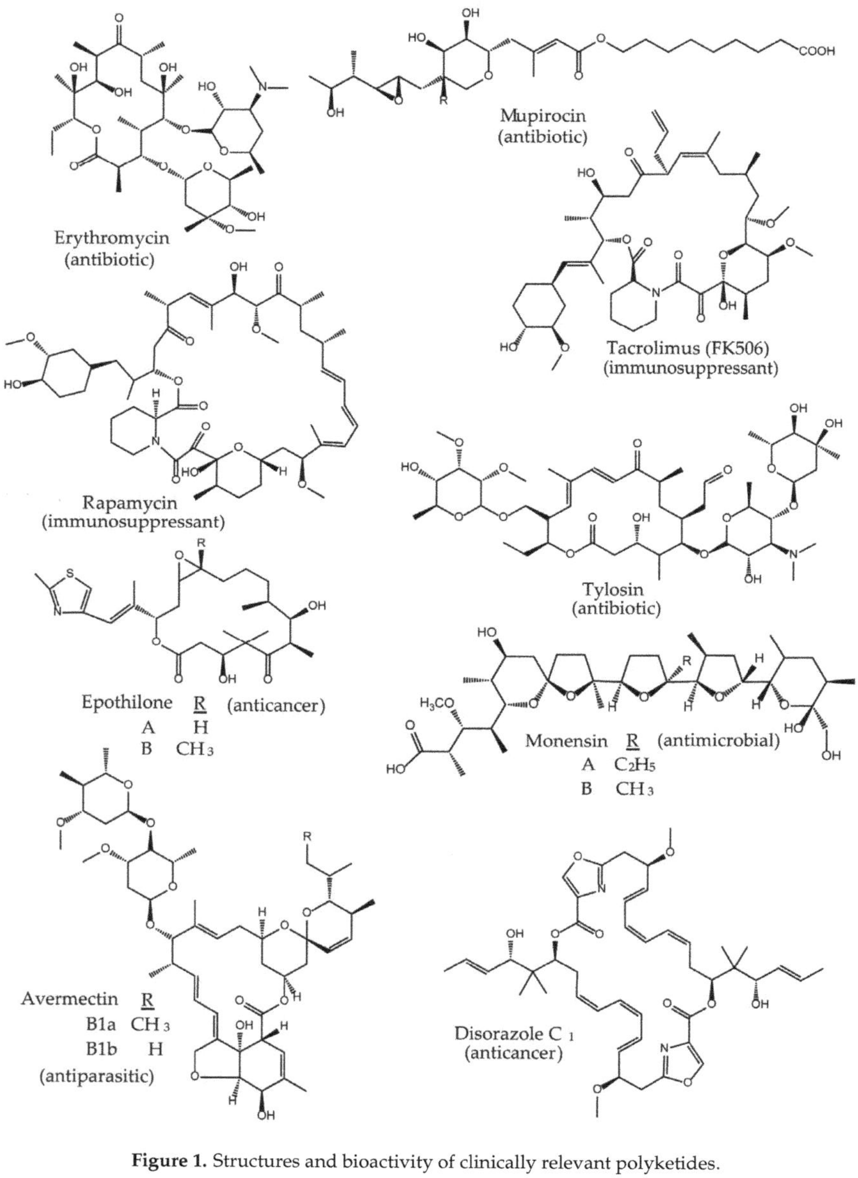

Figure 1. Structures and bioactivity of clinically relevant polyketides.

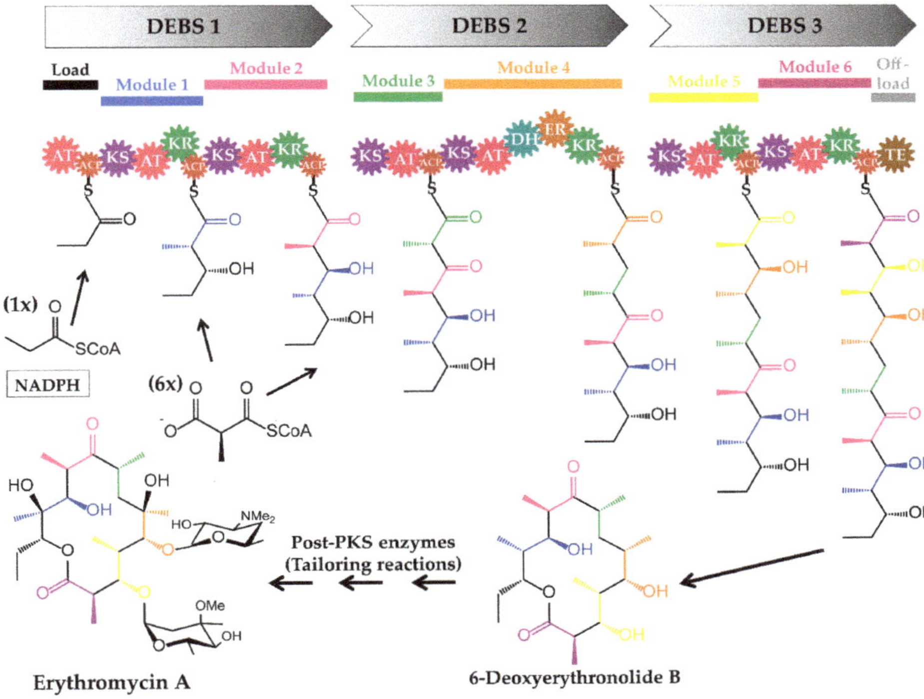

Figure 2. The erythromycin assembly line (DEBS) [20,21]. The prototypical type I modular PKS machinery is composed of three subunits DEBS 1, DEBS 2 and DEBS 3, which are organized into modules. Each module contains a set of domains catalysing one elongation step. The biosynthesis of the chain begins with the loading of the starter unit (derived from propionyl-coenzyme A(-CoA)) onto the loading module (load). Subsequently, the polyketide chain is extended by one extender unit (derived from methylmalonyl-CoA) on each of the "downstream" extension modules. In modules which possess reductive domains (optional), the polyketide intermediate is modified. The synthesized chain is released from the last module (off-load) by the thioesterase domain. Finally, the generated erythromycin intermediate (6-deoxyerythronolide B) is further processed in the post-PKS steps (tailoring). PKS domains: AT, acyltransferase; ACP, acyl carrier protein; KS, ketosynthase; DH, dehydrogenase; KR, ketoreductase; ER, enoylreductase; TE, thioesterase.

The polyketide chain synthesizing enzymes, the polyketide synthases (PKSs), often differ in their composition and organization, and the iterative or non-iterative fashion. Based on those features, the PKSs were classified into "different" groups: modular type I PKSs (including the so-called *cis*-AT (AT-containing) type I PKSs and *trans*-AT (AT-less) type I PKSs), iterative type I PKSs, type II PKSs (iterative), and type III PKSs (iterative (chalcone synthase-like PKSs)) [22]. However, in the past two decades many PKS assembly lines were identified, which do not conform to this classification (non-canonical PKSs) [22–27]. For example, in aureothin or sceliphrolactam biosynthesis, the modular PKSs AurA [28] and SceQ [29], respectively, harbour an iterative module. Because of such discrepancies, the described classification and terminology are currently used with caution, especially for the newly identified, non-canonical PKS pathways.

Modular type I PKSs (e.g., the prototypical 6-deoxyerythronolide B synthase (DEBS) from erythromycin) are multifunctional megaenzymes, equipped with domains catalysing the successive linkage of simple building blocks to a polyketide chain (Figure 2). The PKS domains are organized into units termed modules, of which each harbours a set of domains required for one elongation (and modification) cycle. Thus, a minimal PKS consists of the essential domains: the acyltransferase (AT), the acyl carrier protein (ACP) and the ketosynthase (KS). The AT is either embedded into the PKS (*cis*-AT PKS) [30] or encoded as a separate gene, of which the gene product (discrete AT) complements the AT-less PKS (*trans*-AT PKS) [31,32]. In contrast to PKS-independent acyltransferase enzymes, suchas dihydroxyacetonephosphate acyltransferases [33] or long-chain-alcohol O-fatty-acyltransferases (wax synthases) [34,35], which typically transfer acyl groups to a non-acyl carrier protein (non-ACP) acceptor molecule, the ATs of polyketide assembly lines load simple units (building blocks) derived from thioester-activated precursors onto the ACPs of a PKS.

After the ACP-loading step, the KS domain of the PKS catalyses the decarboxylative Claisen-like condensation of the newly loaded unit and the already existing polyketide chain. Additional optional domains, such as the ketoreductase (KR), dehydratase (DH), enoylreductase (ER) and methyltransferase (MT) process the generated β-ketoacyl thioester intermediate. Finally, the polyketide chain is released from the last module of the assembly line, which is usually catalysed by thioesterase (TE). Tailoring enzymes (if present) further modify the intermediate into the final product (Figure 2).

The structural diversity of polyketides can be assigned to several factors. Those include the length of the polyketide chain, oxidation state and stereochemistry of the β-keto groups, mechanism of termination and chain release, as well as post-PKS tailoring steps [2]. Most of those variations result from the features of the PKS system (type of the PKS, PKS module and domain composition) and modification enzymes. However, the fact that the ATs provide the essential precursors for the polyketide biosynthesis and that they differ in their substrate specificities across the PKS pathways make the ATs to one of the most promising "targets" for engineering of the polyketide assembly lines. Although, the AT-based engineering was successful in some cases, this strategy is still limited by numerous challenges resulting from the complexity of the PKS systems. Therefore, the AT-based engineering is not yet a well-established high-throughput technology. To eliminate or at the very least reduce the engineering bottlenecks and increase the effectiveness of this approach, further knowledge about those complex systems is required. This will contribute to a better understanding of the PKS machineries and enable a more predictable PKS-engineering.

The goal of this review is to present an overview on the AT-targeted PKS engineering attempts and the recently gained knowledge, which renews the hope for programmable PKS modification and the production of new polyketide derivatives. The synopsis includes an introduction into the PKS assembly lines and its AT-domains, a brief summary of the biosynthesis of polyketide precursors, mechanistic and structural insights into PKSs and finally a more detailed description of the AT-based engineering strategies applied to modular type I PKSs and their future perspectives.

2. Polyketide Synthases and the Essential Acyltransferase Domains

The PKSs are the core enzymes of the polyketide biosynthetic machineries. In modular PKSs, the PKS domains are combined into modules. Each module contains one set of domains required for one elongation and potential modification cycle (Figure 2).

In this review, we discuss the AT-based PKS engineering efforts applied to modular type I PKSs. The structure of a polyketide scaffold, synthesized by a modular PKS assembly line reflects the order of the catalytic domains within the PKS-modules, which usually corresponds to the chromosomal order of the underlying genes. This assembly line-like structure of the textbook modular PKSs (*cis*-AT type I PKSs) and the corresponding products was defined as the "collinearity rule" [21,36–38]. It was thereby generally assumed that relying on the sequence encoding the assembly line and the collinear architecture of the modules/domains, including the conserved domain motifs, enable the prediction of the polyketide structure and its stereochemistry [39–41]. However, the previous PKS "module" definition, which was deduced from the organization and function of the domains observed in mammalian fatty acid synthase (KS → AT → DH → ER → KR → ACP → TE; from the KS to the ACP) was reconsidered very recently. In 2017, Zhang et al. reported that processing enzymes co-migrate during assembly line evolution with the KS domain downstream (not upstream) of the ACP [42]. Consequently, the term "module", for modular *cis*-AT PKSs had to be redefined and the polyketide assembly lines updated [43]. According to the newly proposed definition, a PKS module containing all processing domains has the following organization: AT+DH+ER+KR+ACP+KS. Subsequently, 526 ACPs from 33 characterized *trans*-AT PKS assembly lines were analysed using bioinformatic tools [44]. To group the ACPs, "module types" (*a–y*) were defined based on the chemistry they perform and classes of enzymes present in the module. A cladogram of ACPs that belong to defined module types was generated and ACP families, which contain related module types, were specified [44]. The analysis uncovered that ACPs from the same module type generally clade together, reflective of the co-evolution of these domains with their cognate enzymes [44,45]. Moreover, cladograms of KSs upstream and downstream of ACPs revealed that in most of the analysed systems, the KSs downstream of ACPs from the same module type also clade together. However, this was not the case for the KSs upstream of ACPs, which was inconsistent with the traditional definition of a module. Therefore, the authors suggested to update the term "module" also for the *trans*-AT assembly lines [44]. We refer to the fact that a new definition of the PKS module was recently introduced. Nonetheless, to avoid any confusion in this review, we rely on the original publications and use the previously reported module nomenclature whenever a PKS assembly line was characterized according to the old module terminology.

The modularity, together with the structural and functional PKS characteristics further strengthened the idea of exploiting the megaenzymes and their domains for diversification of polyketide molecules. The ATs are particularly eligible for these purposes as those domains are responsible for the selection of the precursors and provision of the building blocks (units) to the PKS. More specifically, the precursor- and occasionally the ACP-specificity of the ATs are features that essentially contribute to structural diversity of polyketides and encourage the AT-based PKS engineering. Albeit there are still gaps in understanding of the complex polyketide assembly lines [46,47], various approaches of AT-engineering, including AT domain swapping, AT site-directed mutagenesis, and AT knockout and complementation by an AT-domain from other modules were reported [48,49] (Table 1 and Figure 3). However, it does not necessarily mean that the engineering potential of PKSs by manipulation of the AT features is exhausted. Recent findings such as updates on PKS module boundaries might boost implementation of these strategies and lead to a new breakthrough in this field. Therefore, it seems to be the right time to summarize the available information on the ATs of PKSs, the AT-based engineering of polyketides and the potential new prospects.

Table 1. Examples of acyltransferase-based polyketide synthase engineering. Abbreviations: acyltransferase (AT); methylmalonyl-coenzyme A (MM-CoA); malonyl-CoA (M-CoA), methoxymalonyl-ACP (MeO-ACP).

Engineering Strategy	Examples			
	Acceptor PKS (Target Assembly Line for the Engineering)	**Tool (e.g., Donor AT or Motif) and Origin (Pathway)**	**Substrate Specificities Native AT (Acceptor PKS)/Used Tool (AT or Motif)**	**References**
AT-swapping	AT from module 1, and 2/erythromycin	"Hyg" AT2 (module 2)/"Hyg" PKS gene, AT (module 14)/rapamycin, Ven AT/unknown PKS-like gene cluster	MM-CoA/M-CoA	[50]
	AT from module 1, and 2/erythromycin	AT (module 2)/rapamycin	MM-CoA/M-CoA	[51,52]
	AT from module 1/erythromycin	AT (module 3)/erythromycin	MM-CoA/MM-CoA	[53]
	AT from module 6/erythromycin	AT (module 6)/rapamycin	MM-CoA/M-CoA	[54]
	AT from module 1/erythromycin	AT (module 2)/epothilone	MM-CoA/M-CoA	[55]
	AT from module 1/erythromycin	AT (module 3)/epothilone	MM-CoA/promiscuous MM, M-CoA	[55]
	AT from module 4/erythromycin	AT (module 5)/niddamycin	MM-CoA/EM-CoA	[56]
	AT from module 6/erythromycin	hydroxymalonate-specifying *fkbA*-AT8/ascomycin (FK520)	MM-CoA/Methoxymalonyl-ACP	[57,58]
	AT from module 1-5 and 7/geldanamycin	AT (module 2 and 14)/rapamycin	MM-CoA, MeO-ACP/M-CoA	[59]
AT-site directed mutagenesis	AT from module 1 and 4/erythromycin	consensus motifs (YASH/HAFH motif)	MM-CoA/M-CoA	[60–62]
	AT from module 6 of/erythromycin	structurally identified relevant residue	MM-CoA/SNAC-derivatives	[63,64]
	AT from module 6/erythromycin	saturation mutagenesis of relevant motifs	MM-CoA/non-natural alkynyl-modified unit	[65]
AT-cross complementation	AT6null (inactivated AT in module 6)/erythromycin	*S. coelicolor* malonyl-CoA: ACP transacylase	MM-CoA/M-CoA	[66]
	AT6null (inactivated AT in module 6)/erythromycin	AT (BryP-AT1)/bryostatin	MM-CoA/M-CoA	[67]
	AT6null (inactivated AT in module 6)/erythromycin	AT (BryP-AT2)/bryostatin	MM-CoA/MM-CoA	[67]

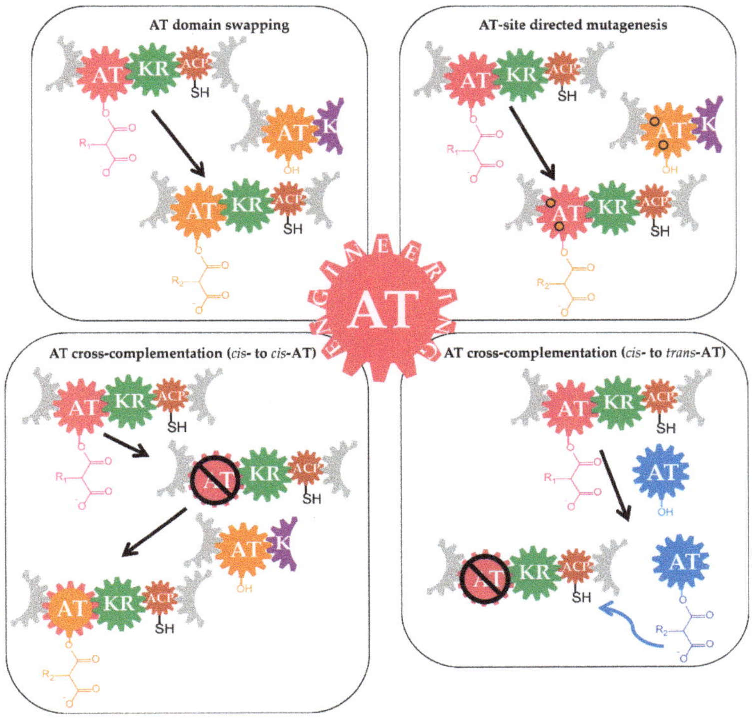

Figure 3. General strategies for acyltransferase-based engineering of polyketide synthases. The acyltransferase (AT)-based engineering of polyketide synthases (PKSs) includes the AT domain swapping, AT-site directed mutagenesis and the AT-cross complementation. In the AT domain swapping approach, the AT (red-purple) of a "target" assembly line is replaced by an AT from another module of the native assembly line or an AT of a foreign pathway (heterologous ATs) (orange). AT-site directed mutagenesis usually involves the identification of conserved motifs, which contribute to substrate specificity of the AT and the mutation of those sites of the AT (red-purple AT with orange circles/black outline). This approach enables the alteration of the AT-substrate specificity. In AT cross-complementation experiments, the AT (red-purple) of a "target" assembly line is inactivated and the PKS module is complemented by a foreign AT (orange or blue). The AT for complementation of the AT-inactivated PKS is either incorporated in place of the native AT (*cis*- to *cis*-AT complementation (orange)) or encoded as separate gene, of which gene product (discrete AT (blue)) complements the assembly line *in trans* (*cis*- to *trans*-AT complementation). Because the outcome of the *cis*- to *cis*-AT complementation is similar to AT-domain substitution or swapping, this strategy is often regarded as such.

2.1. Acyltransferase Substrates and Loading of the Acyl Carrier Protein

ATs are indispensable for polyketide assembly as they select the precursors and load the building blocks (units) for the polyketide biosynthesis onto the ACPs of PKSs. However, the ACPs have to be "prepared" for accepting the unit prior to the loading reaction. This is accomplished in a post-translational modification by the 4-phosphopantetheinyl (Ppant) transferase (PPTase) [68–70]. The enzyme transfers the Ppant arm of coenzyme A (CoA) to the serine residue of the ACP (*apo*-ACP form), which results in the activated ACP (*holo*-ACP form). The ATs of PKSs load the building block onto the Ppant group of the *holo*-ACP. The units for loading of the *holo*-ACP originate from thioester-activated precursors (e.g., CoA- or ACP-linked units such as acyl starter unit, malonyl units or their 2-substituted derivatives). The loading of the ACP in PKSs is accomplished in a two-step reaction [71]. In the first step, the AT recognizes and binds the precursor to its active serine residues, which results in the acyl-*O*-AT intermediate (self-acylation reaction; step 1). Subsequently, the acyl-*O*-AT interacts with the activated ACP (*holo*-ACP) and transfers the unit onto the protein, which leads to the formation of an acyl-*S*-ACP (transacylation reaction; step 2).

The ATs of *cis*-AT PKSs usually load only one type of extender units (e.g., malonate derived from malonyl-CoA or methylmalonate derived from methylmalonyl-CoA) onto its ACP, which strongly indicates that the ATs are substrate-specific at the native polyketide production conditions. However, there are examples of assembly lines where an AT domain transfers different substrates simultaneously to the same ACP, resulting in the production of polyketide mixtures. This is the case, for example, for the AT domain in module five of the monensin PKS (AT5mon). The monensin AT5 accepts both, ethylmalonyl- and methylmalonyl-CoA as substrates, leading to the production of monensins A and B [72,73] at standard fermentation conditions. The inherent promiscuity of the AT5mon was exploited for loading of synthetic non-natural extender units (allylmalonyl-*N*-acetylcysteamine (allylmalonyl-SNAC), propargylmalonyl-SNAC, propylmalonyl-SNAC, butylmalonyl-SNAC, and hexyl-SNAC) onto the cognate ACP in in vivo feeding experiments [74]. Except 2-hexylmalonyl-SNAC, which was toxic to the producer strain, the synthetic building blocks were incorporated in significant amounts and new premonensin derivatives, including propargyl-premonensin, were produced. This suggests that some ATs might accept a wider range of substrates than originally assumed and that limitations of the availability of non-native precursors for the production of polyketide derivatives can be overcome by feeding with synthetic building blocks. Moreover, it makes the idea of heterologous expression of genes or pathways, directly delivering non-native precursors for the production of polyketide derivatives, very promising (examples for this strain engineering approach were also provided in Sections 3.1.1 and 3.1.2).

In the following, we briefly summarized the biosynthetic routes for those precursors, which are frequently used by the ATs for the assembly of polyketide structures.

2.1.1. Provision of Malonyl-CoA and Methylmalonyl-CoA Precursors

Malonyl-CoA and methylmalonyl-CoA are most commonly used and metabolically available precursors for the biosynthesis of polyketides. In general, there are two pathways for the biosynthesis of malonyl- or methylmalonyl-CoA. One route is the carboxylation of the acetyl- and propionyl-CoA, respectively [75,76]. The acetyl-CoA carboxylases (ACCs) carboxylate acetyl-CoA to malonyl-CoA and propionyl-CoA carboxylases (PCCs) convert propionyl-CoA to methylmalonyl-CoA [77,78]. The other route involves the direct conversion of malonate or methylmalonate to the CoA-activated forms by ATP-dependent malonyl-CoA synthetase and homologous enzymes. Malonyl-CoA synthetases, such as MatB have been described for *Streptomyces coelicolor* [79] and *Rhizobium* [80–82]. Hughes and Keatinge-Clay studied the substrate flexibility of the *S. coelicolor* MatB and have shown that this adenylate-forming enzyme is capable of producing most CoA-linked polyketide extender units, as well as pantetheine- and *N*-acetylcysteamine-linked analogs, useful for in vitro PKS studies [79]. It was demonstrated that the methylmalonyl groups ligated by MatB to CoA, pantetheine or *N*-acetylcysteamine were utilized and incorporated into a triketide pyrone by the terminal module of the 6-deoxyerythronolide B synthase (Mod6TE) in vitro [79].

The MatB enzyme of *Rhizobium trifolii* has shown a tolerance for a variety of C2 substituted malonic acids [80] however, in most cases the activity towards the non-native substrates was low. This encouraged the engineering of the malonyl-CoA synthetase and the generation of enzyme mutants, which deliver exotic precursors [65,83–85] for polyketide biosynthesis with improved efficiency.

The incorporation of some of the generated unnatural building blocks into polyketide scaffolds was confirmed in vivo [86–90], which makes the use of the engineered MatB variants attractive for non-chemical polyketide derivatization ("bio-derivatization").

2.1.2. Provision of Ethylmalonyl-CoA and Exotic Alkylmalonyl-CoA Precursors

In addition to the typically used substrates malonyl-CoA and methylmalonyl-CoA, some ATs of PKSs utilize unusual alkylmalonyl-CoA precursors for polyketide assembly. Such exotic precursors are generated by crotonyl-CoA carboxylase/reductase (CCR) and homologues (CCRs were broadly defined as enoyl-CoA carboxylase/reductases (ECRs)) [91–94]. CCRs/ECRs are members of the medium-chain reductase/dehydrogenase (MDR) protein superfamily, which can be found in all three domains of life [95]. The MDR superfamily includes enzymes that reduce either C=O or C=C bonds in α,β-unsaturated carbonyl compounds.

Recent structure-based engineering of the active-site binding pocket of CCRs enabled significant alteration of their catalytic activity towards non-native substrates [96,97]. For example, site-directed mutagenesis of the V350G-site in the CCR enzyme AntE expanded its substrate scope to afford indolylmethylmalonyl-CoA [96].

An alternative strategy to the construction of modified biocatalysts for the provision of unusual polyketide precursors is the screening and identification of independent pathways or novel enzymes, directly delivering the exotic substrates. In 2016 Ray et al. described a CCRC (crotonyl-CoA reductase/carboxylase)-independent mechanism for assembly of unusual PKS precursors, where an acyl-CoA carboxylase (YCC, biotin-dependent enzyme) directly carboxylates medium chain acyl-CoA thioesters to alkylmalonyl-CoA [98].

The provision of diverse precursors is the prerequisite for the exploitation of the AT promiscuity and the production of polyketide derivatives. Malonyl-CoA synthetases (e.g., MatB-like synthetases) and CCR enzymes enable the generation of a variety of substrates, which can be utilized by the ATs. However, for their use in directed production of AT substrates, the enantiomeric selectivity of the ATs needs to be considered [99]. For example, the AT domains of the model assembly line erythromycin (Figure 2) utilize solely the (2*S*)-isomer of methylmalonyl-CoA for each of the extension step [100]. In such cases the application of CCRCs or biotin-dependent carboxylase enzymes, which generate (*S*)-enantiomers might be the better route to obtain 2-substituted malonyl-CoAs, compared to malonyl-CoA synthetases providing (*R*)-enantiomers. Nevertheless, both enantiomeric forms are useful for testing and characterization of the AT-stereochemistry, which is an important determinant for the efficient incorporation of unnatural moieties into polyketide structures.

2.2. *Mechanistic and Structural Insights into Acyltransferases and Polyketide Synthase Modules*

The challenges of polyketide engineering, resulting from the complex modularity and functionality of PKSs became more obvious after some of the initial PKS engineering experiments, including skipping and swapping of domains/modules failed [101–103]. Disruptions of the modular architecture of the PKS, destructions of the PKS polypeptide integrity and/or prevention of protein-protein interface interactions were common reasons for the insufficiency of the modified PKSs. This demonstrated that not all paradigms derived from investigations of model assembly lines can be arbitrary transferred to another PKS biosynthetic pathway for its engineering. Accordingly, detailed understanding of these systems is required and indispensable for rational PKS modification. This encouraged the more comprehensive mechanistic and structural analysis of the PKS assembly lines. Here, we focus on AT-substrate and AT-ACP-KS interactions, and present examples underlining the significance of the interdomain communication for the polyketide biosynthesis.

2.2.1. Substrate Recognition and Acyltransferase-Acyl Carrier Protein Interactions

Typically, the ATs accomplish the loading of the ACP by binding of the precursor and the release of the CoASH group prior to the interaction with the respective ACP domain [86,104]. During the loading process, the acyl-AT intermediate is exposed to a nucleophilic attack of the thiol residue of the ACP's phosphopantetheine moiety. This reaction scenario was termed "ping-pong-bi-bi" mechanism [105]. However, exceptions of this mechanism were observed. For example, the *trans*-AT Dis/Dsz AT from the disorazole pathway exhibited different kinetics, compared to the classical "ping-pong" mechanism. Wong et al. demonstrated that the transacylation in case of the *trans*-AT Dis/Dsz AT depends on the interaction with its ACP before CoASH is released from the building block [106].

The significance of the AT-ACP interactions for the transacylation was also confirmed for other assembly lines (e.g., erythromycin, zwittermicin A or vicenistatin).

In the erythromycin PKS DEBS, protein-protein interactions between AT domains, ACP domains and the linkers that flank AT domains were systematically probed [48,107,108]. The ATs and their cognate ACPs exposed at least 10-fold greater specificity for each other than for heterologous proteins [107]. Moreover, the flanking (N- and C-terminal) linkers of an AT domain contributed to the efficiency and specificity of transacylation, underlining the importance of the linker regions for proper protein-protein interaction [107].

The examples of AT-substrate interactions described above involve ATs which select free precursors and load the building blocks onto ACPs being integrated into the PKS. In addition, AT-substrate recognition and protein-protein "communication" were investigated for the less abundant ACP-linked units (standalone ACPs to which the building block is attached, such as hydroxymalonyl-ACP). In zwittermicin A biosynthesis, the acyltransferase domain of the PKS ZmaA (ZmaA-AT) recognizes the hydroxymalonyl-ACP as substrate and transfers the hydroxymalonyl unit to a downstream ACP via a transacylated AT domain intermediate [109]. The X-ray crystal structure of ZmaA-AT revealed that the ACP itself biases extender unit selection [110]. Furthermore, it indicated that the ACP interaction with the hydrophobic motif promotes secondary structure formation at the binding site and leads to opening of the adjacent substrate pocket lid to allow the binding of the substrate in the active site of the AT [110].

In the biosynthesis of vicenistatin, the AT VinK transfers a very early vicenistatin-intermediate between two ACPs, the standalone ACP VinL and VinP1LdACP (loading module) [111,112]. Thus, it was strongly suggested that the AT VinK needs to distinguish between the ACPs VinL and VinP1LdACP. In a later study, a crystal structure of the AT–ACP (VinK-VinL) complex was obtained, which revealed that Arg153, Met-206, and Arg-299 of VinK interact with the negatively charged helix II region of VinL [113]. The structure of the VinK-VinL complex visualized the interaction between an AT and ACP and provided the first detailed mechanistic insights into ACP recognition by an AT enzyme.

2.2.2. Interactions between the Acyltransferase, Ketosynthase and the Acyl Carrier Protein

The condensation of ACP-bound extender units with the growing polyketide chain is essential for completing the biosynthesis and for the formation of the final product. Those steps (translocation and elongation) are dependent on protein-protein recognition between the AT, KS and the ACP domains [47,114–118]. Regions in the KS domain, KS to AT linkers, and the AT domains of DEBS were described as docking site for the ACP during chain transfer and elongation [118,119]. The protein-substrate and protein-protein interactions cause conformational changes within PKS modules during the catalytic cycle of polyketide biosynthesis. Based on the cryo-EM data obtained for the full-length PikAIII module of pikromycin, Dutta et al. and Whicher et al. described the dynamics of PKS modules as concerted actions to mediate appropriate substrate processing [120,121]. Specifically, the ACP domain is differently positioned after polyketide chain substrate loading onto the active site of the ketosynthase, after extension to the β-keto intermediate, and after β-hydroxy product generation. The conformational rearrangements enable optimal positioning for reductive processing of the polyketide chain elongation intermediate, bound to the ACP [121].

More recently, the synchronized processing of intermediates on a PKS assembly line, was reported as "turnstile mechanism" (chain elongation induces module closure, while chain translocation to the next module reopens the turnstile) [122]. In this study, it was shown that modules, of which the ACP is occupied by an acyl substrate, are not able to load an ACP-borne diketide intermediate onto their KS. The KS was first accepting an acyl chain after the substrate was released from the ACP. However, the exact mechanisms that "communicate" the ACP-loading status to the KS are still unknown. The presumption that the acylation of the KS itself initiates the transfer of the information about the ACP-conformation and ACP-loading state to the KS remains speculative.

2.2.3. Docking Domains: Intersubunit Communication in Polyketide Assembly Lines

The recent mechanistic and structural data on the PKS assembly lines and their modules provided additional details about these complex systems and uncovered some of the problems, which led to the failure of the initial PKS engineering efforts. From the obtained data, it was possible to generate a more nuanced map of modular *cis*-AT PKS structure and function [49,123], including the determination of so called "docking domains" (DDs) at the N and C termini of the PKSs [124]. The N- and C-terminal DDs ensure the correct PKS assembly into a functional multiprotein complex. For example, the polyketide core of erythromycin A is synthesized by three multienzyme subunits DEBS 1, DEBS 2, and DEBS 3, each harbouring two extension modules (Figure 2). This requires a proper inter-modular transfer between ACP and KS domains as well as inter-protein transfer between the *cis*-AT PKSs. A typical "docking domain", which enables the intersubunit communication, was identified in the erythromycin assembly line [124] (e.g., between the C terminus of DEBS 2 and N terminus of DEBS 3). The structure of the DD contains two separate four α-helix bundles, which together mediate not only specific docking interactions, but also promote dimerization of each homodimer [124].

Docking domains were postulated and experimentally confirmed for the less well understood *trans*-AT PKSs [125–127]. In this section, we describe the DDs involved in intersubunit communication of the PKS systems, which must not be confounded with the regions, C-terminal to the KS, which were originally proposed as AT-docking domains for the discrete AT in leinamycin biosynthesis (*trans*-AT PKS pathway) [128,129].

Unlike the *cis*-AT PKSs, in which the DDs for intersubunit communication occur between the PKS modules (the docking is C-terminal to the ACP of an upstream polypeptide and N-terminal to the KS of a downstream polypeptide), *trans*-AT PKSs are disconnected within and between modules. Specifically, the junctions between subunits in *trans*-AT PKSs occur within and between PKS-modules [125,126]. Such C- and N-terminal DDs were observed for VirA and VirFG in the *trans*-AT PKS pathway of virginiamycin [125]. In this study, it was shown that the deletion of the C-terminal partner (VirA ^{C}DD) or the downstream catalytic domain, significantly affects the N-terminal DD (VirFG ^{N}DD). The N-terminal DD (VirFG ^{N}DD) exhibited multiple characteristics of an intrinsically disordered protein [125]. The stability of the protein (VirFG ^{N}DD) was recovered after fusion of the docking domains. Similar results were provided for the *trans*-AT PKS system of macrolactin [125]. Two-helix, pseudosymmetric motifs of the length of about 25-residue, which non-covalently connect domains between and within the PKS modules were identified. The docking domains and their cognate domain were heterologously expressed and purified as stable complexes from *Escherichia coli* [126], indicating their importance, not only for the correct assembly of the polypeptides, but also for the overall protein stability of the PKS complex. Furthermore, the structure and portability of the four-helix bundle docking domains was demonstrated in deletion and swapping experiments [126].

Very recently, a fundamentally different mechanism of intersubunit communication at the KS/DH was reported for the *trans*-AT PKS gladiolin (gbn) [130]. In contrast to the virginiamycin and macrolactin PKSs, which use mutually compatible docking domains at both, the N- and C-termini of the interacting subunits [125,126], the GbnD4 KS domain utilizes a single, largely unstructured docking domain at the C terminus for a direct interaction with the GbnD5 DH domain [130]. The data confirmed that the docking domain is required for the communication of the KS with the ACP, appended to the DH.

Successful engineering of PKS pathways, especially in case of non-canonical assembly lines, requires the identification of crucial interaction regions for the PKS subunits and the elucidation of the mechanism involved in the communication between their domains. The new insights are valuable for the maintenance of proper protein-protein interactions, dynamics, and finally, for the chemical outcome of the modified PKS assembly lines.

3. Strategies of Acyltransferase-Based Polyketide Engineering

The most significant and commonly applied strategies of AT-based engineering of PKS pathways are: the AT domain-swapping, site-specific mutagenesis of the ATs and the cross-complementation of an AT-inactivated PKS by another AT, either from the cognate assembly line or using heterologous ATs (Figure 3 and Table 1).

3.1. Domain-Swapping

In general, there are two strategies to achieve AT domain swapping in a PKS system. The AT domain can be exchanged by another AT or indirectly replaced by an entire module swapping. For both principles, endogenous ATs/modules from the native assembly line or ATs/modules of a foreign pathway (heterologous parts) can be utilized. In cases where the original AT substrate specificity was altered and functional engineered PKSs were generated, new polyketide derivatives are produced (Figure 4).

3.1.1. Acyltransferase Domain Substitution and the Provision of the Required Non-Native Precursors

One of the most prominent AT-based engineering approaches for PKSs and polyketide structures is the substitution of the AT domain with an AT of a different substrate specificity. The best known example of an assembly line to which this strategy was applied, is erythromycin. In the following, examples of AT-swapping in the erythromycin PKS (DEBS) and in a modified variant of the DEBS (DEBS1-TE), and cases of AT-substitution combined with the optimization of precursor supply are presented. Moreover, the replacement of an AT in a non-erythromycin PKS, is described.

The erythromycin PKS DEBS was successfully engineered by replacing the methylmalonyl-CoA specific ATs from modules 1 and 2. For example, malonyl-CoA specific ATs from *Streptomyces hygroscopicus* ATCC 29253 (Hyg-AT2 from module 2 of a type I PKS and RAPS-AT14 from module 14 of the rapamycin PKS) and from *Streptomyces venezuelae* ATCC 15439, (Ven-AT from a PKS-like pathway) were inserted into the DEBS of the wild type strain *Saccharopolyspora erythraea* ER720 [50,131]. The mutants of *Sacch. erythraea* ER720, harbouring the engineered DEBS1 produced novel, bioactive erythromycins (12-desmethyl-12-deoxyerythromycin A, 10-desmethylerythromycin A and 10-desmethyl-12-deoxyerythromycin A) [50].

The engineering of the DEBS-PKS was also successful in heterologous hosts, such as *S. coelicolor* CH999 [132] and *Streptomyces lividans* K4–114 (in both strains the actinorhodin gene cluster was deleted) [133,134], expressing the entire DEBS-PKS. Liu et al. have shown that swapping of the DEBS-AT from module 6 in the *S. coelicolor* CH999 strain, with the AT originating from module 6 of a rapamycin PKS, leads to the production of the antibacterial erythromycin analog 2-nor-6-deoxyerythronolide B [54]. McDaniel et al. used the two heterologous hosts, *S. coelicolor* CH999 and *S. lividans* K4–114 and applied AT-substitution, combined AT and KR swapping, and other engineering strategies to generate a library of >50 macrolides, including examples of analogs with one, two, and three altered carbon centers of the polyketide products [135].

To minimize the DEBS and achieve a premature release of the polyketide chain from the erythromycin PKS, the assembly line's TE domain was relocated to the C-terminus of various ACPs (e.g., fusion of the TE to the C-terminus of the DEBS1 resulted in DEBS1-TE) [136,137]. The minimized PKS DEBS1-TE, which consisted of the first two modules of the erythromycin PKS-complex and the TE domain, immediately became a model system for studying not only the mechanistic issues in polyketide biosynthesis [138,139], but also for the engineering of the DEBS-PKS (e.g., AT-domain substitution). Among other experiments, the DEBS1-TE was used to replace the entire AT domain from module 1 by a heterologous AT derived from module 2 of the rapamycin PKS [51]. Two new lactones (2-methyl-3,5-dihydroxy-*n*-hexanoic acid δ-lacton and 2-methyl-3,5-dihydroxy-*n*-heptanoic acid δ-lacton) were produced by *S. coelicolor* containing the modified DEBS1-TE [51]. Although functional engineered PKSs were obtained, the yields of the products generated by these chimeras often greatly depended on the position of the polyketide chain at which the new building block was incorporated and frequently resulted in significantly lower production of the compounds [50,51]. In contrast, Lau et al., generated a variant of the bimodular DEBS1-TE, in which the AT2 domain was replaced with the malonyl-CoA-specific AT2 domain of the rapamycin PKS that produced 10 mg/L of the expected 2-desmethyl triketide lactone [52]. The productivity of the recombinant *S. coelicolor* strain, containing the engineered DEBS1+TE, was 50% relative to the production levels of the parent triketide lactone ((2*R*,3*S*,4*S*,5*R*)-2,4-dimethyl-3,5-dihydroxy-*n*-heptanoic acid δ-lactone), 20 mg/L [52].

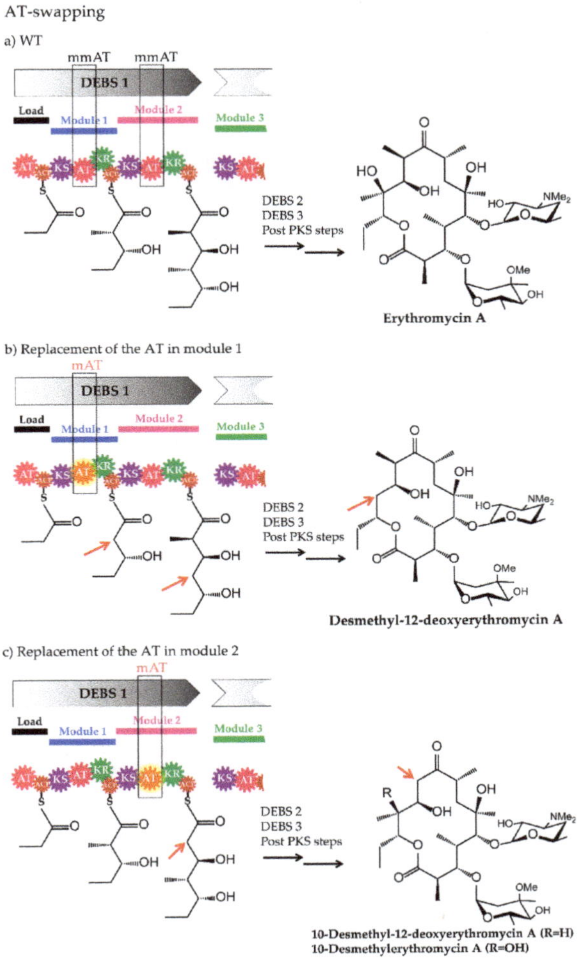

Figure 4. Examples of AT swapping [50]. (**a**) Part of the erythromycin assembly line illustrating DEBS 1 and the final structure of erythromycin A, produced by the wild type strain *Saccharopolyspora erythraea* ER720. (**b**) Mutant in which the AT domain of module 1 was replaced by a malonyl-CoA specific AT (loading malonate onto the ACP). The mutant strain produced the derivative 12-Desmethyl-12-deoxyerythromycin A. (**c**) Mutant in which the AT domain of module 2 was replaced by a malonyl-CoA specific AT (loading malonate onto the ACP). The mutant produced the derivatives 10-desmethyl-12-deoxyerythromycin A and 10-desmethylerythromycin A. Abbreviations: AT, acyltransferase; ACP, acyl carrier protein; KS, ketosynthase; KR, ketoreductase; mmAT: AT loading methylmalonate; mAT: AT loading malonate.

Further engineering attempts of the DEBS1-TE fusion include the replacement of module 2 of DEBS1-TE with the module 12 from rapamycin or module 3 from the erythromycin assembly line [53,60]. In both cases, engineered PKSs were obtained and triketide lactones, in which for example, acetate extender units were incorporated instead of propionate, were identified in the recombinant *Sacch. erythraea* JC2 strain [140] (*Sacch. erythraea* JC2 is a genome reduced derivative of *Sacch. erythraea* NRRL2338 lacking almost all erythromycin PKS genes). In another study the strain *Sacch. erythraea* JC2 was exploited for the replacement of the AT1 (methylmalonyl-CoA specific) in the model PKS DEBS1-TE with the epothilone EpoAT2 (malonyl-CoA specific in epothilone biosynthesis) and EpoAT3 (utilizes malonyl- and methylmalonyl-CoA at native epothilone production conditions) [55]. Functional PKSs, producing the expected triketide lactone compounds, were generated. However, lower yields of total products were detected when compared to DEBS1-TE (2% and 11.5% respectively) [55].

In some cases the alteration of the AT with an AT incorporating a different extender unit is not sufficient as the substrate is not available in the producer and it requires the external supplementation with the precursor or further genetic modification of the strain to overcome this limitation. Such an observation was made for a substitution experiment in which the AT domain of module 4 in erythromycin (eryAT4/methylmalonate-specific) was exchanged with the AT domain from module 5 (nidAT5/ethylmalonate-specific) of the niddamycin PKS [56]. The strain containing the modified PKS produced erythromycin A, but did not synthesize the ethyl-substituted derivative. After supplementation of the culture with ethylmalonate, moderate production of 6-desmethyl-6-ethylerythromycin A (6-ethylEr) was detected in addition to erythromycin A. To eliminate the limitation and improve the intracellular ethylmalonyl-CoA concentration, the crotonyl-CoA reductase (see also Section 2.1.2) from *Streptomyces collinus* [141] was introduced into the *Sacch. erythraea* EAT4, yielding *Sacch. erythraea* EAT4-crr. The *crr*-expressing strain synthesized the derivative 6-ethylErA [56].

Another example of a combined AT-substitution- and precursor supply optimization approach was provided by Kato and co-authors [57]. In their study, the previously designed *S. lividans* K4–114, harbouring a plasmid with genes encoding the erythromycin DEBS was modified: namely, the AT6 domain was replaced by the presumably hydroxymalonate-specifying *fkbA*-AT8 domain [54,58]. This DEBS construction was utilized for a substitution with an AT providing an exotic substrate (methoxymalonate) to the ascomycin assembly line [57]. The *fkbA*-AT8 (originally AT6 in DEBS) of the modified DEBS in *S. lividans* K4–114 was exchanged by the heterologous AT8 domain (loading methoxymalonate onto the PKS) from ascomycin (FK520, synthesized by *S. hygroscopicus*). In addition, a subcluster of five genes (*asm13-17*) from the ansamitocin biosynthetic gene cluster of *Actinosynnema pretiosum* was coexpressed in the modified *S. lividans* K4–114 strain to provide the methoxymalonyl extender unit for the engineered erythromycin PKS. Two novel analogs of erythronolide, 2-desmethyl-2-methoxy-DEB and 2-desmethyl-6-DEB, were produced by this strain [57].

Besides the AT of the erythromycin PKS, other assembly lines were modified by AT-swapping to produce polyketide derivatives. The AT domains in the synthesizing PKS-complex of geldanamycin, a potential anticancer drug, were substituted in six AT swaps (in modules 1, 2, 3, 4, 5, and 7) of the seven GdmPKS modules [59]. The AT-swapping using the RapAT2 domain and/or the RapAT14 domain, both from the rapamycin PKS, resulted in functional PKSs in four (modules 1, 4, 5, and 7) of the six modules. The geldanamycin analogs: 2-desmethyl, 6-desmethoxy, 8-desmethyl, and 14-desmethyl derivatives, including one analog with a four-fold enhanced affinity for its target (chaperone Hsp90, essential for growth of cancer cells), were produced in *S. hygroscopicus* [59].

3.1.2. Examples of Acyltransferase Domain Substitution by Exchanging the Entire Module and the Supply of Non-Native Precursors

The AT substitution resulting from the exchange of entire modules of a PKS assembly line is an indirect strategy and often a "side-effect" of an engineering experiment, primary aiming at alteration of a module and not necessarily at AT-swapping. The attempts of module swapping were recently discussed elsewhere [138,142–147]. Therefore, we provide only a few of the most relevant or recent examples for this engineering strategy.

The exchange of a whole module was successful for loading modules as well as for a chain extension module of a PKS [148–150]. For example, the wide-specificity loading module of the avermectin was introduced in place of the first module of DEBS and the resulting hybrid PKS gene was expressed in *Sacch. erythraea* [149]. Novel erythromycins derived from endogenous branched-chain acid starter units were observed, confirming the functionality of the engineered PKS and the flexibility of the assembly line, which accepted and prolonged the polyketide chain with alternative moieties [149].

Replacing of an entire loading module by a loading module of the avermectin PKS from *Streptomyces avermitilis* or the erythromcyin PKS of *Sacch. erythraea* was also successful for the spinosyn PKS in *Saccharopolyspora spinosa* BIOT-1066 [151]. The resulting strain derivatives of *Sacch. spinosa* BIOT-1066, expressing the hybrid PKS pathways produced the anticipated spinosyn analogs [151]. In a previous study, the inherent promiscuity of the loading module of the avermectin PKS was exploited for the generation of a compound library in *S. avermitilis* mutants, blocked in the biosynthesis of native starter units [152]. Structures including the most effective antiparasitic avermectin derivatives (e.g., dormectin) [153] were isolated. More recently, the avermectin biosynthetic pathway was engineered to provide alternative doramectin producers [154]. In this study, the loading module of the avermectin PKS in a *S. avermitilis* strain was replaced with a cyclohexanecarboxylic (CHC) unique loading module from phoslactomycin [154,155]. Furthermore, a CHC-CoA biosynthetic gene cassette was introduced into the engineered strain to ensure the production of the precursor for directed biosynthesis of doramectin. The obtained strain synthesized higher amounts of doramectin (53 mg/L), relative to the yields detected in case of an external supplementation of the wild type strain with the substrate (CHC) (9 mg/L). However, the quantity was significantly lower when compared to the parental avermectin producer (500 mg/L). Nevertheless, the valuable "target" compound (doramectin) could be derived from the module swapping approach and the potential limitations, leading to reduced production yields, might be eliminated by applying diverse strain engineering methods (mutagenesis, synthetic biology etc.).

In addition to the loading modules, chain extension modules were replaced, as it was the case for DEBS PKS variants. For example, the second module of the bimodular mini-PKS DEBS (loading + DEBS module 1 + module 2) was swapped with cognate modules of the erythromycin assembly line (DEBS module 3 and 6) and heterologous module 5 from rifamycin [150]. The engineered DEBS-PKSs produced the expected triketide lactone in the heterologous expression host *S. coelicolor* CH999 (see also Section 3.1.1). This demonstrates that both approaches (exchanging of whole loading- or chain extension module) can result in engineered PKSs, which are capable of assembling new polyketide derivatives.

3.2. Site-Specific Mutagenesis of Acyltransferases

A different approach to the whole domain swapping, aiming at the substitution of AT-substrate specificity, is the site-directed mutagenesis (Figure 5). Recent advances in sequencing methodology, bioinformatics, structural and synthetic biology contributed to the identification of amino acid signatures of ATs and have a significant impact on the engineering of PKS assembly lines [49,61,74,106,113,138,143,144,148,156–169].

Analysis of the amino acid sequences of ATs resulted in the identification of approximately 100 residues at the C-terminal end of the active-site serine in the analysed ATs, which was assigned to different AT-substrate specificities. Indeed, in most cases, the substrate specificity of the ATs could be predicted based on this motif. This observation encouraged the modification of the AT substrate specificity by exchanging the specific sequence signature. Such strategy was applied, for example, to the model PKS-complex DEBS and its ATs. The methylmalonyl-CoA specific YASH-motif of AT1, AT4 and AT6 of DEBS was altered to the malonyl-CoA specific HAFH (and HASH) [60–62], which resulted in AT variants incorporating both building blocks (methylmalonate and malonate) (Figure 5). Mutagenesis of sequences apart of this motif led to the generation of ATs transferring and loading non-native units onto the PKS however, reduced efficiency was observed [61,62].

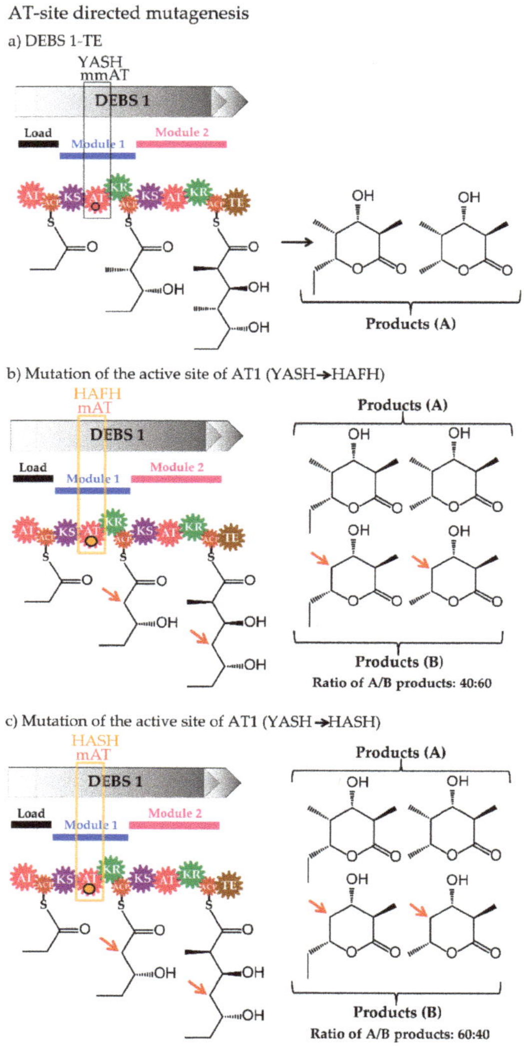

Figure 5. Examples of AT-site directed mutagenesis [60]. (**a**) The fusion of DEBS 1 and TE domain line (DEBS 1-TE) form the erythromycin assembly line. The strain containing the modified fusion DEBS 1-TE produced the respective triketide lactones (products A). (**b**) The fusion of DEBS 1-TE in which the AT of module 1 was modified by exchanging the specificity motif YASH (native, methylmalonyl-CoA) by the HAFH motif (malonyl-CoA). The strain *Saccharopolyspora erythraea* containing the modified DEBS-TE produced triketide lactones A and B (40:60). (**c**) The fusion of DEBS 1-TE in which the AT of module 1 was modified by exchanging the specificity motif YASH by the HASH motif (malonyl-CoA). The strain *Saccharopolyspora erythraea* containing the modified DEBS-TE produced triketide lactones A and B (60:40). Abbreviations: AT, acyltransferase; ACP, acyl carrier protein; KS, ketosynthase; KR, ketoreductase; TE, thioesterase; mmAT: AT loading methylmalonate; mAT: AT loading malonate.

Harvey et al. exploited the features of the loading AT by mutating one of the ATs of DEBS and obtaining a new class of alkynyl- and alkenyl-substituted macrolides (e.g., 15-propargyl erythromycin A) with activities comparable to that of the natural product [170].

In other studies, the prototypical PKS (DEBS) was subjected to mutagenesis and the generated active site-mutant library was screened for substrate selectivity [61,63–65]. Several single amino acid substitutions, having an impact on the selectivity of the PKS, were identified. The substitution Tyr189Arg in DEBS-AT6 inverted the selectivity of the DEBS from its natural substrate toward an alkynyl-modified unit [65].

Bravo-Rodriguez et al. used a combined approach, which involved molecular modelling and mutagenesis of the AT6 of DEBS and showed that the V295A mutation of this AT leads to a wider active site and improves the promiscuity for non-native substrates [61,63,64].

Based on AT amino acid sequences, loading ATs (ATs incorporated into the loading module) usually have relaxed substrate specificity and load the loading module with diverse starter units. For example, the loading AT from the modular PKS of avermectin accepts more than 40 carboxylic acids as substrate for the loading module leading to the biosynthesis of a series of congeners [165]. Using modelling tools and site-directed mutagenesis targeting the active site residues, altered specificity toward a panel of synthetic substrate mimics was achieved [165].

Very recently, high-resolution X-ray crystal structures of a broadly selective AT SpnD-AT of the splenocin *cis*-AT PKS, accepting long aliphatic chains, up to C7-malonyl-CoA and aromatic benzylmalonyl-CoAs were solved [171]. To the best of the authors' knowledge, their work provided first structures of an AT-substrate complex and enabled the understanding of both, the stereoselectivity and the broad substrate specificity of the SpnD-AT [171]. Furthermore, using the structural data, it was possible to mutate key residues of the canonical Ery-AT6 from the erythromycin PKS and "shift" its substrate preference. The modified Ery-AT6 was able to accept diverse bulky extender units [171]. These results suggest that future efforts to expand AT substrate tolerance should target the three important residues, Q150, Y278, S280 in methylmalonyl-CoA specific AT or Q150, H278, F280 in malonyl-CoA specific AT [171].

In addition to the AT of *cis*-AT PKS, the discrete AT from the *trans* AT-PKS pathways [31,32,172] was investigated to identify engineering opportunities. While the *cis*-AT PKSs occasionally harbour ATs providing non-malonate building blocks to the assembly line, the ATs of the *trans*-AT PKSs are mostly malonyl-CoA-specific at natural, non-manipulated production conditions. The KirCII-AT [173] from the kirromycin pathway [174] is an exception. This AT loads a branched precursor (ethylmalonate) onto one particular ACP (KirACP5) at native, non-manipulated conditions and therefore belongs to one of the most promiscuous discrete ATs for regiospecific PKS engineering. To map the interaction epitope KirCII:KirACP5, the ACP was subjected to modelling and alanine scanning mutagenesis, where 61 surface residues were individually mutated to Ala [175]. Additionally, several KirCII-mutants were constructed and tested in vitro. The regions involved in the KirCII:KirACP5 interaction, *trans*-acylation activity were identified. Further in vitro investigation of the substrate specificity of KirCII in ACP loading assays revealed that the AT is able to load KirACP5 with allyl- and propargyl-malonate and to a lesser extent with azidoethyl- and phenyl-malonate [176]. Those precursors are not available in the producer cell and the in vivo production of the respective kirromycin derivatives required external supplementation (feeding) with the substrate [90]. Similar in vitro studies were conducted for the discrete AT (malonyl-CoA specific at native condition) from disorazole biosynthesis. Wong et al. used alanine-scanning mutagenesis of one of the ACPs (ACP1 from DSZS) of the discrete AT and identified a conserved Asp45 residue in ACP1, which is co-responsible for the recognition of this ACP as substrate by the AT enzyme [106].

While the *cis*-ATs, such as the ATs of the erythromycin assembly line are relatively well studied and were manipulated for several times, there is not much known about the substrate flexibility and the exact mechanisms of partner (precursor or ACP) recognition for ATs from other PKS pathways (e.g., those of the *trans*-AT PKSs). The knowledge obtained from manipulations conducted in the past years will contribute to understanding of the less-explored systems and help to overcome the still existing limitations for AT-based engineering of PKS pathways.

3.3. Cross-Acyltransferase Complementation

In principle, an AT-inactivated *cis*-AT PKS can be complemented in two ways. One option is the insertion of an AT domain (synthetic, from another module of the native or derived from heterologous assembly lines) in place of the original domain into the PKS module (*cis*-to-*cis* AT-complementation). Another approach is the complementation of the PKS by introduction of a discrete AT, which provides the building blocks in *trans* (*cis*-to-*trans* AT-complementation) (Figures 3 and 6 and Table 1).

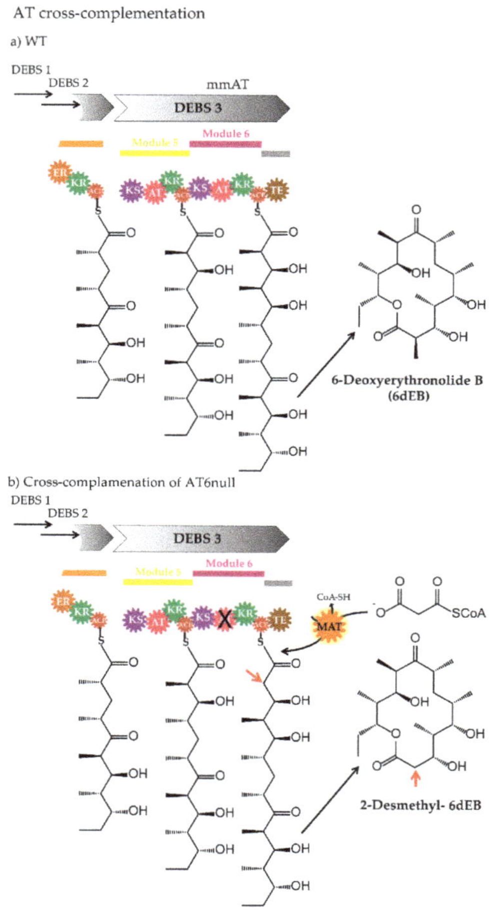

Figure 6. Example of AT cross-complementation [66]. (**a**) Part of the erythromycin biosynthetic pathway illustrating the DEBS complex and the chain intermediates on modules 4, 5 and 6, and the product 6-deoxyerythronolide B (6dEB). (**b**) Part of the erythromycin biosynthetic pathway illustrating the DEBS complex in which the AT6 was inactivated (AT6null) and complemented by a discrete enzyme malonyl-CoA:ACP transacylase (MAT) from *Streptomyces coelicolor*. The strain produced 2-desmethyl-6dEB. Abbreviations: AT, acyltransferase; ACP, acyl carrier protein; KS, ketosynthase; KR, ketoreductase; TE, thioesterase; mmAT: AT loading methylmalonate.

Most of the past AT complementation studies were carried out using the PKSs of the model *cis*-AT PKS assembly line erythromycin from *Sacch. erythraea* and related polyketide pathways. Their modification by deletion of the native AT and integration of a foreign AT, often regarded as domain substitution or swapping, was described in Section 3.1, as well as in diverse reports and reviews [48–50,65,138,142–144,167,168,177–182]. Here, we provide examples of cross-AT complementation, where the functionality of an AT-inactivated PKS (PKS module) was restored by an external heterologous AT protein (e.g., introduction of a heterologous AT into a mutant strain containing an AT-inactivated PKS).

Although the *trans*-AT PKS pathways were overlooked for a long time and are less well characterized than the *cis*-AT PKSs, the number of this type of biosynthetic assembly lines is constantly increasing [32,183]. In most cases, discrete ATs (freestanding ATs or *trans*-ATs) exhibit malonyl-CoA specificity in the native producer strain (e.g., disorazole AT Dis/Dsz [106], bryostatin AT BryP [67], bacillaene AT PksC [184], rhizopodin ATs RizA and RizF [185], and kirromycin AT KirCI [186]) and the diversification of the polyketide chain takes place rather after the elongation by reduction, methylation, and other modifications. However, very few *trans*-AT PKS pathways, which involve discrete ATs with non-malonyl-CoA specificity at native production conditions, were identified. The so far experimentally confirmed examples include the ethylmalonyl-CoA-specific KirCII [173] from kirromycin, the methoxymalonyl-ACP-specific OzmC [187] from oxazolomycins, and the (2*S*)-aminomalonyl-ACP-specific ZmaF [109] required for zwittermicin A biosynthesis. The discrete ATs might exhibit relaxed substrate specificity when the respective precursor is available. This was shown for example for KirCII [90,173,176]. Considering this observation and the fact that the freestanding ATs are encoded by separate genes and act *in trans* to complement the cognate PKSs, their development to a tool for PKS engineering might be easier, compared to the in the ATs embedded in the PKS (*cis*-AT PKSs). In in vitro studies, it was demonstrated that a *trans*-AT from kirromycin (KirCII) and disorazole (Dis/Dsz AT) can complement the AT-null DEBS of the *cis*-AT type [172]. In the past, the AT-null-DEBS module 6 (DEBS, in which the AT domain of module 6 was deleted) was used for similar complementation experiments. For instance, the functionality of the AT-inactivated module was restored after the supplementation with malonyl-CoA:ACP transacylase from *S. coelicolor* (Figure 6), which led to the production of 2-desmethyl-6-dEB [66]. The complementation of AT-null DEBS module 6 was also demonstrated in combination with the AT-domains of the discrete tandem AT BryP from the bryostatin PKS [67] (from ca. *Endobugula sertula*- bacterial symbiont of the marine bryozoan *Bugula neritina*). BryP was also able to catalyse the acyl-transfer onto ACPs of pikromycin and other bryostatin PKS modules [67].

In addition, an in vivo cross-species complementation, using ATs from two *trans*-AT PKSs was successful for the Δ*mmpC*-AT1 mutant from mupirocin [188] (*Pseudomonas fluorescens*). *P. fluorescens* produces a mixture of several pseudomonic acids, named mupirocin [189]. Constructs for BryP-AT1 and the didomain protein BryP-AT1AT2 complementation were introduced into the Δ*mmpC*-AT1 mutant [67]. The complementation of the Δ*mmpC*-AT1 mutant expressing the BryP-AT1 restored the biosynthesis of the main compound of mupirocin, the pseudomonic acid A (PA-A), to approximately 83% of WT production levels.

The efficiency of *trans*-AT complementation is often affected by the identity of the building block and the ACP [172,175]. Thus, the successful complementation and production of efficient modified PKSs still needs a better understanding of the discrete AT-substrate (AT-precursor and AT-ACP) recognition and interaction as well as experimentally verified data, which disclose if the gained engineering knowledge can be transferred to diverse assembly lines or which modifications are required to recover the systems.

4. Advances in Natural Science and Future Perspectives of AT-Based PKS Engineering

The recent progress and advances related to polyketides and their biosynthetic machineries provide valuable and essential know-how, which makes the AT-based engineering of PKS assembly lines even more promising than it was before. Nonetheless, this strategy is not yet a standard, high-throughput technology, which is applicable to any desired PKS enzyme. To enable an extensive and efficient construction of functional polyketide machineries, further analysis and experimentally confirmed insights into the complex systems, including non-canonical PKSs are indispensable. The understanding of the structural, mechanistic, catalytic, biochemical details not only for individual domains, but the whole assembly line would make the engineering and tuning more predictable and executable. Furthermore, directed optimization of the producer cell "hosting" the biosynthetic complex would lead to improvements in the productivity of those engineered systems. Today's sequencing technologies, bioinformatics, structural biology, chemistry, molecular biology, and other in vitro and in vivo tools facilitate the further investigation of the PKS systems and the PKS engineering. Researchers aiming at AT-based PKS engineering will most probably continue to use the established and approved strategies and methodology. Those include in general, the expanding of the pool of precursors [97,98,190] for the ATs, the modification of the AT enzymes under the consideration of the obtained structural and mechanistic insights, AT-domain swapping and cross-complementation of inactivated ATs with ATs from other modules/pathways, and site-directed mutagenesis to exchange the substrate specificity of the AT. Specifically, the information existing in databases [191,192] is incredibly helpful, not only for direct sequence alignments and phylogenetic analysis [193], but also for the development of advanced bioinformatics tools supporting the AT-targeted PKS engineering [194]. The application of such tools allows the identification of new PKS pathways [195] and/or novel types of ATs, the prediction of AT substrate specificity [61,62,158], the analysis of amino acid coevolution in protein sequences [196–198], and even protein interaction surfaces [41,199,200]. Recently, the computational online platform ClusterCAD, which facilitates the selection of natural *cis*-AT PKS parts to design novel chimeric PKSs for the biosynthesis of small PKS-derived compounds, was developed [144,201].

The protein structures of PKS domains [112,127,162,169,202–205], including the ATs [113,161,162,171,184] and whole modules [120,206] enable a more reliable modelling of the architecture of those enzymes and will improve the structure-based PKS engineering. Diverse protein modelling platforms are already available [207–212] and were successfully exploited. For example, the modelling of the erythromycin DEBS AT6 with (2*S*)-methylmalonyl-CoA confirmed the role of the proposed active-site residues and revealed residues important for the substrate binding [61]. The obtained information was used for simulations on mutants with the native methylmalonyl-CoA and non-native malonyl-CoA extender units, which led to the identification of residues prohibiting the binding of the desired substrate. Consequently, the respective sites were mutated and mutants, which were able to utilize the substrate 2-propargylmalonyl-SNAC were generated [61].

In another study, the small-angle X-ray scattering (SAXS) was used to characterize the structure of a module (*apo* module 5 from the VirA subunit) and to identify the positions of domains flanking the ACP and KS in the in the virginamycin PKS [213]. The outcome of this analysis enabled modelling of the complete intersubunit interface in the virginiamycin *trans*-AT PKS system [125].

Those results are encouraging the implementation of the computational tools, which contribute to the better understanding and more efficient engineering of the complex PKS assembly lines. It is very likely that the bioinformatics software tools will become more important due to their intensive use for diverse simulations and analysis of structural details of PKSs and their AT domains, as well as for the design of engineering experiments in the future. Furthermore, the existing databases will be complemented with the knowledge (e.g., structural data on non-canonical ATs) gained from the planned and currently conducted research on PKSs. This will lead to the development of new computational tools (or tool features) and improve the predictability of PKS-variation as well.

In addition, the structural biology technique itself is constantly improving. A new method for the direct delivery of the sample into an X-ray free-electron laser was used for the *trans*-AT from the disorazole PKS [161]. The novel sample extractor efficiently delivered limited quantities of microcrystals directly from the native crystallization solution into the X-ray beam at room temperature. A crystal structure of the discrete AT with resolution of 2.5 Å was obtained [161]. The crystallization of this difficult to handle enzymes and their complexes (e.g., AT-substrate) is highly desired, as it will provide useful information about the ATs and their interaction partners. We speculate that the progress in this field will continue and support the engineering efforts.

Directed evolution [214–218] was successfully applied to generate enzyme libraries. This approach usually uses saturation mutagenesis, error-prone polymerase chain reaction (PCR) or DNA shuffling to generate a library of mutated proteins. Subsequently the protein variants are screened in an adequate assay to identify the most promising mutants. The selected mutants are subjected to another cycle of mutagenesis to "evolve" the enzyme to the favoured protein version. This artificial process of repeating cycles of mutagenesis and selection mimics the evolution in the Nature. Although the high throughput directed evolution of PKSs and the AT enzymes is mainly limited by rapid screening (assays and analytics for detection of the of the enzyme catalysis), progress can be observed [219–221]. The directed evolution screening/selection strategies, which have been employed to improve or modify the functions of nonribosomal peptide synthetases (NRPSs) and PKSs were summarized in a review [222] recently published by Rui and Zhang. Directed evolution strategies enable the identification of residues affecting the enzyme properties, which might be difficult to detect using alternative methods. Thus, the directed evolution approach provides a good alternative for PKS and AT optimization.

In the future, those AT-based PKS engineering strategies and methodologies, which rely on DNA steps, will be supported by advanced cloning methods such as Red/ET system [223–225], the TAR cloning [226,227], USER cloning [228–230], CRISPR-Cas9 [231–235], and other synthetic biology tools [138,143,144,236–238].

Finally, the efficient production of natural and engineered polyketides often requires the tuning of the native producer strain or the development of suitable heterologous expression hosts [182,239,240]. Preferably, diverse actinomycetes [241–245], *E. coli* [246–248], *Bacillus* [249], *Saccharomyces* [250,251] and/or *Aspergillus* strains [252,253] were used for heterologous expression of polyketide pathways in the past. It is very likely, that the existing hosts will be further optimized and specifically adapted to the expression of engineered pathways, including AT-modified PKSs. Most probably, new potential hosts will be identified for the needs of expression of the respective assembly lines and the biosynthesis of the final engineered compound.

5. Conclusions

In the past, many attempts of combinatorial biosynthesis, aiming at engineered PKSs resulted in inactive or inefficient enzymes and/or assembly lines. The newly obtained knowledge about the ATs and their PKS pathways in combination with methodology advancements provide exciting new perspectives for the AT-based PKS engineering. The design and generation of functional and efficient PKS pathways might enable the production of new bioactive compounds of which structures would be extremely difficult to access using traditional synthetic chemistry approaches.

Author Contributions: The review was written and edited by the authors E. M.-K. and W.W.

Funding: The authors and work in their laboratory is supported by the Eberhard Karls Universität Tübingen, the DFG, the BMBF (FKZ 031L 0018A, ERASysApp), and Biovet (Sofia, Bulgaria).

Conflicts of Interest: The authors declare no conflict of interest.

Abbreviations

ACC	acetyl-CoA carboxylase
ACP	acyl carrier protein
AT	acyltransferase
CCR	crotonyl-CoA carboxylase/reductase
CCRC	crotonyl-CoA reductase/carboxylase
CoA	coenzyme A
CRISPR-Cas9	clustered regularly interspersed palindromic repeats; CRISPR associated (Cas) protein (Cas9)
Cryo-EM	cryo-electron microscopy
KS	ketosynthase
DD	docking domains
DEBS	6-deoxyerythromycin B synthase
DH	dehydrogenase
ECR	enoyl-CoA carboxylase/reductase
ER	enoylreductase
KR	ketoreductase
NRPS	nonribosomal peptide synthetase
PCC	propionyl-CoA carboxylase
PKS	polyketide synthase
PPTase	4-phosphopantetheinyl transferase
Red/ET	homologous recombination system based on the Red operon of lambda phage or RecE/RecT from Rac phage
SNAC	*N*-acetylcysteamine
TAR	transformation-associated recombination
TE	thioesterase
USER	uracil-excision based cloning
YCC	acyl-CoA carboxylase

References

1. Cane, D.E.; Walsh, C.T.; Khosla, C. Harnessing the biosynthetic code: Combinations, permutations, and mutations. *Science* **1998**, *282*, 63–68. [CrossRef] [PubMed]
2. Hertweck, C. The biosynthetic logic of polyketide diversity. *Angew. Chem. Int. Ed. Engl.* **2009**, *48*, 4688–4716. [CrossRef] [PubMed]
3. Mahajan, G.B.; Balachandran, L. Antibacterial agents from actinomycetes—A review. *Front. Biosci.* **2012**, *4*, 240–253. [CrossRef]
4. Vannucchi, V. Clinical study of a new antibiotic; erythromycin. *Riv. Crit. Clin. Med.* **1952**, *52*, 128–136. [PubMed]
5. Haight, T.H.; Finland, M. Laboratory and clinical studies on erythromycin. *N. Engl. J. Med.* **1952**, *247*, 227–232. [CrossRef] [PubMed]
6. Wuite, J.; Davies, B.I.; Go, M.; Lambers, J.; Jackson, D.; Mellows, G. Pseudomonic acid: A new topical antimicrobial agent. *Lancet* **1983**, *2*, 394. [CrossRef]
7. Casewell, M.W.; Hill, R.L. Mupirocin ('pseudomonic acid')—A promising new topical antimicrobial agent. *J. Antimicrob. Chemother.* **1987**, *19*, 1–5. [CrossRef] [PubMed]
8. Vezina, C.; Kudelski, A.; Sehgal, S.N. Rapamycin (AY-22,989), a new antifungal antibiotic. I. Taxonomy of the producing streptomycete and isolation of the active principle. *J. Antibiot. (Tokyo)* **1975**, *28*, 721–726. [CrossRef] [PubMed]
9. Sehgal, S.N.; Baker, H.; Vezina, C. Rapamycin (AY-22,989), a new antifungal antibiotic. II. Fermentation, isolation and characterization. *J. Antibiot. (Tokyo)* **1975**, *28*, 727–732. [CrossRef] [PubMed]
10. Kino, T.; Hatanaka, H.; Hashimoto, M.; Nishiyama, M.; Goto, T.; Okuhara, M.; Kohsaka, M.; Aoki, H.; Imanaka, H. FK-506, a novel immunosuppressant isolated from a *Streptomyces*. I. Fermentation, isolation, and physico-chemical and biological characteristics. *J. Antibiot. (Tokyo)* **1987**, *40*, 1249–1255. [CrossRef] [PubMed]

11. Kino, T.; Hatanaka, H.; Miyata, S.; Inamura, N.; Nishiyama, M.; Yajima, T.; Goto, T.; Okuhara, M.; Kohsaka, M.; Aoki, H.; et al. FK-506, a novel immunosuppressant isolated from a *Streptomyces*. II. Immunosuppressive effect of FK-506 in vitro. *J. Antibiot. (Tokyo)* **1987**, *40*, 1256–1265. [CrossRef] [PubMed]
12. Bollag, D.M.; McQueney, P.A.; Zhu, J.; Hensens, O.; Koupal, L.; Liesch, J.; Goetz, M.; Lazarides, E.; Woods, C.M. Epothilones, a new class of microtubule-stabilizing agents with a taxol-like mechanism of action. *Cancer Res.* **1995**, *55*, 2325–2333. [PubMed]
13. Gerth, K.; Bedorf, N.; Hofle, G.; Irschik, H.; Reichenbach, H. Epothilons A and B: Antifungal and cytotoxic compounds from *Sorangium cellulosum* (myxobacteria). Production, physico-chemical and biological properties. *J. Antibiot. (Tokyo)* **1996**, *49*, 560–563. [CrossRef] [PubMed]
14. Paterson, I.; Lam, N.Y.S. Challenges and discoveries in the total synthesis of complex polyketide natural products. *J. Antibiot. (Tokyo)* **2018**, *71*, 215–233. [CrossRef] [PubMed]
15. Mapperson, R.R.; Kotiw, M.; Davis, R.A.; Dearnaley, J.D. The diversity and antimicrobial activity of *Preussia* sp. endophytes isolated from Australian dry rainforests. *Curr. Microbiol.* **2014**, *68*, 30–37. [CrossRef] [PubMed]
16. El-Hawary, S.S.; Mohammed, R.; AbouZid, S.F.; Bakeer, W.; Ebel, R.; Sayed, A.M.; Rateb, M.E. Solamargine production by a fungal endophyte of *Solanum nigrum*. *J. Appl. Microbiol.* **2016**, *120*, 900–911. [CrossRef] [PubMed]
17. Iftime, D.; Kulik, A.; Hartner, T.; Rohrer, S.; Niedermeyer, T.H.; Stegmann, E.; Weber, T.; Wohlleben, W. Identification and activation of novel biosynthetic gene clusters by genome mining in the kirromycin producer *Streptomyces collinus* Tü 365. *J. Ind. Microbiol. Biotechnol.* **2016**, *43*, 277–291. [CrossRef] [PubMed]
18. Teanpaisan, R.; Kawsud, P.; Pahumunto, N.; Puripattanavong, J. Screening for antibacterial and antibiofilm activity in Thai medicinal plant extracts against oral microorganisms. *J. Tradit. Complement Med.* **2017**, *7*, 172–177. [CrossRef] [PubMed]
19. Zhu, J.J.; Yao, S.; Guo, X.; Yue, B.S.; Ma, X.Y.; Li, J. Bioactivity-guided screening of wound-healing active constituents from American cockroach (*Periplaneta americana*). *Molecules* **2018**, *23*, 101. [CrossRef] [PubMed]
20. Cortes, J.; Haydock, S.F.; Roberts, G.A.; Bevitt, D.J.; Leadlay, P.F. An unusually large multifunctional polypeptide in the erythromycin-producing polyketide synthase of *Saccharopolyspora erythraea*. *Nature* **1990**, *348*, 176–178. [CrossRef] [PubMed]
21. Donadio, S.; Staver, M.J.; McAlpine, J.B.; Swanson, S.J.; Katz, L. Modular organization of genes required for complex polyketide biosynthesis. *Science* **1991**, *252*, 675–679. [CrossRef] [PubMed]
22. Shen, B. Polyketide biosynthesis beyond the type I, II and III polyketide synthase paradigms. *Curr. Opin. Chem. Biol.* **2003**, *7*, 285–295. [CrossRef]
23. Müller, R. Don't classify polyketide synthases. *Chem. Biol.* **2004**, *11*, 4–6. [CrossRef] [PubMed]
24. Kellmann, R.; Stuken, A.; Orr, R.J.; Svendsen, H.M.; Jakobsen, K.S. Biosynthesis and molecular genetics of polyketides in marine dinoflagellates. *Mar. Drugs* **2010**, *8*, 1011–1048. [CrossRef] [PubMed]
25. Chen, H.; Du, L. Iterative polyketide biosynthesis by modular polyketide synthases in bacteria. *Appl. Microbiol. Biotechnol.* **2016**, *100*, 541–557. [CrossRef] [PubMed]
26. Shimizu, Y.; Ogata, H.; Goto, S. Type III polyketide synthases: Functional classification and phylogenomics. *ChemBioChem* **2017**, *18*, 50–65. [CrossRef] [PubMed]
27. Shimizu, Y.; Ogata, H.; Goto, S. Corrigendum: Type III polyketide synthases: Functional classification and phylogenomics. *ChemBioChem* **2017**, *18*, 1048–1049. [CrossRef] [PubMed]
28. He, J.; Hertweck, C. Functional analysis of the aureothin iterative type I polyketide synthase. *ChemBioChem* **2005**, *6*, 908–912. [CrossRef] [PubMed]
29. Low, Z.J.; Pang, L.M.; Ding, Y.; Cheang, Q.W.; Le Mai Hoang, K.; Thi Tran, H.; Li, J.; Liu, X.W.; Kanagasundaram, Y.; Yang, L.; et al. Identification of a biosynthetic gene cluster for the polyene macrolactam sceliphrolactam in a *Streptomyces* strain isolated from mangrove sediment. *Sci. Rep.* **2018**, *8*, 1594. [CrossRef] [PubMed]
30. Staunton, J.; Weissman, K.J. Polyketide biosynthesis: A millennium review. *Nat. Prod. Rep.* **2001**, *18*, 380–416. [CrossRef] [PubMed]
31. Musiol, E.M.; Weber, T. Discrete acyltransferases involved in polyketide biosynthesis. *Med. Chem. Comm.* **2012**, *3*, 871–886. [CrossRef]
32. Helfrich, E.J.; Piel, J. Biosynthesis of polyketides by *trans*-AT polyketide synthases. *Nat. Prod. Rep.* **2016**, *33*, 231–316. [CrossRef] [PubMed]

33. Hajra, A.K. Dihydroxyacetone phosphate acyltransferase. *Biochim. Biophys. Acta* **1997**, *1348*, 27–34. [CrossRef]
34. Barney, B.M.; Wahlen, B.D.; Garner, E.; Wei, J.; Seefeldt, L.C. Differences in substrate specificities of five bacterial wax ester synthases. *Appl. Environ. Microbiol.* **2012**, *78*, 5734–5745. [CrossRef] [PubMed]
35. Wenning, L.; Yu, T.; David, F.; Nielsen, J.; Siewers, V. Establishing very long-chain fatty alcohol and wax ester biosynthesis in *Saccharomyces cerevisiae*. *Biotechnol. Bioeng.* **2017**, *114*, 1025–1035. [CrossRef] [PubMed]
36. Caffrey, P. Conserved amino acid residues correlating with ketoreductase stereospecificity in modular polyketide synthases. *ChemBioChem* **2003**, *4*, 654–657. [CrossRef] [PubMed]
37. Minowa, Y.; Araki, M.; Kanehisa, M. Comprehensive analysis of distinctive polyketide and nonribosomal peptide structural motifs encoded in microbial genomes. *J. Mol. Biol.* **2007**, *368*, 1500–1517. [CrossRef] [PubMed]
38. Sugimoto, Y.; Ishida, K.; Traitcheva, N.; Busch, B.; Dahse, H.M.; Hertweck, C. Freedom and constraint in engineered noncolinear polyketide assembly lines. *Chem. Biol.* **2015**, *22*, 229–240. [CrossRef] [PubMed]
39. Reid, R.; Piagentini, M.; Rodriguez, E.; Ashley, G.; Viswanathan, N.; Carney, J.; Santi, D.V.; Hutchinson, C.R.; McDaniel, R. A model of structure and catalysis for ketoreductase domains in modular polyketide synthases. *Biochemistry* **2003**, *42*, 72–79. [CrossRef] [PubMed]
40. Anand, S.; Mohanty, D. Modeling *holo*-ACP:DH and *holo*-ACP:KR complexes of modular polyketide synthases: A docking and molecular dynamics study. *BMC Struct. Biol.* **2012**, *12*, 10. [CrossRef] [PubMed]
41. Yadav, G.; Anand, S.; Mohanty, D. Prediction of inter domain interactions in modular polyketide synthases by docking and correlated mutation analysis. *J. Biomol. Struct. Dyn.* **2013**, *31*, 17–29. [CrossRef] [PubMed]
42. Zhang, L.; Hashimoto, T.; Qin, B.; Hashimoto, J.; Kozone, I.; Kawahara, T.; Okada, M.; Awakawa, T.; Ito, T.; Asakawa, Y.; et al. Characterization of giant modular PKSs provides insight into genetic mechanism for structural diversification of aminopolyol polyketides. *Angew. Chem. Int. Ed. Engl.* **2017**, *56*, 1740–1745. [CrossRef] [PubMed]
43. Keatinge-Clay, A.T. Polyketide synthase modules redefined. *Angew. Chem. Int. Ed. Engl.* **2017**, *56*, 4658–4660. [CrossRef] [PubMed]
44. Vander Wood, D.A.; Keatinge-Clay, A.T. The modules of *trans*-acyltransferase assembly lines redefined with a central acyl carrier protein. *Proteins* **2018**, *86*, 664–675. [CrossRef] [PubMed]
45. Medema, M.H.; Cimermancic, P.; Sali, A.; Takano, E.; Fischbach, M.A. A systematic computational analysis of biosynthetic gene cluster evolution: Lessons for engineering biosynthesis. *PLoS Comput. Biol.* **2014**, *10*, e1004016. [CrossRef] [PubMed]
46. Gulder, T.A.; Freeman, M.F.; Piel, J. The catalytic diversity of multimodular polyketide synthases: Natural product biosynthesis beyond textbook assembly rules. *Top. Curr. Chem.* **2011**. [CrossRef]
47. Khosla, C.; Herschlag, D.; Cane, D.E.; Walsh, C.T. Assembly line polyketide synthases: Mechanistic insights and unsolved problems. *Biochemistry* **2014**, *53*, 2875–2883. [CrossRef] [PubMed]
48. Dunn, B.J.; Khosla, C. Engineering the acyltransferase substrate specificity of assembly line polyketide synthases. *J. R. Soc. Interface* **2013**, *10*, 20130297. [CrossRef] [PubMed]
49. Bayly, C.L.; Yadav, V.G. Towards precision engineering of canonical polyketide synthase domains: Recent advances and future prospects. *Molecules* **2017**, *22*. [CrossRef] [PubMed]
50. Ruan, X.; Pereda, A.; Stassi, D.L.; Zeidner, D.; Summers, R.G.; Jackson, M.; Shivakumar, A.; Kakavas, S.; Staver, M.J.; Donadio, S.; et al. Acyltransferase domain substitutions in erythromycin polyketide synthase yield novel erythromycin derivatives. *J. Bacteriol.* **1997**, *179*, 6416–6425. [CrossRef] [PubMed]
51. Oliynyk, M.; Brown, M.J.; Cortes, J.; Staunton, J.; Leadlay, P.F. A hybrid modular polyketide synthase obtained by domain swapping. *Chem. Biol.* **1996**, *3*, 833–839. [CrossRef]
52. Lau, J.; Fu, H.; Cane, D.E.; Khosla, C. Dissecting the role of acyltransferase domains of modular polyketide synthases in the choice and stereochemical fate of extender units. *Biochemistry* **1999**, *38*, 1643–1651. [CrossRef] [PubMed]
53. Ranganathan, A.; Timoney, M.; Bycroft, M.; Cortes, J.; Thomas, I.P.; Wilkinson, B.; Kellenberger, L.; Hanefeld, U.; Galloway, I.S.; Staunton, J.; et al. Knowledge-based design of bimodular and trimodular polyketide synthases based on domain and module swaps: A route to simple statin analogues. *Chem. Biol.* **1999**, *6*, 731–741. [CrossRef]
54. Liu, L.; Thamchaipenet, A.; Fu, H.; Betlach, M.; Ashley, G. Biosynthesis of 2-nor-6-deoxyerythronolide B by rationally designed domain substitution. *J. Am. Chem. Soc.* **1997**, *119*, 10553–10554. [CrossRef]

55. Petkovic, H.; Sandmann, A.; Challis, I.R.; Hecht, H.J.; Silakowski, B.; Low, L.; Beeston, N.; Kuscer, E.; Garcia-Bernardo, J.; Leadlay, P.F.; et al. Substrate specificity of the acyl transferase domains of EpoC from the epothilone polyketide synthase. *Org. Biomol. Chem.* **2008**, *6*, 500–506. [CrossRef] [PubMed]
56. Stassi, D.L.; Kakavas, S.J.; Reynolds, K.A.; Gunawardana, G.; Swanson, S.; Zeidner, D.; Jackson, M.; Liu, H.; Buko, A.; Katz, L. Ethyl-substituted erythromycin derivatives produced by directed metabolic engineering. *Proc. Natl. Acad. Sci. USA* **1998**, *95*, 7305–7309. [CrossRef] [PubMed]
57. Kato, Y.; Bai, L.; Xue, Q.; Revill, W.P.; Yu, T.W.; Floss, H.G. Functional expression of genes involved in the biosynthesis of the novel polyketide chain extension unit, methoxymalonyl-acyl carrier protein, and engineered biosynthesis of 2-desmethyl-2-methoxy-6-deoxyerythronolide B. *J. Am. Chem. Soc.* **2002**, *124*, 5268–5269. [CrossRef] [PubMed]
58. Reeves, C.D.; Chung, L.M.; Liu, Y.; Xue, Q.; Carney, J.R.; Revill, W.P.; Katz, L. A new substrate specificity for acyl transferase domains of the ascomycin polyketide synthase in *Streptomyces hygroscopicus*. *J. Biol. Chem.* **2002**, *277*, 9155–9159. [CrossRef] [PubMed]
59. Patel, K.; Piagentini, M.; Rascher, A.; Tian, Z.Q.; Buchanan, G.O.; Regentin, R.; Hu, Z.; Hutchinson, C.R.; McDaniel, R. Engineered biosynthesis of geldanamycin analogs for Hsp90 inhibition. *Chem. Biol.* **2004**, *11*, 1625–1633. [CrossRef] [PubMed]
60. Del Vecchio, F.; Petkovic, H.; Kendrew, S.G.; Low, L.; Wilkinson, B.; Lill, R.; Cortes, J.; Rudd, B.A.; Staunton, J.; Leadlay, P.F. Active-site residue, domain and module swaps in modular polyketide synthases. *J. Ind. Microbiol. Biotechnol.* **2003**, *30*, 489–494. [CrossRef] [PubMed]
61. Sundermann, U.; Bravo-Rodriguez, K.; Klopries, S.; Kushnir, S.; Gomez, H.; Sanchez-Garcia, E.; Schulz, F. Enzyme-directed mutasynthesis: A combined experimental and theoretical approach to substrate recognition of a polyketide synthase. *ACS Chem. Biol.* **2013**, *8*, 443–450. [CrossRef] [PubMed]
62. Reeves, C.D.; Murli, S.; Ashley, G.W.; Piagentini, M.; Hutchinson, C.R.; McDaniel, R. Alteration of the substrate specificity of a modular polyketide synthase acyltransferase domain through site-specific mutations. *Biochemistry* **2001**, *40*, 15464–15470. [CrossRef] [PubMed]
63. Bravo-Rodriguez, K.; Klopries, S.; Koopmans, K.R.M.; Sundermann, U.; Yahiaoui, S.; Arens, J.; Kushnir, S.; Schulz, F.; Sanchez-Garcia, E. Substrate flexibility of a mutated acyltransferase domain and implications for polyketide biosynthesis. *Chem. Biol.* **2015**, *22*, 1425–1430. [CrossRef] [PubMed]
64. Klopries, S.; Bravo-Rodriguez, K.; Koopmans, K.R.; Sundermann, U.; Yahiaoui, S.; Arens, J.; Kushnir, S.; Sanchez-Garcia, E.; Schulz, F. Data in support of substrate flexibility of a mutated acyltransferase domain and implications for polyketide biosynthesis. *Data Brief.* **2015**, *5*, 528–536. [CrossRef] [PubMed]
65. Koryakina, I.; Kasey, C.; McArthur, J.B.; Lowell, A.N.; Chemler, J.A.; Li, S.; Hansen, D.A.; Sherman, D.H.; Williams, G.J. Inversion of extender unit selectivity in the erythromycin polyketide synthase by acyltransferase domain engineering. *ACS Chem. Biol.* **2017**, *12*, 114–123. [CrossRef] [PubMed]
66. Kumar, P.; Koppisch, A.T.; Cane, D.E.; Khosla, C. Enhancing the modularity of the modular polyketide synthases: Transacylation in modular polyketide synthases catalyzed by malonyl-CoA:ACP transacylase. *J. Am. Chem. Soc.* **2003**, *125*, 14307–14312. [CrossRef] [PubMed]
67. Lopanik, N.B.; Shields, J.A.; Buchholz, T.J.; Rath, C.M.; Hothersall, J.; Haygood, M.G.; Hakansson, K.; Thomas, C.M.; Sherman, D.H. In vivo and in vitro *trans*-acylation by BryP, the putative bryostatin pathway acyltransferase derived from an uncultured marine symbiont. *Chem. Biol.* **2008**, *15*, 1175–1186. [CrossRef] [PubMed]
68. Ray, T.K.; Cronan, J.E., Jr. Activation of long chain fatty acids with acyl carrier protein: Demonstration of a new enzyme, acyl-acyl carrier protein synthetase, in *Escherichia coli*. *Proc. Natl. Acad. Sci. USA* **1976**, *73*, 4374–4378. [CrossRef] [PubMed]
69. Quadri, L.E.; Weinreb, P.H.; Lei, M.; Nakano, M.M.; Zuber, P.; Walsh, C.T. Characterization of Sfp, a *Bacillus subtilis* phosphopantetheinyl transferase for peptidyl carrier protein domains in peptide synthetases. *Biochemistry* **1998**, *37*, 1585–1595. [CrossRef] [PubMed]
70. Pavlidou, M.; Pross, E.K.; Musiol, E.M.; Kulik, A.; Wohlleben, W.; Weber, T. The phosphopantetheinyl transferase KirP activates the ACP and PCP domains of the kirromycin NRPS/PKS of *Streptomyces collinus* Tü 365. *FEMS Microbiol. Lett.* **2011**, *319*, 26–33. [CrossRef] [PubMed]
71. Cheng, Y.Q.; Coughlin, J.M.; Lim, S.K.; Shen, B. Type I polyketide synthases that require discrete acyltransferases. *Methods Enzymol.* **2009**, *459*, 165–186. [PubMed]

72. Oliynyk, M.; Stark, C.B.; Bhatt, A.; Jones, M.A.; Hughes-Thomas, Z.A.; Wilkinson, C.; Oliynyk, Z.; Demydchuk, Y.; Staunton, J.; Leadlay, P.F. Analysis of the biosynthetic gene cluster for the polyether antibiotic monensin in *Streptomyces cinnamonensis* and evidence for the role of *monB* and *monC* genes in oxidative cyclization. *Mol. Microbiol.* **2003**, *49*, 1179–1190. [CrossRef] [PubMed]
73. Bhatt, A.; Stark, C.B.; Harvey, B.M.; Gallimore, A.R.; Demydchuk, Y.A.; Spencer, J.B.; Staunton, J.; Leadlay, P.F. Accumulation of an *E,E,E*-triene by the monensin-producing polyketide synthase when oxidative cyclization is blocked. *Angew. Chem. Int. Ed. Engl.* **2005**, *44*, 7075–7078. [CrossRef] [PubMed]
74. Bravo-Rodriguez, K.; Ismail-Ali, A.F.; Klopries, S.; Kushnir, S.; Ismail, S.; Fansa, E.K.; Wittinghofer, A.; Schulz, F.; Sanchez-Garcia, E. Predicted incorporation of non-native substrates by a polyketide synthase yields bioactive natural product derivatives. *ChemBioChem* **2014**, *15*, 1991–1997. [CrossRef] [PubMed]
75. Guchhait, R.B.; Polakis, S.E.; Dimroth, P.; Stoll, E.; Moss, J.; Lane, M.D. Acetyl coenzyme A carboxylase system of *Escherichia coli*. Purification and properties of the biotin carboxylase, carboxyltransferase, and carboxyl carrier protein components. *J. Biol. Chem.* **1974**, *249*, 6633–6645. [PubMed]
76. Bramwell, H.; Hunter, I.S.; Coggins, J.R.; Nimmo, H.G. Propionyl-CoA carboxylase from *Streptomyces coelicolor* A3(2): Cloning of the gene encoding the biotin-containing subunit. *Microbiology* **1996**, *142 Pt 3*, 649–655. [CrossRef] [PubMed]
77. Gago, G.; Kurth, D.; Diacovich, L.; Tsai, S.C.; Gramajo, H. Biochemical and structural characterization of an essential acyl coenzyme A carboxylase from *Mycobacterium tuberculosis*. *J. Bacteriol.* **2006**, *188*, 477–486. [CrossRef] [PubMed]
78. Diacovich, L.; Peiru, S.; Kurth, D.; Rodriguez, E.; Podesta, F.; Khosla, C.; Gramajo, H. Kinetic and structural analysis of a new group of acyl-CoA carboxylases found in *Streptomyces coelicolor* A3(2). *J. Biol. Chem.* **2002**, *277*, 31228–31236. [CrossRef] [PubMed]
79. Hughes, A.J.; Keatinge-Clay, A. Enzymatic extender unit generation for in vitro polyketide synthase reactions: Structural and functional showcasing of *Streptomyces coelicolor* MatB. *Chem. Biol.* **2011**, *18*, 165–176. [CrossRef] [PubMed]
80. Pohl, N.L.; Hans, M.; Lee, H.Y.; Kim, Y.S.; Cane, D.E.; Khosla, C. Remarkably broad substrate tolerance of malonyl-CoA synthetase, an enzyme capable of intracellular synthesis of polyketide precursors. *J. Am. Chem. Soc.* **2001**, *123*, 5822–5823. [CrossRef] [PubMed]
81. An, J.H.; Kim, Y.S. A gene cluster encoding malonyl-CoA decarboxylase (MatA), malonyl-CoA synthetase (MatB) and a putative dicarboxylate carrier protein (MatC) in *Rhizobium trifolii*–cloning, sequencing, and expression of the enzymes in *Escherichia coli*. *Eur. J. Biochem.* **1998**, *257*, 395–402. [CrossRef] [PubMed]
82. Kim, Y.S.; Chae, H.Z. Purification and properties of malonyl-CoA synthetase from *Rhizobium japonicum*. *Biochem. J.* **1991**, *273 Pt 3*, 511–516. [CrossRef] [PubMed]
83. Maharjan, S.; Park, J.W.; Yoon, Y.J.; Lee, H.C.; Sohng, J.K. Metabolic engineering of *Streptomyces venezuelae* for malonyl-CoA biosynthesis to enhance heterologous production of polyketides. *Biotechnol. Lett.* **2010**, *32*, 277–282. [CrossRef] [PubMed]
84. Koryakina, I.; Williams, G.J. Mutant malonyl-CoA synthetases with altered specificity for polyketide synthase extender unit generation. *ChemBioChem* **2011**, *12*, 2289–2293. [CrossRef] [PubMed]
85. Crosby, H.A.; Rank, K.C.; Rayment, I.; Escalante-Semerena, J.C. Structure-guided expansion of the substrate range of methylmalonyl coenzyme A synthetase (MatB) of *Rhodopseudomonas palustris*. *Appl. Environ. Microbiol.* **2012**, *78*, 6619–6629. [CrossRef] [PubMed]
86. Xie, X.; Watanabe, K.; Wojcicki, W.A.; Wang, C.C.; Tang, Y. Biosynthesis of lovastatin analogs with a broadly specific acyltransferase. *Chem. Biol.* **2006**, *13*, 1161–1169. [CrossRef] [PubMed]
87. Mo, S.; Kim, D.H.; Lee, J.H.; Park, J.W.; Basnet, D.B.; Ban, Y.H.; Yoo, Y.J.; Chen, S.W.; Park, S.R.; Choi, E.A.; et al. Biosynthesis of the allylmalonyl-CoA extender unit for the FK506 polyketide synthase proceeds through a dedicated polyketide synthase and facilitates the mutasynthesis of analogues. *J. Am. Chem. Soc.* **2011**, *133*, 976–985. [CrossRef] [PubMed]
88. Klopries, S.; Sundermann, U.; Schulz, F. Quantification of N-acetylcysteamine activated methylmalonate incorporation into polyketide biosynthesis. *Beilstein. J. Org. Chem.* **2013**, *9*, 664–674. [CrossRef] [PubMed]
89. Richardson, M.T.; Pohl, N.L.; Kealey, J.T.; Khosla, C. Tolerance and specificity of recombinant 6-methylsalicyclic acid synthase. *Metab. Eng.* **1999**, *1*, 180–187. [CrossRef] [PubMed]

90. Musiol-Kroll, E.M.; Zubeil, F.; Schafhauser, T.; Hartner, T.; Kulik, A.; McArthur, J.; Koryakina, I.; Wohlleben, W.; Grond, S.; Williams, G.J.; et al. Polyketide bioderivatization using the promiscuous acyltransferase KirCII. *ACS Synth. Biol.* **2017**, *6*, 421–427. [CrossRef] [PubMed]
91. Erb, T.J.; Berg, I.A.; Brecht, V.; Muller, M.; Fuchs, G.; Alber, B.E. Synthesis of C5-dicarboxylic acids from C2-units involving crotonyl-CoA carboxylase/reductase: The ethylmalonyl-CoA pathway. *Proc. Natl. Acad. Sci. USA* **2007**, *104*, 10631–10636. [CrossRef] [PubMed]
92. Erb, T.J.; Brecht, V.; Fuchs, G.; Muller, M.; Alber, B.E. Carboxylation mechanism and stereochemistry of crotonyl-CoA carboxylase/reductase, a carboxylating enoyl-thioester reductase. *Proc. Natl. Acad. Sci. USA* **2009**, *106*, 8871–8876. [CrossRef] [PubMed]
93. Chan, Y.A.; Podevels, A.M.; Kevany, B.M.; Thomas, M.G. Biosynthesis of polyketide synthase extender units. *Nat. Prod. Rep.* **2009**, *26*, 90–114. [CrossRef] [PubMed]
94. Ray, L.; Moore, B.S. Recent advances in the biosynthesis of unusual polyketide synthase substrates. *Nat. Prod. Rep.* **2016**, *33*, 150–161. [CrossRef] [PubMed]
95. Riveros-Rosas, H.; Julian-Sanchez, A.; Villalobos-Molina, R.; Pardo, J.P.; Pina, E. Diversity, taxonomy and evolution of medium-chain dehydrogenase/reductase superfamily. *Eur. J. Biochem.* **2003**, *270*, 3309–3334. [CrossRef] [PubMed]
96. Zhang, L.; Mori, T.; Zheng, Q.; Awakawa, T.; Yan, Y.; Liu, W.; Abe, I. Rational control of polyketide extender units by structure-based engineering of a crotonyl-CoA carboxylase/reductase in antimycin biosynthesis. *Angew. Chem. Int. Ed. Engl.* **2015**, *54*, 13462–13465. [CrossRef] [PubMed]
97. Kundert, J.; Gulder, T.A. Extending polyketide structural diversity by using engineered carboxylase/reductase enzymes. *Angew. Chem. Int. Ed. Engl.* **2016**, *55*, 858–860. [CrossRef] [PubMed]
98. Ray, L.; Valentic, T.R.; Miyazawa, T.; Withall, D.M.; Song, L.; Milligan, J.C.; Osada, H.; Takahashi, S.; Tsai, S.C.; Challis, G.L. A crotonyl-CoA reductase-carboxylase independent pathway for assembly of unusual alkylmalonyl-CoA polyketide synthase extender units. *Nat. Commun.* **2016**, *7*, 13609. [CrossRef] [PubMed]
99. Kwan, D.H.; Schulz, F. The stereochemistry of complex polyketide biosynthesis by modular polyketide synthases. *Molecules* **2011**, *16*, 6092–6115. [CrossRef] [PubMed]
100. Marsden, A.F.; Caffrey, P.; Aparicio, J.F.; Loughran, M.S.; Staunton, J.; Leadlay, P.F. Stereospecific acyl transfers on the erythromycin-producing polyketide synthase. *Science* **1994**, *263*, 378–380. [CrossRef] [PubMed]
101. Thomas, I.; Martin, C.J.; Wilkinson, C.J.; Staunton, J.; Leadlay, P.F. Skipping in a hybrid polyketide synthase. Evidence for ACP-to-ACP chain transfer. *Chem. Biol.* **2002**, *9*, 781–787. [CrossRef]
102. Moss, S.J.; Martin, C.J.; Wilkinson, B. Loss of co-linearity by modular polyketide synthases: a mechanism for the evolution of chemical diversity. *Nat. Prod. Rep.* **2004**, *21*, 575–593. [CrossRef] [PubMed]
103. Kalkreuter, E.; Williams, G.J. Engineering enzymatic assembly lines for the production of new antimicrobials. *Curr. Opin. Microbiol.* **2018**, *45*, 140–148. [CrossRef] [PubMed]
104. Khosla, C.; Tang, Y.; Chen, A.Y.; Schnarr, N.A.; Cane, D.E. Structure and mechanism of the 6-deoxyerythronolide B synthase. *Annu. Rev. Biochem.* **2007**, *76*, 195–221. [CrossRef] [PubMed]
105. Joshi, V.C.; Wakil, S.J. Studies on the mechanism of fatty acid synthesis. XXVI. Purification and properties of malonyl-coenzyme A–acyl carrier protein transacylase of *Escherichia coli*. *Arch. Biochem. Biophys.* **1971**, *143*, 493–505. [CrossRef]
106. Wong, F.T.; Jin, X.; Mathews, I.I.; Cane, D.E.; Khosla, C. Structure and mechanism of the *trans*-acting acyltransferase from the disorazole synthase. *Biochemistry* **2011**, *50*, 6539–6548. [CrossRef] [PubMed]
107. Wong, F.T.; Chen, A.Y.; Cane, D.E.; Khosla, C. Protein-protein recognition between acyltransferases and acyl carrier proteins in multimodular polyketide synthases. *Biochemistry* **2010**, *49*, 95–102. [CrossRef] [PubMed]
108. Dunn, B.J.; Cane, D.E.; Khosla, C. Mechanism and specificity of an acyltransferase domain from a modular polyketide synthase. *Biochemistry* **2013**, *52*, 1839–1841. [CrossRef] [PubMed]
109. Chan, Y.A.; Thomas, M.G. Recognition of (2*S*)-aminomalonyl-acyl carrier protein (ACP) and (2*R*)-hydroxymalonyl-ACP by acyltransferases in zwittermicin A biosynthesis. *Biochemistry* **2010**, *49*, 3667–3677. [CrossRef] [PubMed]
110. Park, H.; Kevany, B.M.; Dyer, D.H.; Thomas, M.G.; Forest, K.T. A polyketide synthase acyltransferase domain structure suggests a recognition mechanism for its hydroxymalonyl-acyl carrier protein substrate. *PLoS ONE* **2014**, *9*, e110965. [CrossRef] [PubMed]
111. Shindo, K.; Kamishohara, M.; Odagawa, A.; Matsuoka, M.; Kawai, H. Vicenistatin, a novel 20-membered macrocyclic lactam antitumor antibiotic. *J. Antibiot. (Tokyo)* **1993**, *46*, 1076–1081. [CrossRef] [PubMed]

112. Miyanaga, A.; Cieslak, J.; Shinohara, Y.; Kudo, F.; Eguchi, T. The crystal structure of the adenylation enzyme VinN reveals a unique beta-amino acid recognition mechanism. *J. Biol. Chem.* **2014**, *289*, 31448–31457. [CrossRef] [PubMed]
113. Miyanaga, A.; Iwasawa, S.; Shinohara, Y.; Kudo, F.; Eguchi, T. Structure-based analysis of the molecular interactions between acyltransferase and acyl carrier protein in vicenistatin biosynthesis. *Proc. Natl. Acad. Sci. USA* **2016**, *113*, 1802–1807. [CrossRef] [PubMed]
114. Wu, N.; Tsuji, S.Y.; Cane, D.E.; Khosla, C. Assessing the balance between protein-protein interactions and enzyme-substrate interactions in the channeling of intermediates between polyketide synthase modules. *J. Am. Chem. Soc.* **2001**, *123*, 6465–6474. [CrossRef] [PubMed]
115. Wu, N.; Cane, D.E.; Khosla, C. Quantitative analysis of the relative contributions of donor acyl carrier proteins, acceptor ketosynthases, and linker regions to intermodular transfer of intermediates in hybrid polyketide synthases. *Biochemistry* **2002**, *41*, 5056–5066. [CrossRef] [PubMed]
116. Chen, A.Y.; Schnarr, N.A.; Kim, C.Y.; Cane, D.E.; Khosla, C. Extender unit and acyl carrier protein specificity of ketosynthase domains of the 6-deoxyerythronolide B synthase. *J. Am. Chem. Soc.* **2006**, *128*, 3067–3074. [CrossRef] [PubMed]
117. Weissman, K.J.; Muller, R. Protein-protein interactions in multienzyme megasynthetases. *ChemBioChem* **2008**, *9*, 826–848. [CrossRef] [PubMed]
118. Charkoudian, L.K.; Liu, C.W.; Capone, S.; Kapur, S.; Cane, D.E.; Togni, A.; Seebach, D.; Khosla, C. Probing the interactions of an acyl carrier protein domain from the 6-deoxyerythronolide B synthase. *Protein Sci.* **2011**, *20*, 1244–1255. [CrossRef] [PubMed]
119. Kapur, S.; Chen, A.Y.; Cane, D.E.; Khosla, C. Molecular recognition between ketosynthase and acyl carrier protein domains of the 6-deoxyerythronolide B synthase. *Proc. Natl. Acad. Sci. USA* **2010**, *107*, 22066–22071. [CrossRef] [PubMed]
120. Dutta, S.; Whicher, J.R.; Hansen, D.A.; Hale, W.A.; Chemler, J.A.; Congdon, G.R.; Narayan, A.R.; Hakansson, K.; Sherman, D.H.; Smith, J.L.; et al. Structure of a modular polyketide synthase. *Nature* **2014**, *510*, 512–517. [CrossRef] [PubMed]
121. Whicher, J.R.; Dutta, S.; Hansen, D.A.; Hale, W.A.; Chemler, J.A.; Dosey, A.M.; Narayan, A.R.; Hakansson, K.; Sherman, D.H.; Smith, J.L.; et al. Structural rearrangements of a polyketide synthase module during its catalytic cycle. *Nature* **2014**, *510*, 560–564. [CrossRef] [PubMed]
122. Lowry, B.; Li, X.; Robbins, T.; Cane, D.E.; Khosla, C. A turnstile mechanism for the controlled growth of biosynthetic intermediates on assembly line polyketide synthases. *ACS Cent. Sci.* **2016**, *2*, 14–20. [CrossRef] [PubMed]
123. Khosla, C.; Kapur, S.; Cane, D.E. Revisiting the modularity of modular polyketide synthases. *Curr. Opin. Chem. Biol.* **2009**, *13*, 135–143. [CrossRef] [PubMed]
124. Broadhurst, R.W.; Nietlispach, D.; Wheatcroft, M.P.; Leadlay, P.F.; Weissman, K.J. The structure of docking domains in modular polyketide synthases. *Chem. Biol.* **2003**, *10*, 723–731. [CrossRef]
125. Dorival, J.; Annaval, T.; Risser, F.; Collin, S.; Roblin, P.; Jacob, C.; Gruez, A.; Chagot, B.; Weissman, K.J. Characterization of intersubunit communication in the virginiamycin *trans*-acyl transferase polyketide synthase. *J. Am. Chem. Soc.* **2016**, *138*, 4155–4167. [CrossRef] [PubMed]
126. Zeng, J.; Wagner, D.T.; Zhang, Z.; Moretto, L.; Addison, J.D.; Keatinge-Clay, A.T. Portability and structure of the four-helix bundle docking domains of *trans*-acyltransferase modular polyketide synthases. *ACS Chem. Biol.* **2016**, *11*, 2466–2474. [CrossRef] [PubMed]
127. Gay, D.C.; Gay, G.; Axelrod, A.J.; Jenner, M.; Kohlhaas, C.; Kampa, A.; Oldham, N.J.; Piel, J.; Keatinge-Clay, A.T. A close look at a ketosynthase from a *trans*-acyltransferase modular polyketide synthase. *Structure* **2014**, *22*, 444–451. [CrossRef] [PubMed]
128. Tang, G.L.; Cheng, Y.Q.; Shen, B. Leinamycin biosynthesis revealing unprecedented architectural complexity for a hybrid polyketide synthase and nonribosomal peptide synthetase. *Chem. Biol.* **2004**, *11*, 33–45. [CrossRef] [PubMed]
129. Lohman, J.R.; Ma, M.; Osipiuk, J.; Nocek, B.; Kim, Y.; Chang, C.; Cuff, M.; Mack, J.; Bigelow, L.; Li, H.; et al. Structural and evolutionary relationships of "AT-less" type I polyketide synthase ketosynthases. *Proc. Natl. Acad. Sci. USA* **2015**, *112*, 12693–12698. [CrossRef] [PubMed]

130. Jenner, M.; Kosol, S.; Griffiths, D.; Prasongpholchai, P.; Manzi, L.; Barrow, A.S.; Moses, J.E.; Oldham, N.J.; Lewandowski, J.R.; Challis, G.L. Mechanism of intersubunit ketosynthase-dehydratase interaction in polyketide synthases. *Nat. Chem. Biol.* **2018**, *14*, 270–275. [CrossRef] [PubMed]
131. DeWitt, J.P. Evidence for a sex factor in *Streptomyces erythreus*. *J. Bacteriol.* **1985**, *164*, 969–971. [PubMed]
132. Kao, C.M.; Katz, L.; Khosla, C. Engineered biosynthesis of a complete macrolactone in a heterologous host. *Science* **1994**, *265*, 509–512. [CrossRef] [PubMed]
133. McDaniel, R.; Ebert-Khosla, S.; Hopwood, D.A.; Khosla, C. Engineered biosynthesis of novel polyketides. *Science* **1993**, *262*, 1546–1550. [CrossRef] [PubMed]
134. Ziermann, R.; Betlach, M.C. Recombinant polyketide synthesis in *Streptomyces*: Engineering of improved host strains. *Biotechniques* **1999**, *26*, 106–110. [PubMed]
135. McDaniel, R.; Thamchaipenet, A.; Gustafsson, C.; Fu, H.; Betlach, M.; Ashley, G. Multiple genetic modifications of the erythromycin polyketide synthase to produce a library of novel "unnatural" natural products. *Proc. Natl. Acad. Sci. USA* **1999**, *96*, 1846–1851. [CrossRef] [PubMed]
136. Kao, C.M.; Luo, G.; Katz, L.; Cane, D.E.; Khosla, C. Engineered biosynthesis of a triketide lactone from an incomplete modular polyketide synthase. *J. Am. Chem. Soc.* **1994**, *116*, 11612–11613. [CrossRef]
137. Cortes, J.; Wiesmann, K.E.; Roberts, G.A.; Brown, M.J.; Staunton, J.; Leadlay, P.F. Repositioning of a domain in a modular polyketide synthase to promote specific chain cleavage. *Science* **1995**, *268*, 1487–1489. [CrossRef] [PubMed]
138. Weissman, K.J. Genetic engineering of modular PKSs: From combinatorial biosynthesis to synthetic biology. *Nat. Prod. Rep.* **2016**, *33*, 203–230. [CrossRef] [PubMed]
139. Luo, G.; Pieper, R.; Rosa, A.; Khosla, C.; Cane, D.E. Erythromycin biosynthesis: Exploiting the catalytic versatility of the modular polyketide synthase. *Bioorg. Med. Chem.* **1996**, *4*, 995–999. [CrossRef]
140. Rowe, C.J.; Cortes, J.; Gaisser, S.; Staunton, J.; Leadlay, P.F. Construction of new vectors for high-level expression in actinomycetes. *Gene* **1998**, *216*, 215–223. [CrossRef]
141. Robertsen, H.L.; Musiol-Kroll, E.M.; Ding, L.; Laiple, K.J.; Hofeditz, T.; Wohlleben, W.; Lee, S.Y.; Grond, S.; Weber, T. Filling the gaps in the kirromycin biosynthesis: Deciphering the role of genes involved in ethylmalonyl-CoA supply and tailoring reactions. *Sci. Rep.* **2018**, *8*, 3230. [CrossRef] [PubMed]
142. Weissman, K.J.; Leadlay, P.F. Combinatorial biosynthesis of reduced polyketides. *Nat. Rev. Microbiol.* **2005**, *3*, 925–936. [CrossRef] [PubMed]
143. Maervoet, V.E.T.; Briers, Y. Synthetic biology of modular proteins. *Bioengineered* **2017**, *8*, 196–202. [CrossRef] [PubMed]
144. Yuzawa, S.; Backman, T.W.H.; Keasling, J.D.; Katz, L. Synthetic biology of polyketide synthases. *J. Ind. Microbiol. Biotechnol.* **2018**. [CrossRef] [PubMed]
145. Donadio, S.; Sosio, M. Strategies for combinatorial biosynthesis with modular polyketide synthases. *Comb. Chem. High Throughput Screen* **2003**, *6*, 489–500. [CrossRef] [PubMed]
146. Sun, H.; Liu, Z.; Zhao, H.; Ang, E.L. Recent advances in combinatorial biosynthesis for drug discovery. *Drug Des. Devel. Ther.* **2015**, *9*, 823–833. [PubMed]
147. Smanski, M.J.; Zhou, H.; Claesen, J.; Shen, B.; Fischbach, M.A.; Voigt, C.A. Synthetic biology to access and expand nature's chemical diversity. *Nat. Rev. Microbiol.* **2016**, *14*, 135–149. [CrossRef] [PubMed]
148. Yuzawa, S.; Deng, K.; Wang, G.; Baidoo, E.E.; Northen, T.R.; Adams, P.D.; Katz, L.; Keasling, J.D. Comprehensive in vitro analysis of acyltransferase domain exchanges in modular polyketide synthases and its application for short-chain ketone production. *ACS Synth. Biol.* **2017**, *6*, 139–147. [CrossRef] [PubMed]
149. Marsden, A.F.; Wilkinson, B.; Cortes, J.; Dunster, N.J.; Staunton, J.; Leadlay, P.F. Engineering broader specificity into an antibiotic-producing polyketide synthase. *Science* **1998**, *279*, 199–202. [CrossRef] [PubMed]
150. Gokhale, R.S.; Tsuji, S.Y.; Cane, D.E.; Khosla, C. Dissecting and exploiting intermodular communication in polyketide synthases. *Science* **1999**, *284*, 482–485. [CrossRef] [PubMed]
151. Sheehan, L.S.; Lill, R.E.; Wilkinson, B.; Sheridan, R.M.; Vousden, W.A.; Kaja, A.L.; Crouse, G.D.; Gifford, J.; Graupner, P.R.; Karr, L.; et al. Engineering of the spinosyn PKS: Directing starter unit incorporation. *J. Nat. Prod.* **2006**, *69*, 1702–1710. [CrossRef] [PubMed]
152. Dutton, C.J.; Gibson, S.P.; Goudie, A.C.; Holdom, K.S.; Pacey, M.S.; Ruddock, J.C.; Bu'Lock, J.D.; Richards, M.K. Novel avermectins produced by mutational biosynthesis. *J. Antibiot. (Tokyo)* **1991**, *44*, 357–365. [CrossRef] [PubMed]

153. Goudie, A.C.; Evans, N.A.; Gration, K.A.; Bishop, B.F.; Gibson, S.P.; Holdom, K.S.; Kaye, B.; Wicks, S.R.; Lewis, D.; Weatherley, A.J.; et al. Doramectin—A potent novel endectocide. *Vet. Parasitol.* **1993**, *49*, 5–15. [CrossRef]
154. Palaniappan, N.; Kim, B.S.; Sekiyama, Y.; Osada, H.; Reynolds, K.A. Enhancement and selective production of phoslactomycin B, a protein phosphatase IIa inhibitor, through identification and engineering of the corresponding biosynthetic gene cluster. *J. Biol. Chem.* **2003**, *278*, 35552–35557. [CrossRef] [PubMed]
155. Ghatge, M.; Palaniappan, N.; Das Choudhuri, S.; Reynolds, K. Genetic manipulation of the biosynthetic process leading to phoslactomycins, potent protein phosphatase 2A inhibitors. *J. Ind. Microbiol. Biotechnol.* **2006**, *33*, 589–599. [CrossRef] [PubMed]
156. Gomez-Escribano, J.P.; Alt, S.; Bibb, M.J. Next generation sequencing of actinobacteria for the discovery of novel natural products. *Mar. Drugs* **2016**, *14*, 78. [CrossRef] [PubMed]
157. Blin, K.; Wolf, T.; Chevrette, M.G.; Lu, X.; Schwalen, C.J.; Kautsar, S.A.; Suarez Duran, H.G.; de Los Santos, E.L.C.; Kim, H.U.; Nave, M.; et al. antiSMASH 4.0-improvements in chemistry prediction and gene cluster boundary identification. *Nucleic. Acids. Res.* **2017**, *45*, W36–W41. [CrossRef] [PubMed]
158. Yadav, G.; Gokhale, R.S.; Mohanty, D. Computational approach for prediction of domain organization and substrate specificity of modular polyketide synthases. *J. Mol. Biol.* **2003**, *328*, 335–363. [CrossRef]
159. Khayatt, B.I.; Overmars, L.; Siezen, R.J.; Francke, C. Classification of the adenylation and acyl-transferase activity of NRPS and PKS systems using ensembles of substrate specific hidden Markov models. *PLoS ONE* **2013**, *8*, e62136. [CrossRef] [PubMed]
160. Haydock, S.F.; Aparicio, J.F.; Molnar, I.; Schwecke, T.; Khaw, L.E.; Konig, A.; Marsden, A.F.; Galloway, I.S.; Staunton, J.; Leadlay, P.F. Divergent sequence motifs correlated with the substrate specificity of (methyl)malonyl-CoA:acyl carrier protein transacylase domains in modular polyketide synthases. *FEBS Lett.* **1995**, *374*, 246–248. [CrossRef]
161. Mathews, I.I.; Allison, K.; Robbins, T.; Lyubimov, A.Y.; Uervirojnangkoorn, M.; Brunger, A.T.; Khosla, C.; DeMirci, H.; McPhillips, S.E.; Hollenbeck, M.; et al. The conformational flexibility of the acyltransferase from the disorazole polyketide synthase is revealed by an X-ray free-electron laser using a room-temperature sample delivery method for serial crystallography. *Biochemistry* **2017**, *56*, 4751–4756. [CrossRef] [PubMed]
162. Wagner, D.T.; Zhang, Z.; Meoded, R.A.; Cepeda, A.J.; Piel, J.; Keatinge-Clay, A.T. Structural and functional studies of a pyran synthase domain from a *trans*-acyltransferase assembly line. *ACS Chem. Biol.* **2018**. [CrossRef] [PubMed]
163. Skiba, M.A.; Maloney, F.P.; Dan, Q.; Fraley, A.E.; Aldrich, C.C.; Smith, J.L.; Brown, W.C. PKS-NRPS enzymology and structural biology: Considerations in protein production. *Methods Enzymol.* **2018**, *604*, 45–88. [PubMed]
164. Gay, D.C.; Wagner, D.T.; Meinke, J.L.; Zogzas, C.E.; Gay, G.R.; Keatinge-Clay, A.T. The LINKS motif zippers *trans*-acyltransferase polyketide synthase assembly lines into a biosynthetic megacomplex. *J. Struct. Biol.* **2016**, *193*, 196–205. [CrossRef] [PubMed]
165. Wang, F.; Wang, Y.; Ji, J.; Zhou, Z.; Yu, J.; Zhu, H.; Su, Z.; Zhang, L.; Zheng, J. Structural and functional analysis of the loading acyltransferase from avermectin modular polyketide synthase. *ACS Chem. Biol.* **2015**, *10*, 1017–1025. [CrossRef] [PubMed]
166. Wang, Y.Y.; Bai, L.F.; Ran, X.X.; Jiang, X.H.; Wu, H.; Zhang, W.; Jin, M.Y.; Li, Y.Q.; Jiang, H. Biochemical characterization of a malonyl-specific acyltransferase domain of FK506 biosynthetic polyketide synthase. *Protein Pept. Lett.* **2015**, *22*, 2–7. [CrossRef] [PubMed]
167. Zhang, W.; Liu, J. Recent advances in understanding and engineering polyketide synthesis. *F1000Res* **2016**, *5*. [CrossRef] [PubMed]
168. Cai, W.; Zhang, W. Engineering modular polyketide synthases for production of biofuels and industrial chemicals. *Curr. Opin. Biotechnol.* **2018**, *50*, 32–38. [CrossRef] [PubMed]
169. Herbst, D.A.; Huitt-Roehl, C.R.; Jakob, R.P.; Kravetz, J.M.; Storm, P.A.; Alley, J.R.; Townsend, C.A.; Maier, T. The structural organization of substrate loading in iterative polyketide synthases. *Nat. Chem. Biol.* **2018**, *14*, 474–479. [CrossRef] [PubMed]
170. Harvey, C.J.; Puglisi, J.D.; Pande, V.S.; Cane, D.E.; Khosla, C. Precursor directed biosynthesis of an orthogonally functional erythromycin analogue: Selectivity in the ribosome macrolide binding pocket. *J. Am. Chem. Soc.* **2012**, *134*, 12259–12265. [CrossRef] [PubMed]

171. Li, Y.; Zhang, W.; Zhang, H.; Tian, W.; Wu, L.; Wang, S.; Zheng, M.; Zhang, J.; Sun, C.; Deng, Z.; et al. Structural basis of a broadly selective acyltransferase from the polyketide synthase of splenocin. *Angew Chem. Int. Ed. Engl.* **2018**, *57*, 5823–5827. [CrossRef] [PubMed]
172. Dunn, B.J.; Watts, K.R.; Robbins, T.; Cane, D.E.; Khosla, C. Comparative analysis of the substrate specificity of *trans*- versus *cis*-acyltransferases of assembly line polyketide synthases. *Biochemistry* **2014**, *53*, 3796–3806. [CrossRef] [PubMed]
173. Musiol, E.M.; Hartner, T.; Kulik, A.; Moldenhauer, J.; Piel, J.; Wohlleben, W.; Weber, T. Supramolecular templating in kirromycin biosynthesis: The acyltransferase KirCII loads ethylmalonyl-CoA extender onto a specific ACP of the *trans*-AT PKS. *Chem. Biol.* **2011**, *18*, 438–444. [CrossRef] [PubMed]
174. Weber, T.; Laiple, K.J.; Pross, E.K.; Textor, A.; Grond, S.; Welzel, K.; Pelzer, S.; Vente, A.; Wohlleben, W. Molecular analysis of the kirromycin biosynthetic gene cluster revealed beta-alanine as precursor of the pyridone moiety. *Chem. Biol.* **2008**, *15*, 175–188. [CrossRef] [PubMed]
175. Ye, Z.; Musiol, E.M.; Weber, T.; Williams, G.J. Reprogramming acyl carrier protein interactions of an Acyl-CoA promiscuous *trans*-acyltransferase. *Chem. Biol.* **2014**, *21*, 636–646. [CrossRef] [PubMed]
176. Koryakina, I.; McArthur, J.; Randall, S.; Draelos, M.M.; Musiol, E.M.; Muddiman, D.C.; Weber, T.; Williams, G.J. Poly specific *trans*-acyltransferase machinery revealed *via* engineered acyl-CoA synthetases. *ACS Chem. Biol.* **2013**, *8*, 200–208. [CrossRef] [PubMed]
177. Hans, M.; Hornung, A.; Dziarnowski, A.; Cane, D.E.; Khosla, C. Mechanistic analysis of acyl transferase domain exchange in polyketide synthase modules. *J. Am. Chem. Soc.* **2003**, *125*, 5366–5374. [CrossRef] [PubMed]
178. Kim, W.; Lee, D.; Hong, S.S.; Na, Z.; Shin, J.C.; Roh, S.H.; Wu, C.Z.; Choi, O.; Lee, K.; Shen, Y.M.; et al. Rational biosynthetic engineering for optimization of geldanamycin analogues. *ChemBioChem* **2009**, *10*, 1243–1251. [CrossRef] [PubMed]
179. Katz, L. The DEBS paradigm for type I modular polyketide synthases and beyond. *Methods Enzymol.* **2009**, *459*, 113–142. [PubMed]
180. Weber, T.; Charusanti, P.; Musiol-Kroll, E.M.; Jiang, X.; Tong, Y.; Kim, H.U.; Lee, S.Y. Metabolic engineering of antibiotic factories: New tools for antibiotic production in actinomycetes. *Trends Biotechnol.* **2015**, *33*, 15–26. [CrossRef] [PubMed]
181. Robbins, T.; Liu, Y.C.; Cane, D.E.; Khosla, C. Structure and mechanism of assembly line polyketide synthases. *Curr. Opin. Struct. Biol.* **2016**, *41*, 10–18. [CrossRef] [PubMed]
182. Barajas, J.F.; Blake-Hedges, J.M.; Bailey, C.B.; Curran, S.; Keasling, J.D. Engineered polyketides: Synergy between protein and host level engineering. *Synth. Syst. Biotechnol.* **2017**, *2*, 147–166. [CrossRef] [PubMed]
183. Song, L.; Jenner, M.; Masschelein, J.; Jones, C.; Bull, M.J.; Harris, S.R.; Hartkoorn, R.C.; Vocat, A.; Romero-Canelon, I.; Coupland, P.; et al. Discovery and biosynthesis of gladiolin: A *Burkholderia gladioli* antibiotic with promising activity against *Mycobacterium tuberculosis*. *J. Am. Chem. Soc.* **2017**, *139*, 7974–7981. [CrossRef] [PubMed]
184. Cuskin, F.; Solovyova, A.S.; Lewis, R.J.; Race, P.R. Crystallization and preliminary X-ray analysis of the bacillaene synthase *trans*-acting acyltransferase PksC. *Acta Crystallogr. Sect. F Struct. Biol. Cryst. Commun.* **2011**, *67*, 464–466. [CrossRef] [PubMed]
185. Pistorius, D.; Muller, R. Discovery of the rhizopodin biosynthetic gene cluster in *Stigmatella aurantiaca* Sg a15 by genome mining. *ChemBioChem* **2012**, *13*, 416–426. [CrossRef] [PubMed]
186. Musiol, E.M.; Greule, A.; Hartner, T.; Kulik, A.; Wohlleben, W.; Weber, T. The AT(2) domain of KirCI loads malonyl extender units to the ACPs of the kirromycin PKS. *ChemBioChem* **2013**, *14*, 1343–1352. [CrossRef] [PubMed]
187. Zhao, C.; Coughlin, J.M.; Ju, J.; Zhu, D.; Wendt-Pienkowski, E.; Zhou, X.; Wang, Z.; Shen, B.; Deng, Z. Oxazolomycin biosynthesis in *Streptomyces albus* JA3453 featuring an "acyltransferase-less" type I polyketide synthase that incorporates two distinct extender units. *J. Biol. Chem.* **2010**, *285*, 20097–20108. [CrossRef] [PubMed]
188. El-Sayed, A.K.; Hothersall, J.; Cooper, S.M.; Stephens, E.; Simpson, T.J.; Thomas, C.M. Characterization of the mupirocin biosynthesis gene cluster from *Pseudomonas fluorescens* NCIMB 10586. *Chem. Biol.* **2003**, *10*, 419–430. [CrossRef]

189. El-Sayed, A.K.; Hothersall, J.; Thomas, C.M. Quorum-sensing-dependent regulation of biosynthesis of the polyketide antibiotic mupirocin in *Pseudomonas fluorescens* NCIMB 10586. *Microbiology* **2001**, *147*, 2127–2139. [CrossRef] [PubMed]
190. Wang, J.; Wang, C.; Song, K.; Wen, J. Metabolic network model guided engineering ethylmalonyl-CoA pathway to improve ascomycin production in *Streptomyces hygroscopicus* var. *ascomyceticus*. *Microb. Cell Fact.* **2017**, *16*, 169. [CrossRef] [PubMed]
191. Sayers, E.W.; Barrett, T.; Benson, D.A.; Bolton, E.; Bryant, S.H.; Canese, K.; Chetvernin, V.; Church, D.M.; Dicuccio, M.; Federhen, S.; et al. Database resources of the National Center for Biotechnology Information. *Nucleic. Acids. Res.* **2012**, *40*, D13–D25. [CrossRef] [PubMed]
192. Coordinators, N.R. Database resources of the National Center for Biotechnology Information. *Nucleic. Acids. Res.* **2018**, *46*, D8–D13. [CrossRef] [PubMed]
193. Boddy, C.N. Bioinformatics tools for genome mining of polyketide and non-ribosomal peptides. *J. Ind. Microbiol. Biotechnol.* **2014**, *41*, 443–450. [CrossRef] [PubMed]
194. Jenke-Kodama, H.; Dittmann, E. Bioinformatic perspectives on NRPS/PKS megasynthases: Advances and challenges. *Nat. Prod. Rep.* **2009**, *26*, 874–883. [CrossRef] [PubMed]
195. Blin, K.; Medema, M.H.; Kottmann, R.; Lee, S.Y.; Weber, T. The antiSMASH database, a comprehensive database of microbial secondary metabolite biosynthetic gene clusters. *Nucleic Acids Res.* **2017**, *45*, D555–D559. [CrossRef] [PubMed]
196. Sandler, I.; Abu-Qarn, M.; Aharoni, A. Protein co-evolution: How do we combine bioinformatics and experimental approaches? *Mol. Biosyst.* **2013**, *9*, 175–181. [CrossRef] [PubMed]
197. Suplatov, D.; Sharapova, Y.; Timonina, D.; Kopylov, K.; Svedas, V. The visualCMAT: A web-server to select and interpret correlated mutations/co-evolving residues in protein families. *J. Bioinform. Comput. Biol.* **2017**, *16*, 1840005. [CrossRef] [PubMed]
198. De Oliveira, S.H.; Shi, J.; Deane, C.M. Comparing co-evolution methods and their application to template-free protein structure prediction. *Bioinformatics* **2017**, *33*, 373–381. [CrossRef] [PubMed]
199. Milano, T.; Paiardini, A.; Grgurina, I.; Pascarella, S. Type I pyridoxal 5′-phosphate dependent enzymatic domains embedded within multimodular nonribosomal peptide synthetase and polyketide synthase assembly lines. *BMC Struct. Biol.* **2013**, *13*, 26. [CrossRef] [PubMed]
200. Sagar, M.; Pandey, N.; Qamar, N.; Singh, B.; Shukla, A. Domain analysis of 3 Keto Acyl-CoA synthase for structural variations in *Vitis vinifera* and *Oryza brachyantha* using comparative modelling. *Interdiscip. Sci.* **2015**, *7*, 7–20. [CrossRef] [PubMed]
201. Eng, C.H.; Backman, T.W.H.; Bailey, C.B.; Magnan, C.; Garcia Martin, H.; Katz, L.; Baldi, P.; Keasling, J.D. ClusterCAD: A computational platform for type I modular polyketide synthase design. *Nucleic Acids Res.* **2018**, *46*, D509–D515. [CrossRef] [PubMed]
202. Annaval, T.; Rudolf, J.D.; Chang, C.Y.; Lohman, J.R.; Kim, Y.; Bigelow, L.; Jedrzejczak, R.; Babnigg, G.; Joachimiak, A.; Phillips, G.N., Jr.; et al. Crystal structure of thioesterase SgcE10 supporting common polyene intermediates in 9- and 10-membered enediyne core biosynthesis. *ACS Omega* **2017**, *2*, 5159–5169. [CrossRef] [PubMed]
203. Weissman, K.J. Uncovering the structures of modular polyketide synthases. *Nat. Prod. Rep.* **2015**, *32*, 436–453. [CrossRef] [PubMed]
204. Weissman, K.J. Correction: Uncovering the structures of modular polyketide synthases. *Nat. Prod. Rep.* **2017**, *34*, 1035. [CrossRef] [PubMed]
205. Edwards, A.L.; Matsui, T.; Weiss, T.M.; Khosla, C. Architectures of whole-module and bimodular proteins from the 6-deoxyerythronolide B synthase. *J. Mol. Biol.* **2014**, *426*, 2229–2245. [CrossRef] [PubMed]
206. Smith, J.L.; Skiniotis, G.; Sherman, D.H. Architecture of the polyketide synthase module: Surprises from electron cryo-microscopy. *Curr. Opin. Struct. Biol.* **2015**, *31*, 9–19. [CrossRef] [PubMed]
207. Guex, N.; Peitsch, M.C.; Schwede, T. Automated comparative protein structure modeling with SWISS-MODEL and Swiss-PdbViewer: A historical perspective. *Electrophoresis* **2009**, *30*, S162–S173. [CrossRef] [PubMed]
208. Bertoni, M.; Kiefer, F.; Biasini, M.; Bordoli, L.; Schwede, T. Modeling protein quaternary structure of homo- and hetero-oligomers beyond binary interactions by homology. *Sci. Rep.* **2017**, *7*, 10480. [CrossRef] [PubMed]
209. Wells, J.N.; Bergendahl, L.T.; Marsh, J.A. Computational modelling of protein complex structure and assembly. *Methods Mol. Biol.* **2018**, *1764*, 347–356. [PubMed]

210. Mertens, H.D.; Svergun, D.I. Structural characterization of proteins and complexes using small-angle X-ray solution scattering. *J. Struct. Biol.* **2010**, *172*, 128–141. [CrossRef] [PubMed]
211. Khater, S.; Gupta, M.; Agrawal, P.; Sain, N.; Prava, J.; Gupta, P.; Grover, M.; Kumar, N.; Mohanty, D. SBSPKSv2: Structure-based sequence analysis of polyketide synthases and non-ribosomal peptide synthetases. *Nucleic Acids Res.* **2017**, *45*, W72–W79. [CrossRef] [PubMed]
212. Diminic, J.; Zucko, J.; Ruzic, I.T.; Gacesa, R.; Hranueli, D.; Long, P.F.; Cullum, J.; Starcevic, A. Databases of the thiotemplate modular systems (CSDB) and their in silico recombinants (r-CSDB). *J. Ind. Microbiol. Biotechnol.* **2013**, *40*, 653–659. [CrossRef] [PubMed]
213. Davison, J.; Dorival, J.; Rabeharindranto, H.; Mazon, H.; Chagot, B.; Gruez, A.; Weissman, K.J. Insights into the function of *trans*-acyl transferase polyketide synthases from the SAXS structure of a complete module. *Chem. Sci.* **2014**, *5*, 3081–3095.
214. Porter, J.L.; Rusli, R.A.; Ollis, D.L. Directed evolution of enzymes for industrial biocatalysis. *ChemBioChem* **2016**, *17*, 197–203. [CrossRef] [PubMed]
215. Li, G.; Wang, J.B.; Reetz, M.T. Biocatalysts for the pharmaceutical industry created by structure-guided directed evolution of stereoselective enzymes. *Bioorg Med. Chem.* **2017**, *26*, 1241–1251. [CrossRef] [PubMed]
216. Barley, M.H.; Turner, N.J.; Goodacre, R. Recommendations on the implementation of genetic algorithms for the directed evolution of enzymes for industrial purposes. *ChemBioChem* **2017**, *18*, 1087–1097. [CrossRef] [PubMed]
217. Sun, Z.; Wikmark, Y.; Backvall, J.E.; Reetz, M.T. New concepts for increasing the efficiency in directed evolution of stereoselective enzymes. *Chemistry* **2016**, *22*, 5046–5054. [CrossRef] [PubMed]
218. Kataoka, M.; Miyakawa, T.; Shimizu, S.; Tanokura, M. Enzymes useful for chiral compound synthesis: Structural biology, directed evolution, and protein engineering for industrial use. *Appl. Microbiol. Biotechnol.* **2016**, *100*, 5747–5757. [CrossRef] [PubMed]
219. Hughes, A.J.; Tibby, M.R.; Wagner, D.T.; Brantley, J.N.; Keatinge-Clay, A.T. Investigating the reactivities of a polyketide synthase module through fluorescent click chemistry. *Chem. Commun. (Camb)* **2014**, *50*, 5276–5278. [CrossRef] [PubMed]
220. Dorrestein, P.C.; Bumpus, S.B.; Calderone, C.T.; Garneau-Tsodikova, S.; Aron, Z.D.; Straight, P.D.; Kolter, R.; Walsh, C.T.; Kelleher, N.L. Facile detection of acyl and peptidyl intermediates on thiotemplate carrier domains via phosphopantetheinyl elimination reactions during tandem mass spectrometry. *Biochemistry* **2006**, *45*, 12756–12766. [CrossRef] [PubMed]
221. Gao, X.; Xie, X.; Pashkov, I.; Sawaya, M.R.; Laidman, J.; Zhang, W.; Cacho, R.; Yeates, T.O.; Tang, Y. Directed evolution and structural characterization of a simvastatin synthase. *Chem. Biol.* **2009**, *16*, 1064–1074. [CrossRef] [PubMed]
222. Rui, Z.; Zhang, W. Engineering biosynthesis of non-ribosomal peptides and polyketides by directed evolution. *Curr. Top. Med. Chem.* **2016**, *16*, 1755–1762. [CrossRef] [PubMed]
223. Su, C.; Zhao, X.Q.; Wang, H.N.; Qiu, R.G.; Tang, L. Seamless stitching of biosynthetic gene cluster containing type I polyketide synthases using Red/ET mediated recombination for construction of stably co-existing plasmids. *Gene* **2015**, *554*, 233–240. [CrossRef] [PubMed]
224. Liu, Q.; Shen, Q.; Bian, X.; Chen, H.; Fu, J.; Wang, H.; Lei, P.; Guo, Z.; Chen, W.; Li, D.; et al. Simple and rapid direct cloning and heterologous expression of natural product biosynthetic gene cluster in *Bacillus subtilis* via Red/ET recombineering. *Sci. Rep.* **2016**, *6*, 34623. [CrossRef] [PubMed]
225. Bian, X.; Huang, F.; Stewart, F.A.; Xia, L.; Zhang, Y.; Muller, R. Direct cloning, genetic engineering, and heterologous expression of the syringolin biosynthetic gene cluster in *E. coli* through Red/ET recombineering. *ChemBioChem* **2012**, *13*, 1946–1952. [CrossRef] [PubMed]
226. Wu, N.; Huang, H.; Min, T.; Hu, H. TAR cloning and integrated overexpression of 6-demethylchlortetracycline biosynthetic gene cluster in *Streptomyces aureofaciens*. *Acta Biochim. Biophys. Sin. (Shanghai)* **2017**, *49*, 1129–1134. [CrossRef] [PubMed]
227. Kouprina, N.; Larionov, V. Transformation-associated recombination (TAR) cloning for genomics studies and synthetic biology. *Chromosoma* **2016**, *125*, 621–632. [CrossRef] [PubMed]
228. Hansen, B.G.; Salomonsen, B.; Nielsen, M.T.; Nielsen, J.B.; Hansen, N.B.; Nielsen, K.F.; Regueira, T.B.; Nielsen, J.; Patil, K.R.; Mortensen, U.H. Versatile enzyme expression and characterization system for *Aspergillus nidulans*, with the *Penicillium brevicompactum* polyketide synthase gene from the mycophenolic acid gene cluster as a test case. *Appl. Environ. Microbiol.* **2011**, *77*, 3044–3051. [CrossRef] [PubMed]

229. Geu-Flores, F.; Nour-Eldin, H.H.; Nielsen, M.T.; Halkier, B.A. USER fusion: A rapid and efficient method for simultaneous fusion and cloning of multiple PCR products. *Nucleic Acids Res.* **2007**, *35*, e55. [CrossRef] [PubMed]
230. Hansen, N.B.; Lubeck, M.; Lubeck, P.S. Advancing USER cloning into simpleUSER and nicking cloning. *J. Microbiol. Methods* **2014**, *96*, 42–49. [CrossRef] [PubMed]
231. Lee, N.C.; Larionov, V.; Kouprina, N. Highly efficient CRISPR/Cas9-mediated TAR cloning of genes and chromosomal loci from complex genomes in yeast. *Nucleic Acids Res.* **2015**, *43*, e55. [CrossRef] [PubMed]
232. Zhang, M.M.; Wong, F.T.; Wang, Y.; Luo, S.; Lim, Y.H.; Heng, E.; Yeo, W.L.; Cobb, R.E.; Enghiad, B.; Ang, E.L.; et al. CRISPR-Cas9 strategy for activation of silent *Streptomyces* biosynthetic gene clusters. *Nat. Chem. Biol.* **2017**, *13*, 607–609. [CrossRef] [PubMed]
233. Huang, H.; Zheng, G.; Jiang, W.; Hu, H.; Lu, Y. One-step high-efficiency CRISPR/Cas9-mediated genome editing in *Streptomyces*. *Acta Biochim. Biophys. Sin. (Shanghai)* **2015**, *47*, 231–243. [CrossRef] [PubMed]
234. Zeng, H.; Wen, S.; Xu, W.; He, Z.; Zhai, G.; Liu, Y.; Deng, Z.; Sun, Y. Highly efficient editing of the actinorhodin polyketide chain length factor gene in *Streptomyces coelicolor* M145 using CRISPR/Cas9-CodA(sm) combined system. *Appl. Microbiol. Biotechnol.* **2015**, *99*, 10575–10585. [CrossRef] [PubMed]
235. Tong, Y.; Robertsen, H.L.; Blin, K.; Weber, T.; Lee, S.Y. CRISPR-Cas9 toolkit for actinomycete genome editing. *Methods Mol. Biol.* **2018**, *1671*, 163–184. [PubMed]
236. Siegl, T.; Tokovenko, B.; Myronovskyi, M.; Luzhetskyy, A. Design, construction and characterisation of a synthetic promoter library for fine-tuned gene expression in actinomycetes. *Metab. Eng.* **2013**, *19*, 98–106. [CrossRef] [PubMed]
237. Luo, Y.; Zhang, L.; Barton, K.W.; Zhao, H. Systematic identification of a panel of strong constitutive promoters from *Streptomyces albus*. *ACS Synth. Biol.* **2015**, *4*, 1001–1010. [CrossRef] [PubMed]
238. Phelan, R.M.; Sachs, D.; Petkiewicz, S.J.; Barajas, J.F.; Blake-Hedges, J.M.; Thompson, M.G.; Reider Apel, A.; Rasor, B.J.; Katz, L.; Keasling, J.D. Development of next generation synthetic biology tools for use in *Streptomyces venezuelae*. *ACS Synth. Biol.* **2017**, *6*, 159–166. [CrossRef] [PubMed]
239. Fang, L.; Guell, M.; Church, G.M.; Pfeifer, B.A. Heterologous erythromycin production across strain and plasmid construction. *Biotechnol. Prog.* **2018**, *34*, 271–276. [CrossRef] [PubMed]
240. Huang, J.; Yu, Z.; Li, M.H.; Wang, J.D.; Bai, H.; Zhou, J.; Zheng, Y.G. High level of spinosad production in the heterologous host *Saccharopolyspora erythraea*. *Appl. Environ. Microbiol.* **2016**, *82*, 5603–5611. [CrossRef] [PubMed]
241. Zhang, X.; Lu, C.; Bai, L. Conversion of the high-yield salinomycin producer *Streptomyces albus* BK3-25 into a surrogate host for polyketide production. *Sci. China Life Sci.* **2017**, *60*, 1000–1009. [CrossRef] [PubMed]
242. Liu, C.; Zhang, J.; Lu, C.; Shen, Y. Heterologous expression of galbonolide biosynthetic genes in *Streptomyces coelicolor*. *Antonie Van Leeuwenhoek* **2015**, *107*, 1359–1366. [CrossRef] [PubMed]
243. Baltz, R.H. *Streptomyces* and *Saccharopolyspora* hosts for heterologous expression of secondary metabolite gene clusters. *J. Ind. Microbiol. Biotechnol.* **2010**, *37*, 759–772. [CrossRef] [PubMed]
244. Jung, W.S.; Kim, E.; Yoo, Y.J.; Ban, Y.H.; Kim, E.J.; Yoon, Y.J. Characterization and engineering of the ethylmalonyl-CoA pathway towards the improved heterologous production of polyketides in *Streptomyces venezuelae*. *Appl. Microbiol. Biotechnol.* **2014**, *98*, 3701–3713. [CrossRef] [PubMed]
245. Pokhrel, A.R.; Dhakal, D.; Jha, A.K.; Sohng, J.K. Herboxidiene biosynthesis, production, and structural modifications: Prospect for hybrids with related polyketide. *Appl. Microbiol. Biotechnol.* **2015**, *99*, 8351–8362. [CrossRef] [PubMed]
246. Zhang, H.; Fang, L.; Osburne, M.S.; Pfeifer, B.A. The continuing development of *E. coli* as a heterologous host for complex natural product biosynthesis. *Methods Mol. Biol.* **2016**, *1401*, 121–134. [PubMed]
247. Sun, L.; Zeng, J.; Zhang, S.; Gladwin, T.; Zhan, J. Effects of exogenous nutrients on polyketide biosynthesis in *Escherichia coli*. *Appl. Microbiol. Biotechnol.* **2015**, *99*, 693–701. [CrossRef] [PubMed]
248. Yuzawa, S.; Kim, W.; Katz, L.; Keasling, J.D. Heterologous production of polyketides by modular type I polyketide synthases in *Escherichia coli*. *Curr. Opin. Biotechnol.* **2012**, *23*, 727–735. [CrossRef] [PubMed]
249. Kumpfmuller, J.; Methling, K.; Fang, L.; Pfeifer, B.A.; Lalk, M.; Schweder, T. Production of the polyketide 6-deoxyerythronolide B in the heterologous host *Bacillus subtilis*. *Appl. Microbiol. Biotechnol.* **2016**, *100*, 1209–1220. [CrossRef] [PubMed]
250. Bond, C.; Tang, Y.; Li, L. *Saccharomyces cerevisiae* as a tool for mining, studying and engineering fungal polyketide synthases. *Fungal. Genet. Biol.* **2016**, *89*, 52–61. [CrossRef] [PubMed]

251. Vickery, C.R.; Cardenas, J.; Bowman, M.E.; Burkart, M.D.; Da Silva, N.A.; Noel, J.P. A coupled in vitro/in vivo approach for engineering a heterologous type III PKS to enhance polyketide biosynthesis in *Saccharomyces cerevisiae*. *Biotechnol. Bioeng.* **2018**, *15*, 1394–1402. [CrossRef] [PubMed]
252. Ma, Z.; Li, W.; Zhang, P.; Lyu, H.; Hu, Y.; Yin, W.B. Rational design for heterologous production of aurovertin-type compounds in *Aspergillus nidulans*. *Appl. Microbiol. Biotechnol.* **2018**, *102*, 297–304. [CrossRef] [PubMed]
253. Gressler, M.; Hortschansky, P.; Geib, E.; Brock, M. A new high-performance heterologous fungal expression system based on regulatory elements from the *Aspergillus terreus* terrein gene cluster. *Front. Microbiol.* **2015**, *6*, 184. [CrossRef] [PubMed]

antibiotics

MDPI

Review

Production of β-Lactamase Inhibitors by *Streptomyces* Species

Daniela de Araújo Viana Marques [1,*], Suellen Emilliany Feitosa Machado [2], Valéria Carvalho Santos Ebinuma [3], Carolina de Albuquerque Lima Duarte [4], Attilio Converti [5] and Ana Lúcia Figueiredo Porto [6]

1 Campus Serra Talhada, University of Pernambuco, Avenida Custódio Conrado, 600, AABB, Serra Talhada, Pernambuco 56912-550, Brazil
2 Department of Antibiotics, Federal University of Pernambuco, Avenida da Engenharia, 2° andar, Cidade Universitária, Recife, Pernambuco 50740-600, Brazil; suellen_feitosa_@hotmail.com
3 Department of Bioprocesses and Biotechnology, School of Pharmaceutical Sciences, São Paulo State University (UNESP), Rodovia Araraquara-Jaú/Km 01, Araraquara 14800-903, Brazil; valeriac@fcfar.unesp.br
4 Campus Garanhuns, University of Pernambuco, Rua Capitão Pedro Rodrigues, 105, São José, Garanhuns, Pernambuco 55295-110, Brazil; carolina.albuquerque@upe.br
5 Department of Civil, Chemical and Environmental Engineering, Chemical Pole, University of Genoa, Via Opera Pia 15, 16145 Genoa, Italy; converti@unige.it
6 Department of Morphology and Animal Physiology, Federal Rural University of Pernambuco, Av. Dom Manoel de Medeiros, Recife, Pernambuco 52171-900, Brazil; analuporto@yahoo.com.br
* Correspondence: daniela_viana@yahoo.com.br; Tel.: +55-81-995047420

Received: 30 May 2018; Accepted: 12 July 2018; Published: 17 July 2018

Abstract: β-Lactamase inhibitors have emerged as an effective alternative to reduce the effects of resistance against β-lactam antibiotics. The *Streptomyces* genus is known for being an exceptional natural source of antimicrobials and β-lactamase inhibitors such as clavulanic acid, which is largely applied in clinical practice. To protect against the increasing prevalence of multidrug-resistant bacterial strains, new antibiotics and β-lactamase inhibitors need to be discovered and developed. This review will cover an update about the main β-lactamase inhibitors producers belonging to the *Streptomyces* genus; advanced methods, such as genetic and metabolic engineering, to enhance inhibitor production compared with wild-type strains; and fermentation and purification processes. Moreover, clinical practice and commercial issues are discussed. The commitment of companies and governments to develop innovative strategies and methods to improve the access to new, efficient, and potentially cost-effective microbial products to combat the antimicrobial resistance is also highlighted.

Keywords: actinobacteria; β-lactamase; resistance; antibiotic; β-lactamase inhibitor

1. Introduction

Infectious diseases, which are caused by bacteria, viruses, parasites, or fungi, persist to be the main cause of mortality worldwide. Despite the success of antibiotics and advances in their production and purification, children and old people are affected by bacterial infections that cause approximately 17 million deaths per year. One of the main reasons for this occurrence is the increasing prevalence of antibiotic-resistant strains [1].

Penicillin was the first antibiotic discovered by Alexander Fleming in 1928, and since then, these compounds have been essential for healthcare [2]. Although the first antibiotic producer discovered was a *Penicillium* strain, many others have this potential, and since 1942, the *Streptomyces* genus has been known for its extraordinary ability to produce secondary metabolites, mainly antibiotics; nearly two-thirds of which occur naturally. This genus is one of the largest genera as it contains over 500 species, described along with a variety of species recognized as Actinomycetes [3,4].

Streptomyces are among the most versatile soil microorganisms, given their high metabolite production rate and the large variety of biotransformations. Overall, intracellular mechanisms control the accumulation of metabolites, which depends on process variables, types of nutrients, their concentrations, and operating conditions in submerged culture [5,6]. Therefore, the study and selection of appropriate culture medium composition is essential to ensure high productivity and low costs of the production process [7]. Additionally, many natural antibiotics must be purified after the production process through cost-effective methods that enable one to recover the final product at the highest level of purity and yield [8].

The first *Streptomyces* species used in industrial antibiotic production were *S. griseus* and *S. venezuelae*, which allowed one to obtain streptomycin and chloromycetin, respectively [8]. Since then, a large number of antibiotics (over 55% of available antibiotics) produced by the genus *Streptomyces* were detected between 1945 and 1978 [9].

The first important antibiotic used in clinical practice was penicillin G (benzylpenicillin), a β-lactam compound that attracted the interest of researchers to develop other derivatives. The β-lactam ring present in the structure of this drug, which is typical of this class of antibiotics, acts by linking intimately to the penicillin-binding proteins (transpeptidases), affecting cell wall biosynthesis in both Gram-negative and Gram-positive bacteria [10,11].

Afterwards, as a result of quick replication, recombination, and the high mutation rate of bacteria, resistance to β-lactam antibiotics emerged among the β-lactamase-producing organisms. These enzymes act directly on the β-lactam ring, inactivating the antibiotic. The first β-lactamase was a plasmid-mediated enzyme [12], named TEM from the name (Temoniera) of a patient in which the enzyme-producing *E. coli* strain was isolated. Subsequently, a plasmid with similar biochemical properties was detected and named TEM-2. Faced with this fact, the efforts to discover inhibitors able to inactivate β-lactamases began around 1970. Among them, clavulanic acid, which was obtained by screening of natural products, and possesses a similar β-lactam ring, was found to be a potent inhibitor of staphylococcal penicillinases and other plasmid-encoded penicillinases present in enteric bacteria, including TEM [13].

Nevertheless, antibiotic resistance was not over, and other bacteria began to produce similar enzymes; thus, the main challenge would be discovering novel inhibitors with activity against a broad spectrum of enzymes from multiple classes [10]. Genetic engineering has been an alternative tool to achieve this aim; thanks to a better knowledge of the expression of regulatory proteins in mutant organisms, it suggested that antibiotic production might be influenced by these regulatory events [14].

In accordance with the points explained above, this review provides a summary of the β-lactamase inhibitors produced by *Streptomyces* species. Herein, we emphasize some topics such as antibiotic resistance, β-lactamases inhibitor producers, production and purification processes, use of β-lactamase inhibitors in clinical practice, and commercial aspects.

2. Antibiotic Resistance

As previously mentioned, antibiotics produced by *Streptomyces* spp. were detected in 1942 with the discovery of streptomycin, and since then, researchers intensified the search for similar antibiotics within this genus [15].

However, as a mechanism of resistance, bacteria started to produce enzymes with the capacity to hydrolyze β-lactam antibiotics (β-lactamases), hence decreasing their efficiency. Meanwhile, penicillin-resistant strains of *Staphylococcus aureus* and *Streptococcus pneumoniae* emerged, which led to the proposition that the resistance mechanism could be intrinsically associated with the genomes of these bacteria [16–20]. The same phenomenon was observed with Gram-negative bacteria belonging to the *Neisseria* genus [21] and other streptomycin-resistant bacteria [22]. Biofilm formation by microorganisms is also allied with chronic, recurrent human infections and resistance [23]. The spread of resistant strains is linked to the migration of people as well, who dissipate resistant strains among population in remote communities where the use of antibiotics is very restricted [24].

In a bacterial infection, antibiotics act in two stages: in the former, a large part of the bacterial cells are killed, while in the latter, a few of them stay alive and are called persistent bacteria. If the antibiotic is withdrawn, these survivors begin a new multiplication, leading to a slow elimination of infection and recurrent bacterial infections. This phenomenon, which is called persistent sub-minimum inhibitory concentration (MIC), is observed in most bacteria. For instance, tuberculosis, which is caused by *Mycobacterium tuberculosis*, is treated for at least six months, and when the treatment is not continuous, recurrence of infection is commonly observed. Although different from resistant bacteria, the persistent ones are of great concern to the medical field, and the study and knowledge of the mechanisms of this phenomenon are of paramount importance [25].

These sub-MICs suggest a greater selection of resistance, increase bacterial mutation rates, cause phenotypic and genotypic variability, and affect the biofilm formation. In addition, they promote the maintenance of horizontal transmission of resistance genes [26] and can be especially important in multispecies communities, where even small changes in species interaction can have cascading effects [27].

Many Gram-negative bacteria, such as *Haemophilus influenza*; *Klebsiella pneumonia*; *Acinetobacter baumanii*; *Pseudomonas aeruginosa*; *Enterobacter* spp.; *Escherichia coli*; *Serratia* spp.; *Proteus* spp.; *Providencia* spp.; *Helicobacter pylori*; *Salmonella* spp.; *Neisseria gonorrhoeae*; *Shigella* spp.; and some Gram-positive ones such as *Enterococcus faecium*, *S. aureus*, and *S. pneumoniae*, were classified by the World Health Organization (WHO) as "ESKAPE" pathogens, with an extremely important role in the antimicrobial resistance era [28,29].

In addition, another interesting theory was that, unlike antibiotic-producing fungi, *Streptomyces* species defend themselves from antimicrobial attach. These microorganisms possess a self-resistance mechanism to avert the suicide, which made the antimicrobial resistance appear as a natural event preceding the current selective pressure of clinical antibiotic use [30]. The main mechanisms involved in β-lactam antibiotic resistance are inactivation of β-lactamases and modification of penicillin-binding proteins (PBPs). Furthermore, about 90 years after penicillin discovery, β-lactam-based therapies have been largely used against some illness. It is known that the chemical structures of drugs have greatly been modified in the recent past, and this action has become the tactic for most of currently-available antibiotics, which are known as second-, third-, and fourth-generation antibiotics [31,32]. Despite this progress, the indiscriminate use of antibiotics may be responsible for a microbial war in the near future [33].

Beta-Lactam antibiotics were also the first natural antibacterial compounds to be successfully developed and modified by the pharmaceutical industry [34]. This class of compounds has in common a β-lactam four-membered ring associated with different structures (Figure 1). They act by inhibiting bacterial cell wall peptidoglycan biosynthesis, thereby leading to cell lysis and death [35,36]. Specifically, β-lactam antibiotics bind to the acylate active site of PBPs, which is essential for the bacteria cell wall biosynthesis [37].

Beta-Lactam antibiotics have been widely employed for several years to treat infectious diseases because of their high specificity and strong killing efficiency [37]; however, bacterial capacity to develop apparatuses of resistance to β-lactam drugs is one of the big problems of health care [37,38]. Among the mechanisms of antibiotic resistance, we can cite the production of efflux pumps, modification or reduced production of outer membrane porins (in Gram-negative bacteria), alteration of PBPs (the molecular target of β-lactams), and production of β-lactamases [39]. In this review, we will focus on the last resistance mechanism and the mode to overcome this problem by means of biotechnology.

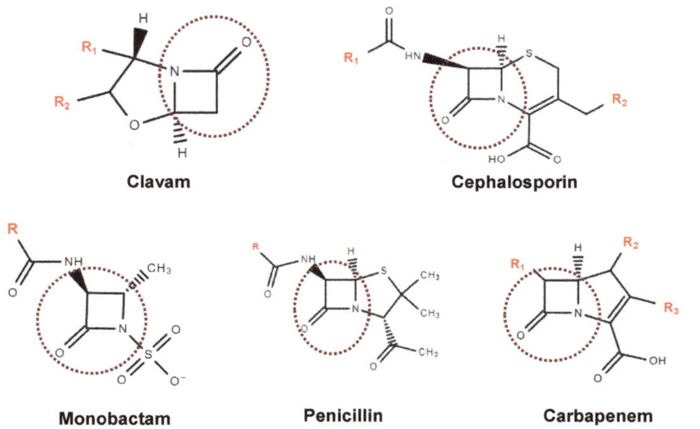

Figure 1. Backbone composition of β-lactam structures with the β-lactam ring highlighted (Structures drawn in ChemSpider).

3. Producers of β-Lactamases

Because β-lactamases act by breaking the amide bond present in the β-lactam ring of penicillins, they inactivate the drug biological activity, resulting in the inability of antibiotics to kill bacterial cells [40]. The microbial production of these enzymes strongly depends on cell structure. Gram-negative bacteria, with their inner and outer cell membranes separated from the periplasmic space, might be considered more developed microorganisms than the Gram-positive ones, which possess only a rigid cell wall composed of several peptidoglycan layers. As a result, Gram-negative bacteria produce high amounts of β-lactamases compared with the Gram-positive ones. Beta-Lactamase production by the former group can occur by induction (direct interaction of a β-lactam with the microorganism regulatory system) or constitutively, while that by Gram-positive bacteria is not yet well understood [37]. Among the Gram-negative bacteria, the β-lactamase production level can vary greatly; as an example, the active β-lactamase level in some enteric bacteria can exceed 4% of total soluble protein [41].

The enzyme characteristics are quite dependent on the original host. More than 1600 proteins with capacity to hydrolyze β-lactam rings have already been described [42], which, on one hand, demonstrates the diversity of β-lactamase structure [43], but from the other, hinders understanding of their action mechanism [42]. Efforts have been made to classify these enzymes, which were mainly based on their molecular structure (full nucleotide, amino acid sequence, and conserved motifs), functional characteristics (including substrate and inhibition profiles), or variances in structural features [43]. According to Ambler classification, β-lactamases are grouped in classes A, B, C, and D. Classes A, C, and D include enzymes that use serine in their active site to hydrolyze the substrate forming the acyl enzyme, while those belonging to class B are metallo-β-lactamases that require a bivalent ion, usually Zn^{2+}, for the hydrolysis [10,44].

On the other hand, Bush, Jacoby, and Medeiros in 1995 [43] classified β-lactamases in four groups based on their functional features: (1) cephalosporinases not fully inhibited by clavulanic acid; (2) penicillinases, cephalosporinases, and broad-spectrum β-lactamases inhibited by active site-directed inhibitors of β-lactamases; and (3) metallo-β-lactamases that are able to hydrolyze the main β-lactam antibiotics, but are scarcely inhibited by typical inhibitors of β-lactamases.

Later, Bush and Jacoby [44] expanded this classification to include subgroups with more recently discovered β-lactamases. As an example, the group 2d comprises the OXA enzymes (oxacillinases) that can hydrolyze cloxacillin and oxacillin [44]. Table 1 summarizes the groups of β-lactamases, while deeper description can be found in the work of Bush and Jacoby [44].

Table 1. Classification of β-lactamases based on Bush and Jacoby [44].

Functional Group According Bush, Jacoby, and Medeiros *		Ambler Class *	Hydrolyzed β-Lactam Antibiotics
1		C	Cephalosporins, benzylpenicillin, cephamycins
1e		C	Ceftazidime and often other oxyimino-β-lactams
2	2a	A	Benzylpenicillin
	2b	A	Benzylpenicillin and cephalosporins
	2be	A	Cefotaxime, ceftazidime, ceftriaxone, cefepime, aztreonam
	2br	A	Resistance to clavulanic acid, sulbactam, and tazobactam
	2ber	A	Oxyimino-β-lactams combined with resistance to clavulanic acid, sulbactam, and tazobactam
	2c	A	Carbenicillin
	2ce	A	Carbenicillin, cefepime, and cefpirome
	2d	D	Cloxacillin or oxacillin
	2de	D	Cloxacillin or oxacillin, and oxyimino-β-lactams
	2df	D	Cloxacillin or oxacillin and carbapenems
	2e	A	Cephalosporins inhibited by clavulanic acid
	2f	A	Carbapenems, oxyimino-β-lactams, cephamycins
3		B	Carbapenems
4		Unknown	Carbapenems

According to the work of Laxminarayan et al. published in 2016, around 214,000 neonatal sepsis deaths worldwide are related to resistant pathogens each year [45]. To overcome the problem of antimicrobial resistance due to β-lactamase production, the pharmaceutical industry has employed two strategies: (i) improvement of β-lactam antibiotics resistant to these enzymes (as an example, the expanded-spectrum cephalosporins and carbapenems); and (ii) the application of β-lactamase inhibitors (BLIs) associated with a β-lactam antibiotic [34]. However, the first strategy has not demonstrated to be so effective because bacteria have developed the capacity to produce enzymes with enhanced features such as extended-spectrum β-lactamases (ESBLs) [40] and carbapenemases [45]. Therefore, application of BLIs may be a valid approach to overcome the problem of antimicrobial resistance.

4. Inhibitors of β-Lactamases

Although resembling β-lactam antibiotic structure, BLIs are compounds with weak antibacterial activity. Thus, the therapeutic strategy is to co-administer BLIs with penicillins and cephalosporins [36] in two different ways: (i) employing substrates that bind the enzyme reversibly and/or irreversibly, forming unfavorable steric interactions; or (ii) developing mechanism-based or irreversible "suicide inhibitors" as clavulanic acid (CA), sulbactam, and tazobactam (Figure 2) [10,46].

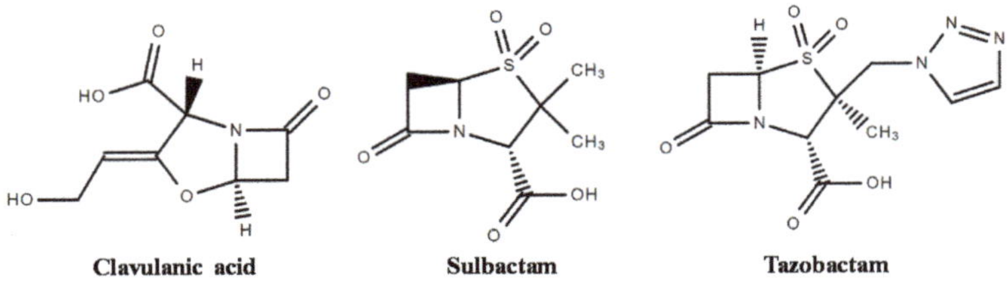

Figure 2. Chemical structure of β-lactamase inhibitors (Structures drawn in ChemSpider).

CA is a natural drug isolated in 1976 from *Streptomyces clavuligerus* fermentation broth [41], while sulbactam and tazobactam are BLIs developed by synthetic route in 1980 [40]. CA binds irreversibly with the serine hydroxyl group present in the active site of the enzyme, producing a stable acylated intermediate and inactivating the enzyme. As a result, the antibiotic co-administered with CA performs its main action [47]. Augmentin™ (amoxicillin and potassium clavulanate) [40] and Timentin™ (ticarcillin and potassium clavulanate) [47], both produced by GlaxoSmithKline [10], are examples of combinations available in the market. Drugs combined with CA have proven clinical efficacy against several bacteria (both Gram-negative and Gram-positive) as described by Drawz and Bonomo [10]. CA is associated with other β-lactam antibiotics in its salt form because of its instability under several conditions, such as acidic or alkaline conditions, and in the presence of salts [36].

Sulbactam and tazobactam have inhibitory spectra and mechanisms similar to those of CA [40]. Sulbactam is available on the medicine market in combination with ampicillin (Unasyn™ produced by Pfizer) and cefoperazone (Magnex™ produced by Pfizer) [10]. Sulbactam has advantages compared with other BLIs because it has its own activity against some *Acinetobacter baumannii* strains and does not induce class I (Ampc) chromosomal β-lactamases in Enterobacteriaceae [48]. Moreover, sulbactam combinations have not demonstrated strong selective pressures for ESBL-producing Enterobacteriaceae and vancomycin-resistant enterococci [48].

Tazobactam combined with piperacillin (Zosyn™ produced by Wyeth) has proven clinical efficacy against Gram-positive and Gram-negative pathogens [10]. Tazobactam was also associated with Ceftolozane being an antibacterial agent with potent activity against Gram-negative bacteria, including drug-resistant *P. aeruginosa* and many ESBL-producing Enterobacteriaceae [49]. Another association used to improve the spectrum of antibiotic action is that of piperacillin–tazobactam and vancomycin [50].

Although clavulanic acid, tazobactam, and sulbactam are commercially available BLIs, they exert inhibitory effect against serine-β-lactamases, mainly belonging to classes A, C, and D, which means that they are not inhibitors of metallo-β-lactamases [34]. Regarding metallo-β-lactamases inhibitors, the most important are the following: (i) thiol derivates, such as thiomandelic acids, which bind at the hydrophobic pocket of the enzyme active site and bind to, or interfere with, the bonding network between the hydrolytic water and the Zn^{2+} ions [10]; (ii) dicarboxylate, that is, succinic acid, which binds at the dinuclear metal center using one carboxylate to form a monodentate bridge between both zinc ions, and the second carboxylate to bridge between Zn^{2+} and a conserved Lys residue [51]; (iii) trifluoromethyl ketones and alcohols, the tetrazole portion of whose molecule interacts directly with the active-site Zn^{2+} ions; (iv) carbapenem analogs; (v) tricyclic natural products; and (vi) penicillin derivatives with C-6-mercaptomethyl substituent [10].

Considering that the bacterial resistance is a health care problem, the development of new β-lactamase inhibitors or improvement of already-available BLIs is essential. Meropenem-RPX7009 and Biapenem-RPX7009 by Rempex Pharmaceuticals that are both boronate β-lactamase inhibitors are still in development phase [42]. Although the first citation of boronic acids as serine-β-lactamases inhibitors was already reported many years ago, they have recently been explored as the next generation of pan-β-lactamase inhibitors. Trigonal boron (III) compounds behave as Lewis acids and are prone to react with nucleophiles, resulting in tetrahedral covalent adducts able to resist enzymatic hydrolysis [34]. Another example is Avibactam, which is a diazabycloootanone approved by the U.S. Food and Drug Administration (FDA) to be used in combination with Ceftazidime (AstraZeneca Pharmaceuticals, Forest–Cerexa, Actavis–Allergan) [42] to treat complicated intra-abdominal infections (cIAI), complicated urinary tract infections (cUTI), and hospital-acquired pneumonia (HAP) [52].

The most part of β-lactamase inhibitors considered in the present review and available in the market is produced through synthetic routes. The exception is CA, which is produced by *Streptomyces* strains by fermentation. However, the development of biotechnology/bioprocesses in the last decades has shown a new perspective to develop new pharmaceutical products or improve already-established processes. Thus, we will present old and new panoramas to overcome the question of antibiotic resistance using biotechnology.

5. Producers of β-Lactamases Inhibitors

Streptomyces spp. are aerobic, filamentous, Gram-positive bacteria, which resemble fungi and are obtained naturally in different environments. These microorganisms are able to create a chain of spores developing a multicellular complex that, after the sporulation phase, forms hyphae with multinucleated mycelium [53]. These morphological characteristics make *Streptomyces* spp. the most adaptable microorganisms found in the soil, given their broad spectrum of metabolite produced and their biotransformation processes.

The discovery of β-lactamase inhibitors introduced new ways to overcome the problem of antibiotic resistance, through drug research and development [54]. However, as they also contain the same β-lactam ring, they are as susceptible to time-limited application as the β-lactam antibiotics [55]. Resistance to the β-lactam/β-lactamase inhibitor combinations especially of Gram-negative microorganisms is an old and clinically difficult situation [11,56] that has stimulated the use of computation-based design methods to develop new β-lactamase inhibitors [54].

As previously mentioned, sulbactam and tazobactam are obtained by synthetic route [54], while clavulanic acid is a secondary metabolite naturally produced by the actinomycete *S. clavuligerus* [57] that belongs to the class of clavams to which it gives its name.

The discovery of clavams, in which an oxygen substitutes a sulfur of penicillins and cephalosporins, occurred during a screening of microorganisms looking for natural products able to inhibit β-lactamases [58,59]. They are structurally related to CA, which, however, is the only one exerting β-lactamase inhibitory activity. Such an activity may be related to the peculiar 3R, 5R stereochemistry of CA, while all the other clavams have a 3S, 5S stereochemistry, although some of them have antibacterial or antifungal characteristics [57,60].

According to Nobary and Jensen [61], *S. clavuligerus* produces the following 5S clavam compounds in addition to CA: 2-hydroxymethylclavam, 2-formyloxymethylclavam, clavam-2-carboxylic acid, and alanylclavam (Figure 3). Even though not clinically effective, these compounds can be used as precursors to produce CA through a different metabolic route.

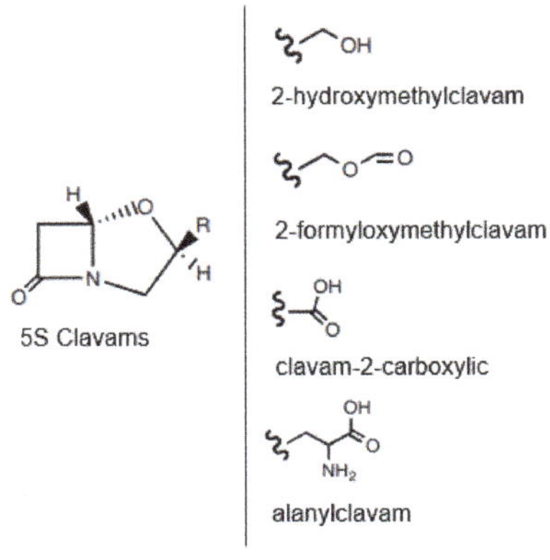

Figure 3. Chemical structure of 5S clavam compounds obtained from *Streptomyces clavuligerus*. (Adapted from Nobary and Jensen [61]).

CA was identified later in other *Streptomyces* species, namely *Streptomyces jumonjinensis* and *Streptomyces katsurahamanus*, while a great variety of *Streptomyces* species showed the capacity to produce other clavam metabolites with structures similar to CA (clavaminic acid, valclavam, and clavamycins) [62]. More recently, genome-sequencing projects have resulted in deposits of DNA sequences that suggest that CA biosynthetic capability exists in a wider range of strains, including *Streptomyces flavogriseus* ATCC 33331 and *Saccharomonospora viridis* DSM 43017 [61]. The number of species that produce non-CA clavams overpasses that of CA producers, so the ability to biosynthesize and distribute CA is rather limited [62].

Studies investigating biosynthetic ways and genes correlated with the production of CA and other clavams have been performed [63–65]. CA biosynthesis involves early and late stages including oxidative reactions catalyzed by 2-oxoglutarate dependent dioxygenases. The "early" steps are largely accepted, but those of the "later" fraction are not understood or are even unknown. Gene clusters for the biosynthesis of CA, 5S clavams, and their paralogs have been identified in *S. clavuligerus* genome sequence. According to Tahlan et al. [66], *ceaS2*, *bls2*, *pah2*, *cas2*, and *oat2* are genes involved in clavaminic acid biosynthesis, thus contributing to both CA and 5S clavams productions, while other genes are only involved in that of CA.

CA obtained by fermentation is then recovered from the medium and purified several times. This production has been the subject of intensive research in recent years because of its clinical and commercial importance [61]. According to Costa and Badino [67] and Ünsaldı et al. [65], CA fermentation processes provide low CA concentrations, which hinders its large-scale chemical synthesis. Moreover, the commercial CA feasibility is restricted because of its elaborate production process [68]. Thus, it is important to get higher production rates of this valuable compound applying more powerful fermentations.

According to Viana-Marques et al. [69], certain nutrients present in the culture medium (such as C, N, P, S sources and salts) and compounds produced in the biosynthetic steps can enhance CA production. Additionally, there are many ways to increase CA production such as optimization of bioreactor operation, conditions of agitation and aeration, and medium composition [70].

After fermentation, the CA-containing fermented broth is clarified by centrifugation and/or filtration [71], and the clarified broth subjected to primary extraction steps involving liquid–liquid extraction followed by adsorption procedures. We will discuss the production and purification process to obtain β-lactamases inhibitors later.

The production of new antibiotics, and investigating and manipulating the biosynthesis routes, represent an important, efficient, and sustainable tool [60], with a chance to discover novel mechanisms of action [72]. To promote the production of new compounds, many strategies have been employed such as cloning and heterologous expression of biosynthetic gene clusters, affecting the regulatory ways, varying culture conditions, and co-culturing two or more organisms together [73].

5.1. Natural Microorganisms

The search for bioactive metabolites like novel antibiotics produced by microorganisms for potential use in several industrial applications, mainly in the agricultural and pharmaceutical fields, has become more significant because of the progress of drug/multi-drug resistance in most of pathogenic microbes [74]. Even though several antibiotics used in the clinical practice and agriculture were produced by Streptomycetes, many others were found from natural products produced by this genus and were developed into precious therapeutic agents [62,75].

As mentioned above, *Streptomyces* members are Gram-positive bacteria that grow in different environments with a filamentous form, such as fungi. Soil is the main niche in which they were isolated and identified as prolific producers of effective bioactive compounds other than antibiotics such as antivirals, anticancers, anti-hypertensives, and immunosuppressives [1,74].

Streptomyces species were considered as the major producers of bioactive compounds for the biotechnology industry [76]. Genome sequence surveys on various actinomycetes indicate that each bacterium is able to produce about 10-fold more secondary metabolites than the wild type selected by screening analysis before the availability of the genome sequence data. This suggests that actinomycetes are still promising sources of novel bioactive compounds [77].

According to Chater [75], *Streptomyces* are being explored even more intensively in the hope that they will help extensively to the provision of new therapeutic products to face the global threat of antibiotic resistance among pathogenic bacteria, as well as to the supply of other bioactive agents for medical purposes.

More than 23,000 bioactive secondary metabolites obtained from microorganisms have been found, and over 10,000 of them are produced by actinomycetes [78]. In general, the genome sequencing of actinomycetes isolated from soil, or of marine and plant origin, has resulted in the identification of a range of clusters for secondary metabolites, in the quantity of about 20 to 30 clusters per genome. This report reveals unexpected truths in the discovery of new useful agents [79].

Of the 20 (or more) secondary metabolites produced by *S. clavuligerus*, many have clinical importance, like cephamycin C (CephC) and CA, which are synthesized simultaneously through different metabolic routes, rigorously controlling intra- and extracellular factors [80,81].

S. clavuligerus is a great model for analyzing the relations involved in the biosynthesis of secondary metabolites because of its productive diversity [82]; for this reason, it has been used in metabolic engineering to obtain new strains able to overproduce CA. Particularly, amplification of biosynthetic genes encoding specific enzymes can conduct to an earlier and more rational method to upgrade strains with higher antibiotic production [62].

5.2. *Genetically-Modified Microorganisms*

Large-scale obtention of antibiotics by microbial fermentation has been the ground of the industry since the development of penicillin in the 1940s. Amounts of these products are nowadays very prominent after years of strong improvement of strategies as such mutagenesis and selection. These strategies, which contribute for strain enhancement, were early adopted for the penicillin strain. For example, Olano et al. reported in 2008 a penicillin production by *Penicillium chrysogenum* higher than 70 g/L, while in 1949, the original strain produced only 60 mg/L, which represents more than a 1000-fold increase [83].

Random mutagenesis and selection techniques are frequently used to obtain the strain most suitable for industrial fermentations, aiming to get high amounts of secondary metabolites [82]. This advance can be realized through over-expression of overall regulators, pathway-specific regulators, or biosynthetic genes. However, genetic engineering attempts to create high-yield strains of a specific product have rarely been successful; thus, enhancing production is still considered a challenge [64].

Furthermore, the deployment of new technologies such as DNA sequencing, transcription profiling, genomics, proteomics, metabolomics, transcriptomics, and metabolite profiling have offered new chances to engineer strains for obtaining high yields of natural products [83].

The improvement of microbial species is also an important way to decrease the production costs of industrial fermentations. Now, the mutation and selection technique is used frequently with success; however, it is very slow and labor-intensive [64].

According to Li and Townsend [82], the creation of a new generation of highly performant strains by this approach may take at least five years. Techniques of molecular genetics have been improved, and the possibility to change existing ways or make non-native pathways has progressed fast (Table 2). Advances in genetics, transcriptional analysis, proteomics, metabolic reconstructions, and metabolic flux analysis offer genetic engineering the chance to enhance the approaches for strain improvement in a targeted way.

Table 2. Genetic methods applied to improve the production of antibiotics and β–lactamase inhibitors (Adapted from Adrio and Demain [84]).

Genetic Method	Secondary Metabolite
Protoplast fusion	Penicillin G, cephalosporin C, cephamycin C, clavulanic acid, indolizomycin, rifamycins
Metabolic engineering	Antibiotics (penicillin G, cephalosporin C, cephamycin C, clavulanic acid, semisynthetic cephalosporins)
Transposition	Daptomycin, tylosin
Combinatorial biosynthesis	Erythromycins, tetracenomycins, tylosin, spiramycins
Genome mining	Echinosporamicin-type antibiotics, antifungal compounds (ECO-02301), and so on

Most models of CA overproduction have come from manipulation of genes encoding biosynthetic enzymes or transcriptional regulators [62]. CA is obtained by submerged fermentation and then purified from the fermented broth in different ways; however, the most common protocol is centrifugation for cell separation, followed by liquid–liquid extraction with organic solvents and/or adsorption techniques and, finally, chromatography techniques [85].

6. Production of β-Lactamase Inhibitors

6.1. Biosynthesis of Clavulanic Acid

Studies carried out by Higgens and Kastner [86] and Brown et al. [87] were the first reports on the production of β-lactamase inhibitors by *S. clavuligerus* ATCC27064 and their recovery from the fermented broth. Although more than 40 years have passed since then, their strategy remains one of the most frequently applied to produce important drugs for medicinal use [88].

The biosynthetic pathway to produce CA (Figure 4) has not been fully elucidated, although many intermediates and enzymes have already been isolated. Nonetheless, isotope studies, purification, and characterization of the enzymes involved in the process, together with genetic studies, have contributed to clarify its biosynthetic pathway. Arginine (C5 precursor) and glutaraldehyde-3-phosphate (C3 precursor) were identified as two important precursors for clavulanic acid in *S. clavuligerus* [89,90].

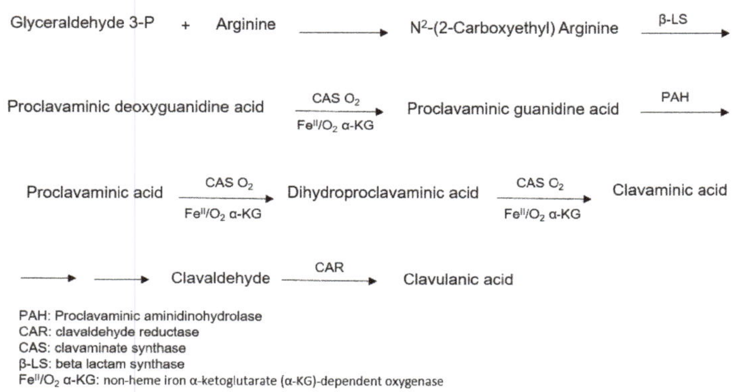

Figure 4. Scheme of clavulanic acid biosynthesis. (Adapted from Oliveira et al. [89,90]).

6.2. Production Process

After the discovery of β-lactamase inhibitors with antibacterial activity reported by Brown et al. [87], many studies have been carried out in recent decades regarding the production processes of β-lactamases inhibitors [70]. Usually, the industrial production of CA is almost entirely based on *S. clavuligerus* cultivation in complex medium [91]. Because of the clinical importance of this compound, the increase in CA production has been the focus of several studies. Strategies to enhance CA production include optimization of batch or fed-batch operation [92,93], temperature [94–96], agitation and aeration [60,93], and medium composition [70,94,95], as well as the selection of new microbial strains [97].

The productivity of microbial metabolites is closely related to the submerged culture process. The selection of the most suitable medium composition is of primary importance to increase the productivity and decrease the cost of any bioprocess [7]. It is well known that extracellular microbial CA production is greatly influenced by medium components, especially carbon and nitrogen sources [70,94], salts composition [70,91,98], and pH [93,94,99]. However, no single medium has been established to optimize CA production by different strains, because each organism requires different conditions for maximum production, which is mainly controlled by intracellular effectors [100].

6.2.1. Carbon Sources

Several carbon sources (glycerol, starch, sucrose, and lipid) have been used to produce CA [70,95,101–103]. Glycerol plays an important role in CA production as the C3 precursor of the molecule [104]. The formation of the C3 precursor did in fact appear to be the rate-limiting step of CA synthesis, while excess arginine, which is the C5 precursor, failed to increase CA production [105]. Bellão et al. [80] investigated the effect of carbon source and feeding conditions on the productions of CA and cephamycin C (CephC) by *S. clavuligerus*. In the experimental range studied, glycerol feeding conditions did not influence maximum CephC production (566.5 mg/L), whereas maximum CA concentration (1022 mg/L) was strongly dependent on culture conditions. These results are consistent with those reported by Saudagar and Singhal [93], who obtained higher amounts of clavulanic acid using glycerol in the production medium.

Wang et al. [106] achieved maximum CA production in shake flask batch cultivation at a glycerol concentration of 15.0 g/L, while Teodoro et al. did so in fed-batch culture at 120 g/L [107]. In a fed-batch study carried out in a 10 L bioreactor to determine the influence of glycerol feeding on CA production by *S. clavuligerus*, the highest CA production (1.6 g/L) was obtained using a glycerol concentration of 180 g/L, highlighting a positive effect of glycerol on CA biosynthesis [102]. Many authors have observed that a glycerol concentration above 15 g/L inhibited batch CA biosynthesis [94,100,108]. Viana et al. [94] observed a decrease in CA production by *Streptomyces* DAUFPE 3060 when glycerol concentration was raised from 5 to 10 g/L. The better performance of the fed-batch process compared with the batch one is likely due to its capacity to prevent substrate inhibition and metabolite repression, besides controlling the growth rate and prolonging the stationary phase.

The preferred carbon source for CA production are lipids because of the inability of *S. clavuligerus* to utilize simple carbohydrates such as glucose. Oils are preferred in terms of energy, as they contain approximately 2.4 times the energy of glucose [57]. Several authors have reported that oils may stimulate CA production [93,99,109–113]. Lee and Ho [109] identified palm and palm-kernel oils as the most suitable carbon sources for growth of *S. clavuligerus* and CA production. Large et al. [110] reported maximum CA production (80 mg/L) in a production medium containing unspecified lipids (C16 and C18 unsaturated and saturated fatty acids). Maranesi et al. [111], studying the use of vegetable oil in CA production by *S. clavuligerus* ATCC 27064, found the highest CA concentration (753 mg/L) in a medium containing 30 g/L soybean oil. Saudagar and Singhal [93] obtained similar CA concentrations in media containing palm oil and soybean oil.

Kim et al. [113] investigated the effect of oils on cell growth and CA production during *S. clavuligerus* NRRL 3585 fermentation. Triolein, whose fatty acid is oleic acid only, was the best oil source for CA production, but free fatty acids generated from oil hydrolysis affected both CA production and cell growth. In the same work, the authors screened for *S. clavuligerus* mutants resistant to high oleic acid concentrations and identified a mutant (*S. clavuligerus* OL13) whose oleic acid minimum inhibitory concentration (MIC = 2.1 g/L) was much higher than that of *S. clavuligerus* NRRL 3585 (0.4 g/L). Not only cell growth was improved, but also maximum CA concentration (1,950 mg/L) was approximately twice as high as that of the parent strain.

Efthimiou et al. [112] described an increase in CA production when 47 mg/L olive oil was used instead of 25 mg/L glycerol as the sole carbon source. In a similar study on the effects of different vegetable oil-based media on cell growth and CA production during *S. clavuligerus* ATCC 27064 cultivation, Salem-Berkhit et al. observed that three out of eight tested oils supported CA production [99] and that the olive oil-containing medium ensured a CA concentration twice as high as glycerol-containing medium.

The use of vegetable oils as the sole carbon source can support bacterial growth and enhance CA production, but a careful choice of the oil is essential to prevent affecting the CA yield [112]. These findings can be explained with the residual oil levels in culture medium [114] and the high oxygen requirement for oil metabolism. Residual oil levels may lead to problems associated with the increased medium viscosity and warrant additional downstream processing [93].

Comparing glycerol and sucrose as the sole carbon source, Lee and Ho [109] observed no production of CA in a glycerol-containing medium, but high production in sucrose-containing medium. Similar findings were reported by Ives and Bushell [105], who observed no CA production in glycerol-containing C-limited medium. Another study by Thakur et al. [101] demonstrated that the addition of dextrin or glycerol as the sole carbon source neither improved nor decreased CA production. However, two studies reported a totally different observation, that is, a glycerol-containing basal medium allowed for a maximum CA level (348.5 mg/L) about twice as high as a starch-based medium [80,93].

Additionally, two other studies by Saudagar and Singhal [93] and Chen et al. [100] revealed a biphasic dose response of glycerol, whereby CA production was inhibited at either too high or too low concentrations.

6.2.2. Nitrogen Sources

Soybean derivatives (flour, protein isolate, and meal) have been used as nitrogen sources for CA production [7,67,70,94,95,97,106,107]. To perform a screening of medium constituents for fermentative CA production by *S. clavuligerus*, Rodrigues et al. [70] performed fermentations using soybean protein isolate (SPI) and soybean flour (SF) as the primary nitrogen source, and obtained higher CA concentration (437 mg/L) with the former ingredient. On the other hand, with SF, CA production remained steady for a long time likely because this ingredient induced the release of extracellular proteases by *S. clavuligerus*, which hydrolyzed it during the growth phase, providing a steady supply of essential nutrients to the microorganism [115].

Being a by-product of oil extraction, SF has been recognized as a potentially useful and cost-effective ingredient. It consists of approximately 40% proteins and is rich in other organic and inorganic compounds, thus being a good candidate for a culture medium [116]. According to Chen et al. [102], soybean derivatives, such as soy meal flour and soybean protein hydrolyzates, are excellent components of media for CA production because they contain arginine, the precursor of CA.

Viana et al. [94], in their attempt to investigate the effect of SF concentration on CA production by the new isolate *Streptomyces* DAUFPE 3060, observed maximum CA production (494 mg/L) at the highest level (20 g/L). After optimization by response surface methodology, Marques et al. [95] achieved, with the same strain, a maximum CA concentration as high as 629 mg/L using 40 g/L SF.

Ortiz et al. [7] carried out a study on the influence of the type of soybean derivatives as nitrogen sources on CA production by *S. clavuligerus*. Using two different media, one containing 20 g/L SF and the other 20 g/L SPI, they obtained the highest CA production (698 mg/L) with the former ingredient. On the other hand, Teodoro et al. [107], who investigated the effect of SPI level on CA production, achieved the highest CA yield (380 mg/L) at intermediate SPI concentration (20 g/L corresponding to 2.95 g/L total N).

6.2.3. Amino Acids as Supplements in Basal Medium

Arginine and ornithine exert a concentration-dependent stimulation of CA production, and both amino acids are effectively incorporated into the CA molecule [117–119]. The investigation of the role of amino acids as nitrogen sources in CA production began in 1986 [117]. Since then, several studies focusing on the effects of amino acids, mainly arginine and ornithine, have been performed.

The incorporation of arginine into the CA molecule does not establish ornithine as a direct precursor, because the enzyme ornithinecarbamoyl transferase of *S. clavuligerus* exhibits arginase activity, which converts arginine to ornithine [120]. However, Valentine et al. [121] used blocking mutants in the *argF* and *argG* genes, which were unable to convert ornithine into arginine, even though they found a great incorporation of arginine in the CA molecule and a poor incorporation of ornithine.

This demonstrates that arginine is the direct precursor of CA and indicates that arginase activity does not produce sufficient ornithine to incorporate into CA.

Townsend and Ho [119] suggested arginine and pyruvate as CA precursors. However, studies have demonstrated that exogenous ornithine, rather than arginine, effectively enhances CA production, provided that there is a sufficient amount of C3 precursor [120]. A 270% increase in CA production was observed intermittently feeding glycerol and ornithine compared with the batch cultivation in shake flasks, and a 150% increase compared with cultures with glycerol and arginine feeding or when only glycerol was fed [122]. Teodoro et al. [102], who investigated the influence of glycerol and ornithine feeding on CA production by *S. clavuligerus* in batch bioreactor, observed an increase in CA productivity, but a small decrease in CA concentration, in the presence of ornithine. Rodrigues et al. [70] observed that glutamate and ornithine negatively affected CA production, while arginine and threonine had no influence.

6.2.4. Salts in Basal Medium

Compounds containing phosphorus, magnesium, and iron are also included in culture media used for CA production [70,106]. Rodrigues et al. [70] published a study on the nutritional requirements of *S. clavuligerus* for CA production where they demonstrated that ferrous sulfate is an essential ingredient of the fermentation medium because the enzymes involved in CA biosynthesis are Fe^{2+}-dependent. A medium containing ferrous sulfate allowed for a CA concentration of 437 mg/L, while a formulation without this salt yielded only 41 mg/L.

Phosphate is a crucial growth-limiting nutrient that regulates the synthesis of antibiotics belonging to different groups; therefore, industrial production of antibiotics is carried out at growth-limiting concentrations of inorganic phosphate [57]. Saudagar and Singhal [98] observed that the optimum concentration of KH_2PO_4 for CA production (878 mg/L) was 10 mM. According their results, higher CA values are expected at lower temperatures and KH_2PO_4 concentrations.

6.2.5. Effect of pH

It has been reported that one of the important characteristics of *S. clavuligerus* is its strong dependence on the extracellular pH for cell growth and CA production [93,94,99]. Viana et al. [94], using a fractional factorial design to investigate the influence of the initial medium pH on CA production by *Streptomyces* DAUFPE 3060, observed the highest CA concentration (494 mg/L) at pH 6.0, while Saudagar and Singhal [93], using a L_{25} orthogonal array, identified pH 7.0 and 7.5 as the optimum pH values for CA production (500 mg/L) and cell growth (140 mg/L nucleic acid), respectively. The marked decrease in the CA yield out of this pH range suggested the occurrence of CA degradation under either acidic or alkaline conditions.

6.2.6. Extractive Fermentation of Clavulanic Acid

CA fermentation processes still present problems such as low CA concentrations [7,92,93,106]; thus, it is essential to search for more effective fermentation methods able to increase the production yield of this valuable compound. Moreover, CA recovery is based on a relatively complex downstream protocol, including successive liquid–liquid extraction steps with organic solvents and a final chromatographic step, which results in low purification yields [68,123]. Therefore, the search for new environmentally friendly (lower amounts of organic solvents) purification strategies is fundamental to achieve higher yields and lower costs. Liquid–liquid extraction in aqueous two-phase systems (ATPS) is an interesting alternative for this purpose [47].

Typical ATPS are the aqueous two-phase polymer systems (ATPPS), formed by two polymers or one polymer and a salt, which separate into two immiscible phases above their critical solubility. Such systems consist of a light phase (top phase), rich in one polymer, and a heavy phase (bottom phase), rich in the second polymer or the salt [124]. Because of the low cost and large difference in hydrophobicity of the two phases, ATPPS, mainly poly(ethylene glycol) (PEG)/salt systems, have been widely used to separate enzymes [125], other proteins [126], and antibiotics [47,68,127].

Some advantages of applying ATPS to whole culture media are the increase in recovery yields, feasibility for continuous operation, reduction of the number of steps, and decrease in the process costs because of joining clarification and partial purification [127]. For most of the ATPS constituted by PEG and salt, the antibiotic is commonly separated into the PEG-rich top phase, whereas the by-products such as amino acids and peptides are discarded into the salt-rich bottom phase [128].

An alternative process has been proposed, which integrates fermentation and extraction with the aim to improve the rate of product formation and production costs. Extractive fermentation or in situ product recovery provides a technological solution to overcome the limitations of product inhibition and low product titer, which are typical drawbacks of biotechnological processes. The concept, as the name suggests, involves integration of an extractive step as the first stage of downstream processing, to remove the product simultaneously to its synthesis. In extractive fermentation using ATPS, cells are considered to be immobilized in one of the phases or at the interface, and the required product to partition into the other phase by proper handling of the system [128].

The advantages of extractive fermentation also include reduction of the toxic effect of product on microbial growth [129] and extended fermentation time [130]. Moreover, continuous product removal during the entire fermentation minimizes the temperature- and pH-dependent degradation of product [131], by reducing its exposure to such damaging conditions. This is particularly beneficial for labile products like CA.

Viana Marques et al. [127] carried out an optimization study according to a 2^2-central composite design to investigate the influence of four variables, specifically, PEG molar mass, PEG and phosphate concentrations, and agitation intensity, on CA extractive fermentation with a PEG/phosphate ATPS. Whereas CA partitioned towards the PEG-rich top phase, cells positioned at the interface. Moreover, it was found that 25% PEG with molecular weight of 8000 g/mol and phosphate salts at 240 rpm were the best conditions for the extractive fermentation, leading to the best results in terms of CA partition coefficient ($K = 8.2$), yield in the PEG-rich phase ($Y = 93\%$), and productivity ($P = 5.3$ mg/L.h).

Panas et al. [132], who attempted the purification of CA produced by *S. clavuligerus* via submerged fermentation using different ATPS, obtained the highest CA recovery yield (64.91%) and purification factor (22.70) with PEG-600/sodium polyacrylate-8000 and PEG-600/cholinium chloride, respectively. These results support the use of these systems as effective techniques to purify CA from fermented broth in a single partitioning step.

7. β-Lactamase Inhibitors in Clinical Practice

Immediately after the discovery of β-lactam antibiotics as natural antibacterial compounds, they were effectively modified by the pharmaceutical industry for clinical purposes. However, they even constitute an effective group of antibiotics. With the advance of technologies through rapid bacterial genome sequencing, improvements in protein structure determination, and advanced genetic engineering, many industries started to modify the antibiotic structures leading to the second, third, and fourth generations of these drugs. Nevertheless, the problem of antimicrobial resistance persisted, and combined drug therapies arose [31,32].

Mycobacteria antibiotic resistance is multifactorial, including enzyme inactivation and cell wall impenetrability [133]. Additionally, because this bacteria group developed genetic and epigenetic resistance through selection [132], treatment options for mycobacteria have been restricted and inert for over thirty years [134]. An example is the rapid and global diffusion of genes of strains with multidrug resistance such as NDM-1-producing *K. pneumoniae* [135].

Co-therapy with β-lactam antibiotics and β-lactamase inhibitors, with or without a β-lactam ring, has shown some success (Table 3) [136–138]. Oxyimino-cephalosporins or carbapenems are still potentially very reliable and represent a validated strategy to overcome antimicrobial resistance in Gram-negative and Gram-positive bacteria.

Meanwhile, taking into account that Avibactam (and derivatives) are inhibitors of both class A and class C (and some of class D) serine enzymes, new approaches and new targets are essential to diversify treatment options. The approval of Vabomere (meropenem/vaborbactam) or Avycaz (ceftazidime-avibactam) demonstrates that novel combinations could lead to an amenable successful clinical development.

Table 3. Combinations of β-lactamase inhibitors and β-lactam antibiotics of clinical use (Adapted from Docquier and Mangani et al. [34] and Bush [41]).

<table>
<tr><th>β–Lactamase Inhibitor</th><th>β–Lactam Antibiotic</th><th>Development Status</th><th>References</th></tr>
<tr><td>Clavulanic acid</td><td>Amoxicillin</td><td>Approved by FDA, EMA</td><td>[139]</td></tr>
<tr><td rowspan="2">Sulbactam</td><td>Ampicillin</td><td>Approved by FDA, EMA</td><td>[140]</td></tr>
<tr><td>ETX2514</td><td>Phase 1 trials completed in 2017</td><td>[34]</td></tr>
<tr><td rowspan="3">Tazobactam</td><td>Piperacillin</td><td>Approved by FDA, EMA</td><td>[141]</td></tr>
<tr><td>Ceftolozane</td><td>Approved in 2014 by FDA and in 2015 by EMA</td><td>[34,142–146]</td></tr>
<tr><td>Cefepime</td><td>Used in Asia</td><td>[147]</td></tr>
<tr><td rowspan="3">Avibactam</td><td>Ceftazidime</td><td>Approved in 2015 by FDA and in 2016 by EMA</td><td>[34,148–159]</td></tr>
<tr><td>Aztreonam</td><td>Phase 2 in progress</td><td>[149,153,160,161]</td></tr>
<tr><td>Ceftaroline</td><td>Phase 2 in progress</td><td>[149,159,162,163]</td></tr>
<tr><td>Relebactam (MK7655)</td><td>Imipenem
(+ cilastatin)</td><td>Phase 3, cUTI (in progress)
Phase 2 cIAI (completed)</td><td>[164,165]</td></tr>
<tr><td>Vaborbactam (RPX7009)</td><td>Meropenem</td><td>Approved in 2017 by FDA</td><td>[34,166]</td></tr>
<tr><td>AAI101</td><td>Cefepime</td><td>Phase 2 initiated in 2017</td><td>[34,167,168]</td></tr>
<tr><td>RG6080 (OP0595, FPI-1459)</td><td>Unknown</td><td>Phase 1 complete</td><td>[165]</td></tr>
</table>

FDA, U.S. Food and Drug Administration; EMA, European Medicines Agency; cUTI, complicated urinary tract infection; cIAI, complicated intra-abdominal infection.

Boronic acids, which have recently been discovered as a new class of pan-β-lactamase inhibitors, are selective, devoid of toxicity, and able to inhibit both serine- and metallo-β-lactamases. There is a perspective that these inhibitors could be promising and available for clinical use [34].

Phosphonic and phosphinic acids, which contain an inert C–P bond, constitute a group of bioactive small molecules with great pharmaceutical potential. Among the most known examples are fosfomycin (the only FDA-approved antibiotic to treat acute cystitis during pregnancy) and fosmidomycin (a potent antimalarial agent), as well as glyphosate and phosphinothricin (widely used herbicides). However, the use of phosphonic acids as antibiotics/herbicides led to the occurrence of multiple mechanisms of resistance to them [169].

Possible uses of kinase inhibitors in another field, like immune response to bacterial or viral infections, is being investigated in preclinical studies. The receptor tyrosine kinase inhibitor, approved by the FDA, affects *M. tuberculosis* growth through increased lysosomal targeting and suppression of signal transducer and activator of transcription activation [170]. These drugs could also have the ability to inhibit the survival of other microbes and the replication of viruses and, consequently, to decrease the resistance to drugs in patients with infections [171].

8. Commercial Use

Despite the emerging need for new antibiotics, there is no advanced research into new antimicrobial drugs. In 2017, only 39 antibiotics were in stages I and III of progress, an amount insufficient to meet the current and planned clinical demand. Further, considering these 39 drugs, only 13 (33%) will be turn into a marketable drug [172]. Because of growing antibiotic resistance and multiresistance, the mortality rate is expected to reach 10 million by 2050, with an estimated economic cost of US $100 trillion [173]. As an example, 250,000 people die every year from the drug-resistant tuberculosis, and only 52% of patients worldwide are successful in treatment. And yet, just two novel antibiotics to treat this disease have reached the market in 70 years.

At first glance, although large companies have some advantages in producing new antibiotics, such as an established research method, sophisticated tools for dosage study, and fast approval by regulatory agencies, they give priority to the development of other drugs for the treatment of other diseases. This occurs because the profit of antibiotic investment is low, representing a market of US $45 billion, losing only to drugs used in cardiovascular and central nervous diseases [174].

A further issue is the antibiotic price when it reaches the market. The competition is great with generic drugs that have a far lower price. Therefore, in some instances, large companies transfer responsibility to small business to develop these drugs; for example, this happens with daptomycin, produced by Cubist and licensed by Lilly [175,176].

The United States and Europe are joining forces to reach the purpose of producing 10 to 15 new antibiotics every decade. This initiative is part of a government program that aims to make US $10–$30 billion available in the market over the next 10 years [177], which would be a real war against drug resistance. Nevertheless, the main obstacle to these strategies is their implementation, as it often implies public acquisition of the rights to distribute the antibiotic that poses a significant risk to companies and major upfront public costs to support this burden.

9. Future Perspectives

Nowadays, multidrug resistance is the biggest concern of government agencies and companies that develop antibiotics. Multidrug resistant tuberculosis affects half a million people every year, requires two years of treatment with success in only few cases, and is frequently observed in sites with low human development index. However, many of the novel drugs that are arriving in the market are not directed to infections caused by antibiotic-resistant pathogens. Additionally, according to Pew Charitable Trust studies, only 31% of drugs in advanced stages of clinical trials are effective against an ESKAPE pathogen, and only 33% against the multidrug resistant ones [178].

There are companies and entities around the world that are encouraged to solve the problem of antibiotic resistance. The Global Antibiotic Research and Development Partnership (GARDP), a non-profit organization, is designed to fund research and commercialization of antibiotics. Between 2016–2018, it received funding of more than €5 million for research on new antimicrobial drugs [179].

The isolation of new microorganisms from different environments, such as actinomycetes and more specifically Streptomycetes, able to produce new substances, may considerably accelerate the development of new treatments. Additionally, genetic engineering application and proteomic techniques could generate high performing strains able to produce new bioactive compounds and, consequently, new powerful antibiotics [65].

Besides that, most of the research did not take into account the interactions among bacterial species, which are used to live in communities in the natural environment. In addition, the leakage of antibiotics into natural environments has the potential to radically alter the evolution of resistance along with the microbial dynamics and structure of the communities. Faced with this, the important collaboration of the Amazon Biotechnology Center (CBA) and TB Alliance in the search for new antibiotic producers, and funds to support for the final phase of clinical trials and regulation of using antibiotics, respectively, are a hope [1].

The risk of "super bugs" resistant to relatively all licensed antibiotics may rise in the future; therefore, constant worldwide surveillance for multidrug-resistant bacteria is urgently required.

Because of this, strong emphasis on collaboration between companies and governments encourages synergy across the search of new antibiotics and the antibiotic market [1].

10. Conclusions

This review focused on a general discussion on the production of β-lactamase inhibitors by the members of the genus *Streptomyces*. Antibiotic resistance has been around for some time and has grown quite a bit as a result of increased mutations in genes encoding enzymes such as β-lactamases, which are responsible for inactivation of β-lactams antibiotics. The classification of these enzymes was reorganized because of the appearance of new genes and, consequently, new enzymes. In view of the fact that antibiotics used in clinical practice at the time were no longer very effective, β-lactamase inhibitors have arisen to circumvent the serious problem constituted by the spread of many extended-spectrum β-lactamases.

Clavulanic acid (CA), the first and most important β-lactamase inhibitor used in clinical practice to date, was obtained naturally by *Streptomyces clavuligerus*, while the other two most known inhibitors, Tazobactam and Sulbactam, are of synthetic origin. The CA industrial production has been extensively studied. Several sources of carbon, nitrogen, vitamins, amino acids, and salts were tested in batch or fed-batch and small or large-scale cultivations. The influence of physicochemical parameters such as pH, temperature, agitation, and aeration on fermentation was also investigated for the purpose of improving CA production and reducing time and costs. The most used purification processes are chromatographic methods, aqueous two-phase extraction using polymers and salts, and extraction systems using solvents. Extractive fermentation is a promising emerging technology that integrates the production and extraction stages, thus increasing productivity and reducing time.

Although *Streptomyces* spp. are known to be excellent producers of antibiotics and β-lactamase inhibitors, rapid and global diffusion of the genes of strains with multidrug resistance is a concern; thus, it is still a considerable challenge to improving production and reducing costs. In this respect, genetic engineering may make it possible to construct combinations of genes capable of coding new, hitherto unknown antibiotics; what we might call hybrid antibiotics.

Recently, bacteria named "ESKAPE" by WHO have emerged, which are extremely important pathogens that spread high levels of resistance across the world. Faced with this, the search for new producers of antibiotics and β-lactamase inhibitors and combined drug therapies are emergency alternatives. For this to occur consistently, companies and government must join forces to develop new low-priced antibiotics to solve the problem of antibiotic resistance.

Author Contributions: All authors conceived, drafted and revised the manuscript together.

Funding: This research received no external funding.

Conflicts of Interest: The authors declare no conflict of interest.

References

1. Procópio, R.E.; Silva, I.R.; Martins, M.K.; Azevedo, J.L.; Araújo, J.M. Antibiotics produced by *Streptomyces*. *Braz. J. Infect. Dis.* **2012**, *16*, 466–471. [CrossRef] [PubMed]
2. Radlinski, L.; Conlon, B.P. Antibiotic efficacy in the complex infection environment. *Curr. Opin. Microbiol.* **2018**, *42*, 19–24. [CrossRef] [PubMed]
3. Bhattacharyya, B.; Pal, S.; Sen, S. Anitibiotic production by *Streptomyces hygroscopicus* $D_{1.5}$: Cultural effect. *Rev. Microbiol.* **1998**, *29*, 254–257. [CrossRef]
4. Mohanraj, G.; Sekar, T. Isolation and screening of actinomycetes from marine sediments for their potential to produce antimicrobials. *Int. J. Life Sci. Pharm. Res.* **2013**, *2*, 115–126.
5. Kirk, S.; Avogmpme-Rossa, C.A.; Bushell, M.E. Growth limiting substrate affects antibiotic production and associated metabolic flues in *Streptomyces clavuligerus*. *Biotechnol. Lett.* **2000**, *22*, 1803–1809. [CrossRef]
6. Gouveia, E.R.; Baptista-Neto, A.; Badino, A.C., Jr.; Hokka, C.O. Optimisation of medium composition for clavulanic acid production by *Streptomyces clavuligerus*. *Biotechnol. Lett.* **2001**, *23*, 157–161. [CrossRef]

7. Ortiz, S.C.A.; Hokka, C.O.; Badino, A.C., Jr. Utilization of soybean derivatives on clavulanic acid production by *Streptomyces clavuligerus*. *Enzyme Microb. Technol.* **2007**, *40*, 1071–1077. [CrossRef]
8. Ehrlich, J.; Bartz, Q.R.; Smith, R.M.; Joslyn, D.A.; Burkholder, P.R. Chloromycetin, a new antibiotic from a soil actinomycete. *Science* **1947**, *106*, 417. [CrossRef] [PubMed]
9. Sethi, S.; Kumar, R.; Gupta, S. Antibiotic production by microbes isolated from soil. *Int. J. Pharm. Sci. Res.* **2013**, *4*, 2967–2973. [CrossRef]
10. Drawz, S.M.; Bonomo, R.A. Three Decades of β-Lactamase Inhibitors. *Clin. Microbiol. Rev.* **2010**, *23*, 160–201. [CrossRef] [PubMed]
11. Bush, K.; Bradford, P.A. β-Lactams and β-Lactamase Inhibitors: An Overview. *Cold Spring Harbor Perspect. Med.* **2016**, *6*, 1–22. [CrossRef] [PubMed]
12. Bradford, P.A. Extended-spectrum β-lactamases in the 21st century: Characterization, epidemiology, and detection of this important resistance threat. *Clin. Microbiol. Rev.* **2001**, *14*, 933–951. [CrossRef] [PubMed]
13. Simpson, I.N.; Harper, P.B.; O'Callaghan, C.H. Principal beta–lactamases responsible for resistance to beta-lactam antibiotics in urinary tract infections. *Antimicrob. Agents Chemother.* **1980**, *17*, 929–936. [CrossRef] [PubMed]
14. Borodina, I.; Siebring, J.; Zhang, J.; Smith, C.P.; van Keulen, G.; Dijkhuizen, L.; Nielsen, J. Antibiotic Overproduction in *Streptomyces coelicolor* A3(2) Mediated by Phosphofructokinase Deletion. *J. Biol. Chem.* **2008**, *283*, 25186–25199. [CrossRef] [PubMed]
15. Watve, M.G.; Tickoo, R.; Jog, M.M.; Bhole, B.D. How many antibiotics are produced by the genus *Streptomyces*? *Arch. Microbiol.* **2001**, *176*, 386–390. [CrossRef] [PubMed]
16. Abraham, E.P.; Chain, E. An enzyme from bacteria able to destroy penicillin. *Rev. Infect. Dis.* **1988**, *10*, 677–678. [CrossRef] [PubMed]
17. Plough, H.H. Penicillin resistance of Staphylococcus aureus and its clinical implications. *Am. J. Clin. Pathol.* **1945**, *15*, 446–451. [CrossRef] [PubMed]
18. Bondi, A., Jr.; Dietz, C.C. Penicillin resistant staphylococci. *Proc. Soc. Exp. Biol. Med.* **1945**, *60*, 55–58. [CrossRef] [PubMed]
19. Eriksen, K.R. Studies on induced resistance to penicillin in pneumococcus type I. *Acta Pathol. Microbiol. Scand.* **1945**, *22*, 398–405. [CrossRef] [PubMed]
20. Gots, J.S. Production of extracellular penicillin-inactivating substances associated with penicillin resistance in *Staphylococcus aureus*. *Proc. Soc. Exp. Biol. Med.* **1945**, *60*, 165–168. [CrossRef] [PubMed]
21. Miller, C.P.; Bohnhoff, M. Studies on the action of penicillin; development of penicillin resistance by gonococcus. *Proc. Soc. Exp. Biol. Med.* **1945**, *60*, 354–356. [CrossRef] [PubMed]
22. Waksman, S.A.; Reilly, H.C.; Schatz, A. Strain specificity and production of antibiotic substances: V. Strain resistance of bacteria to antibiotic substances, especially streptomycin. *Proc. Natl. Acad. Sci. USA* **1945**, *31*, 157–164. [CrossRef] [PubMed]
23. Hassan, A.; Usman, J.; Kaleem, F.; Omair, M.; Khalid, A.; Iqbal, M. Evaluation of different detection methods of biofilm formation in the clinical isolates. *Braz. J. Infect. Dis.* **2011**, *15*, 305–311. [CrossRef]
24. Allen, H.K.; Donato, J.; Wang, H.H.; Cloud-Hansen, K.A.; Davies, J.; Handelsman, J. Call of the wild: Antibiotic resistance genes in natural environments. *Nat. Rev. Microbiol.* **2010**, *8*, 251–259. [CrossRef] [PubMed]
25. Petchiappan, A.; Chatterji, D. Antibiotic Resistance: Current Perspectives. *ACS Omega* **2017**, *2*, 7400–7409. [CrossRef]
26. Andersson, D.I.; Hughes, D. Microbiological effects of sublethal levels of antibiotics. *Nat. Rev. Microbiol.* **2014**, *12*, 465–478. [CrossRef] [PubMed]
27. Hiltunen, T.; Virta, M.; Laine, A.L. Antibiotic resistance in the wild: An eco-evolutionary perspective. *Philos. Trans. R. Soc. B Biol. Sci.* **2017**, *19*, 1–7. [CrossRef] [PubMed]
28. World Health Organization. *Antibacterial Agents in Clinical Development: An Analysis of the Antibacterial Clinical Development Pipeline, Including Tuberculosis*; WHO: Geneva, Switzerland, 2017.
29. Brown, E.D.; Wright, G.D. Antibacterial drug discovery in the resistance era. *Nature* **2016**, *529*, 336–343. [CrossRef] [PubMed]

30. Bignell, D.R.D.; Seipke, R.F.; Huguet-Tapia, J.C.; Chambers, A.H.; Parry, R.J.; Loria, R. *Streptomyces scabies* 87-22 contains a coronafacic acid-like biosynthetic cluster that contributes to plant–microbe interactions. *Mol. Plant-Microbe Interact.* **2010**, *23*, 161–175. [CrossRef] [PubMed]
31. Fernandes, P. Antibacterial discovery and development—The failure of success? *Nat. Biotechnol.* **2006**, *24*, 1497–1503. [CrossRef] [PubMed]
32. Fischbach, M.A.; Walsh, C.T. Antibiotics for emerging pathogens. *Science* **2009**, *325*, 1089–1093. [CrossRef] [PubMed]
33. Ogawara, H. Self-resistance in Streptomyces, with Special Reference to β-Lactam Antibiotics. *Molecules* **2016**, *21*, 605. [CrossRef] [PubMed]
34. Docquier, J.D.; Mangani, S. An update on β-lactamase inhibitor discovery and development. *Drug Resist. Updat.* **2018**, *36*, 13–29. [CrossRef] [PubMed]
35. Coulthurst, S.J.; Barnard, A.M.L.; Salmond, G.P.C. Regulation and biosynthesis of carbapenem antibiotics in bacteria. *Nat. Rev. Microbiol.* **2005**, *3*, 295–306. [CrossRef] [PubMed]
36. Santos, V.C.; Pereira, J.F.B.; Haga, R.B.; Rangel-Yagui, C.O.; Teixeira, J.A.C.; Converti, A.; Pessoa, A., Jr. Stability of clavulanic acid under variable pH, ionic strength and temperature conditions. A new kinetic approach. *Biochem. Eng. J.* **2009**, *45*, 89–93. [CrossRef]
37. Zeng, X.; Lin, J. Beta-lactamase induction and cell wall metabolism in Gram-negative bacteria. *Front. Microbiol.* **2013**, *4*, 128. [CrossRef] [PubMed]
38. Cohen, M.L. Changing patterns of infectious disease. *Nature* **2000**, *406*, 762–767. [CrossRef] [PubMed]
39. Munita, J.M.; Arias, C.A. Mechanisms of antibiotic resistance. *Microbiol. Spectr.* **2016**, *4*, 1–37.
40. Wang, W.-J.; Wang, Q.; Zhang, Y.; Lu, R.; Zhang, Y.-L.; Yang, K.-W.; Lei, J.E. Characterization of β-lactamase activity using isothermal titration calorimetry. *Biochim. Biophys. Acta* **2017**, *1861*, 2031–2038. [CrossRef] [PubMed]
41. Bush, K. A resurgence of β-lactamase inhibitor combinations effective against multidrug-resistant Gram-negative pathogens. *Int. J. Antimicrob.* **2015**, *46*, 483–493. [CrossRef] [PubMed]
42. Papp-Wallace, K.M.; Bonomo, R.A. New β-Lactamase inhibitors in the clinic. *Infect. Dis. Clin. N. Am.* **2016**, *30*, 441–464. [CrossRef] [PubMed]
43. Bush, K.; Jacoby, G.A.; Medeiros, A. Functional classification scheme for β-lactamases and its correlation with molecular structure. *Antimicrob. Agents Chemother.* **1995**, *39*, 1211–1233. [CrossRef] [PubMed]
44. Bush, K.; Jacoby, G.A. Updated functional classification of beta-lactamases. *Antimicrob. Agents Chemother.* **2010**, *54*, 969–976. [CrossRef] [PubMed]
45. Laxminarayan, R.; Matsoso, P.; Pant, S.; Brower, C.; Røttingen, J.-A.; Klugman, K.; Davies, S. Access to effective antimicrobials: A worldwide challenge. *Lancet* **2016**, *387*, 168–175. [CrossRef]
46. Bush, K. β-Lactamase inhibitors from laboratory to clinic. *Clin. Microbiol. Rev.* **1988**, *1*, 109–123. [CrossRef] [PubMed]
47. Silva, M.S.C.; de Carvaiho Santos-Ebinuma, V.; Lopes, A.M.; de Oliveira Rangel-Yagui, C. Dextran sulfate/Triton X two-phase micellar systems as an alternative first purification step for clavulanic acid. *Fluid Phase Equilib.* **2015**, *399*, 80–86. [CrossRef]
48. Akova, M. Sulbactam-containing b-lactamase inhibitor combinations. *Clin. Microbiol. Infect.* **2008**, *14*, 185–188. [CrossRef] [PubMed]
49. Jacqueline, C.; Howland, K.; Chesnel, L. In vitro activity of ceftolozane/tazobactam in combination with other classes of antibacterial agents. *J. Glob. Antimicrob. Resist.* **2017**, *10*, 326–329. [CrossRef] [PubMed]
50. Balci, C.; Uzun, O.; Arici, M.; Hayran, S.A.; Yuce, D.; Unal, S. Nephrotoxicity of piperacillin-tazobactam combined with vancomycin: Should it be a concern? *Int. J. Antimicrob. Agents* **2018**, in press. [CrossRef] [PubMed]
51. Fast, W.; Sutton, L.D. Metallo-β-lactamase: Inhibitors and reporter substrates. *Biochim. Biophys. Acta* **2013**, *1834*, 1648–1659. [CrossRef] [PubMed]
52. Ball, M.; Boyd, A.; Ensor, G.J.; Evans, M.; Golden, M.; Linke, S.R.; Milne, D.; Murphy, R.; Telford, A.; Kalyan, Y.; et al. Development of a manufacturing route to avibactam, a β-lactamase inhibitor. *Org. Process Res. Dev.* **2016**, *20*, 1799–1805. [CrossRef]
53. Ohnishi, Y.; Ishikawa, J.; Hara, H.; Suzuki, H.; Ikenoya, M.; Ikeda, H.; Yamashita, A.; Hattori, M.; Horinouchi, S. Genome sequence of the streptomycin-producing microorganism *Streptomyces griséus* IFO 13350. *J. Bacteriol.* **2008**, *190*, 4050–4060. [CrossRef] [PubMed]

54. Drawz, S.M.; Papp-Wallace, K.M.; Bonomo, R.A. New β-lactamase inhibitors: A therapeutic renaissance in an MDR world. *Antimicrob. Agents Chemother.* **2014**, *58*, 1835–1846. [CrossRef] [PubMed]
55. Grigorenko, V.G.; Andreeva, I.P.; Rubtsova, M.Y.; Deygen, I.M.; Antipin, R.L.; Majouga, A.G.; Egorov, A.M.; Beshnova, D.A.; Kallio, J.; Hackenberg, C.; et al. Novel non-β-lactam inhibitor of b-lactamase TEM-171 based on acylated phenoxyaniline. *Biochimie* **2017**, *132*, 45–53. [CrossRef] [PubMed]
56. Thomson, J.M.; Distler, A.M.; Bonomo, R.A. Overcoming resistance to β-lactamase inhibitors: Comparing Sulbactam to novel inhibitors against clavulanate resistant SHV enzymes with substitutions at Ambler position 244. *Biochemistry* **2007**, *46*, 11361–11368. [CrossRef] [PubMed]
57. Saudagar, P.S.; Survase, S.A.; Singhal, R.S. Clavulanic acid: A review. *Biotechnol. Adv.* **2008**, *26*, 335–351. [CrossRef] [PubMed]
58. McGowan, S.J.; Bycroft, B.W.; Salmond, G.P.C. Bacterial production of carbapenems and clavams: Evolution of β-lactam antibiotic pathways. *Trends Microbiol.* **1998**, *6*, 203–208. [CrossRef]
59. Özcengiz, G.; Demain, A.L. Recent advances in the biosynthesis of penicillins, cephalosporins and clavams and its regulation. *Biotechnol. Adv.* **2013**, *31*, 287–311. [CrossRef] [PubMed]
60. Hamed, R.B.; Gomez-Castellanos, J.R.; Henry, L.; Ducho, C.; McDonough, M.A.; Schofield, C.J. The enzymes of β-lactam biosynthesis. *Nat. Prod. Rep.* **2013**, *30*, 21–107. [CrossRef] [PubMed]
61. Nobary, S.G.; Jensen, S.E. A comparison of the clavam biosynthetic gene clusters in *Streptomyces antibioticus* Tü1718 and *Streptomyces clavuligerus*. *Can. J. Microbiol.* **2012**, *58*, 413–425. [CrossRef] [PubMed]
62. Song, J.Y.; Jensen, S.L.; Lee, K.J. Clavulanic acid biosynthesis and genetic manipulation for its overproduction. *Appl. Microbiol. Biotechnol.* **2010**, *88*, 659–669. [CrossRef] [PubMed]
63. Ramirez-Malule, H.; Restrepo, A.; Cardona, W.; Junne, S.; Neubauer, P.; Rios-Estepa, R. Inversion of the stereochemical configuration (3*S*,5*S*)-clavaminic acid into (3*R*,5*R*)-clavulanic acid: A computationally-assisted approach based on experimental evidence. *J. Theor. Biol.* **2016**, *395*, 40–50. [CrossRef] [PubMed]
64. Qin, R.; Zhong, C.; Zong, G.; Fua, J.; Pang, X.; Cao, G. Improvement of clavulanic acid production in *Streptomyces clavuligerus* F613-1 by using a claR-neo reporter strategy. *Electron. J. Biotechnol.* **2017**, *28*, 41–46. [CrossRef]
65. Ünsaldı, E.; Kurt-Kızıldogan, A.; Voigt, B.; Becher, D.; Ozcengiz, G. Proteome-wide alterations in an industrial clavulanic acid producing strain of *Streptomyces clavuligerus*. *Synth. Syst. Biotechnol.* **2017**, *2*, 39–48. [CrossRef] [PubMed]
66. Tahlan, K.; Anders, C.; Wong, A.; Mosher, R.H.; Beatty, P.H.; Brumlik, M.J.; Griffin, A.; Hughes, C.; Griffin, J.; Barton, B.; et al. 5S clavam biosynthetic genes are located in both the clavam and paralog gene clusters in *Streptomyces clavuligerus*. *Chem. Biol.* **2007**, *14*, 131–142. [CrossRef] [PubMed]
67. Costa, C.L.L.; Badino, A.C. Overproduction of clavulanic acid by extractive fermentation. *Electron. J. Biotechnol.* **2015**, *18*, 154–160. [CrossRef]
68. Haga, R.B.; Santos-Ebinuma, V.C.; Silva, M.S.C.; Pessoa, A., Jr.; Rangel-Yagui, C.O. Clavulanic acid partitioning in charged aqueous two-phase micellar systems. *Sep. Purif. Technol.* **2013**, 273–278. [CrossRef]
69. Viana-Marques, D.A.; Santos-Ebinuma, V.C.; Pessoa, A., Jr.; Porto, A.L.; Torres, B.R.; Converti, A. Effect of aeration and agitation on extractive fermentation of clavulanic acid by using aqueous two-phase system. *Biotechnol. Prog.* **2016**, *32*, 1444–1452. [CrossRef] [PubMed]
70. Rodrigues, K.C.S.; Souza, A.T.; Badino, A.C.; Pedrolli, D.B.; Cerri, M.O. Screening of medium constituents for clavulanic acid production by *Streptomyces clavuligerus*. *Braz. J. Microbiol* **2018**, in press. [CrossRef] [PubMed]
71. Forte, M.B.S.; Taviot-Guého, C.; Leroux, F.; Rodrigues, M.I.; Maugeri Filho, F. Clavulanic acid separation on fixed bed columns of layered double hydroxides: Optimization of operating parameters using breakthrough curves. *Process Biochem.* **2016**, *51*, 509–516. [CrossRef]
72. Kumar, P.S.; Duraipandiyan, V.; Ignacimuthu, S. Isolation, screening and partial purification of antimicrobial antibiotics from soil *Streptomyces* sp. SCA 7. *Kaohsiung J. Med. Sci.* **2014**, *30*, 435–446. [CrossRef] [PubMed]
73. Charusanti, P.; Fong, N.L.; Nagarajan, H.; Pereira, A.R.; Li, H.J.; Abate, E.A.; Su, Y.; Gerwick, W.H.; Palsson, B.O. Exploiting adaptive laboratory evolution of streptomyces clavuligerus for antibiotic discovery and overproduction. *PLoS ONE* **2012**, *7*, e033727. [CrossRef] [PubMed]
74. Singh, V.; Haque, S.; Singh, H.; Verma, J.; Vibha, K.; Singh, R.; Jawed, A.; Tripathi, C.K.M. Isolation, screening, and identification of novel isolates of Actinomycetes from India for antimicrobial applications. *Front. Microbiol.* **2016**, *7*, 1921. [CrossRef] [PubMed]

75. Chater, K.F. Recent advances in understanding *Streptomyces*. *F1000Research* **2016**, *5*, 2795. [CrossRef] [PubMed]
76. Al husnan, L.A.; Alkahtani, M.D.F. Molecular Identification of *Streptomyces* producing antibiotics and their antimicrobial activities. *Ann. Agric. Sci.* **2016**, *61*, 251–255. [CrossRef]
77. Weber, T.; Charusanti, P.; Musiol-Kroll, E.M.; Jiang, X.; Tong, Y.; Kim, H.U.; Lee, S.Y. Metabolic engineering of antibiotic factories: New tools for antibiotic production in actinomycetes. *Trends Biotechnol.* **2015**, *33*, 15–26. [CrossRef] [PubMed]
78. Balachandran, C.; Duraipandiyan, V.; Arund, Y.; Sangeetha, B.; Emi, N.; Al-Dhabi, N.A.; Ignacimuthu, S.; Inaguma, Y.; Okamoto, A.; Perumal, P.T. Isolation and characterization of 2-hydroxy-9,10-anthraquinone from *Streptomyces olivochromogenes* (ERINLG-261) with antimicrobial and antiproliferative properties. *Rev. Bras. Farmacogn.* **2016**, *26*, 285–295. [CrossRef]
79. Martínez-Burgo, Y.; Álvarez-Álvarez, R.; Pérez-Redondo, R.; Liras, P. Heterologous expression of *Streptomyces clavuligerus* ATCC 27064 cephamycin C gene cluster. *J. Biotechnol.* **2014**, *186*, 21–29. [CrossRef] [PubMed]
80. Bellão, C.; Antonio, T.; Araujo, M.L.G.C.; Badino, A.C. Production of clavulanic acid and cephamycin C by *Streptomyces clavuligerus* under different fed-batch conditions. *Braz. J. Chem. Eng.* **2013**, *30*, 257–266. [CrossRef]
81. Cavallieri, A.P.; Baptista, A.S.; Leite, C.A.; Araujo, M.L.G.C. A case study in flux balance analysis: Lysine, a cephamycin C precursor, can also increase clavulanic acid production. *Biochem. Eng. J.* **2016**, *112*, 42–53. [CrossRef]
82. Li, R.; Townsend, C.A. Rational strain improvement for enhanced clavulanic acid production by genetic engineering of the glycolytic pathway in *Streptomyces clavuligerus*. *Metab. Eng.* **2006**, *8*, 240–252. [CrossRef] [PubMed]
83. Olano, C.; Lombó, F.; Méndez, C.; Salas, J.A. Improving production of bioactive secondary metabolites in actinomycetes by metabolic engineering. *Metab. Eng.* **2008**, *10*, 281–292. [CrossRef] [PubMed]
84. Adrio, J.L.; Demain, A.L. Recombinant organisms for production of industrial products. *Bioeng. Bugs* **2010**, *1*, 116–131. [CrossRef] [PubMed]
85. Hirata, D.B.; Oliveira, J.H.H.L.; Leão, K.V.; Rodrigues, M.I.; Ferreira, A.G.; Giulietti, M.; Barboza, M.; Hokka, C.O. Precipitation of clavulanic acid from fermentation broth with potassium 2-ethyl hexanoate salt. *Sep. Purif. Technol.* **2009**, *66*, 598–605. [CrossRef]
86. Higgens, C.E.; Kastner, R.E. *Streptomyces clavuligerus* sp. nova b-lactam antibiotic producer. *Int. J. Syst. Evol. Microbiol.* **1971**, *21*, 326–331. [CrossRef]
87. Brown, A.G.; Butterworth, D.; Cole, M.; Hanscomb, G.; Hood, J.D.; Reading, C.; Rolinson, G.N. Naturally occurring beta-lactamase inhibitors with antibacterial activity. *J. Antibiot.* **1976**, *29*, 668–669. [CrossRef] [PubMed]
88. Schmidt, F.R. Optimization and scale up of industrial fermentation processes. *Appl. Microbiol. Biotechnol.* **2005**, *68*, 425–435. [CrossRef] [PubMed]
89. Oliveira, J.H.L.; Granato, A.C.; Hirata, D.B.; Hokka, C.O.; Barboza, M. Clavulanic acid and cephamicin C: A perspective of the biosynthesis, isolation and action mechanism. *Quim. Nova* **2009**, *32*, 2142–2150. [CrossRef]
90. Townsend, C.A. New reactions in clavulanic acid biosynthesis. *Curr Opin Chem Biol.* **2002**, *6*, 583–589. [CrossRef]
91. Ser, H.L.; Law, J.W.F.; Chaiyakunapruk, N.; Jacob, S.A.; Palanisamy, U.D.; Chan, K.G.; Goh, B.H.; Lee, L.H. Fermentation conditions that affect clavulanic acid production in *Streptomyces clavuligerus*: A systematic review. *Front. Microbiol.* **2016**, *7*, 522. [CrossRef] [PubMed]
92. Domingues, L.C.G.; Teodoro, J.C.; Hokka, C.O.; Badino, A.C.; Araujo, M.L.G.C. Optimisation of the glycerol-to-ornithine molar ratio in the feed medium for the continuous production of clavulanic acid by *Streptomyces clavuligerus*. *Biochem. Eng. J.* **2010**, *53*, 7–11. [CrossRef]
93. Saudagar, P.S.; Singhal, R.S. Optimization of nutritional requirements and feeding strategies for clavulanic acid production by *Streptomyces clavuligerus*. *Bioresour. Technol.* **2007**, *98*, 2010–2017. [CrossRef] [PubMed]
94. Viana, D.A.; Carneiro-Cunha, M.N.; Araújo, J.M.; Barros-Neto, B.; Lima-Filho, J.L.; Converti, A.; Pessoa-Júnior, A.; Porto, A.L.F. Screening of variables influencing the clavulanic acid production by *Streptomyces* DAUFPE 3060 strain. *Appl. Biochem. Biotechnol.* **2010**, *160*, 1797–1807. [CrossRef] [PubMed]

95. Marques, D.A.V.; Carneiro da Cunha, M.N.; Araújo, J.M.; Lima-Filho, J.L.; Converti, A.; Pessoa, A., Jr.; Porto, A.L.F. Optimization of clavulanic acid production by *Streptomyces* DAUFPE 3060 by response surface methodology. *Braz. J. Microbiol.* **2011**, *42*, 658–667. [CrossRef] [PubMed]
96. Costa, C.L.L.; Badino, A.C. Production of clavulanic acid by *Streptomyces clavuligerus* in batch cultures without and with glycerol pulses under different temperature conditions. *Biochem. Eng. J.* **2012**, *69*, 1–7. [CrossRef]
97. Marques, D.A.V.; Santos-Ebinuma, V.C.; Oliveira, P.M.S.; Lima, G.M.S.; Araújo, J.M.; Lima-Filho, J.L.; Converti, A.; Pessoa-Júnior, A.; Porto, A.L.F. Screening of wild type *Streptomyces* isolates able to overproduce clavulanic acid. *Braz. J. Microbiol.* **2014**, *45*, 919–928. [CrossRef]
98. Saudagar, P.S.; Singhal, R.S. A statistical approach using L25 orthogonal array method to study fermentative production of clavulanic acid by *Streptomyces clavuligerus* MTCC1142. *Appl. Biochem. Biotechnol.* **2007**, *136*, 345–359. [CrossRef] [PubMed]
99. Salem-Berkhit, M.M.; Alanazi, F.K.; Alsarra, I.A. Improvement and enhancement of clavulanic acid production in *Streptomyces clavuligerus* using vegetable oils. *Afr. J. Biotechnol.* **2010**, *9*, 6806–6812.
100. Chen, K.C.; Lin, Y.H.; Tsai, C.M.; Hsieh, C.H.; Houng, J.Y. Optimization of glycerol feeding for clavulanic acid production by *Streptomyces clavuligerus* with glycerol feeding. *Biotechnol. Lett.* **2002**, *24*, 455–458. [CrossRef]
101. Thakur, R.; Roy, M.K.; Dutta, N.N.; Bezbaruah, R.L. Coordinate production of cephamycin c and clavulanic acid by *Streptomyces clavuligerus*. *Indian J. Exp. Biol.* **1999**, *37*, 1031–1033. [PubMed]
102. Teodoro, J.C.; Baptista-Neto, A.; Araujo, M.L.G.D.C.; Hokka, C.O.; Badino, A.C. Influence of glycerol and ornithine feeding on clavulanic acid production by *Streptomyces clavuligerus*. *Braz. J. Chem. Eng.* **2010**, *27*, 499–506. [CrossRef]
103. Vasconcelos, E.S.; Lima, V.A.; Goto, L.S.; Cruz-Hernández, I.L.; Hokka, C.O. Clavulanic acid production by the MMS 150 mutant obtained from wild type *Streptomyces clavuligerus* ATCC 27064. *Braz. J. Microbiol.* **2014**, *44*, 1049–1057. [CrossRef] [PubMed]
104. Baggaley, K.H.; Brown, A.G.; Schofield, C.J. Chemistry and biosynthesis of clavulanic acid and other clavams. *Nat. Prod. Rep.* **1997**, *14*, 309–333. [CrossRef] [PubMed]
105. Ives, P.R.; Bushell, M.E. Manipulation of the physiology of clavulanic acid production in *Streptomyces clavuligerus*. *Microbiology* **1997**, *143*, 3573–3579. [CrossRef] [PubMed]
106. Wang, Y.H.; Yang, B.; Ren, J.; Dong, M.L.; Liang, D.; Xu, A.L. Optimization of medium composition for the production of clavulanic acid by *Streptomyces clavuligerus*. *Process Biochem.* **2005**, *40*, 1161–1166. [CrossRef]
107. Teodoro, J.C.; Baptista-Neto, A.; Cruz-Hernandez, I.L.; Hokka, C.O.; Badino, A.C. Influence of feeding conditions on clavulanic acid production in fed-batch cultivation with medium containing glycerol. *Appl. Microbiol. Biotechnol.* **2006**, *72*, 450–455. [CrossRef] [PubMed]
108. Romero, J.; Liras, P.; Martin, J.F. Dissociation of cephamycin and clavulanic acid biosynthesis in *Streptomyces clavuligerus*. *Appl. Microbiol. Biotechnol.* **1984**, *20*, 318–325. [CrossRef]
109. Lee, P.C.; Ho, C.C. Production of clavulanic acid and cephamycin C by *Streptomyces clavuligerus* in palm-oil medium. *World J. Microbiol. Biotechnol.* **1996**, *12*, 73–75. [CrossRef] [PubMed]
110. Large, K.P.; Ison, A.P.; Williams, D.J. The effect of agitation rate on lipid utilisation and clavulanic acid production in *Streptomyces clavuligerus*. *J. Biotechnol.* **1998**, *63*, 111–119. [CrossRef]
111. Maranesi, G.L.; Baptista-Neto, A.; Hokka, C.O.; Badino, A.C. Utilization of vegetable oil in the production of clavulanic acid by *Streptomyces clavuligerus* ATCC 27064. *World J. Microbiol. Biotechnol.* **2005**, *21*, 509–514. [CrossRef]
112. Efthimiou, G.; Thumser, A.E.; Avignone−Rossa, C.A. A novel finding that *Streptomyces clavuligerus* can produce the antibiotic clavulanic acid using olive oil as a sole carbon source. *J. Appl. Microbiol.* **2008**, *105*, 2058–2064. [CrossRef] [PubMed]
113. Kim, S.J.; Kim, J.O.; Shin, C.H.; Park, H.W.; Kim, C.W. An approach to strain improvement and enhanced production of clavulanic acid in *Streptomyces clavuligerus*. *Biosci. Biotech. Biochem.* **2009**, *73*, 160–164. [CrossRef] [PubMed]
114. Stowell, J.D. The application of oils and fats in antibiotic processes. In *Carbon Substrates in Biotechnology*; Stowell, J.D., Beardsmore, A.J., Keevil, C.W., Woodward, J.R., Eds.; IRL Press: Oxford, UK, 1987; pp. 139–159.
115. Mayer, A.F.; Deckwer, W.D. Simultaneous production and decomposition of clavulanic acid during *Streptomyces clavuligerus* cultivations. *Appl. Microbiol. Biotechnol.* **1996**, *45*, 41–46. [CrossRef] [PubMed]

116. Joo, H.S.; Chang, C.S. Production of protease from a new alkalophilic *Bacillus* sp. I-312 grow on soybean meal: Optimization and some properties. *Process Biochem.* **2005**, *40*, 1263–1270. [CrossRef]
117. Romero, J.; Liras, P.; Martín, J.F. Utilization of ornithine and arginine as specific precursors of clavulanic acid. *Appl. Environ. Microbiol.* **1986**, *52*, 892–897. [PubMed]
118. Romero, J.; Liras, P.; Martín, J.F. Isolation and biochemical characterization of *Streptomyces clavuligerus* mutants in the biosynthesis of clavulanic acid and cephamycin C. *Appl. Microbiol. Biotechnol.* **1988**, *27*, 510–516. [CrossRef]
119. Townsend, C.A.; Ho, M.F. Biosynthesis of clavulanic acid: Origin of C_3 unit. *J. Am. Chem. Soc.* **1985**, *107*, 1065–1066. [CrossRef]
120. De la Fuente, J.L.; Martin, J.F.; Liras, P. New type of hexameric ornithine carbamoyltransferase with arginase activity in the cephamycin producers *Streptomyces clavuligerus* and *Nocardia lactamdurans*. *Biochem. J.* **1996**, *15*, 173–179.
121. Valentine, B.P.; Bailey, C.R.; Doherty, A.; Morris, J.; Elson, S.W.; Baggaley, K.H.; Nicholson, N.H. Evidence that arginine is a later metabolic intermediate than ornithine in the biosynthesis of clavulanic acid by *Streptomyces clavuligerus*. *Chem. Soc. Chem. Commun.* **1993**, *15*, 1210–1211. [CrossRef]
122. Chen, K.C.; Lin, Y.H.; Wu, J.Y.; Hwang, S.C.J. Enhancement of clavulanic acid production in *Streptomyces clavuligerus* with ornithine feeding. *Enzyme Microb. Technol.* **2003**, *32*, 152–156. [CrossRef]
123. Barboza, M.; Almeida, R.M.; Hokka, C.O. Kinetic studies on clavulanic acid recovery by ion exchange chromatography. *Bioseparation* **2001**, *10*, 221–227. [CrossRef] [PubMed]
124. Albertsson, P.A. *Partition of Cell Particles and Macromolecules*, 3rd ed.; Wiley: New York, NY, USA, 1986.
125. Lima, C.A.; Freitas, A.C.V., Jr.; Lima Filho, J.L.; Converti, A.; Viana Marques, D.A.; Carneiro-da-Cunha, M.G.; Porto, A.L.F. Two-phase partitioning and partial characterization of a collagenase from Penicillium aurantiogriseum URM4622: Application to collagen hydrolysis. *Biochem. Eng. J.* **2013**, *75*, 64–71. [CrossRef]
126. Saravanan, S.; Rao, J.R.; Nair, B.U.; Ramasami, T. Aqueous two-phase poly(ethylene glycol)–poly(acrylic acid) system for protein partitioning: Influence of molecular weight, pH and temperature. *Process Biochem.* **2008**, *43*, 905–911. [CrossRef]
127. Viana Marques, D.A.; Pessoa-Júnior, A.; Lima-Filho, J.L.; Converti, A.; Perego, P.; Porto, A.L.F. Extractive fermentation of clavulanic acid by *Streptomyces* DAUFPE 3060 using aqueous two-phase system. *Biotechnol. Prog.* **2011**, *27*, 95–103. [CrossRef] [PubMed]
128. Silva, C.S.; Cuel, M.F.; Barreto, V.O.; Kwong, W.H.; Hokka, C.O.; Barboza, M. Separation of clavulanic acid from fermented broth of aminoacids by an aqueous two-phase system and ion-exchange adsorption. *New Biotechnol.* **2012**, *29*, 428–431. [CrossRef] [PubMed]
129. Lu, C.; Dong, J.; Yang, S.T. Butanol production from wood pulping hydrolysate in na integrated fermentation-gas stripping process. *Bioresour. Technol.* **2013**, *143*, 467–475. [CrossRef] [PubMed]
130. Lee, J.C.; Park, H.R.; Park, D.J.; Lee, H.B.; Kim, Y.B.; Kim, C.J. Improved production of teicoplanin using adsorbent resin in fermentations. *Lett. Appl. Microbiol.* **2003**, *37*, 196–200. [CrossRef] [PubMed]
131. Bersanetti, P.A.; Almeida, R.M.; Barboza, M.; Araújo, M.L.G.; Hokka, C.O. Kinetic studies on clavulanic acid degradation. *Biochem. Eng. J.* **2005**, *23*, 31–36. [CrossRef]
132. Panas, P.; Lopes, C.; Cerri, M.O.; Ventura, S.P.; Santos-Ebinuma, V.C.; Pereira, J.F. Purification of clavulanic acid produced by *Streptomyces clavuligerus* via submerged fermentation using polyethylene glycol/cholinium chloride aqueous two-phase systems. *Fluid Phase Equilib.* **2017**, *450*, 42–50. [CrossRef]
133. Smith, T.; Wolff, K.A.; Nguyen, L. Molecular biology of drug resistance in *Mycobacterium tuberculosis*. *Curr. Top. Microbiol. Immunol.* **2013**, *374*, 53–80. [CrossRef] [PubMed]
134. Cole, S.T. Inhibiting Mycobacterium tuberculosis within and without. *Philos. Trans. R. Soc. B Biol. Sci.* **2016**, *371*, 1–8. [CrossRef] [PubMed]
135. Wlodarchak, N.; Teachout, N.; Procknow, R.; Beczkiewicz, J.; Schaenzer, A.; Satyshur, K.; Pavelka, M.; Zuercher, B.; Drewry, D.; Sauer, J.D.; et al. Repurposed kinase inhibitors and β-lactams as a novel therapy for antibiotic resistant. *BioRxiv* **2017**. [CrossRef]
136. Diacon, A.H.; van der Merwe, L.; Barnard, M.; von Groote-Bidlingmaier, F.; Lange, C.; Garcia-Basteiro, A.L.; Sevene, E.; Ballell, L.; Barros-Aguirre, D. Beta-Lactams against tuberculosis—New trick for an old dog? *N. Engl. J. Med.* **2016**, *375*, 393–394. [CrossRef] [PubMed]

137. Ramon-Garcia, S.; Gonzalez Del Rio, R.; Villarejo, A.S.; Sweet, G.D.; Cunningham, F.; Barros, D.; Ballell, L.; Mendoza-Losana, A.; Ferrer-Bazaga, S.; Thompson, C.J. Repurposing clinically approved cephalosporins for tuberculosis therapy. *Sci. Rep.* **2016**, *6*, 34293. [CrossRef] [PubMed]
138. Veziris, N.; Truffot, C.; Mainardi, J.L.; Jarlier, V. Activity of carbapenems combined with clavulanate against murine tuberculosis. *Antimicrob. Agents Chemother.* **2011**, *55*, 2597–2600. [CrossRef] [PubMed]
139. Nitsch-Osuch, A.; Kuchar, E.; Zycinska, K.; Gyrczuk, E.; Miskiewicz, K.; Korzeniewski, K. Implementation of hospital's antibiotic policy decreases antimicrobial use in the general pediatric ward. *Adv. Exp. Med. Biol.* **2015**, *857*, 67–74. [CrossRef] [PubMed]
140. Urban, C.; Go, E.; Mariano, N.; Berger, B.J.; Avraham, I.; Rubin, D.; Rahal, J.J. Effect of sulbactam on infections caused by imipenem-resistant *Acinetobacter calcoaceticus* biotype *anitratus*. *J. Infect. Dis.* **1993**, *167*, 448–451. [CrossRef] [PubMed]
141. Brunetti, L.; Poustchi, S.; Cunningham, D.; Toscani, M.; Nguyen, J.; Lim, J.; Ding, Y.; Nahass, R.G. Clinical and economic impact of empirical extended-infusion piperacillin–Tazobactam in a community medical center. *Ann. Pharmacother.* **2015**, *49*, 754–760. [CrossRef] [PubMed]
142. Farrell, D.J.; Flamm, R.K.; Sader, H.S.; Jones, R.N. Antimicrobial activity of ceftolozane–Tazobactam tested against Enterobacteriaceae and *Pseudomonas aeruginosa* with various resistance patterns isolated in U.S. hospitals (2011–2012). *Antimicrob. Agents Chemother.* **2013**, *57*, 6305–6310. [CrossRef] [PubMed]
143. Sutherland, C.A.; Nicolau, D.P. Susceptibility profile of ceftolozane/tazobactam and other parenteral antimicrobials against *Escherichia coli*, *Klebsiella pneumoniae*, and *Pseudomonas aeruginosa* from US hospitals. *Clin. Ther.* **2015**, *37*, 1564–1571. [CrossRef] [PubMed]
144. Sader, H.S.; Farrell, D.J.; Castanheira, M.; Flamm, R.K.; Jones, R.N. Antimicrobial activity of ceftolozane/tazobactam tested against *Pseudomonas aeruginosa* and Enterobacteriaceae with various resistance patterns isolated in European hospitals (2011–12). *J. Antimicrob. Chemother.* **2014**, *69*, 2713–2722. [CrossRef] [PubMed]
145. Estabrook, M.; Bussell, B.; Clugston, S.L.; Bush, K. In vitro activity of ceftolozane-tazobactam as determined by broth dilution and agar diffusion assays against recent U.S. *Escherichia coli* isolates from 2010 to 2011 carrying CTX-M-type extended-spectrum β-lactamases. *J. Clin. Microbiol.* **2014**, *52*, 4049–4052. [PubMed]
146. Solomkin, J.; Hershberger, E.; Miller, B.; Popejoy, M.; Friedland, I.; Steenbergen, J.; Yoon, M.; Collins, S.; Yuan, G.; Barie, P.S.; et al. Ceftolozane/tazobactam plus metronidazole for complicated intra-abdominal infections in an era of multidrug resistance: Results from a randomized, double-blind, phase 3 trial (ASPECT-cIAI). *Clin. Infect. Dis.* **2015**, *60*, 1462–1471. [CrossRef] [PubMed]
147. Kaur, R.; Gautam, V.; Singhal, L.; Ray, P. Antimicrobial activity of cefepime-tazobactam combination tested against clinical isolates of Enterobacteriaceae. *J. Antibiot.* **2014**, *67*, 603–604. [CrossRef] [PubMed]
148. Lahiri, S.D.; Mangani, S.; Jahic, H.; Benvenuti, M.; Durand-Reville, T.F.; De Luca, F.; Ehmann, D.E.; Rossolini, G.M.; Alm, R.A.; et al. Molecular basis of selective inhibition and slow reversibility of avibactam against class D carbapenemases: A structure-guided study of OXA-24 and OXA-48. *ACS Chem. Biol.* **2015**, *10*, 591–600. [CrossRef] [PubMed]
149. Li, H.; Estabrook, M.; Jacoby, G.A.; Nichols, W.W.; Testa, R.T.; Bush, K. In vitro susceptibility of characterized β-lactamase-producing strains tested with avibactam combinations. *Antimicrob. Agents Chemother.* **2015**, *59*, 1789–1793. [CrossRef] [PubMed]
150. Levasseur, P.; Girard, A.M.; Miossec, C.; Pace, J.; Coleman, K. In vitro antibacterial activity of the ceftazidime–avibactam combination against Enterobacteriaceae, including strains with well characterized β-lactamases. *Antimicrob. Agents Chemother.* **2015**, *59*, 1931–1934. [CrossRef] [PubMed]
151. Castanheira, M.; Mills, J.C.; Costello, S.E.; Jones, R.N.; Sader, H.S. Ceftazidime-avibactam activity tested against Enterobacteriaceae isolates from U.S. hospitals (2011 to 2013) and characterization of β-lactamase-producing strains. *Antimicrob. Agents Chemother.* **2015**, *59*, 3509–3517. [CrossRef] [PubMed]
152. Berkhout, J.; Melchers, M.J.; van Mil, A.C.; Nichols, W.W.; Mouton, J.W. In vitro activity of ceftazidime-avibactam combination in in vitro checkerboard assays. *Antimicrob. Agents Chemother.* **2015**, *59*, 1138–1144. [CrossRef] [PubMed]
153. Wang, X.; Zhang, F.; Zhao, C.; Wang, Z.; Nichols, W.W.; Testa, R.; Li, H.; Chen, H.; He, W.; Wang, Q.; et al. In vitro activities of ceftazidime-avibactam and aztreonam-avibactam against 372 Gram-negative bacilli collected in 2011 and 2012 from 11 teaching hospitals in China. *Antimicrob. Agents Chemother.* **2014**, *58*, 1774–1778. [CrossRef] [PubMed]

154. Sader, H.S.; Castanheira, M.; Flamm, R.K.; Farrell, D.J.; Jones, R.N. Antimicrobial activity of ceftazidime- avibactam against Gram-negative organisms collected from U.S. medical centers in 2012. *Antimicrob. Agents Chemother.* **2014**, *58*, 1684–1692. [CrossRef] [PubMed]
155. Flamm, R.K.; Sader, H.S.; Farrell, D.J.; Jones, R.N. Ceftazidime-avibactam and comparator agents tested against urinary tract isolates from a global surveillance program (2011). *Diagn. Microbiol. Infect. Dis.* **2014**, *80*, 233–238. [CrossRef] [PubMed]
156. Flamm, R.K.; Farrell, D.J.; Sader, H.S.; Jones, R.N. Ceftazidime/avibactam activity tested against Gram-negative bacteria isolated from bloodstream, pneumonia, intra-abdominal and urinary tract infections in US medical centres (2012). *J. Antimicrob. Chemother.* **2014**, *69*, 1589–1598. [CrossRef] [PubMed]
157. Sader, H.S.; Castanheira, M.; Flamm, R.K.; Mendes, R.E.; Farrell, D.J.; Jones, R.N. Ceftazidime/avibactam tested against Gram-negative bacteria from intensive care unit (ICU) and non-ICU patients, including those with ventilator-associated pneumonia. *Int. J. Antimicrob. Agents* **2015**, *46*, 53–59. [CrossRef] [PubMed]
158. Sader, H.S.; Castanheira, M.; Mendes, R.E.; Flamm, R.K.; Farrell, D.J.; Jones, R.N. Ceftazidime-avibactam activity against multidrug-resistant *Pseudomonas aeruginosa* isolated in U.S. medical centers in 2012 and 2013. *Antimicrob. Agents Chemother.* **2015**, *59*, 3656–3659. [CrossRef] [PubMed]
159. Testa, R.; Canton, R.; Giani, T.; Morosini, M.I.; Nichols, W.W.; Seifert, H.; Stefanik, D.; Rossolini, G.M.; Nordmann, P. In vitro activity of ceftazidime, ceftaroline and aztreonam alone and in combination with avibactam against European Gram-negative and Gram-positive clinical isolates. *Int. J. Antimicrob. Agents* **2015**, *45*, 641–646. [CrossRef] [PubMed]
160. Biedenbach, D.J.; Kazmierczak, K.; Bouchillon, S.K.; Sahm, D.F.; Bradford, P.A. In vitro activity of aztreonam–avibactam against a global collection of Gram-negative pathogens from 2012 and 2013. *Antimicrob. Agents Chemother.* **2015**, *59*, 4239–4248. [CrossRef] [PubMed]
161. Livermore, D.M.; Mushtaq, S.; Warner, M.; Zhang, J.; Maharjan, S.; Doumith, M.; Woodford, N. Activities of NXL104 combinations with ceftazidime and aztreonam against carbapenemase-producing Enterobacteriaceae. *Antimicrob. Agents Chemother.* **2011**, *55*, 390–394. [CrossRef] [PubMed]
162. Sader, H.S.; Farrell, D.J.; Flamm, R.K.; Jones, R.N. Activity of ceftaroline and comparator agents tested against *Staphylococcus aureus* from patients with bloodstream infections in US medical centres (2009–2013). *J. Antimicrob. Chemother.* **2015**, *70*, 2053–2056. [CrossRef] [PubMed]
163. Flamm, R.K.; Farrell, D.J.; Sader, H.S.; Jones, R.N. Antimicrobial activity of ceftaroline combined with avibactam tested against bacterial organisms isolated from acute bacterial skin and skin structure infections in United States medical centers (2010–2012). *Diagn. Microbiol. Infect. Dis.* **2014**, *78*, 449–456. [CrossRef] [PubMed]
164. Blizzard, T.A.; Chen, H.; Kim, S.; Wu, J.; Bodner, R.; Gude, C.; Imbriglio, J.; Young, K.; Park, Y.W.; Ogawa, A.; et al. Discovery of MK-7655, a β-lactamase inhibitor for combination with Primaxin®. *Bioorg. Med. Chem. Lett.* **2014**, *24*, 780–785. [CrossRef] [PubMed]
165. Morinaka, A.; Tsutsumi, Y.; Yamada, M.; Suzuki, K.; Watanabe, T.; Abe, T.; Furuuchi, T.; Inamura, S.; Sakamaki, Y.; Mitsuhashi, N.; et al. OP0595, a new diazabicyclooctane: Mode of action as a serine β-lactamase inhibitor, antibiotic and β-lactam 'enhancer'. *J. Antimicrob. Chemother.* **2015**, *70*, 2779–2786. [CrossRef] [PubMed]
166. Lapuebla, A.; Abdallah, M.; Olafisoye, O.; Cortes, C.; Urban, C.; Quale, J.; Landman, D. Activity of meropenem combined with RPX7009, a novel β-lactamase inhibitor, against Gram-negative clinical isolates in New York City. *Antimicrob. Agents Chemother.* **2015**, *59*, 4856–4860. [CrossRef] [PubMed]
167. Crandon, J.L.; Nicolau, D.P. Comparative potency of cefepime (FEP) and cefepime/AAI101 (FEP/AAI101) against highly resistant Enterobacteriaceae. In Proceedings of the 54th Interscience Conference on Antimicrobial Agents and Chemotherapy (ICAAC), Washington, DC, USA, 5–9 September 2014.
168. Crandon, J.L.; Nicolau, D.P. In vivo activities of simulated human doses of cefepime and cefepime–AAI101 against multidrug-resistant Gram-negative Enterobacteriaceae. *Antimicrob. Agents Chemother.* **2015**, *59*, 2688–2694. [CrossRef] [PubMed]
169. Chekan, J.R.; Cogan, D.P.; Nair, S.K. Molecular basis for resistance against phosphonate antibiotics and herbicides. *MedChemComm* **2016**, *7*, 28–36. [CrossRef] [PubMed]
170. Sogi, M.K.; Lien, K.A.; Johnson, J.R.; Krogan, N.J.; Stanley, S.A. The tyrosine kinase inhibitor Gefitinib restricts *Mycobacterium tuberculosis* growth through increased lysosomal biogenesis and modulation of cytokine signaling. *ACS Infect. Dis.* **2017**, *3*, 564–574. [CrossRef] [PubMed]

171. Pandey, R.; Kapur, R. Kinase inhibitors in clinical practice: An expanding world. *J. Allergy Clin. Immunol.* **2017**, *141*, 522–524. [CrossRef] [PubMed]
172. Stern, S.C.S.; Franken, L.; Voller, S.; Rentmeister, H.; Grosch, B. *Breaking through the Wall: A Call for Concerted Action on Antibiotics Research and Development*; The Boston Consulting Group: Berlin, Germany, 2017.
173. The Review on Antimicrobial Resistance. *Antimicrobial Resistance: Tackling a Crisis for the Health and Wealth of Nations*; The Review on Antimicrobial Resistance: London, UK, 2014; pp. 1–20.
174. Bush, K. Antibacterial drug discovery in the 21st century. *Clin. Microbiol. Infect.* **2004**, *10*, 10–17. [CrossRef] [PubMed]
175. Projan, S.J. Why is big pharma getting out of antibacterial drug discovery? *Drug Discov. Today* **2003**, *6*, 427–430. [CrossRef]
176. Boggs, A.F.; Miller, G.H. Antibacterial drug discovery: Is small pharma the solution? *Clin. Microbiol. Infect.* **2004**, *10*, 32–36. [CrossRef] [PubMed]
177. Ferraro, J.S.; Towse, A.; Mestre-Ferrandiz, J. *Incentives for New Drugs to Tackle Anti-Microbial Resistance*; Office of Health Economics: London, UK, 2017.
178. The Pew Charitable Trusts. *Antibiotics Currently in Clinical Development*; The Pew Charitable Trusts: Philadelphia, PA, USA, 2017.
179. Simpkin, V.L.; Renwick, M.J.; Kelly, R.; Mossialos, E. Incentivising innovation in antibiotic drug discovery and development: Progress, challenges and next steps. *J. Antibiot.* **2017**, *70*, 1087–1096. [CrossRef] [PubMed]

antibiotics

Review

Actinomycetes, an Inexhaustible Source of Naturally Occurring Antibiotics

Yōko Takahashi * and Takuji Nakashima

Kitasato Institute for Life Sciences, Kitasato University, 5-9-1 Shirokane, Minato-ku, Tokyo 108-8641, Japan; takuji@lisci.kitasato-u.ac.jp

* Correspondence: ytakaha@lisci.kitasato-u.ac.jp

Received: 17 April 2018; Accepted: 23 May 2018; Published: 24 May 2018

Abstract: Global public health faces a desperate situation, due to the lack of effective antibiotics. Coordinated steps need to be taken, worldwide, to rectify this situation and protect the advances in modern medicine made over the last 100 years. Work at Japan's Kitasato Institute has been in the vanguard of many such advances, and work is being proactively tailored to promote the discovery of urgently needed antimicrobials. Efforts are being concentrated on actinomycetes, the proven source of most modern antibiotics. We devised a novel physicochemical screening mechanism, whereby simple physico-chemical properties, in conjunction with related detection methods, such as LC/MS, LC/UV, and polarity, could be used to identify or predict new compounds in a culture broth, simply by comparing results with existing databases. New compounds are isolated, purified, and their structure determined before being tested for any bioactivity. We used lyophilized actinomycete strains from the Kitasato Microbial Library, most more than 35 years old, and found 330 strains were producers of useful bioactive substances. We also tested organisms found in fresh samples collected in the complex environments from around plant roots, as well as from sediments of mangrove forests and oceans, resulting in the discovery of 36 novel compounds from 11 actinomycete strains. A compound, designated iminimycin, containing an iminium ion in the structure was discovered from the culture broth of *Streptomyces griseus* OS-3601, which had been stored for a long time as a streptomycin-producing strain. This represented the first iminium ion discovery in actinomycetes. Compounds with a cyclopentadecane skeleton containing 5,6-dihydro-4-hydroxyl-2-pyrone ring and tetrahydrofuran ring, designated mangromicins, were isolated from the culture broth of *Lechevalieria aerocolonigenes* K10-0216 obtained from sediment in a mangrove forest. These structures are extremely unique among natural compounds. From the same culture broth, new steroid compounds, named K10-0216 KA and KB, and other new compounds having a thiazole and a pyridine ring, named pyrizomicin A and B, were discovered. New substances can be found from actinomycetes that have been exhaustively studied. Novel compounds with different skeletons can be found from a single broth of one strain. The sought after new antibiotics will arise from continued exploitation of the actinomycetes, especially rare actinomycetes. Work on new organisms and samples should be augmented by re-examination of known actinomycetes already in storage. New research should also be carried out on the manipulation of culture media, thereby stimulating actinomycete strains to produce novel chemicals. The establishment of wide-ranging international research collaborations will facilitate and expedite the efficient and timely discovery and provision of bioactive compounds to help maintain and promote advances in global public health.

Keywords: actinomycetes; secondary metabolites; novel compounds; physicochemical screening; physical and chemical properties; structural diversity; biological activity

1. Introduction

The use of chemicals to maintain or improve human health is as old as recorded history. In recent times, the discovery and development of chemicals to kill or overcome bacteria or pathogens is regarded as one of the most significant medical achievements of the 20th century, and countless millions of human lives have been saved as a result. Natural products have been and remain the mainstay of medical treatments. Chemicals produced in nature, or compounds based on them, accounted for 65% of the 1211 small molecule drugs approved by the United States Food and Drug Administration (FDA) in the 34 years from 1981 to 2014 [1]. The wide and diverse range of microbial primary or secondary metabolites that possess potent and sometimes unique bioactivity, coupled with the enormous and as yet relatively untapped potential and promise they offer, will heavily influence and drive forward future antibiotic research, while simultaneously emphasizing the importance of prioritizing natural products discovery over the manufacturing of synthetic compounds [2].

Penicillin was first used for human treatment in 1942, and it revolutionized the treatment of bacterial infections. It has since saved hundreds of millions of lives, as well as galvanizing the search for similar antibacterial or antimicrobial chemicals. As a result of the worldwide research effort, there was a flood of new antibiotics identified throughout the 1950s and 1960s, with the approval of several distinct novel classes of efficacious antibiotics for human use. However, since that "Golden Age", the number of new antibiotics registered has steadily declined, and very few new classes of antibiotics have reached the marketplace and clinical use. In reality, scientific and economic factors will likely delay the appearance of any new antibiotics. From preclinical testing to approval for human use takes 10–15 years, and the costs involved are prohibitive [3,4]. A recent analysis suggests that, in 2014, the actual cost of driving a new compound from concept to the marketplace was in excess of $1.3 billion [5]. Furthermore, approximately 1 in 1000 potential drugs proceed to clinical trials, and then almost 90% fail in the human testing phase. For example, in the antibiotic field, the prevailing dangerous lack of new antibiotics, coupled with the loss of effectiveness of many already being widely used, threatens a return to the pre-antibiotic era, and the reversal of the gains made in global public health during the 20th century, accompanied by the potential loss of millions of lives.

The world is fast waking up to this dreadful scenario, and in 2015, the World Health Assembly endorsed the Global Action Plan on Antimicrobial Resistance [6]. This committed all member states to prepare national action plans and take proactive steps to promote the discovery, development, and sustainable exploitation of new antibiotics, especially those with novel modes of action. These commitments were reiterated by the United Nations General Assembly in 2016 [7].

So how can we find the urgently needed drugs? We firmly believe that actinomycetes will prove to be the primary source of the desperately needed biological substances over the next 2–3 decades, and this publication contains substantial evidence to support that point of view.

The actinomycetes are a heterogeneous group of Gram-positive bacteria with high guanine (G) and cytosine (C) content in their DNA. They are extremely diverse, with at least 350 genera known to date. They constitute one of the largest bacterial phyla, and are ubiquitous in aquatic and terrestrial ecosystems. Most (especially the streptomycetes are saprophytic, soil-dwelling organisms, but they are also found in fresh and salt water, and the air. They are typically present in soil at densities of 10^6 to 10^9 cells per gram of soil, with streptomycetes accounting for over 95% of all actinomycete strains isolated from soil [8]. Many species are harmless to animals and higher plants, while some are important pathogens.

The actinomycetes, particularly species from the genus *Streptomyces*, have proved to be a tremendous high-impact source of valuable chemicals. They have yielded many clinically essential antimicrobial compounds, including streptomycin, actinomycin, and streptothricin [9]. Besides streptomycin (discovered in 1944 from *Streptomyces griseus*), other examples of the success of this traditional discovery research approach are chloramphenicol (1947, *S. venezuelae*), tetracycline (1948, *S. rimosus*), erythromycin (1952, *Saccharopolyspora erythraea*), leucomycin (1952, *S. kitasatoensis*), and vancomycin (1956, *S. orientales*). Additionally, in 1963, gentamicin was discovered, isolated from *M. purpurea*, a member of the *Micromonospora*. This triggered the search for new compounds from the so-called "rare actinomycetes", which are lower frequency of isolation than members of the genus *Streptomyces* that are well isolated from soil. Compounds from the rare organisms include teicoplanin (1978, *Actinoplanes teichomyceticus*), fortimicin (1977, *M. olivoasterospora*), rosamicin (1972, *M. rosaria*), and nocardicin (1976, *Nocardia uniformis*). Incidentally, salinosporamide A, which holds promise for development of an anticancer drug, is produced by a strain of the genus *Salinispora*, a rare actinomycete isolated from a heat-treated marine sediment sample [10].

Approximately two-thirds of all known antibiotics are produced by actinomycetes, predominantly by *Streptomyces* [8]. It is believed that the actinomycetes are the source of some 61% of all microorganism-derived bioactive substances so far discovered [11], with 16% of the total originating from the "rare actinomycetes", mostly from the *Micromonosporaceae*, with additional smaller contributions from the *Pseudonocardiaceae* and *Thermomonosporaceae*. This suggests that rare actinomycetes are a valuable source of novel compounds, and that improved isolation strategies are required to increase the frequency in which they are isolated.

2. Historical Discovery of Novel Compounds from Actinomycetes by the Kitasato Ōmura-Drug Discovery Group

The Kitasato Institute has, since its inception, concentrated its investigations on soil dwelling microbes, particularly the actinomycetes, as a potential source of bioactive compounds. Up until the mid-1970s, the singular universally employed discovery process involved identifying microorganisms in soil (or other) samples, culturing them and then testing any primary or secondary metabolites or other chemicals they produced to identify predetermined bioactivity that would meet a human need.

Decades of success in our exploration of the actinomycetes is exemplified by the discovery in the Kitasato Institute by Satoshi Ōmura in the early-1970s of *Streptomyces avermectinius* (synonym *S. avermitilis*) MA-4680^{T}, the microbe which produces the avermectins [12,13]. The avermectin derivative, ivermectin, is perhaps the world's greatest, most effective, and safest drug for the treatment and prevention of a diverse range of human diseases and conditions [14]. The importance and significance of the discovery and development of these compounds was recognized by the 2015 Nobel Prize in Physiology or Medicine being awarded to Prof. Ōmura and Prof. William C. Campbell of Merck & Company, Inc., Kenilworth, NJ, USA, representing the industrial partner which has become essential for the discovery, development, production, marketing, and distribution process of all modern-day antibiotics. The award citation stated "William C. Campbell and Satoshi Ōmura discovered a new drug, avermectin, the derivatives of which have radically lowered the incidence of River Blindness and Lymphatic Filariasis, as well as showing efficacy against an expanding number of other parasitic diseases" [15,16]. The 2015 award was the third Nobel Prize given for discovery of an antibiotic, following those for penicillin (for Fleming, Florey, and Chain in 1945) and for streptomycin (for Waksman in 1952), the man who first coined the term "antibiotic".

The discovery of ivermectin arose because of Ōmura's unwavering belief that microorganisms are a limitless source of useful chemical compounds—"Microbes do not produce useless metabolites: we just have little knowledge of their usefulness for mankind" [17,18], and because the partnership he set up between his group and Merck scientists were looking for specific anthelmintic compounds. Although ivermectin has proved to be a multifaceted, extremely effective chemical with a wide range of impacts, the original bioactivity screening focused almost predominantly on looking for an anthelmintic. Hence, that was what was found.

In the early-1970s, Ōmura decided to introduce an innovative new approach to drug discovery, namely to simply identify novel chemicals with no fixed goal in mind, carry out preliminary assays and evaluations, catalog and store both the chemicals and the producing microorganisms, and make the chemicals available for others to assay for all variety of bioactivity, or for use as biological or chemical reagents. This novel process was referred to as physicochemical (PC) screening.

As members of the Kitasato Institute for Life Sciences' Drug Discovery Group, with Satoshi Ōmura as team leader, we have long and extensive experience in the search for novel compounds derived from microorganisms. Our cohesive integrated research program now encompasses the following three foci:

1. Isolation of microorganisms, identification, and microbial cultivation.
2. Discovery of substances from microbial metabolites.
3. Optimization of compounds by organic synthesis.

Latterly, our isolation work has been significantly refocused. We are now investigating existing but hitherto underutilized actinomycetes which exist in storage. We have also switched our attention from soil dwelling microbes to the exploration of microorganisms living in the complex environments found in the immediate vicinity of plant roots. We quickly discovered that whereas more than 90% of actinomycetes isolated from soil are *Streptomyces* strains, the rare actinomycetes dominate in strains isolated from plant roots. Currently, some 642 strains of actinomycetes have been isolated from 16 plant root locations, about 80% of which are rare. Two new genera (*Phytohabitans sufuscus* and *Rhizocola hellebori*) plus seven new species have, so far, been proposed through taxonomic study of these strains [19–21].

In our case, the discovery of useful microbial chemicals has been facilitated and accelerated by employing a two-pronged approach. Initially, using the traditional method attempting to acquire a new compound with a preconceived specific biological activity; more recently by identifying any and all novel substances by detecting and exploiting the basic physical and chemical properties and structural features of compounds. This bifurcated approach, both mechanisms of which are ongoing in Kitasato University, has led to the discovery of over 500 compounds, most of which were found using the original method [22].

In the mid-1970s, PC screening was introduced initially using Dragendorff's reagent to identify nitrogen-containing compounds (alkaloids) which would cause a simple, visible color change. Staurosporine [23] was discovered as the first indolocarbazole compound from the culture broth of *Saccharothrix aerocolonigenes* subsp. *staurosporeus* AM 2282^{T} [24] (renamed *Lentzea albida* in 2002 [25]) in 1977, using this method. We initially determined that the compound possessed antifungal properties, and demonstrated a hypotensive effect. Nine years after discovery, in 1986, another research group discovered that staurosporine was a nanomolar inhibitor of protein kinases, as assessed by the prevention of ATP binding to the kinase [26,27]. This interesting biological activity stimulated an explosion in exploratory research for selective protein kinase inhibitors by numerous laboratories and pharmaceutical companies worldwide, staurosporine becoming the parent compound for many of today's highly-successful anticancer agents. This example helps to illustrate that all substances produced by microbes may be of great benefit, and that they should be examined for potential use in all forms of human endeavor, especially for use in modern medicine, and that they should be made available for exhaustive testing and use by all, wherever practical and possible.

We now routinely search for novel chemicals from actinomycetes by analyzing a range of physico-chemical properties, such as LC/MS, LC/UV and polarity. It is now possible to predict whether a new substance is present by analyzing results and comparing with existing databases. This approach has so far identified some 36 novel compounds (including analogs) [28]. In this report, we describe these results, and discuss the ability of actinomycetes to produce a wide spectrum of novel chemicals, as well as draw attention to the diversity of metabolites that a single microbial strain can provide for us.

3. Novel Compounds Discovered by Physicochemical (PC) Screening of Cultured Broths of Actinomycetes

The novel compounds derived from actinomycetes discovered through our PC screening during the past eight years are displayed in Table 1. The compounds are accompanied by the name of the producing microorganism, their original source, the primary biological activity of the compound and relevant publications. The PC screening procedure was carried out as follows.

After cultivation in 10 mL of several kinds of preset media, an equivalent amount of ethanol was added, the ingredients were thoroughly mixed, the cells were then disrupted, and the ethanol extract was subjected to PC screening.

After LC/MS and LC/UV analysis, each peak recorded was compared with known data from the Dictionary of Natural Products, and our own database. A peak was predicted to be a novel substance. When this was the case, we scaled up the culture and isolated and purified the target compound using column chromatography and preparative HPLC. After obtaining the unique compound, its structure was determined by high-resolution mass spectrometry, NMR, etc. The new compounds underwent preliminary bioassays, either in-house or in established collaborative research projects with other groups.

Identification of the strains being cultured was carried out using morphological characteristics, chemical composition in cells, and phylogenetic analysis based on 16S rRNA gene sequences.

The Kitasato Ōmura-Drug Discovery Group has already discovered avermectin [14], staurosporine [23], herbimycin [29], setamycin [30], and lactacystin [31] from secondary metabolites of actinomycetes [22]. The actinomycete strains producing these compounds (as well as strains producing a variety of other chemicals) have all been catalogued, freeze-dried, and stored in the Kitasato Microbial Library (KML). In an effort to respond to the urgent global demand for new antibiotics, we have recently revived the KML strains to confirm their viability, the continuance of compound production, and the reliability of the preservation process. Survival rates and compound productivity maintenance rates have been good, but specific data in this respect will be reported elsewhere. During this work, PC screening was carried out on culture broths of 330 strains, resulting in the discovery of several new compounds (No. 1 to No. 3 in Table 1).

With respect to the three entries in question, the name of the original compound and retention period by lyophilization are stated, all three having been stored for 35 years or more. The compounds recently discovered, namely bisoxazolomycin [32], the iminimycins [33,34], and the nanaomycins [35,36], would probably not have been detected by an assay system seeking a specific bioactive property, the compounds being found as a direct result of PC screening. Discovery of the iminimycins and nanaomycins are described in detail below in Section 4.2.

With respect to new isolates (No. 4 to No. 13 in Table 1), actinomycete strains isolated from around the roots of plants (No. 4 to No. 7), sediment from mangrove forests (No. 8 to No. 10), sea sediment (No. 11), and soil samples (No. 12 & No. 13) are listed. Actinoallolides [37], hamuramicins [38], spoxazomicins [39,40], and trehangelins [41,42] were discovered from endophytic actinomycete strains, and these are classified as rare actinomycetes. The mangromicins [43–45], K10-0216 KA and KB [46], and pyrizomicins [47], which have differing core structures, were found in a culture broth of *Lechevalieria aerocolonigenes* K10-0216 isolated from sediment from mangroves. Mumiamicin [48] was found in an actinomycete strain isolated from sea sediment, while sagamilactam [49] and the dipyrimicins [50] originated in actinomycete strains from soil. In Section 5.1, we describe, in detail, the discovery of other compounds, notably the trehangelins from *Polymorphospora rubra* K07-0510 and compounds from *Lechevalieria aerocolonigenes* K10-0216.

Assays of the 36 compounds, involving collaboration with other research groups, led to the discovery of varying bioactivity, as shown in Table 1. These results help demonstrate the usefulness and cost/time effectiveness of PC screening, as well as the potential diversity of metabolites produced by a single microorganism. The outcome clearly demonstrates that, as Prof Ōmura rightly opines, "microorganisms are a treasure trove of new natural products".

Table 1. Compounds discovered by physicochemical screening of actinomycete strains (March 2018).

No.	Compound	Producing Microorganism	Source	Biological Activity	References
1	Bisoxazolomycin	*Streptomyces subflavus* subsp. *irumaensis* AM-3603	* KML Irumamycin producing strain, ** 36 years	Antibacterial	[32]
2	Iminimycin A & B	*Streptomyces griseus* OS-3601	* KML, Streptomycin producing strain, ** 43 years	Antibacterial	[33,34]
3	Nanaomycin F–H	"*Streptomyces rosa* subsp. *Notoensis*" OS-3966	* KML Nanaomycin producing strain, ** 36 years	Inhibitor of Epithelial-Mesenchymal Transition induced cells	[35,36]
4	Actinoallolide A–E	*Actinoallomurus fulvus* MK10-036	Roots of *Capsicum frutescents* in Thailand	Antitrypanosomal	[37]
		A. fulvus K09-0307	Roots of mondo grass in Saitama Pref., Japan		
5	Hamuramicin A & B	*Allostreptomyces* sp. K12-0794	Roots of fern in Hamura city, Tokyo, Japan	Antibacterial	[38]
6	Spoxazomicin A–C	*Streptosporangium oxazolinicum* K07-0460^T	Roots of orchid in Iriomote Island, Japan	Antitrypanosomal	[39,40]
7	Trehangelin A–C	*Polymorphospora rubra* K07-0510	Roots of orchid in Iriomote Island, Japan	Anti-Lipid peroxidation	[41,42]
				Enhanced production of collagen	
8	Mangromicin A–I	*Lechevalieria aerocolonigenes* K10-0216	Sediment from mangrove forest in Iriomote Island, Japan	Antitrypanosomal	[43–45]
				Antioxidative	
9	K10-0216 KA & KB	*Lechevalieria aerocolonigenes* K10-0216	Sediment from mangrove forest in Iriomote Island, Japan	Inhibitory effect on the lipid accumulation	[46]
10	Pyrizomicin A & B	*Lechevalieria aerocolonigenes* K10-0216	Sediment from mangrove forest in Iriomote Island, Japan	Antimicrobial	[47]
11	Mumiamicin	*Mumia* sp. YSP-2-79	Sea sediment, Namako Pond in Kagoshima Pref., Japan	Antibacterial	[48]
				Antioxidative	
12	Sagamilactam	*Actinomadura* sp. K13-0306	Soil, Kanagawa Pref., Japan	Cytotoxicity	[49]
				Antitrypanosomal	
13	Dipyrimicin A & B	*Amycolatopsis* sp. K16-0194	Soil, Okinawa Pref., Japan	Antibacterial	[50]

* KML: Kitasato Microbial Library, ** length of preservation by lyophilization; No. 1–3: Compounds from the KML; No. 4–13: Compounds from fresh isolates (No. 4–7; Roots of plants, No. 8–10; Sediment of mangrove forest, No. 11; Marine sediment, Nos. 12 & 13; Soil).

4. Novel Compounds Discovered from the Kitasato Microbial Library (KML)

4.1. Iminimycin A by Streptomyces griseus OS-3601, a Streptomycin Producing Strain

KML strain OS-3601 (Figure 1) was isolated from a soil sample collected at Aso, Kumamoto prefecture, Japan, in 1971, and identified as *Streptomyces griseus*, which produces streptomycin. As part of our screening of 330 preserved actinomycete strains, using four different liquid production media, a unique metabolite, predicted to be a new compound, was observed in an extract of a culture of this strain grown on defatted wheat germ medium. The metabolite was not observed following growth in the other three production media. The compound produced by strain OS-3601 showed an 242.1910 *m/z* $[M + H]^+$ and maximal absorption at 269 and 282 nm. Purification from the culture broth of strain OS-3601 yielded a new iminium compound, designated iminimycin A (Figure 1) [33], which was strongly supported by an IR absorption spectrum, indicative of the presence of an iminium ion. Several plant-derived compounds containing an iminium ion are known [51] but, to our knowledge, this is the first compound arising from an actinomycete source.

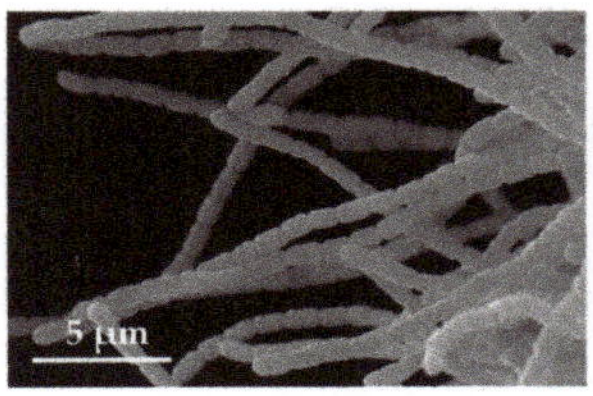

Streptomyces griseus **OS-3601**

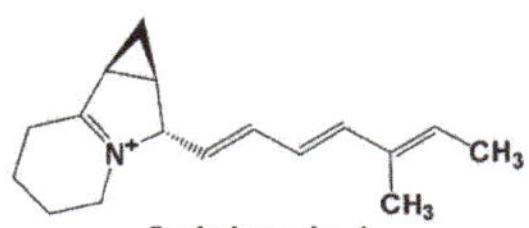

Iminimycin A

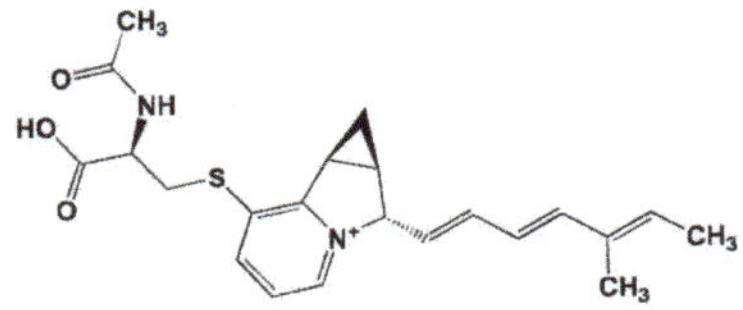

Iminimycin B

Figure 1. Scanning electron micrograph of the aerial spore chain of the streptomycin-producing strain *Streptomyces griseus* OS-3601 (**Left**) and structures of iminimycin A and B discovered from the culture broth (**Right**).

An unidentified compound with physical and chemical properties similar to iminimycin A was observed in the octadecyl silyl fraction lacking iminimycin A; the compound having an 399.1736 *m/z* $[M + H]^+$ and maximal absorption at 269 and 282 nm. Purification of this compound revealed a new indolizine alkaloid, designated iminimycin B (Figure 1) [34], that possessed *N*-acetylcysteine and pyridinium moieties, instead of the iminium moiety of iminimycin A. Iminimycin A and B both show antimicrobial activity against Gram-positive and Gram-negative bacteria. Since 1943, when Selman A. Waksman discovered streptomycin from the secondary metabolite of *Streptomyces griseus*, about 200 compounds have been reported from strains identified as *Streptomyces griseus* [9].

Among the actinomycetes, *Streptomyces griseus* is one of the organisms most frequently isolated from soil samples, and has been studied extensively. It was therefore surprising that our PC screening allowed us to find new substances from this strain, demonstrating, yet again, the unmatched ability of microorganisms, especially actinomycetes, to produce a plethora of chemicals.

One chemical component of the culture broth of strain OS-3601 was predicted to be novel from analysis of HPLC and LC/MS data gathered through PC screening. Our prediction proved true following isolation, purification, and structure determination.

Mapping of the genome of streptomycin-producing *Streptomyces griseus* has already been completed [52], and this information also indicates that production of the novel substance is predictable. So why has it not been discovered before? It is conceivable that productivity is low, or that it has been overlooked because it has been masked by the cycloheximide produced in large quantities at the same time in a lipid soluble fraction, or because the compound is inherently unstable and short-lived. We believe that prediction of the existence of new compounds can be envisaged through a combination of an enquiring mind, technological progress, and the creation and use of more sophisticated equipment. Indeed, it can be said that Prof. Ōmura's belief that "microorganisms are infinite resources" and that "microbial research has really just begun" are an accurate depiction of the current situation.

4.2. Nanaomycins F, G and H Derived from Nanaomycins A–E Produced by "Streptomyces rosa subsp. notoensis" OS-3966

Nanaomycins A, B, C, D, and E produced by "*Streptomyces rosa* subsp. *notoensis*" OS-3966 (Figure 2a,b) [53–55] were isolated from soil sampled in Nanao City, Japan. The nanaomycins contain a naphthoquinone skeleton and were found to have antimycoplasma properties. Nanaomycin A (Figure 2b) was developed as a therapeutic agent for cattle dermatophytosis in 1980 [56]. It generates semiquinone radicals by forming double bonds at positions 4a and 10a, damaging DNA in the process, consequently displaying antibacterial and antifungal activity.

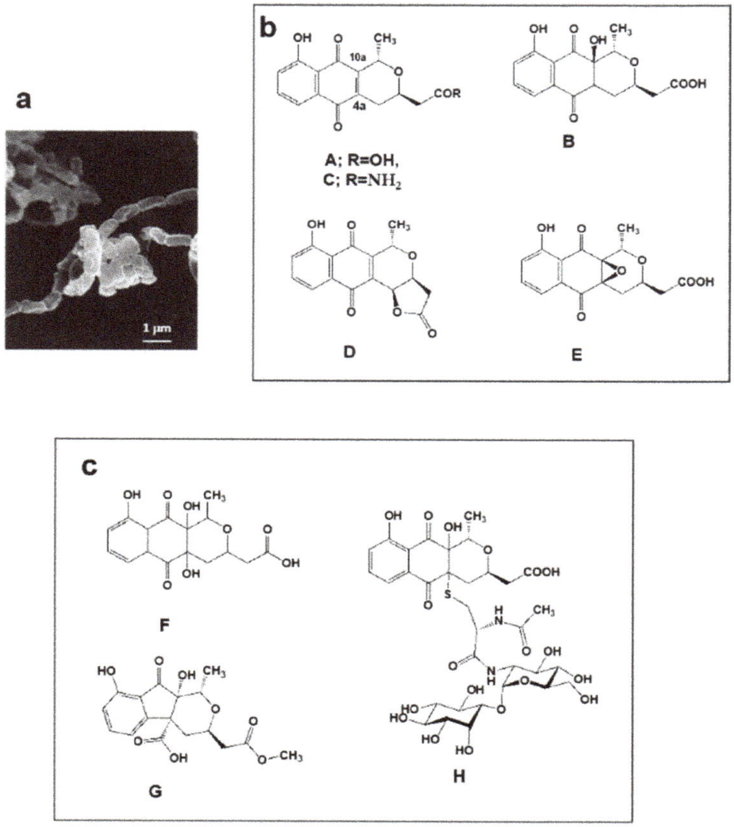

Figure 2. Nanaomycin-producing strain "*Streptomyces rosa* subsp. *notoensis*" OS-3966 and new analogs discovered by PC screening from a culture broth. (**a**) Scanning electron micrograph of aerial spore chain of "*S. rosa* subsp. *notoensis*" OS-3966 grown on agar medium; (**b**) Nanaomycins A–E discovered as antibacterial and antifungal substances during 1974–1979; (**c**) New analogs (F–H) discovered from the culture broth via PC screening.

We undertook PC screening using the culture broths of stored freeze-dried nanaomycin-producing strain OS-3966. Two new compounds appeared in the EtOH extract obtained from culturing in defatted wheat germ production medium. The first compound had 337.0917 *m/z* $[M + H]^+$ and maximal absorption at 231, 248 (sh), 267 (sh), and 347 nm; the second compound had 351.1075 *m/z* $[M + H]^+$ and maximal absorption at 225, 256 (sh), and 308 nm. Purification of these compounds revealed they were two new nanaomycin analogs. The first, which we named nanaomycin F (Figure 2c) [35], is a 4a-hydroxyl analog of nanaomycin B (Figure 2b). The second, which we named nanaomycin G (Figure 2c) [35], has a unique 1-indanone skeleton fused with a tetrahydropyran ring. During chromatographic purification of these compounds, another new compound was obtained, nanaomycin H (Figure 2c) [36]. Structure elucidation of nanaomycin H showed it to be a pyranonaphthoquinone with a mycothiol moiety. Our assays detected no antibacterial or antifungal activity in these three compounds. The reason for this seems to be that the radical-generating ability disappeared due to the reduction of the double bond of the quinone skeleton. It was found that the compounds inhibited epithelial-mesenchymal transition, inducing proliferation of mammalian cells [57]. These compounds would not have been found using an assay simply targeting bioactivity. Incidentally, the production medium of nanaomycin A to E is mainly composed of glycerol and soybean meal [53], whereas nanaomycins F, G, and H were produced on a medium containing mainly soluble starch and defatted wheat germ [35,36]. This raises the interesting possibility that simply changing the composition of culture media may allow the discovery of new compounds. These results showed to support OSMAC (one strain, many compounds) approach [58].

5. Novel Compounds Found from Fresh Isolates

5.1. Novel Substances, Trehangelins Found from Metabolites of the Plant-Derived Rare Actinomycete Polymorphospora Rubra K07-0510

Microbes isolated from plant root environments were cultured in each of four production media, with the resulting broths being subjected to PC screening. Strain K07-0510, when grown in one of the production media, yielded a peak (indicating a new compound) that was predicted based on spectrometric data.

Strain K07-0510 was isolated from the roots of an orchid collected on Iriomote Island, Okinawa, Japan. Short spore chains were formed, and spores with a smooth surface were cylindrical in shape (Figure 3). Whole-cell hydrolysates contained *meso*-DAP (diaminopimelic acid). The 16S rRNA gene sequence was determined and analyzed using the EzTaxon-e database (present name EzBioCloud) [59] to reveal a 99.9% similarity with *Polymorphospora rubra* TT97-42^T. On the basis of the morphological and cultural properties and 16S rRNA gene sequence analyses, strain K07-0510 was identified as *Polymorphospora. rubra*.

A new predicted peak showing 507.2087 *m/z* and maximal absorption at 216 nm was also found in the culture broth of strain K07-0510. Purification of this compound eventually identified three new compounds, which were named trehangelin A, B, C (Figure 3) [41,42]. These compounds were separated from the culture broth of strain K07-0510 by ethyl acetate extraction, followed by silica gel and ODS column chromatography, with final purification by HPLC. Eighteen liters of culture broth yielded 59 mg of trehangelin A as the major component, along with 4.4 mg and 1.3 mg of the minor components trehangelin B and C, respectively.

Structural analysis revealed that two molecules of angelic acid were bound to one molecule of trehalose. Trehangelin A, the main compound, binds angelic acid to the 3,3′ positions of trehalose, while B and C does so to the 3,2′ and 4,4′ positions, respectively. After substance isolation, preliminary bioassays were carried out, and it was found that trehangelin A and C inhibited erythrocyte hemolysis by photooxidation. Trehangelin A and C, which have symmetric structures, showed more potent inhibition than ascorbic acid. In addition, it has been found that the compounds facilitate cytoprotective action and accumulation of procollagen type I C-peptide in a cell culture assay [60], and further research is currently in progress. Our group is now close to elucidating the genetic basis of the biosynthesis of the trehangelins [42].

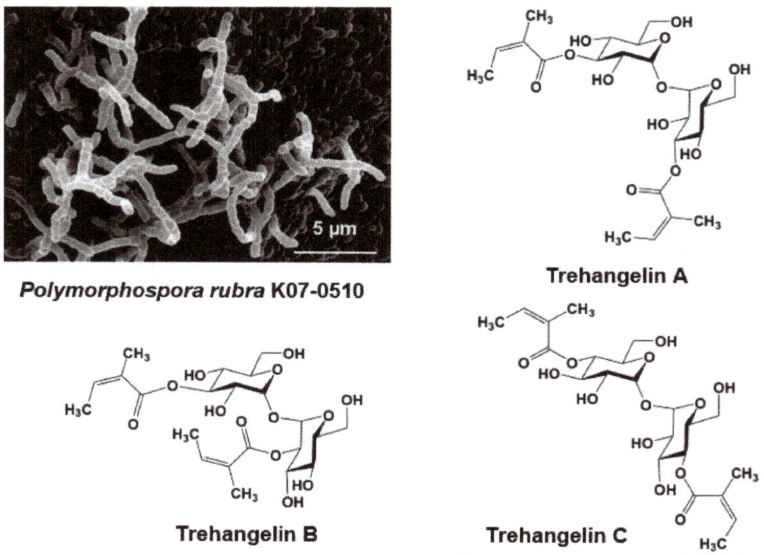

Figure 3. Scanning electron micrograph of the aerial spore chain of the trehangelin-producing strain *Polymorphospora rubra* K07-0510 and structures of trehangelin A, B, and C.

Angelic acid is known to occur in many plants, particularly chamomile, and has been used as a tonic and sedative to treat a variety of minor complaints, but reports of it being produced by microorganisms are extremely rare. It is, thus, interesting to note that a microbe in a plant root environment also produces the compound.

5.2. New Compounds Produced by Lechevalieria Aerocolonigenes K10-0216, Mangromycin A-I, K10-0216 KA & KB, and Pyrizomicin A & B

Lechevalieria aerocolonigenes K10-0216 (Figure 4a) is a rare actinomycete isolated from sediment collected in a mangrove forest in Iriomote Island, Okinawa Prefecture, Japan in 2011. In the culture broth of this strain, we found 13 new compounds.

Mangrove trees growing in brackish water are known to be a rich source of microorganisms, and many rare actinomycetes have been discovered in such environments [61]. We isolated 65 actinomycetes strains from five samples collected from a mangrove forest. Simple use of 16S rRNA gene sequences resulted in identification of 44 strains of the genus *Micromonospora*, as well as a few from the *Actinomadura* and *Verrucosispora*. The so-called rare actinomycetes accounted for 83% of organisms isolated.

Strain K10-0216 grown on an inorganic salt–starch agar medium produced sparse white aerial mycelia that formed characteristic clumps of interwoven hyphae, as shown in Figure 4a. The 16S rRNA gene sequence, which was determined and analyzed using the EzTaxon-e database [59], demonstrated a 99.8% similarity to *Lechevalieria aerocolonigenes* ISP 5034^T. On the basis of the morphological and cultural properties and 16S rRNA gene sequence analyses, strain K10-0216 was identified as *Lechevalieria aerocolonigenes*.

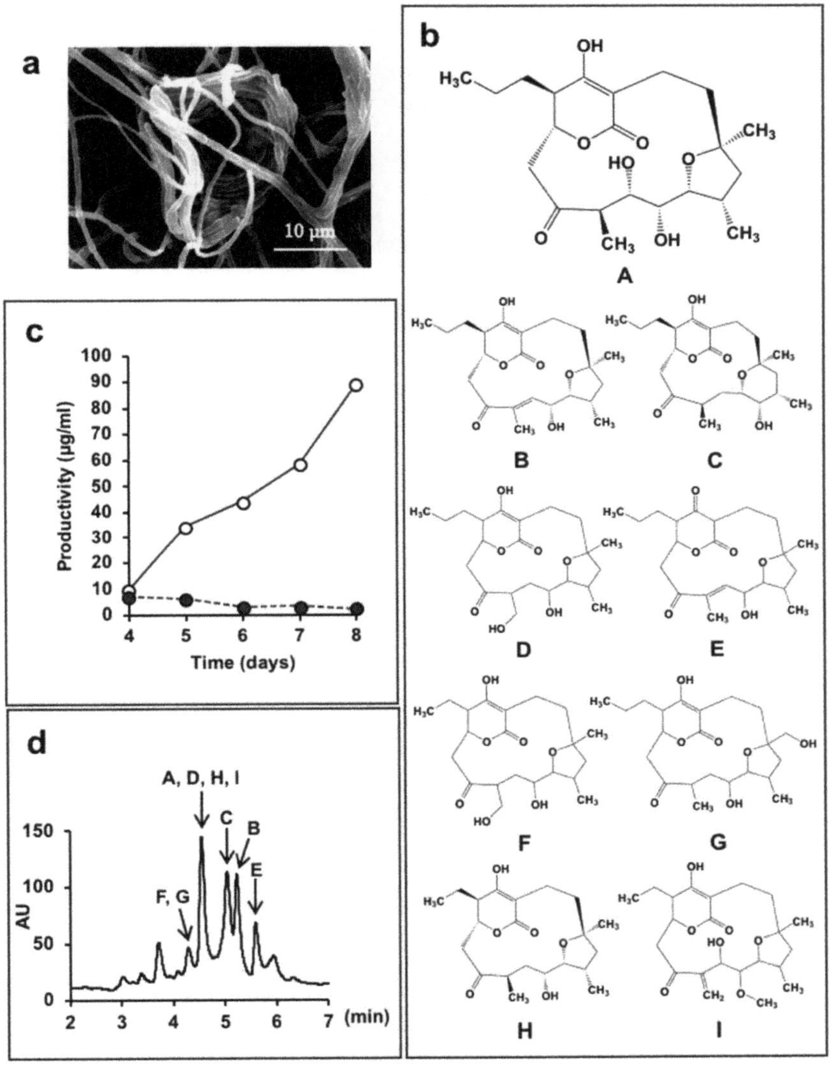

Figure 4. *Lechevalieria aerocolonigenes* K10-0216 and the mangromicins discovered from the culture broth. (**a**) Scanning electron micrograph of a clump of interwoven hyphae of *L. aerocolonigenes* K10-0216 grown on inorganic salt–starch agar at 27 °C for four weeks; (**b**) Structure of mangromicin A–I; (**c**) Productivity of mangromicin A. Black circle: Basic medium; soluble starch 2.0(%), defatted wheat germ 1.0, glycerol 0.5, dry yeast 0.3, $CaCO_3$ 0.5, meat extract 0.5. Open circle: Improved medium; soluble starch 5.0(%), defatted wheat germ 1.0, glycerol 0.5, dry yeast 1.0, $CaCO_3$ 0.5, meat extract 0.0. (**d**) HPLC analysis of the mangromicins. Chromatographic separation was undertaken using a MonoBis (3.2 × 150 mm, Kyoto Monotech Co., Ltd., Kyoto, Japan) at 40 °C. With regard to gradient elution, solvent A was water with 0.1% formic acid, and solvent B was methanol with 0.1% formic acid. The gradient elution was 0–10 min and 5–100% B. The flow rate was 0.5 mL/min, the injection volume was 5 µL, and detection occurred at 254 nm using a photodiode array detector.

We found that the culture broth contained predicted novel substances with molecular weights of 410 or 392, and maximal absorption at 251 nm or at 236 nm, respectively. We named these compounds mangromicin A and B [43]. However, these peaks were not obtained in a jar fermenter culture. A total of 500 mL Erlenmeyer flasks containing 100 mL culture medium were used to obtain the target substances that were isolated and purified. Structural analysis revealed a unique structure of a cyclopentadecane skeleton containing a 5,6-dihydro-4-hydroxyl-2-pyrone ring and a tetrahydrofuran ring (Figure 4b). Both substances displayed antitrypanosomal activity, and a patent application was filed. Subsequently, we experimented with different production media in order to obtain a large amount of the target compounds. As a result, by changing the concentration of soluble starch in the medium ingredients from 2.0% to 5.0%, and dry yeast 0.3% to 1.0%, the amount of mangromicin A increased dramatically, from 0.24 μg/mL to 88.6 μg/mL (Figure 4c). Eventually, nine analogs were obtained from 15 L of culture solution using this procedure (Figure 4b) [44,45]. The HPLC profile is shown in Figure 4d. After isolation and purification, preliminary bioassays identified DPPH radical scavenging activity and, in particular, NO scavenging was potent [44,45].

Furthermore, from the same culture broth, we also discovered four novel compounds (Figure 5). Two contained a steroid skeleton (named K10-0216 KA and KB) that inhibited lipid accumulation in 3T3-L1 adipocytes [46], while the others displayed a thiazole and a pyridine ring (named pyrizomicin A and B) that showed antibacterial activity [47].

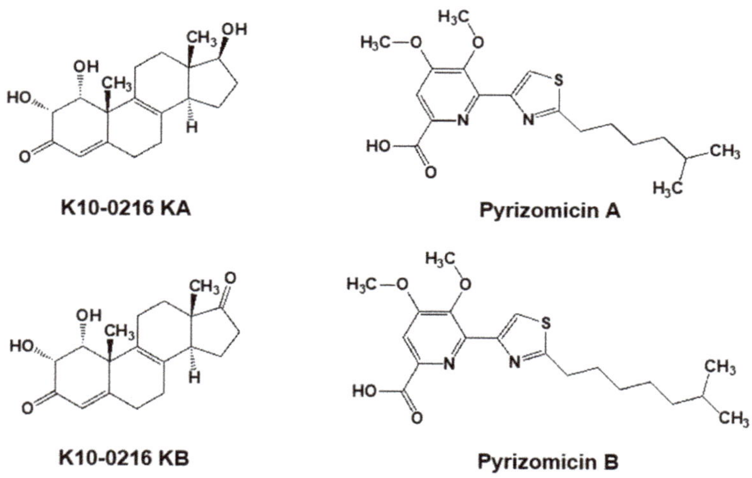

Figure 5. Two steroid compounds (K10-0216 KA and KB) and two compounds containing thiazole and pyridine (pyrizomicin A and B) from *L. aerocolonigenes* K10-0216 (Figure 4a).

As described above, 13 compounds, including three with different skeletons, were found from a single culture broth of one microorganism. This reinforces our belief that it is highly likely to be able to identify and acquire new compounds simply by changing the culture method and production medium.

6. Conclusions

This report describes compounds derived from actinomycete strains which were found via PC screening between 2011–2017 by the Kitasato Ōmura-Drug Discovery Group. Iminimycins and nanaomycins were found from culture broths of conserved KML strains, while the trehangelins from *P. rubra* K07-0510 and new compounds, mangromicins, K10-0216 KA and KB, pyrizomicin A and B, from *L. aelocolonigenes* K10-0216, originated in fresh isolates. Our results illustrate that PC screening can unearth novel compounds that are not likely to be discovered through traditional targeted bioassay systems. Genomic analysis of actinomycetes has found that more than 30 secondary metabolite biosynthetic genes may be involved in chemical production, depending on the strain. However, knowing the genetic production mechanism may not necessarily make obtaining the compound easier.

This report also shows that it is possible to obtain a novel compound from actinomycete species, such as *Streptomyces griseus*, that has already undergone long and intensive study. In addition, as in the case of *L. aerocolonigenes* K10-0216, a variety of compounds with significantly different skeletons can be obtained from a single culture broth using only one microbe strain.

Our work also demonstrates that the aptly classified "rare actinomycetes" remain a unique and, so far, relatively unexploited source of potentially useful chemicals. This is supported by work that found that strains of the genus *Actinoallomurus* (a rare actinomycete) have high secondary metabolite production capacity [62]. We also found five actinoallolide analogs [37] (shown in Table 1) produced by *A. fulvus* MK 10-036 and *A. fulvus* K 09-0307, later discovering two more new compounds together with seven known compounds from an *A. fulvus* K09-0307 broth.

Microorganisms can present us with interesting compounds with unique structures which our current scientific knowledge and expertise cannot easily predict nor easily replicate. For example, the mangromicins have unique skeletons which subsequently attracted attention in the field of organic synthesis. Organic chemists tried to devise a total synthesis of mangromicin A, and finally managed to achieve it via a complicated and lengthy 30 step process [63], whereas *L. aerocolonigenes* K10-0216 produces the compound naturally during culturing.

To illustrate the immeasurable scope for success in this respect, it has been reported that 99% of microorganisms in nature have not yet been isolated [64]. Furthermore, our results suggest that interesting compounds can be found even from *Streptomyces* strains that are thought to have been exhausted, simply by devising new identification methods or culture conditions. It is therefore essential to revisit existing actinomycetes, common and rare, and comprehensively examine them for new compounds.

Traditionally, the search for useful natural products has been advanced using approaches that focus on specific biological activity and target molecules. In this approach, there is a clearly defined goal. However, with this method alone, it is likely that many of the microorganisms isolated have been exploited and then discarded without fully utilizing their abilities. It is clearly difficult to devise a wide variety of screening systems and, certainly, almost impossible for all these systems to be operational in a single institution. Consequently, extensive, multifaceted research collaborations will need to be established to work towards full and comprehensive testing of all chemicals, be they newly isolated from existing compound libraries or from new sources. Naturally, many obstacles will need to be overcome, including protection of Intellectual Property Rights, transfer of technology, and capacity building aspects. None of these should be insurmountable, especially so if the goal of getting as many new bioactive substances in the shortest time possible is to be achieved.

We remain firm in our commitment to discover as many new compounds as possible by exploiting the ability of microorganisms, especially the actinomycetes, to produce such attractive substances. Moreover, we will continue to follow a twin-pronged approach to this task, while striving to devise other alternative screening methods, and adopt any other measures that could help expedite the research and development process.

Author Contributions: Y.T. and T.N. designed and performed the experiments, analyzed the data, and wrote the paper

Funding: This research was funded by the Institution for Fermentation, Osaka (IFO).

Acknowledgments: We deeply appreciate the guidance and support of Satoshi Ōmura (Distinguished Emeritus Kitasato Ōmura-Drug Discovery Group director, Kitasato University). And for his constant emphasis of the importance of isolation, culture and classification of microbes—to "Learn from microorganisms," and "Be grateful for microorganisms." We are grateful to all members of the Kitasato Ōmura-Drug Discovery Group, especially to Atsuko Matsumoto for management of the actinomycete strains and to Kazuro Shiomi, Toshiaki Sunazuka, Masato Iwatsuki and Kazuhiko Ōtoguro for structure determinations and bioassay work, and to Yuki Inahashi, Hirotaka Matsuo and Suga Takuya for their efforts in seeking new compounds by PC screening and the subsequent isolation, purification and structure determination. We would also like to extend our thanks to Toru Kimura, Rei Miyano, Yoshiyuki Kamiya, Shoko Izuta and many other researcher staff and students of the Kitasato Institute for Life Sciences for their collegiate support. We also thank Jun Nakanishi of the National Institute for Materials Science for assistance with biological activity evaluation. We would like to express our deepest gratitude to the IFO for their financial support.

Conflicts of Interest: The authors declare no conflict of interest.

References

1. Newman, D.J.; Cragg, G.M. Natural Products as sources of new drugs from 1981 to 2014. *J. Nat. Prod.* **2016**, *79*, 629–661. [CrossRef] [PubMed]
2. Bérdy, J. Thought and facts about antibiotics: Where we are now and where we are heading. *J. Antibiot.* **2012**, *65*, 385–395. [CrossRef] [PubMed]
3. Van Norman, G.A. Drugs, Devices, and the FDA: Part 1 an Overview of Approval Processes for Drugs. *JACC Basic Transl. Sci.* **2016**, *1*, 170–179. [CrossRef]
4. Morgan, S.; Grootendorst, P.; Lexchin, J.; Cunningham, C.; Greyson, D. The cost of drug development: A systematic review. *Health Policy* **2011**, *100*, 4–17. [CrossRef] [PubMed]
5. Dimasi, J.A.; Grabowski, G.H.; Hansen, R.W. Innovation in the pharmaceutical industry: New estimates of R&D costs. *J. Health Econ.* **2016**, *47*, 20–33. [PubMed]
6. World Health Organization. *Global Action Plan on Antimicrobial Resistance*; World Health Organization: Geneva, Switzerland, 2015.
7. World Health Organization. *United Nations General Assembly High-Level Meeting on Antimicrobial Resistance*; United Nations: New York, NY, USA, 2016.
8. Barka, E.A.; Vatsa, P.; Sanchez, L.; Gaveau-Vaillant, N.; Jacquard, C.; Meier-Kolthoff, J.P.; Klenk, H.P.; Clément, C.; Ouhdouch, Y.; van Wezel, G.P. Taxonomy, physiology, and natural products of Actinobacteria. *Microbiol. Mol. Biol. Rev.* **2015**, *80*, 1–43. [CrossRef] [PubMed]
9. Waksman, S.A.; Schatz, A.; Reynolds, D.M. Production of antibiotic substances by actinomycetes. *Ann. N. Y. Acad. Sci.* **2010**, *1213*, 112–124. [CrossRef] [PubMed]
10. Felling, R.H.; Buchanan, G.O.; Mincer, T.J.; Kauffman, C.A.; Jensen, P.R.; Fenical, P.R. Salinosporamide A: A highly cytotoxic proteasome inhibitor from a novel microbial source, a marine bacterium of the new genus *Salinospora*. *Angew. Chem. Int. Ed.* **2003**, *42*, 355–357. [CrossRef] [PubMed]
11. Tiwari, K.; Gupta, R.K. Rare actinomycetes: A potential storehouse for novel antibiotics. *Crit. Rev. Biotechnol.* **2012**, *32*, 108–132. [CrossRef] [PubMed]
12. Burg, R.W.; Miller, B.M.; Baker, J.; Birnbaum, E.E.; Currie, S.A.; Hartman, R.R.; Kong, Y.L.; Monaghan, R.L.; Olsen, G.; Putter, I.; et al. Avermectins, new family of potent anthelmintic agents: Producing organism and fermentation. *Antimicrob. Agents Chemother.* **1979**, *15*, 361–367. [CrossRef] [PubMed]
13. Takahashi, Y.; Matsumoto, A.; Seino, A.; Ueno, J.; Iwai, Y.; Ōmura, S. *Streptomyces avermectinius* sp. nov., an avermectin-producing strain. *Int. J. Syst. Evolut. Microbiol.* **2002**, *52*, 2163–2168.
14. Campbell, W.C.; Fisher, M.H.; Stapley, E.O.; Albers-Schönberg, G.; Jacobs, T.A. Ivermectin: A potent new antiparasitic agent. *Science* **1983**, *221*, 823–828. [CrossRef] [PubMed]
15. Crump, A.; Ōmura, S. Ivermectin, 'wonder drug' from Japan: The human use perspective. *Proc. Jpn. Acad. Ser. B Phys. Biol. Sci.* **2011**, *87*, 13–28.
16. Crump, A. Ivermectin: Enigmatic multifaceted 'wonder' drug continues to surprise and exceed expectations. *J. Antibiot.* **2017**, *70*, 495–505. [CrossRef] [PubMed]
17. The Nobel Assembly at Karolinska Institute, The Nobel Prize in Physiology or Medicine 2015. Available online: https://www.nobelprize.org/nobel_prizes/medicine/prize_awarder/ (accessed on 24 May 2018).
18. Satoshi Ōmura-Nobel Lecture: "A Splendid Gift from the Earth: The Origins and Impact of Avermectin". Nobelprize.org. Available online: http://www.nobelprize.org/nobel_prizes/medicine/laureates/2015/omura-lecture.html (accessed on 24 May 2018).
19. Takahashi, Y. Continuing fascination of exploration in natural substances from microorganisms. *Biosci. Biotechnol. Biochem.* **2017**, *81*, 6–12. [CrossRef] [PubMed]
20. Matsumoto, A.; Takahashi, Y. Endophytic actinomycetes: Promising source of novel bioactive compounds. *J. Antibiot.* **2017**, *70*, 514–519. [CrossRef] [PubMed]

21. Matsumoto, A. Potential of endophytic actinomycetes for agricultural applications. *Agric. Biotechnol.* **2018**, *2*, 180–185.
22. Ōmura, S. *Splendid Gifts from Microorganisms—The Achievements of Satoshi Ōmura and Collaborators*, 5th ed.; Kitasato Institute for Life Sciences, Kitasato University: Tokyo, Japan, 2015.
23. Ōmura, S.; Iwai, Y.; Hirano, A.; Nakagawa, A.; Awaya, J.; Tsuchiya, H.; Takahashi, Y.; Masuma, R. A new alkaloid AM-2282 of *Streptomyces* origin. *J. Antibiot.* **1977**, *30*, 275–282. [CrossRef] [PubMed]
24. Takahashi, Y.; Shinose, M.; Seino, A.; Iwai, Y.; Ōmura, S. Transfer of staurosporine-producing strain *Streptomyces staurosporeus* AM-2282 to the genus *Saccharothrix* as, *Saccharothrix aerocolonigenes* (labeda 1986) subsp. *staurosporeus* subsp. nov. *Actinomycetologica* **1995**, *9*, 19–26. [CrossRef]
25. Xie, Q.; Wang, Y.; Huang, Y.; Wu, Y.; Ba, F.; Liu, Z. Description of *Lentzea flaviverrucosa* sp. nov. and transfer of the type strain of *Saccharothrix aerocolonigenes* subsp. *staurosporea* to *Lentzea albida*. *Int. J. Syst. Evolut. Microbiol.* **2002**, *52*, 1815–1820.
26. Ōmura, S.; Sasaki, Y.; Iwai, Y.; Takeshima, H. Staurosporine, a potentially important gift from a microorganism. *J. Antibiot.* **1995**, *48*, 535–548. [CrossRef] [PubMed]
27. Nakano, H.; Ōmura, S. Chemical biology of natural indolocarbazole products: 30 years since the discovery of staurosporine. *J. Antibiot.* **2009**, *62*, 17–26. [CrossRef] [PubMed]
28. Nakashima, T.; Takahashi, Y.; Ōmura, S. Search for new compounds from Kitasato microbial library by physicochemical screening. *Biochem. Pharmacol.* **2017**, *134*, 42–55. [CrossRef] [PubMed]
29. Ōmura, S.; Iwai, Y.; Takahashi, Y.; Sadakane, N.; Nakagawa, A.; Oiwa, R.; Hasegawa, Y.; Ikai, T. Herbimycin, a new antibiotic produced by a strain of *Streptomyces*. *J. Antibiot.* **1979**, *32*, 255–261. [CrossRef] [PubMed]
30. Ōmura, S.; Otoguro, K.; Nishikiori, T.; Oiwa, R.; Iwai, Y. Setamycin, a new antibiotic. *J. Antibiot.* **1981**, *34*, 1253–1256. [CrossRef] [PubMed]
31. Ōmura, S.; Fujimoto, K.; Otoguro, K.; Matsuzaki, K.; Moriguchi, R.; Tanaka, H.; Sasaki, Y. Lactacystin, a novel microbial metabolite, induces neuritogenesis of neuroblastoma cells. *J. Antibiot.* **1991**, *44*, 113–116. [CrossRef] [PubMed]
32. Koomsiri, W.; Inahashi, Y.; Kimura, T.; Shiomi, K.; Takahashi, Y.; Ōmura, S.; Thamchaipenet, A.; Nakashima, T. Bisoxazolomycin A: A new natural product from '*Streptomyces subflavus* subsp. *Irumaensis*' AM-3603. *J. Antibiot.* **2017**, *70*, 1142–1145. [PubMed]
33. Nakashima, T.; Miyano, R.; Iwatsuki, M.; Shirahata, T.; Kimura, T.; Asami, Y.; Kobayashi, Y.; Shiomi, K.; Petersson, G.A.; Takahashi, Y.; Ōmura, S. Iminimycin A, the new iminium metabolite produced by *Streptomyces griseus* OS-3601. *J. Antibiot.* **2016**, *69*, 611–615. [CrossRef] [PubMed]
34. Nakashima, T.; Miyano, R.; Matsuo, H.; Iwatsuki, M.; Shirahata, T.; Kobayashi, Y.; Shiomi, K.; Petersson, G.A.; Takahashi, Y.; Ōmura, S. Absolute configuration of iminimycin B, the new indolizidine alkaloid from *Streptomyces griseus* OS-3601. *Tetrahedron Lett.* **2016**, *57*, 3284–3286. [CrossRef]
35. Nakashima, T.; Boonsnongcheep, P.; Kimura, T.; Iwatsuki, M.; Sato, N.; Nonaka, K.; Prathanturarug, S.; Takahashi, Y.; Ōmura, S. New compounds, nanaomycin F and G, discovered by physicochemical screening from a culture broth of *Streptomyces rosa* subsp. *notoensis* OS-3966. *J. Biosci. Bioeng.* **2015**, *120*, 596–600. [PubMed]
36. Nakashima, T.; Kimura, T.; Miyano, R.; Matsuo, H.; Hirose, T.; Kimishima, A.; Nonaka, K.; Iwatsuki, M.; Nakanishi, J.; Takahashi, Y.; et al. Nanaomycin H: A new nanaomycin analog. *J. Biosci. Bioeng.* **2017**, *123*, 765–770. [CrossRef] [PubMed]
37. Inahashi, Y.; Iwatsuki, M.; Ishiyama, A.; Matsumoto, A.; Hirose, T.; Oshita, J.; Sunazuka, T.; Watanalai, P.W.; Takahashi, Y.; Kaiser, M.; et al. Actinoallolides A-E, New anti-trypanosomal macrolides, produced by an endophytic actinomycete, *Actinoallomurus fulvus* MK10-036. *Org. Lett.* **2015**, *17*, 864–867. [CrossRef] [PubMed]
38. Suga, T.; Kimura, T.; Inahashi, Y.; Iwatsuki, M.; Nonaka, K.; Také, A.; Matsumoto, A.; Takahashi, Y.; Ōmura, S.; Nakashima, T. Hamuramicins A and B, 22-membered macrolides, produced by an endophytic actinomycete *Allostreptomyces* sp. K12-0794. *J. Antibiot.* **2018**. [CrossRef] [PubMed]
39. Inahashi, Y.; Matsumoto, A.; Ōmura, S.; Takahashi, Y. *Streptosporangium oxazolinicum* sp. nov., a novel endophytic actinomycete producing new antitrypanosomal antibiotics, spoxazomicins. *J. Antibiot.* **2011**, *64*, 297–302. [CrossRef] [PubMed]

40. Inahashi, Y.; Iwatsuki, M.; Ishiyama, A.; Namatame, M.; Nishihara-Tsukashima, A.; Matsumoto, A.; Hirose, T.; Sunazuka, T.; Yamada, H.; Otoguro, K.; et al. Spoxazomicins A-C, novel antitrypanosomal alkaloids produced by an endophytic actinomycete, *Streptosporangium oxazolinicum* K07-0460^{T}. *J. Antibiot.* **2011**, *64*, 303–307. [CrossRef] [PubMed]
41. Nakashima, T.; Okuyama, R.; Kamiya, Y.; Matsumoto, A.; Iwatsuki, M.; Inahashi, Y.; Yamaji, K.; Takahashi, Y.; Ōmura, S. Trehangelins A, B and C, novel photo-oxidative hemolysis inhibitors produced by an endophytic actinomycete, *Polymorphospora rubra* K07-0510. *J. Antibiot.* **2013**, *66*, 311–317. [CrossRef] [PubMed]
42. Inahashi, Y.; Shiraishi, T.; Palm, K.; Takahashi, Y.; Ōmura, S.; Kuzuyama, T.; Nakashima, T. Biosynthesis of Trehangelin in *Polymorphospora rubra* K07-0510: Identification of methabolic pathway to angelyl-CoA. *ChemBioChem* **2016**, *17*, 1442–1447. [CrossRef] [PubMed]
43. Nakashima, T.; Iwatsuki, M.; Ochiai, J.; Kamiya, Y.; Nagai, K.; Matsumoto, A.; Ishiyama, A.; Otoguro, K.; Shiomi, K.; Takahashi, Y.; et al. Mangromicins A and B: Structure and antitrypanosomal activity of two new cyclopentadecane compounds from *Lechevalieria aerocolonigenes* K10-0216. *J. Antibiot.* **2014**, *67*, 253–260. [CrossRef] [PubMed]
44. Nakashima, T.; Kamiya, Y.; Iwatsuki, M.; Takahashi, Y.; Ōmura, S. Mangromicins, six new anti-oxidative agents isolated from a culture broth of the actinomycete, *Lechevalieria aerocolonigenes* K10-0216. *J. Antibiot.* **2014**, *67*, 533–539. [CrossRef] [PubMed]
45. Nakashima, T.; Kamiya, Y.; Iwatsuki, M.; Sato, N.; Takahashi, Y.; Ōmura, S. Mangromicin C, a new analog of mangromicin. *J. Antibiot.* **2015**, *68*, 220–222. [CrossRef] [PubMed]
46. Nakashima, T.; Kamiya, Y.; Yamaji, K.; Iwatsuki, M.; Sato, N.; Takahashi, Y.; Ōmura, S. New steroidal compounds from an actinomycete strain, *Lechevalieria aerocolonigenes* K10-0216. *J. Antibiot.* **2015**, *68*, 348–350. [CrossRef] [PubMed]
47. Kimura, T.; Inahashi, Y.; Matsuo, H.; Suga, T.; Iwatsuki, M.; Shiomi, K.; Takahashi, Y.; Ōmura, S.; Nakashima, T. Pyrizomicin A and B: Structure and bioactivity of new thiazolyl pyridines from *Lechevalieria aerocolonigenes* K10-0216. *J. Antibiot.* **2018**, *71*, 606–608. [CrossRef] [PubMed]
48. Kimura, T.; Tajima, A.; Inahashi, Y.; Iwatsuki, M.; Kasai, H.; Mokudai, T.; Niwano, Y.; Shiomi, K.; Takahashi, Y.; Ōmura, S.; et al. Mumiamicin: Structure and bioactivity of a new furan fatty acid from *Mumia* sp. YSP-2-79. *J. Gen. Appl. Microbiol.* **2018**, *64*, 62–67. [CrossRef] [PubMed]
49. Kimura, T.; Iwatsuki, M.; Asami, Y.; Ishiyama, A.; Hokari, R.; Otoguro, K.; Matsumoto, A.; Sato, N.; Shiomi, K.; Takahashi, Y.; et al. Anti-trypanosoma compound, sagamilactam, a new polyene macrocyclic lactam from *Actinomadura* sp. K13-0306. *J. Antibiot.* **2016**, *69*, 818–824. [CrossRef] [PubMed]
50. Izuta, S.; Kosaka, S.; Kawai, M.; Miyano, R.; Matsuo, H.; Matsumoto, A.; Nonaka, K.; Takahashi, Y.; Ōmura, S.; Natashima, T. Dipyrimicins A and B, microbial compounds interacted with ergosterolresin, from *Amycolatopsis* sp. K16-0194. *J. Antibiot.* **2018**, *71*, 535–537. [CrossRef] [PubMed]
51. Bringmann, G.; Kajahn, I.; Reichert, M.; Pedersen, S.E.; Faber, J.H.; Gulder, T.; Brun, R.; Christensen, S.B.; Ponte-Sucre, A.; Moll, H.; et al. Ancistrocladinium A and B, the first N, C-coupled naphthyldihydroisoquinoline alkaloids, from a Congolese Ancistrocladus species. *J. Org. Chem.* **2006**, *71*, 9348–9356. [CrossRef] [PubMed]
52. Ohnishi, Y.; Ishikawa, J.; Hara, H.; Suzuki, H.; Ikenoya, M.; Ikeda, H.; Yamashita, A.; Hattori, M.; Horinouchi, S. Genome sequence of the streptomycin-producing microorganism *Streptomyces griseus* IFO 13350. *J. Bacteriol.* **2008**, *190*, 4050–4060. [CrossRef] [PubMed]
53. Ōmura, S.; Tanaka, H.; Koyama, Y.; Oiwa, R.; Katagiri, M. Nanaomycins A and B, new antibiotics produced by a strain of *Streptomyces*. *J. Antibiot.* **1974**, *27*, 363–365. [CrossRef] [PubMed]
54. Tanaka, H.; Marumo, H.; Nagai, T.; Okada, M.; Taniguchi, K. Nanaomycins, new antibiotics produced by a strain of *Streptomyces*. III. A new component, nanaomycin C, and biological activities of nanaomycin derivatives. *J. Antibiot.* **1975**, *28*, 925–930. [CrossRef] [PubMed]
55. Kasai, M.; Shirahata, K.; Ishii, S.; Mineura, K.; Marumo, H.; Tanaka, H.; Ōmura, S. Structure of nanaomycin E, a new nanaomycin. *J. Antibiot.* **1979**, *32*, 442–445. [CrossRef] [PubMed]
56. Kitaura, K.; Araki, Y.; Marumo, H. The therapeutic effect of nanaomycin A against experimental *Trichophyton mentagrophytes* infection in guinea pigs (author's transl). *Jpn. J. Antibiot.* **1980**, *33*, 728–732. [PubMed]
57. Ōmura, S.; Takahashi, Y.; Nakashima, T.; Matsumoto, A.; Nakanishi, J.; Matsuo, H. Epithelial-Mesenchymal Transition Induced Cell Proliferation Inhibitor. Patent No. PCT/JP2017/034819, 26 September 2016.

58. Bode, H.B.; Bethe, B.; Höfs, R.; Zeeck, A. Big Effects from Small Changes: Possible Ways to Explore Nature's Chemical Diversity. *ChemBioChem* **2002**, *3*, 619–627. [CrossRef]
59. Yoon, S.H.; Ha, S.M.; Kwon, S.; Lim, J.; Kim, Y.; Seo, H.; Chun, J. Introducing EzBioCloud: A taxonomically united database of 16S rRNA gene sequences and whole-genome assemblies. *Int. J. Syst. Evol. Microbiol.* **2017**, *67*, 1613–1617. [PubMed]
60. Kamiya, Y.; Enomoto, Y.; Matsukuma, S.; Nakashima, T.; Takahashi, Y.; Omura, S. Study on anti-aging effects of trehangelin, a new trehalose compound. In Proceedings of the 136th Annual Meeting of the Pharmaceutical Society of Japan, Yokohama, Japan, 26–29 March 2016; p. 137.
61. Azman, A.S.; Othman, I.; Velu, S.S.; Chan, K.G.; Lee, L.H. Mangrove rare actinobacteria: Taxonomy, natural compound, and discovery of bioactivity. *Front. Microbiol.* **2015**, *20*, 856. [CrossRef] [PubMed]
62. Pozzi, R.; Simone, M.; Mazzetti, C.; Maffioli, S.; Monciardini, P.; Cavaletti, L.; Bamonte, R.; Sosio, M.; Donadio, S. The genus *Actinoallomurus* and some of its metabolites. *J. Antibiot.* **2011**, *64*, 133–139. [CrossRef] [PubMed]
63. Takada, H.; Yamada, T.; Hirose, T.; Ishihara, T.; Nakashima, T.; Takahashi, Y.; Ōmura, S.; Sunazuka, T. Total synthesis and determination of the absolute configuration of naturally occurring mangromicin A, with potent antitrypanosomal activity. *Org. Lett.* **2017**, *19*, 230–233. [CrossRef] [PubMed]
64. Amann, R.I.; Ludwig, W.; Schleifer, K.H. Phylogenetic identification and in situ detection of individual microbial cells without cultivation. *Microbiol. Rev.* **1995**, *59*, 143–169. [PubMed]

antibiotics

MDPI

Review

Concepts and Methods to Access Novel Antibiotics from Actinomycetes

Joachim J. Hug [1,2], Chantal D. Bader [1,2], Maja Remškar [1,2], Katarina Cirnski [1,2] and Rolf Müller [1,2,*]

1 Department Microbial Natural Products, Helmholtz-Institute for Pharmaceutical Research Saarland (HIPS), Helmholtz Centre for Infection Research (HZI) and Department of Pharmacy, Saarland University, Campus E8.1, 66123 Saarbrücken, Germany; joachim.hug@helmholtz-hzi.de (J.J.H.); chantal.bader@helmholtz-hzi.de (C.D.B.); maja.remskar@helmholtz-hzi.de (M.R.); katarina.cirnski@helmholtz-hzi.de (K.C.)

2 German Center for Infection Research (DZIF), Partner Site Hannover-Braunschweig, 38124 Braunschweig, Germany

* Correspondence: rolf.mueller@helmholtz-hzi.de; Tel.: +49-681-98806-3000

Received: 12 April 2018; Accepted: 17 May 2018; Published: 22 May 2018

check for updates

Abstract: Actinomycetes have been proven to be an excellent source of secondary metabolites for more than half a century. Exhibiting various bioactivities, they provide valuable approved drugs in clinical use. Most microorganisms are still untapped in terms of their capacity to produce secondary metabolites, since only a small fraction can be cultured in the laboratory. Thus, improving cultivation techniques to extend the range of secondary metabolite producers accessible under laboratory conditions is an important first step in prospecting underexplored sources for the isolation of novel antibiotics. Currently uncultured actinobacteria can be made available by bioprospecting extreme or simply habitats other than soil. Furthermore, bioinformatic analysis of genomes reveals most producers to harbour many more biosynthetic gene clusters than compounds identified from any single strain, which translates into a silent biosynthetic potential of the microbial world for the production of yet unknown natural products. This review covers discovery strategies and innovative methods recently employed to access the untapped reservoir of natural products. The focus is the order of actinomycetes although most approaches are similarly applicable to other microbes. Advanced cultivation methods, genomics- and metagenomics-based approaches, as well as modern metabolomics-inspired methods are highlighted to emphasise the interplay of different disciplines to improve access to novel natural products.

Keywords: metagenomics; rare actinomycetes; dereplication; metabolomics; genome mining; natural products

1. Introduction

Antimicrobial resistance (AMR) gains more and more public attention as one of the biggest threats to prevention and treatment of an increasing number of infections. In January 2018, the WHO released a report on global antibiotic resistance crisis. For this report, the new Global Antimicrobial Surveillance System (GLASS) implemented in May 2015 was used to support standardised AMR surveillance globally. The main objective of that study has been tracking the resistance-related issues of medicines in use for treatment of hospital and community acquired infections across the 52 participating countries. The survey focus is on *Escherichia coli*, *Klebsiella pneumoniae*, *Staphylococcus aureus*, *Streptococcus pneumoniae*, *Salmonella* spp., *Neisseria gonorrhoeae*, *Shigella* spp. and *Acinetobacter baumanii*. According to the report, *N. gonorrhoeae* is evolving into a superbug, as it is resistant to 3rd generation cephalosporins and fluoroquinolones and is classified by the WHO Priority Pathogens List of antibiotic-resistant bacteria. *Salmonella* spp., *S. aureus* and *N. gonorrhoeae* were prioritised for research and development of new and effective antibiotic treatments, while *A. baumanii*, *E. coli* and *K. pneumoniae* are classified as critical priority and *Shigella* spp. along with *S. pneumonaiae* are classified as medium priority [1]. Obviously, new lead compounds exhibiting lack of cross-resistance with known antibiotics are urgently searched for.

Between 1950 and 1960—the so-called "golden age of antibiotics" [2]—microbiology and chemistry went hand in hand; bacteria were cultivated and their secondary metabolites extracted to yield a large number of compounds with antibacterial activity. Among these bacteria the genus *Streptomyces* has provided numerous novel bioactive molecules, not just antibiotics but also antifungals, antiprotozoals and antivirals [3]. However, the discovery of novel drugs derived from streptomycete secondary metabolites has continued to steadily decrease since, reflected in a 30% drop of natural product-based drugs in clinical studies between 2001 and 2008 [4]. Natural product research was widely abandoned by pharmaceutical industry due to severe rediscovery issues; in addition, challenging isolation of new producers and low production concentrations contribute to the more difficult access to novel scaffolds with promising activity using well-known natural product sources [4]. In response to the growing needs of medicine for new antimicrobial compounds, the focus in natural products discovery shifted towards underexplored habitats, especially marine and extreme environments. Encouragingly, rare genera of actinomycetes—in general considered as non-streptomycetes—and novel *Streptomyces* isolates found in these habitats have already afforded the discovery of novel antimicrobials with unique chemical moieties, confirming that microbial natural products are still a promising source for drug discovery [5–8]. In particular, non-streptomycetes belonging to genera such as *Micromonospora*, *Nocardia*, *Actinomadura*, *Actinoplanes*, *Streptoverticilllium* and *Saccharopolyspora* have been found to produce chemically unique antibiotics featuring potent activities [8], such as abyssomicins [9] and proximicins [10] from *Verrucosispora* strains. Consequently, isolation of natural products from rare actinomycetes already led to the discovery of approximately 2250 new bioactive secondary metabolites until 2005 [11,12]. Bioinformatic analysis of genomes from long known actinomycetes—for instance, the first whole genome sequenced actinomycete *Streptomyces coelicolor* A3 (2) [13] known to produce only a few secondary metabolites until then—revealed that many secondary metabolite genes are most likely not expressed under conventional laboratory conditions or their corresponding products cannot be detected with widely used analytical methods. This discrepancy translates into a "silent" potential of the microbial world for the production of yet unknown natural products. However, despite these encouraging insights, the majority of nature's microbial biosynthetic potential will remain elusive, if natural product research continues to rely exclusively on a classical culture-dependent discovery platform. Efforts must be intensified to exploit the majority of bacteria not cultured under laboratory conditions to date, and some suitable concepts and methods for that purpose have emerged recently [14].

Outline of this Review

In light of the numerous available review articles covering microbial natural products research—including highly specialised articles diving into specific sub-topics in much detail—we here aim to present a summary of selected technical and conceptual proceedings in the field of natural product research during the last decade. We focus on examples from the world of the actinomycetes although conceptual developments using alternative producers have significant impact on this field [15–19]. The idea of exploring new habitats harbouring rare actinomycetes and their subsequent isolation and cultivation utilising novel cultivation methods serves as a starting point in Section 2. If large-scale cultivation is difficult to achieve or even not possible, culture-independent approaches are required to extend the number of microorganisms accessible for the discovery of novel natural products. Section 3 therefore outlines examples of natural product discovery by metagenomics and heterologous expression of selected pathways as an alternative means to investigate the uncultured environmental diversity and access its hidden natural product potential. Metagenomics mainly relies on prior examination of biosynthetic gene clusters with *in silico* bioinformatics tools. Hence, Section 4 deals with genomics-based approaches; in particular, genome mining to access silent biosynthetic pathways in cultured actinobacteria will be highlighted. Section 5 emphasises developments in analytical chemistry and the increasing use of metabolomics-inspired methods by combining data-driven and experimental approaches. The latter approaches today allow for the prioritisation of minute amounts of compounds that escaped detection previously.

2. Exploring New Habitats

Because of over-exploitation of terrestrial streptomycetes in almost all pharmaceutical and agrochemical companies as well as numerous academic groups, in recent years, the search for new bioactive molecules has moved towards their marine relatives [20], taxa found in the extreme environments [21], endophytic species [22] and on non-*Streptomyces* actinomycetes. Especially marine ecosystems represent one of the richest and underexploited habitats with great microbial diversity showing potential for discovery of novel and chemically diverse antimicrobial compounds [23]. Progress in the beginning of this millennium in the field of selective isolation methods has been achieved, improving access to these diverse actinomycetes. Since *Streptomyces* are the most dominant genus within the actinomycete group, it is necessary to increase chances for the isolation of desirable new species (enrichment) by reducing undesirable background, i.e., previously isolated species (pretreatment) [24]. A combination of thermal pretreatment, beneficial to promote the spore germination and improve the selective isolation of rare actinomycetes and chemical techniques to eliminate bacterial contaminants has proven successful for isolation of some rare actinomycetes [24]. An advanced method of pretreatment applies polyvalent bacteriophages specifically targeting different unwanted background bacteria [25]. Traditional enrichment is based on simple trial and error methods using different chemical and physical treatments to unearth the uncultured bacterial majority. Chaxalactin A–C production—a rare class of 22-membered macrolactone polyketides produced by *Streptomyces* strain C34 which was isolated from the Chilean hyper-arid Atacama Desert soil [26,27]—was highly dependent on the culture medium used for its growth, highlighting the influence of the culture medium on growth and secondary metabolism of actinomycetes. Other recent examples for the empiric progress in isolation of actinomycetes is the halotolerant *Nocardiopsis* sp. HR-4 strain producing (-)-7-deoxy-8-*O*-methyltetrangomycin [28] and the non-extremophilic *Streptomyces* spp. ERI-26 producing 6,6^1-bis(1,5,7-trihydroxy-3-hydroxylmethylanthraquinone) [29,30]. The reason bacteria are remaining "unculturable" is that the growth under laboratory conditions fails to mimic essential aspects of the environment, e.g., the absence of competitors [31]. One common method for simulation of environmental conditions uses diffusion chambers to enable the exchange of nutrients and growth factors within the two chambers while separating the isolates of interest from their competitors. Recovery rates up to 40% have been achieved, comparing the number of growing microcolonies by microscopic counts of cells in the initial inoculum, whereas the same inoculum on standard petri dishes yielded only a very low recovery rate of 0.05% [32]. A further developed and miniaturised *in situ* cultivation platform based on diffusion chambers is used for an isolation chip (iChip). The iChip is composed of several hundred miniature diffusion chambers, of which each is inoculated with an individual environmental cell [33]. To capture on average a single cell, the through-holes of the iChip plate are immersed into a suspension of mixed environmental cells. Covered with an upper and lower plate, the assembled iChip provides a miniature diffusion chamber for each single cell (Figure 1). This technology platform has shown the great potential for upscaling the throughput and improvement of the isolation of previously uncultured bacteria in general. The polycyclic xanthone antibiotics neocitreamicins from *Nocardia* strain G0655 [34] along with other examples described in literature, are showing that simulation of the original environment accomplished the isolation of novel strains [35]. The glycosylated macrolactams NOVO3 and NOVO4 from the strains *Streptosporangium* P1532 and *Amycolatopsis* Z0363, respectively [36], the ribosomal 16-membered cyclic peptide lassomycin from *Lentzea kentuckyensis* [37], and the antibiotic teixobactin from the new β-proteobacterial species *Elftheria terrae* [38] have been isolated by using the iChip technology.

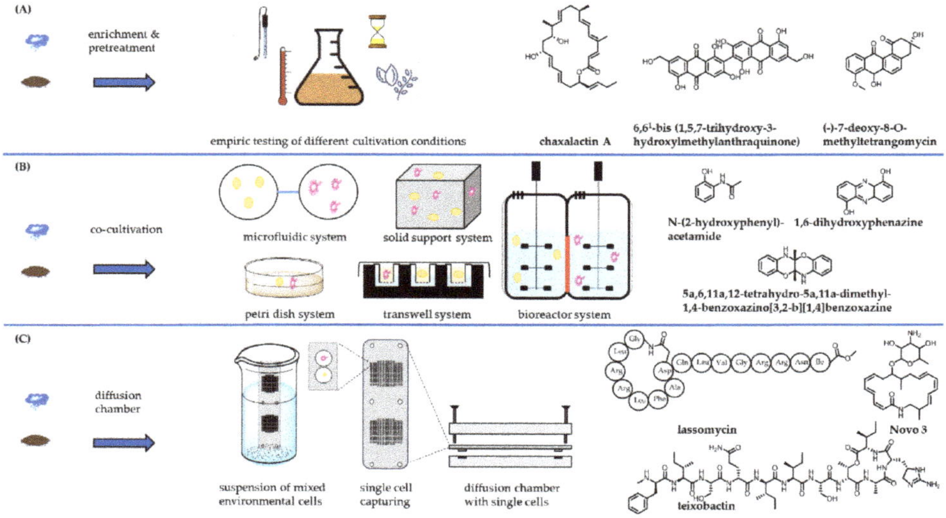

Figure 1. Scheme of isolation strategies, partly adapted from Nichols et al. [33] and Goers et al. [39]. (**A**) Soil sample or marine sample undergoes enrichment and/or pretreatment to increase the chance to isolate new species and/or reduce undesirable background from previously isolated strains. In general, empirical established methods for fermentation varies in incubation time, media composition, additives, pH and temperature to enable growth of desirable strains. Chaxalactin A produced by *Streptomyces* sp. C34 [26,27], 6,6^1-bis (1,5,7-trihydroxy-3-hydroxylmethylanthraquinone) produced by *Streptomyces* spp. ERI-26 [29,30] and (-)-7-deoxy-8-*O*-methyltetrangomycin from *Nocardiopsis* sp. HR-4 [28] are examples for novel metabolites found using this conventional method. (**B**) Soil sample or marine sample is co-cultivated with other microorganisms to promote culturable isolates or to stimulate the secondary metabolism. Co-cultivation is categorised in microfluidic systems, petri dish co-culture systems, co-cultures on solid supports, co-culture systems using bioreactors and transwell systems [39]. Co-cultivation of *Actinokineospora* sp. EG49 and *Nocardiopsis* sp. RV163 induces the biosynthesis of three natural products namely *N*-(2-hydroxyphenyl)-acetamide, 1,6-dihydroxyphenazine and 5a,6,11a,12-tetrahydro-5a,11a-dimethyl[1,4]benzoxazino[3,2-b][1,4]benzoxazine [40]. (**C**) Soil sample or marine sample is used to create a suspension of mixed environmental cells. The isolation chip (iChip) plate is immersed into this suspension to capture (on average) a single cell. Covered with an upper and lower plate, the assembled iChip provides a miniature diffusion chamber for each single cell [33]. NOVO 3 [36], the cyclic peptide lassomycin [37] and the antibiotic teixobactin [38] were isolated from strains obtained by the iChip technology.

These case studies demonstrated how environmental conditions facilitate access to the diversity of yet uncultured microorganisms [32]. An alternative method for separation of uncultured bacteria is encapsulation of single bacterial cells in microdroplets of solidified agarose [41–43]. To acquire insights into metabolic requirements a metatranscriptomic approach can be utilised. High-throughput sequencing of RNA transcripts enables to access expressed genes of growing bacteria that are not culturable in large-scale. Following this metatranscriptomics-based method, an uncultured *Rikenella*-like bacterium in the leech gut was found to utilise mucin as a carbon and energy source [44].

A different approach to simulate authentic circumstances in the environment comprises co-cultivation of different species derived from the same environment, demonstrating significant influence on the production of secondary metabolites [45–47]. The main co-cultivation systems can be categorised by the type of technology used in microfluidic systems, petri dish co-culture systems, co-cultures on solid supports, co-culture systems using bioreactors and transwell systems for co-cultivation, allowing asymmetric levels of separation between various numbers of populations [39]. Co-cultivation of *Actinokineospora* sp. EG49 and *Nocardiopsis* sp. RV163 induces the biosynthesis of three natural products namely *N*-(2-hydroxyphenyl)-acetamide, 1,6-dihydroxyphenazine and 5a,6,11a,12-tetrahydro-5a,11a-dimethyl[1,4]benzoxazino[3,2-b][1,4]benzoxazine [40]. A hypothesised model of mechanisms on the molecular level tries to explain the reason for the necessity of co-cultivation; i.e., neighbouring bacteria are decreasing oxidative, metabolic and environmental stress, provide common nutrients or other growth factors and siderophores in order to bind and solubilise Fe^{3+} [31]. The Black Queen Hypothesis, a theory of reductive evolution, tries to explain how selection leads to helper-dependent bacterial isolates, which lost the ability to perform essential processes on their own. In conclusion the theory states that the "loss of a costly, leaky function is selectively favoured at the individual level and will proceed until the production of public goods is just sufficient to support the equilibrium community" [48]. Similar observations hold true for endosymbiotic marine bacteria, responsible for the prolific production of secondary metabolites [49]. Among the underexplored habitats, analysis of marine ecosystems has flourished (Figure 2). New actinomycetes were isolated from marine sediments, water columns at various depths and even from marine fauna. Screening of 20 marine samples from marine environments in Egypt, employing different methods for selective isolation, has yielded 112 *Streptomyces* isolates. Up to 76% of the isolates displayed activities against tested Gram-positive pathogens, up to 41% against Gram-negative pathogens, 32% against yeasts and 28% against fungi. Although depending on the type of media used, 28 isolates showed activity against methicillin-resistant *S. aureus* (MRSA), i.e., 25% of all obtained isolates. In addition, all *Streptomyces* strains showed significant nematicidal activity against second stage (J_{2}) juveniles of the root-knot nematode *Meloidogyne incognita*, making them a potential candidate for use in agriculture [50]. However, the value of these observations has to be confirmed as the molecular structures of the bioactive compounds have yet to be elucidated. The one strain–many compounds (OSMAC) approach is a strategy to overcome the limited expression of biosynthetic pathways in microbes. Prioritised strains are cultured in a variety of different media compositions and fermentation conditions to maximise the amount and diversity of produced compounds [51]. The use of OSMAC approach has enabled the recovery of a new thiopeptide antibiotic, TP-1161, from marine sediment derived *Nocardiopsis* isolate. The antimicrobial profile revealed no anti-Gram-negative activity; however, the activity against Gram-positive clinical isolates was comparable to the one of vancomycin [52]. Another example for the OSMAC approach is the isolation of the anti-MRSA and anti-vancomycin-resistant *Enterococcus* armeniaspirols produced by *Streptomyces armeniacus* DSM 19369. The armeniaspirols are only produced by the "well-characterised" *S. armeniacus* when cultivated on a malt-containing medium [53].

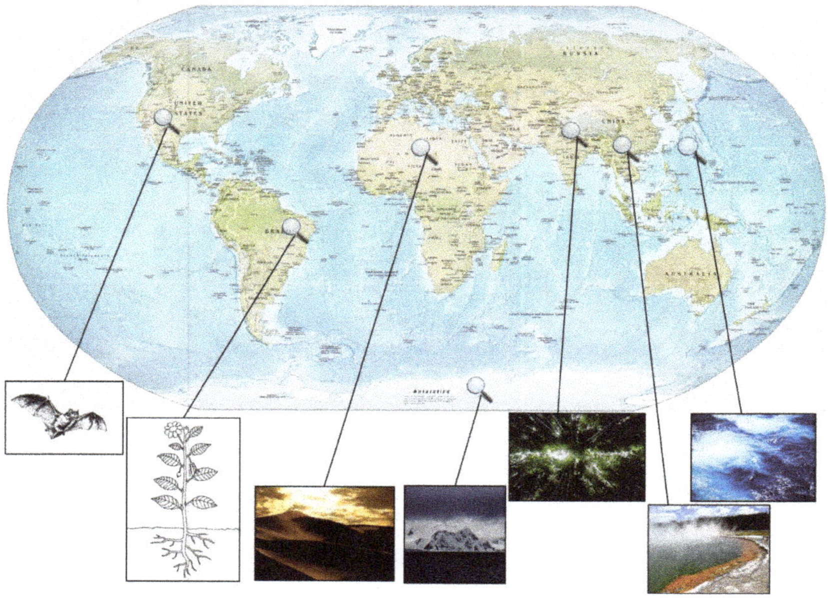

Figure 2. Underexploited habitats of actinobacteria attracted more attention for microbial natural product discovery. Currently, oceans [54], deserts [55], mountains [30] and Antarctica [56] ranges together with hot springs [57] and endophytes [58] and symbionts [59] are focuses of the search for new bioactive compounds.

One of the more recent examples is the isolation and cultivation of the abyssal actinobacterium *Pseudonocardia carboxydivorans* M-227, isolated at 3000 m depth in water column of the Avilés submarine canyon. It produces the two new antibiotics branimycin B and C (Figure 3). Branimycins are structurally related to nargenicins and are mainly active against *S. aureus*. However, both compounds were found to be moderately active against not only *S. aureus* (MIC values 32–64 μg/mL) but also Gram-negative bacteria (MIC values 20–80 μg/mL). It is important to point out that they exhibited activity against MRSA, several clinical *S. aureus* isolates, *E. faecalis* and even *H. influenzae* ATCC 49247 [60]. Contrarily, the novel γ-butyrolactones ghanamycin A and B, isolated from a marine-derived strain *Streptomyces ghanaensis* TXC6-16, only displayed weak anti-Gram-negative activity against *Pseudomans syringae*, *Erwinia* sp. and selected plant pathogens such as *F. oxysporum*, *A. solani* and *P. oryzae* [61]. Lobophorin E and F, new spirotetronate antibiotics, exhibited moderate antimicrobial activity against *B. thuringiensis* SCSIO BT01, *S. aureus* ATCC 29213 and *E. faecalis* ATCC 29212. Interestingly lobophorin F also exhibits moderate cytotoxic activities on human tumour cell lines SF-268, MCF-7 and NCI-233 H460, emphasising the influence of chemical modifications on the observed bioactivity [62]. Spithioneine A and B are structurally unique bohemamine-type pyrrolizidine alkaloids from a marine isolate *Streptomyces spinoverrucosus*. They are especially interesting as they incorporate ergothioneine into a polyketide. Ergothioneine exhibits a wide variety of activities, such as inhibition of oxidative stress, promotion of neuronal differentiation and metal ion chelation, so it will be interesting to see the antimicrobial profile of these novel compounds incorporating this chemical moiety [63].

The question if marine derived small molecules chemically differ from those produced by terrestrial strains remains a matter of debate among natural product researchers. The famous abyssomicin was first thought to be exclusively produced by a marine streptomycete. However, it was later shown to be produced by a terrestrial streptomycete [9,64–66] as well. Probably the best known example that marine and terrestrial actinomycetes are producing similar lead compounds is shown by salinosporamide A from *Salinispora tropica* [67], featuring high structural similarity to the cinnabaramides A−G, isolated from the terrestrial *Streptomyces* sp. DSM 15324 [68]. Recently isolated ilamycins, produced by a *Streptomyces atratus* SCSIO ZH16, a strain isolated from South China deep-sea sediment (Figure 2) [54], carry two unusual building blocks, L-3-nitro-tyrosine and L-2-amino-4-hexenoic acid. New ilamycins produced by a genetically modified strain show very good anti-tuberculosis activity, with the MIC values against *M. tuberculosis* H37Rv in the single-digit nanomolar range (9.8 nM) [54].

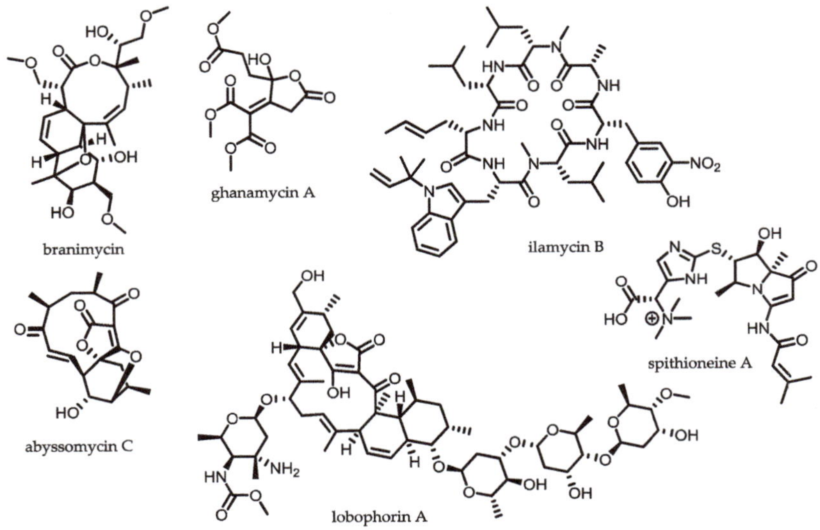

Figure 3. Novel natural products deriving from new habitats.

2.1. Extreme Environments as a Rich Source for Novel Strains

Soil samples from extreme environments such as deserts and samples collected from higher altitudes offer potential to isolate many previously not cultured bacteria. Soil sample from Indian mountains uncovered a new *Streptomyces* sp. isolate, ERI-26, which produces a novel anthraquinone (Figure 2) [30]. The compound exhibits weak antibacterial activity (anti-Gram-positive and anti-Gram-negative activity >62.5 μg/mL) and moderate to good antifungal activity, with MIC values as low as 3.9 μg/mL [29]. Screening of the Egyptian desert revealed that 32 out of 75 isolated strains were active against the tested pathogenic organisms and thus showing that these extremophiles are metabolically active. According to the antimicrobial activities, 12.5% of the isolates were active against Gram-positive as well as Gram-negative bacteria and yeast strains [7]. However, without correlation of these observed activities to any known or unknown compounds, the authentic benefit of these findings is hard to evaluate. Non-*Streptomyces* actinomycete *Saccharotrix* SA198 isolated from Saharian soil has also lead to the discovery of two novel antibiotics A4 and A5 (Figure 2) [55]. Both exhibited good anti-Gram-positive and anti-Gram-negative activity (MIC values between 10 and 50 μg/mL), as well as very good antifungal profile (MIC values between 1–5 μg/mL), but no activity against yeast [55]. Furthermore, the non-*Streptomyces* actinobacterium isolate LAM143cG3 was obtained from an underexplored sebkha lake (dry lakebed consisting primarily of salt) of Kenadsa (Bechar, Southwestern Algeria). Based on the chemical and morphological characteristics, this isolate was identified as a member of the genus *Spirillospora*. Agar diffusion tests showed that crude extracts are primarily and only moderately active on Gram-positive bacteria and have only weak activity on

Gram-negative bacteria [69]. Similar antibiotic bioactivity profile was exhibited by the halotolerant *Nocardiopsis* sp. HR-4 recently isolated from salt lake soil sample in Algerian Sahara [28], however the activities were not connected to known or novel natural products. One of the more promising compounds is actinomadurol, also isolated from a rare actinomycete *Actinomadura* strain KC 191 as it exhibits potent activity against *S. aureus*, *K. rhizophila*, and *P. hauseri* (MIC = 0.39–0.78 µg/mL) [70]. Actinomadurol is not inhibiting the growth of various tested cancer cell lines, even at high concentrations (IC_{50} > 100 µg/mL). Genera *Brevibacterium*, *Gordonia*, *Micromonospora*, *Arthrobacter*, *Demetria*, *Rhodococcus*, *Janibacter*, *Leifsonia*, *Dermacoccus*, *Kocuria*, *Lapillicoccus*, *Microbacterium* and *Nocardioides* were present in soil samples from Antarctica (Figure 2) [56]. These isolates were tested against following pathogens: *C. albicans* ATCC 10231^T, *S. aureus* ATCC 51650^T, MRSA ATCC BAA-44^T and *P. aeruginosa* ATCC 10145^T. Of all the present genera *Brevibacterium* showed the best antimicrobial activity, three isolates were active against *C. albicans*, one against MRSA and one against *S. aureus* [56].

Recently, thermal water sources have been explored in search for new antimicrobial entities. Analysis of the Tengchong hot springs revealed predominantly the presence of the genus *Streptomyces* (Figure 2) [57]. However, with the presence of *Actinomadura*, *Microbispora*, *Micromonospora*, *Nocardiopsis*, *Nonomuraea*, *Promicromonospora*, *Pseudonocardia* and *Verrucosispora*, a high number of rare non-*Streptomyces* actinobacteria could be found. Most of the isolates had a weak activity against *A. baumanii* and some against *M. luteus* [57]. Much stronger antimicrobial profile was exhibited by *Streptomyces* sp. Al-Dhabi-1, collected from Tharban hot spring in Saudi Arabia. Ethyl-acetate extract showed moderate to weak activity against tested pathogens, both Gram-positive and Gram-negative as well as fungi with MIC values <1 mg/mL [71]. Other recently explored habitats are cave systems populated with actinobacteria as its highly diverse microbial community may exhibit antagonistic properties against pathogens. Search for antifungal agents against *P. destructans*, causing white-nose syndrome in bats, led to the identification of 15 novel *Streptomyces* species, encouraging the further investigation of caves (Figure 2) [59] as valuable source for novel actinomycetes producing potent antimicrobials [59]. Unfortunately, most of the studies mentioned do not connect crude extract activity to known or novel natural products and are thus difficult to interpret regarding the potential for new antibiotics. Extreme environments may harbour many antibiotic natural products produced by a plethora of microorganisms, but solely investigate bioactivities without structural correlation to secondary metabolites is not beneficial for the discovery of potential new antibiotics.

2.2. Endophytic Actinomycetes

Endophytic actinomycetes represent a potential repository of novel bioactive compounds as there are almost 300,000 plant species, each hosting one or more types of endophyte [72]. Discovery of new compounds is however hindered by their slow growth and level of production, as it is usually significantly lower than the level of production of *Streptomyces* strains. The genus *Streptomyces* represents only 26% of all actinomycetes found in various plant tissues or in the rhizosphere. Cultured broth of 333 strains from plant roots and 137 strains from rhizospheric soil showed that isolated actinomycetes do have a potential as new sources of antimicrobials. Antimicrobial activity against *B. subtilis* KB-211 was present in 8.5% of isolates from plant roots, 9.1% against *K. rhizophila* KB-212, 5.7% against *X. campestris* pv. *oryzae* KB-88, 3.4% against *M. racemosus* KF-223, 1.7% against *C. albicans* and nothing is active against *E. coli* [73]. In general, the rhizosphere isolates are highlighting more isolates with antimicrobial activity. Overall, 20.3% of the rhizosphere isolates were active against *B. subtilis* KB-211, 17.1% were active against *K. rhizophila* KB-212, 8.9% were active against *X. campestris pv. oryzae* KB-88, 21.1% were active against *M. racemosus* KF-223, 9.8% were active against *C. albicans* and 8.1% were active against *E. coli* [73]. Bioprospecting rhizosphere isolates from the Caatinga biome (northeast Brazil) (Figure 2) [58] gave similar results as the above-mentioned example, where only 16% of isolates showed some activity against tested microorganisms. The most promising isolate was further investigated and crude ethanolic extract of *Streptomyces parvulus* D1.129 inhibited the growth of *S. aureus* and *B. subtilis* at 0.97 µg/mL. Despite the low percentage of isolates with antimicrobial properties, taxonomic difference and genome sequencing data strongly indicate the potential of these strains to produce novel bioactive compounds [58]. *Streptomyces* sp. strain RTd22—isolated from the Mexican sunflower *Tithonia diversifolia*—shows potential as a producer of novel natural products. Although analysis of the genome revealed an abundance of biosynthetic gene clusters (BGCs) not connected to known natural products, until now no novel bioactive compound was isolated from this strain [74].

2.3. Symbiotic Actinomycetes

Bacteria and insects are ubiquitously distributed and some species have naturally developed a symbiotic relationship guided through a co-evolutionary adaptation, where the bacteria are located in specialised anatomical compartments. It is known that attine ants live in a tripartite mutualism with the fungi *Leucoagaricus gongylophorus*, which provides food to the ants, and with antibiotic-producing actinomycetes. Attine ants are hypothesised to acquire actinobacteria from the soil, selecting and maintaining those species that produce useful antibiotics as a way of biocontrolling the fungal counterparts [75]. Similar findings were discovered for wasp—*Streptomyces* and termite—*Streptomyces* symbiosis. Sceliphrolactam was found to be produced by a *Streptomyces* strain isolated from a wasp exoskeleton and inhibits the growth of amphotericin B resistant *C. albicans* [76] (Figure 4). A termite-associated actinomycete, *Amycolatopsis* sp. M39, produces glycosylated polyketide macrolactams, namely the macrotermycins. They exhibit very strong activity against *S. aureus* and *B. subtilis* and a moderate activity against *C. albicans* and *S. cerevisiae* [77]. Another termite-associated natural product, natalamycin A, exhibits a very high and broad-range antifungal activity [78]. *Pseudonocardia* spp. isolated from exoskeletons from several genera of ants were found to produce gerumycins, piperazic acid-containing cyclic depsipeptides that selectively inhibit a fungal pathogen *Escovopsis* [79]. A strain from *Pseudonocardia* spp. isolated from ant colonies harboured a plasmid encoding the biosynthetic gene cluster of a rebeccamycin analogue. Rebeccamycin was originally isolated from *Lechevalieria aerocolonigenes* (*Nocardia aerocolonigenes*) [80]. Rebeccamycin and several analogues have a very pronounced antitumour activity as they inhibit the mammalian topoisomerase I [81]. *Streptomyces* sp. CLI250 isolated from a fungus-growing ant yielded an unusual peptide tryptorubin A that harbours a few unique moieties, such as the linkage between two tryptophans in *para* position to the indole nitrogen. Another unusual structural feature is the linkage between tyrosine aromatic rings to the indole nitrogen of tryptophan, which feature has not been reported in any other natural product [82].

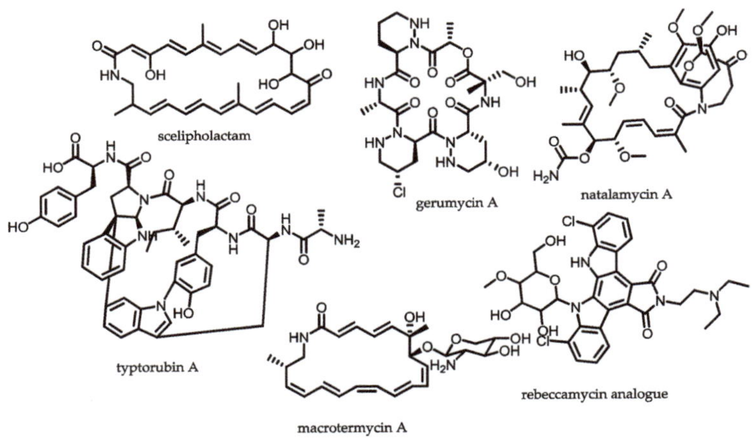

Figure 4. Novel natural products from symbiotic actinobacteria.

2.4. Conclusions

Underexplored habitats represent a largely untapped repository of compounds with unique chemical structures. Overall, new actinomycete isolates frequently exhibit moderate to strong activity against Gram-positive bacteria and fungi. Active compounds against Gram-negative bacteria remain a scarce finding in most habitats whereas the likelihood of finding anti-Gram-negative active natural products appears higher in rich, strongly diverse terrestrial communities. This is probably due to increased selective pressure as a consequence of the struggle of microbes to prevail in their fight for nutrients and territory in densely populated niches. In extreme environments, Nature itself already stamped out a large portion of the competition by making the habitat difficult to live in; therefore, it was previously assumed that the need of remaining microbes to utilise secondary metabolites is greatly decreased [83]. However, recent investigations propose that extremophile microorganisms are in fact able to produce diverse secondary metabolites [84]. The picture is admittedly incomplete since only few extremophiles have been systematically screened for the production of secondary metabolites; nevertheless findings from those that have been analysed are encouraging [84]. Access to products from extremophiles is also limited due to the lack of knowledge on how to isolate and cultivate them under laboratory conditions. As part of increased efforts to overcome this bottleneck, technologies such as the iChip platform can complement empirical improvements in isolation and fermentation of previously uncultured bacteria. Such novel microbiology-based methods are still underrepresented in most natural product discovery pipelines. It is worth taking into consideration that current discovery pipelines are more focused on exploiting new non-extreme habitats to obtain rare actinomycetes, which can be seen as a strategy to circumvent the need for extraordinary cultivation conditions. Although this approach cannot unlock access to extremophiles and uncultured microorganisms it is understandably the preferred approach implemented in many laboratories: as soon as microorganisms can be maintained under laboratory conditions, it is possible to make use of genome mining, apply genetic tools for the activation of silent biosynthetic gene clusters and scrutinise secondary metabolomes using a panel of sensitive analytical methods.

Taken together, access to the full complement of novel natural products within a single microorganism is only achievable through combination of available methods from multiple disciplines. Even without maintaining a constant supply of novel actinomycete isolates, these combined approaches can increase the availability of novel natural products. Considering that the majority of microorganisms cannot be cultured under laboratory conditions, the development of culture-independent approaches are indispensable before their metabolites can be investigated. Thus, eliminating the inherent bottleneck of cultivation is the distinguishing idea of metagenomic approaches, which are based on prior examination of biosynthetic gene clusters with *in silico* bioinformatics tools to enable heterologous expression of selected pathways originated from uncultured microorganisms.

3. Metagenomic Approach to Exploit the Uncultured Bacterial Majority

Since to date only an estimated 1% of bacteria are cultured under laboratory conditions according to 16S ribosomal RNA (rRNA) [14], novel methods are required to access the greater diversity of natural products, circumventing the limitations of traditional culture-based approaches. Metagenomics as a culture-independent approach is based on DNA recovery directly from the environment (eDNA) to gain access to the hidden reservoir of secondary metabolite encoding sequences [85]. Metagenomic approaches in combination with classical culture-based approaches already became an important means to obtain deep insights into the microbial response to contamination or bioremediation techniques [86]. For example, microbial communities were used as environmental biosensors for nitrate [87] and uranium [87,88] or oil [89] contamination by probing 16S rRNA sequencing using high-throughput screening to determine the taxonomic composition of the microbial community. Since several key taxa are indicative of a particular contaminant, this metagenomic screening enables monitoring the presence and extent of contamination in the environment [86]. Therefore, this part of the review focusses on the recent developments in applied metagenomics employed in natural product research and aims to give an account of such culture-independent methods and involved technologies for accessing secondary metabolites from actinobacteria.

3.1. The Metagenomic Screening Workflow

In general, there are two main strategies applying metagenomic methods. The older strategy is relying on functional screening of individual eDNA clones and carried out with the help of huge cosmid libraries. Clones are typically selected by phenotypic readout such as visually detectable signals, bioactivity or more advanced screening procedures such as reporters/biosensors or through specific catalytic assays. The more recent strategy is based on biosynthetic considerations, relying on DNA sequence similarity of conserved biosynthetic genes to selectively get access to potential BGCs within the environmental sample. *In silico* selection of potentially interesting BGCs and subsequent heterologous expression can be used to finally access novel compounds [90]. Figure 5 summarises the main steps in the metagenomics workflow for both methods. Environmental samples are collected from ecologically and geographically diverse environments. In the classical functional metagenomic screening approach, metagenomic libraries are created by extracting, cloning and ligating environmental DNA (eDNA) into a shuttle vector and transformed subsequently into appropriate heterologous hosts. The metagenomic library is screened either for observable phenotypes or for the presence of target DNA sequence. Positive clones are recovered from the metagenomic library and the eDNA insert is sequenced. The antibacterial pigment violacein was isolated from soil metagenomic libraries by applying the classic functional metagenomic screening pipeline [91]. In the targeted sequence-based metagenomic screening approach, crude eDNA is obtained from environmental sample and screened by polymerase chain reaction (PCR amplicons specific for sequences within BGCs. DNA sequence tags are phylogenetically organised, evaluated for biosynthetic origin, compared to reference database and biosynthetic gene clusters are reassembled *in silico*. A metagenomic library is generated from environmental samples harbouring the *in silico* reassembled BGC of interest. Environmental DNA is extracted and screened for the specific sequence of interest. For both approaches, the novel BGCs are finally assembled and modified for heterologous expression in appropriate host and the produced natural product is isolated and structurally elucidated. An example of the targeted sequence-based metagenomic screening is shown by the proteasome inhibitor landepoxin A, which was discovered by comparing ketosynthase (KS) domain sequence tags to the KS domains from known epoxyketone biosynthetic gene clusters to identify epoxyketone proteasome inhibitor derivatives from metagenomic samples [92].

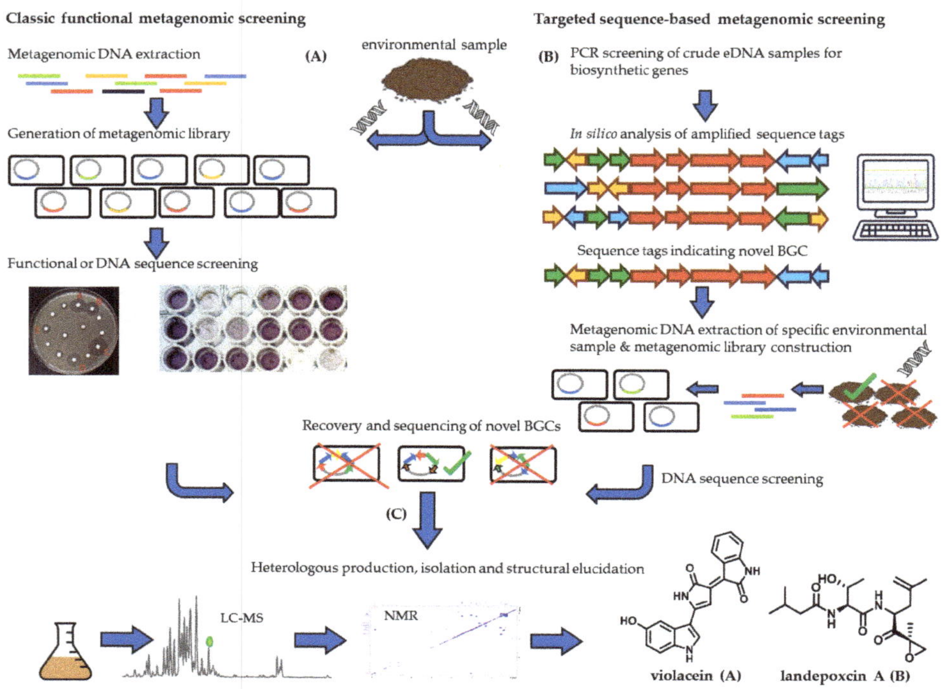

Figure 5. Scheme of metagenomic discovery pipelines adapted from Katz et al. (2016) [93]. Environmental samples are collected from ecologically and geographically diverse environments. (**A**) The classical functional metagenomic screening is based on the generation of metagenomic libraries by using appropriate heterologous hosts. Afterwards the metagenomic library is screened either for observable phenotypes or for the presence of target DNA sequence. (**B**) Crude eDNA is obtained from environmental sample and screened by PCR amplicons specific for sequences within biosynthetic machinery. DNA sequences tags are used to obtain *in silico* reassembled biosynthetic gene clusters (BGCs). A metagenomic library is generated from environmental samples harbouring the *in silico* reassembled BGC of interest. Environmental DNA (eDNA) is extracted and screened for the specific sequence of interest. (**C**) Novel BGCs are finally assembled and modified for heterologous expression in appropriate host and the produced natural product is isolated and structurally elucidated. Violacein is a natural product obtained by the approach (**A**) [91], and landepoxin is a natural product obtained by approach (**B**) [92].

One of the major obstacles in the field of metagenomics is the usage of the model host organism *E. coli* with limited heterologous expression capabilities of complex natural products and the limited insert size of the cosmid-based libraries. These limited heterologous expression capabilities—in particular for secondary metabolites with corresponding huge modular biosynthetic architecture such as polyketides, nonribosomal peptides and hybrids of these—are not surprising, since *E. coli* is in contrast to actinobacteria and myxobacteria not a prolific producer of secondary metabolites [94]. Circumventing these major obstacles will be the key to improving novel natural products output from the metagenomics approach in the future. The usage of broad-host-range shuttle vectors allows effective eDNA clone shuttling to test heterologous expression in taxonomically diverse organisms [95]. Another approach to overcome limited heterologous expression capabilities of natural products in *E. coli* is to increase the transcription and translation of exogenous DNA by changing the host organism. Compared to *Streptomyces*, *E. coli* contains only half the number of different

RNA polymerase sigma factor species. Furthermore, *Streptomyces* has been shown to express more genes deriving from metagenomic libraries than *E. coli* [96], revealing it as a potent host organism for heterologous expression. Because *E. coli* remains one of the few organisms that are featured with easy genetic manipulation as well as fast growth, additional measures than changing the host organism was taken into consideration to overcome its limitations. Overexpression of the missing sigma factor σ^{54} led to the production of oxytetracycline during heterologous expression of the previously silent oxytetracyline BGC originated from *Streptomyces rimosus* in *E. coli* [97]. This is just one powerful example demonstrating the utility of modified *E. coli* strains as heterologous platforms for metagenomic derived BGCs, indicating that even codon usage bias, starter and extender units availability are surmountable obstacles [98–100]. In particular, *E. coli* has proven to be an useful host for the heterologous expression of short biosynthetic pathways such as ribosomally synthesised and post-translationally modified peptides (RiPPs) [101], since the underlying biosynthetic machinery—the ribosome—is provided by the primary metabolism.

Even with a suitable broad host expression platform at hand, heterologous expression remains the critical part in the metagenomic workflow [102,103]. It is a common finding after reassembling of the whole BGC that most of the unmodified BGCs remain silent. To overcome this issue, several different non-targeted approaches are currently utilised such as simulation of environmental conditions through microcolony cultivation [104], co-cultivation [105] and the use of histone deacetylase (HDAC) inhibitors [106]. Targeted approaches to promote heterologous expression comprise the replacement of inactive native promoters through strong artificial or active native promoters, functional in the heterologous host either upstream of positive regulatory elements or individual biosynthetic genes. The operon structure of BGCs requires often the replacement of several promoters to enable heterologous expression; for example the anti-proliferative compounds lazarimide A and B deriving from a silent eDNA BGC, could only be expressed after yeast homologous recombination was employed to perform multiplex promoter exchange [107]. Another case study demonstrates that every gene from the eDNA derived BGC has to be controlled by a separate inducible promoter to finally produce erdasporine, a novel carboxy-indolocarbazole containing tryptophan dimer [108].

3.2. Direct Functional Metagenomic Screening

Functional metagenomic screening is mainly but not exclusively based on phenotypic readouts such as visual detection and chromatographic separation, which is attributed to production of new small molecules. Another simple strategy is testing for growth inhibition against pathogenic microorganisms in a top agar overlay assay [109]. More advanced approaches use reporter/biosensor-based screens such as metabolite-regulated expression (METREX) and substrate-induced gene expression screening (SIGEX). METREX screening can be used to screen for production of quorum-sensing inhibitors able to bind to the LuxR transcriptional activator, which induces expression of a target gene. This leads to accumulation of a green fluorescent protein (GFP) used as reporting system for positive cosmid clones inducing bacterial quorum sensing [110]. SIGEX is based on the substrate-induced expression of a catabolic gene with the similar reporting system for positive cosmid clones such as METREX [111]. Both methods can be used for discovering new genes and gene products that would not have been detected with a sequence-based approach, since no prior knowledge of the gene and gene product is necessary. However, the outcome of natural products discovered by functional metagenomics is to date rather disappointing with only few successful examples such as the antibacterial pigments violacein [91], indirubin with its isomer indigo [112], *N*-acyltyrosines [113] and the turbomycins [114] as shown in Figure 6. All of these examples show simple non-actinomycte derived secondary metabolites, synthesised by very few biosynthetic enzymes. Hence, functional metagenomic screening is obviously failing to access the complexity of actinomycete secondary metabolism.

Enzymatic activity unique or predominantly present in secondary metabolism can be used for advanced functional phenotypic screening to improve the identification of functional BGCs [115,116].

Owen et al. [117] described for example a screening approach based on the single module nonribosomal peptide synthetase (NRPS) *bps*A as reporter gene that synthesises the coloured indigoidine from L-glutamine through activation by 4-phosphopantetheinyl transferase (PPTase)-containing clones. *PPTase* genes are often located in close proximity to a BGC, however most of the BGCs are remote from their required *PPTase* gene, since most microorganisms possess only one *PPTase* gene in their genome, which is involved in secondary metabolism [118]. Wherefore indigoidine production in general is a powerful indicator to identify positive eDNA clones containing BGCs deriving from NRPS or polyketide synthase (PKS) pathway machinery. Five years after inventing this functional *bpsA* gene expression-type BGC screening method, Brady and co-workers screened a soil eDNA library hosted in *S. albus::bpsA ΔPPTase* and identified clones containing NRPS, PKS and NRPS-PKS biosynthetic gene clusters, resulting in the rediscovery of myxochelin A the biosynthesis of which is dependent on a very simple NRPS [119]. This study is a proof of concept for an innovative functional screening method albeit it suffers from a rediscovery issue based on the DNA-sequence paralleling the traditional culture-based high rediscovery rate based on the metabolic level. Furthermore, the relatively simple structure with its underlying plausible biosynthetic pathway, consisting of four biosynthetic proteins (with only one NRPS module) [120], is questioning the general applicability of this method for more complex biosynthetic pathways, since many BGCs are exceeding the typical insert size of a cosmid preventing the functional detection.

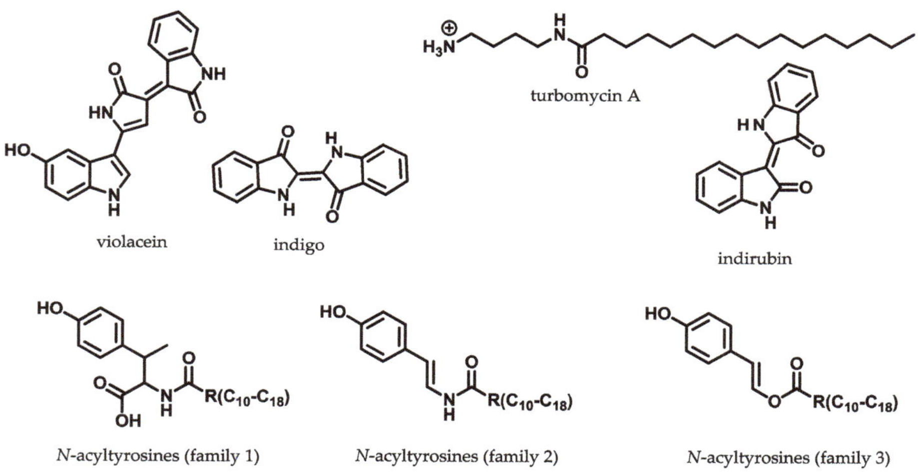

Figure 6. Example of natural products obtained by direct functional metagenomic screening.

3.3. Sequence-Based Metagenomic Discovery Efforts

In contrast to functional metagenomic screening, the identification of potential BGCs through the sequence-based metagenomics approach relies on *in silico* analysis of sequence tags representing microbial genomes for the initial screening process instead of attempting heterologous expression. The complex mixtures of PCR amplicons consisting of numerous domain fragments from potential biosynthetic gene clusters deriving from environmental samples are defining the term of Natural Product Sequence Tags (NPSTs). Unlike whole genome sequences, those datasets are consisting of much shorter DNA stretches, which hampers the identification of biosynthetic gene clusters. Therefore, conventional tools in the field of natural product research such as antiSMASH [121,122] are less suitable for analysing those datasets. Predictive tools such as Surveyor of Natural Product Diversity (eSNAPD) [123] and Natural Product Domain Search (NaPDos) [124] are based on the phylogenetic relationships of sequence tags which enables searching for both close and distant relatives in the large datasets [125]. The advantage of eSNAPD to organise NPST is the requirement of little computational power and the comparable low cost of sequencing small PCR amplicons in contrast to whole genomes. One nuisance with sequence-based metagenomic discovery is the enormous size and density of soil metagenomes in soil samples, where a single sample can contain up to 10^5 unique genomes [93]. Furthermore, metagenomic DNA libraries are sparsely populated with biosynthetic genes of interest, since only a fractional amount of each genome is devoted to secondary metabolism [126]. Therefore, gene cluster enrichment strategies can be used to simultaneously reduce the size and increase the density of biosynthetic sequence tags within eDNA libraries. Employing PCR in sequence-based approaches using degenerated PCR primers to amplify biosynthetic genes of interest was shown to be more sensitive in identifying variants of biosynthetic domain sequences with known architecture than shotgun sequencing [127]. However, these amplicon sequencing-based methods are inherently biased, with regards to detect standard gene organisation such as RiPPs, hybrid NRPS-PKS and nucleoside antibiotic sequences [101,128]. In general, modern sequence-based methods are centred on amplifying conserved biosynthetic enzymes via degenerated primers to construct metagenomic libraries rather than analyse randomly crude eDNA since the majority of DNA sequences are not associated with biosynthetic pathways. A promising conserved biosynthetic gene utilised to find novel metagenomic-based natural products is for example the KS-β [129]. Identification of tailoring and other enzymes occasionally involved in biosynthesis of natural products such as sulfotransferases [130], esterases [131], lipases [132], β-galactosidases [133] and isonitrile synthases [134] could also be utilised to obtain novel natural products such as new sulfated glycopeptide derivatives [130] or (*E*)-3-(2-isocyanovinyl)-1*H*-indole [134] from metagenomic samples. However, those specialised screening approaches of individual biosynthetic sequences are not offering a fully systematic method for accessing the complex entirety of available metagenomes, since these enzymes are also distributed in primary metabolism, and are not common for secondary metabolism, in particular esterases and lipases.

Brady and co-workers could obtain the rare tryptophan dimers hydroxysporine and reductasporine [135], the cytotoxic anthracycline arimetamycin A [136], the tetracyclic MRSA-active antibiotic tetarimycin A [137], the antibiotic cyclic lipopeptides malacidins A and B [138] and the epoxyketone proteasome inhibitors clarepoxcins A–E and landepoxcins A and B [92] with the sequence-based metagenomic approach (Figure 7). These natural products were obtained by using *Streptomyces albus* as heterologous host, underlining this strain as a gifted heterologous host in particular for actinomycete derived BGCs [139]. Recent developments of utilising different so-called broad host heterologous expression platforms for microbial natural product biosynthetic pathways—in particular various hosts from different phylogenetic background such as myxobacterial microorganisms—was described by Müller et al. [102]. The calcium-dependent antibiotics malacidins A and B were discovered by using degenerated primers targeting NRPS adenylation domains to generate amplicons from an arrayed collection of eDNA isolated from 2000 unique soil samples. The malacidins are structurally distinct in comparison to known calcium-dependent antibiotics such as daptomycin and friulimicin, which emphasises the sequence-guided metagenomic pipeline to probe for new congeners of the natural product family. In contrast, the tryptophan dimers were discovered by screening diverse soil samples using degenerate primers for the presence of unique chromopyrrolic acid synthase (CPAS) sequences. This approach demonstrates the large-scale sequence tag-based pre-filtering of diverse environmental samples as input for the metagenomic pipeline to push the discovery of novel BGCs [135]. It is also possible to use NPST libraries to recover natural product genes from metagenomic libraries to obtain through the unique sequence label (barcode) the BGC of interest and the specific sub pool of the library [125].

However, many BGCs are exceeding the typical insert size of a cosmid, and thus several overlapping DNA cosmid clones spanning the complete biosynthetic gene cluster have to be reassembled into a bacterial artificial chromosome (BAC) by the frequently used method of transformation-associated recombination (TAR). Afterwards, the reassembled BGC can be transferred from yeast into different bacterial strains for heterologous expression. Consequently, this approach is rather a heterologous expression based on a metagenomic library than an expression of metagenomic samples [140].

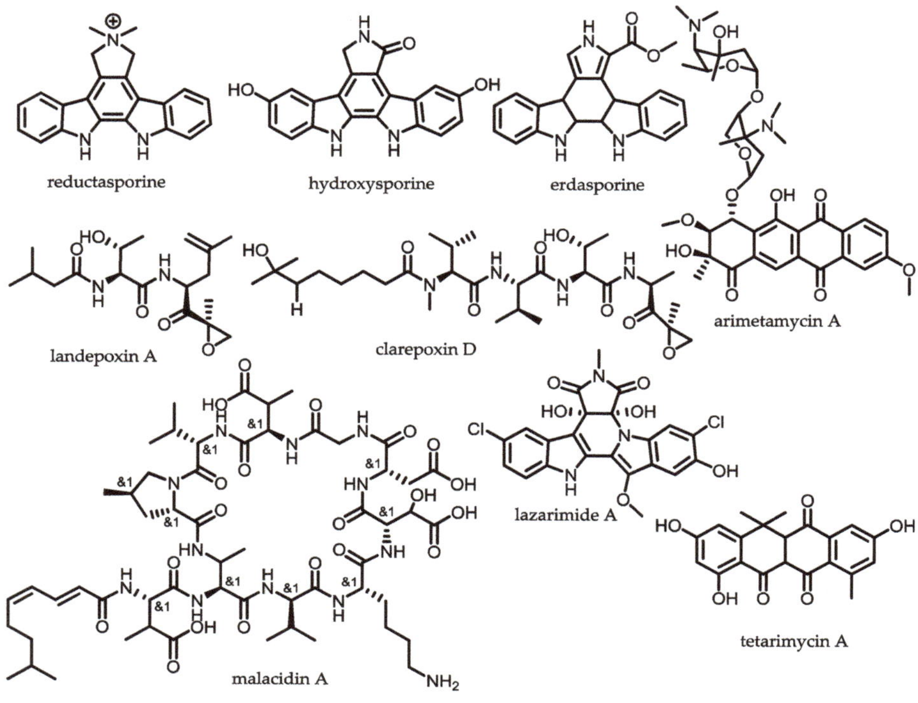

Figure 7. Example of natural products obtained by sequence-based metagenomic approaches.

3.4. Metagenomics for the Assessment of Marine Endophytes

Besides the extensively investigated ecology of soil-dwelling bacteria, the marine environment is still less investigated in terms of natural product research (see Section 2.1) [12]. The major drawback of marine natural product discovery is that cultivation of bacteria is frequently challenging and production of compounds of interest is often not possible in sufficient amounts for structure elucidation [141]. Occasionally, serendipitous discovery of alternative producers of marine natural products might enhance the production and help to characterise subsequently the underlying biosynthetic pathway. One example are the bengamides, a class of natural products that have been characterised as inhibitors of methionine aminopeptidases, emphasising the potential as anticancer compounds [142]. Previously, they have been assumed to exclusively derive from sponge *Jaspis* cf. *coraciae* until they have been found in the terrestrial myxobacterium *Myxococcus virescens* ST200611 (DSM 15898) which allowed studies to be conducted on the biosynthesis of bengamides, their heterologous expression, and the self-resistance mechanism of their producer [143,144]. However when no alternative producer can be found, metagenomic approaches are playing a crucial role to access the diversity of marine natural products by identifying biosynthetic gene clusters for the heterologous production of secondary metabolites [109]. This approach can provide information about the origin of biosynthetic gene clusters, as compounds thought to be produced by sponges, have been later on elucidated to originate from their bacterial symbionts [145]. One example are the unusual antifungal peptides microsclerodermins, which have been reported to derive exclusively from the marine genera *Microscleroderma* and *Theonella* [146,147]. Recent findings, however, revealed the terrestrial alternative producers *Jahnella* sp. MSr9139 and *Sorangium cellulosum* So ce 38 producing microsclerodermins and pedeins along with additional derivatives [148]. The genomic sequence for these two myxobacterial producers enabled to propose a conclusive biosynthetic model for the underlining pathway. Combined with recent metagenomic studies emphasising myxobacteria as putative sponge symbionts, these outcomes provide evidence that a terrestrial bacterial symbiont might be the real biosynthetic source of this "marine" natural product [149]. Considering that seawater bacteria are presumably 10-fold less represented in cultured isolates compared to soil-dwelling bacteria [150], culture-independent methods in marine environment could be expected to access a larger untapped reservoir of chemical and biological diversity of natural products than soil-dwelling bacteria. The following section briefly highlights recently conducted metagenomic case studies, which were guided by previously characterised structure or associated biological activity. The first and second example describes the biosynthetic characterisation through metagenomics of the endosymbiotic derived polytheonamides and calyculin A, the last case study highlights the recent discovery of the anti-HIV lanthipeptides, the divamides, which exemplifies the convergence of traditional functional screening and sequence-based metagenomics (Figure 8).

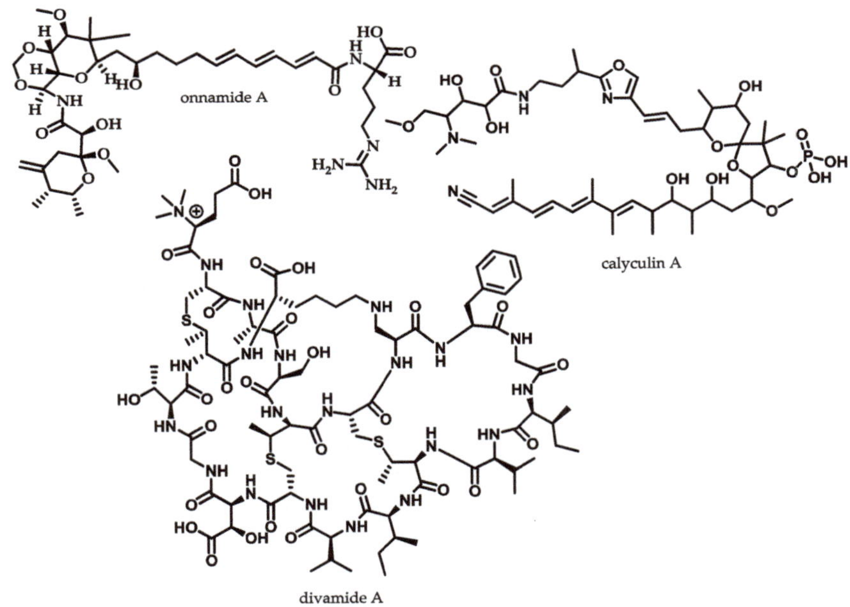

divamide A

Figure 8. Example of marine natural products with elucidated biosynthesis achieved through metagenomic approach.

The marine sponge *T. swinhoei* contains numerous uncultivated bacterial symbionts and is a prolific source of bioactive secondary metabolites [151]. One of those marine natural products with remarkable structure is the 48-residue large peptide polytheonamide, incorporating 13 nonproteinogenic amino acids. Therefore, it was assumed for years that the marine cytotoxic polytheonamide originates from a nonribosomal peptide biosynthesis, despite the gigantic size of 48-residue NRPS machinery [152]. To clarify the biosynthetic origin of the polytheonamides, a semi-nested PCR approach with designed primers specific for precursor peptide consisting of proteinogenic L-configured amino acids was conducted from *T. swinhoei* metagenomic sample [153]. The assembled sequenced amplicons revealed 11 clustered genes, with seven open reading frames forming an operon. Furthermore, several bacterial transposition elements and Shine–Dalgarno sequences were found; these findings not only confirm the ribosomal origin but also suggest that the polytheonamides are produced by bacterial endosymbionts [154]. Further studies of the sponge *T. swinhoei* involving single-cell genomics assisted by enriched bacterial fractionation, fluorescence assisted cell sorting and whole genome amplification combined with pathway specific PCR, revealed the *Entotheonella* spp. bacterial symbiont as the native producer. Similarly, *Candidatus Entotheonella* spp. was assigned as the non-cultured bacterial symbiont of *T. swinhoei*, responsible for the production of onnamides [155] (Figure 8).

Calyculin A first isolated from the marine sponge *Discodermia calyx* in 1986, is a PKS-NRPS hybrid cytotoxic natural product [156]. Since further calyculin-related natural products have been isolated from different sponges, it was assumed that associated bacterial symbionts are producing those compounds [157]. Recently the biosynthetic gene cluster was identified by a metagenomic approach comparable to the methods described for the identification of the BGCs of onnamides and polytheonamides. The PKS-NRPS hybrid origin of calyculin leads to the design of primer screening trans-acyl transferase (AT)-type KS, adenylation domains and 3-hydroxy-3-methylglutaryl-coenzyme A synthase (HMGS)-like motifs using *D. calyx* metagenomic DNA as a template. In total 250,000 clones of a metagenomic library of *D. calyx* total DNA had to be controlled applying a sophisticated pooling strategy [158], before the calyculin BGC could be assigned [159]. Furthermore, the biosynthetic gene cluster was used as a probe, employed catalysed reporter deposition-fluorescence *in situ* hybridisation (CARD-FISH) and laser microdissection was revealing a filamentous bacterium with 97% identity in the 16S rRNA sequence to the *Candidatus Entotheonella* symbiont from the *T. swinhoei* sponges.

Organic extract of the Prochloron-harbouring tunicate *Didemnum molle* (E11-036) from the Eastern Field of Papua New Guinea represented a promising hit in an anti-HIV assay. However further assay-guided fractionation and spectroscopic analysis could not reveal the complete chemical structure of the natural product responsible for the observed antiviral activity, due the very limited sample size. The natural product could at least preliminarily be determined as a novel peptide containing the modified amino acids lanthionine and *N*-trimethylglutamate, as well as the partial peptide sequence "GTTR". The acquired assembled metagenome was searched for the amino acid motif GTTR and for overrepresentation of thioether cross-linked amino acids (cysteine, threonine and serine at the C-terminal end) from which lanthipeptides derive their name [160], yielding one putative divamide BGC. Combined *in silico* analysis of the biosynthetic pathway and re-examination of the acquired NMR (nuclear magnetic resonance spectroscopy) data culminated in the structure elucidation of the divamides. Moreover, heterologous expression of different divamide BGCs originating from the symbiotic cyanobacteria *Prochloron didemni* in *E. coli* confirmed the predicted structure and the biosynthetic pathway [161]. The discovery of the divamides highlights the synergy of different strategies for the search of novel bioactive natural products whereas biological activity within an ecologically relevant system is screened first by functional screening, with the resulting chemical and metagenomic approaches following promising biology. The scarcity of isolated material is circumvented by applying NMR structure elucidation combined with metagenomics and synthetic biology to fully characterise newly discovered natural products and their underlying biosynthesis.

The last case study demonstrates the usage and value of metagenomic approaches in terms of biosynthetic gene cluster identification for the discovery of new natural products from marine endophytes. The existing examples to date have afforded mainly PKS, NRPS or PKS-NRPS hybrids, assembled by gigantic modular biosynthetic pathway machineries. The high mutual similarity of corresponding domain types, in particular KS and AT domains in PKS and condensation (C) and adenylation (A) domains in NRPS, greatly facilitates the *in silico* identification of genes encoding these gigantic microbial biosynthetic pathway machineries, as long as they follow the so-called collinearity rule [162]. In contrast, other types of biosynthetic pathways such as RiPP and nucleoside antibiotics may be much smaller and less obviously clustered than multifunctional NRPS and PKS gene clusters and thus more difficult to identify. However, in terms of heterologous expression, the smaller RiPP BGCs are much easier to express than PKS, NRPS or PKS-NRPS hybrids with the underlying advanced biosynthetic machinery. In terms of structural features, they nevertheless may be at least equally interesting. Additionally, the biosynthetic characterisation of the polytheonamides is demonstrating the use of modern single cell separation and sequencing technologies, which might accelerate the future perspective of metagenomic approaches.

3.5. Sequence Boom: Potential of Next Generation Sequencing and Single-Cell Genomics

With affordable sequencing costs for whole genome sequencing, it became obvious that genome sequencing of single cells would give new perspectives of genetic analysis. In the field of metagenomics, single-cell microorganism sequencing enables genome assembly of new phyla and grant new insights into the microbial dark matter as it can be used for the "non-culturable" bacteria [163], such as described for *T. swinhoei* above. Furthermore, single-cell genomics provide a valuable advantage by reducing the complexity of the genomic signal through the physical separation of cells or chromosomes. However, obtaining the whole genome from a single DNA molecule harboured by individually isolated cells is a technically challenging procedure. Especially the guanine-cytosine (GC) rich genomes of actinobacteria and myxobacteria, both prolific sources of natural products [164,165], are emphasising the problems of whole genome sequencing techniques relying on short reads (typically in the range of several hundred base pairs) [166]. The four major technical challenges in the field of single-cell genomics are efficient isolation of individual cells, amplification of the genome of the specific cell to circumvent the scarcity of genomic material, sequencing of the previously amplified DNA amplicons representing the genome sequence and the evaluation of acquired data [163]. Three whole genome amplification (WGA) methods are commonly applied now: pure PCR-based methods such as degenerate oligonucleotide primed PCR (DOP-PCR), isothermal based methods such as multiple displacement amplification (MDA) and hybrid methods of both such as multiple annealing and looping based amplification cycles (MALBAC) and PicoPLEX. PicoPLEX is based on an initial isothermal preamplification followed by PCR amplification of the amplicons generated during the first step. Among the WGA methods, DOP-PCR achieves higher coverage uniformity with a low physical coverage of the genome. MDA on the contrary exhibits greater genome coverage with lower coverage uniformity of the genome and the hybrid methods MALBAC and PicoPLEX are characterised by coverage and uniformity [163]. A convergent method approach of single-cell genomic sequencing based on MDA and metagenomic library screening accelerated the identification of the apratoxin A biosynthetic gene cluster, a potent cytotoxic compound [167]. Furthermore, the genomes of two chemically distinct *Entotheonella* symbionts, the producers of onnamides and polytheonamides, could be reassembled by using combined single cell isolation and MDA based whole genome amplification [155].

Nevertheless, a truly novel natural product was not discovered to date using single-cell isolation combined with MDA-based whole genome amplification. Single-cell genomics has the potential to assemble the genomes of species that are present at low frequencies in metagenomic samples [167], as well as to produce assemblies of genomes of completely uncharacterised microorganisms. In conclusion, this interesting technology is highlighting great promises and single-cell genomic sequencing might advance the field of natural product research, however the downstream limitations of the metagenomic workflow remain challenging for the discovery of novel compounds.

3.6. Conclusions and Future Considerations

Since metagenomics is a relatively new technology, further advances are necessary to overcome major inherent bottlenecks not only for functional but also for sequence-guided approaches. Brady and co-workers could uncover novel natural products such as the calcium-dependent antibiotics malacidins A and B or the epoxyketone proteasome inhibitors clarepoxcins and landepoxcins via metagenomic approaches. In the field of marine natural product research, metagenomic approaches have been used to identify the biosynthetic pathways of pharmaceutically relevant compounds. Nevertheless, both approaches could not have been successfully conducted without fundamental biosynthetic knowledge. Besides the successful discovery of novel natural products, there are still some key limitations which have to be circumvented in the future [168]. One of the main bottlenecks of metagenomics is the ability to isolate DNA from soil sample for the generation of metagenomics libraries, since samples are inhabited by a great variety of microorganisms and collecting samples containing eDNA harbouring novel biosynthetic pathways is time consuming and requires fundamental ecological background knowledge. Substantial improvement has already been achieved in terms of extraction of high molecular weight DNA and elimination of environmental inhibitors by synchronous coefficient of drag alteration (SCODA) [169,170], indirect DNA extraction through microbial cell separation and formamide treatment [171]. Another limitation arises from the heterogeneous origin of metagenomic libraries, which prevents the taxonomical correlation of biosynthetic gene cluster or bioactivity to the corresponding microorganism. Managing the size and complexity of metagenomic datasets will require for the long-term automated genome mining tools and pattern recognition based algorithms [172]. Nevertheless, these data processing obstacles might be overcome in the near future, since the field of bioinformatics is evolving rapidly; therefore, the final heterologous expression of (reassembled) BGCs remains the major limitation in the near future. Very often promising biosynthetic gene clusters functional in the natural producers remain silent in the heterologous host. Different codon usage, rare tRNAs, tightly regulated promoters, toxicity and stability are only a few of the diverse obstacles. Expressing BGCs from underrepresented bacterial taxonomic origin is still challenging, since even so-called broad host heterologous expression platforms are not always capable of producing the corresponding natural products. Therefore, further progress in understanding the biosynthetic logic of assembly lines found by metagenome mining, including ways to achieve heterologous expression, will help to overcome this inherent bottleneck.

4. Genome Mining: Current Reality and Future Promise of the Post-Genomic Era

With first whole genome sequence of the model microorganism *Streptomyces coelicolor* A3 (2) [13] published in 2002, it became evident that actually little was known about the so-called well-studied strain in terms of secondary metabolism. By that time, *S. coelicolor* was known to produce actinorhodin, methylenomycin, calcium-dependent antibiotic and undecylprodigiosin [173–176]. The last of these compounds was discovered already in 1985. Development of antiSMASH [177] in 2011, by now the most commonly used tool for automatic genomic identification and analysis of biosynthetic gene clusters [121,178], allowed the prediction of additional BGCs located in the genome of *S. coelicolor*, indicating potential for production of hitherto unseen metabolites. This section therefore gives an overview about how biosynthetic gene clusters can be accessed for the discovery of novel natural products. Besides prediction of biosynthetic gene clusters and possible read-outs of *in silico* analysis, the impact of genetically minimised strains for heterologous expression and different approaches to target silent gene clusters will be discussed in detail.

4.1. Biosynthetic Gene Cluster Prediction and Targeted Activation of BGCs

Every experimental method for the assessment of a biosynthetic gene cluster relies on *in silico* analysis of the bacterial genome. NRPS-precursors and modular polyketide synthases assembly lines can be predicted by NRPSpredictor2 [179] or the Stachelhaus code [180], by domain organisation and their predicted substrate specificity [181], respectively, both algorithms being integrated into antiSMASH. Bioinformatic prediction of not yet seen natural product structures based on biosynthetic genes is a powerful tool for follow-up analysis, such as genome–metabolome correlation. However, structural prediction of natural products solely based on *in silico* data is still not reliable. Targeted activation or knockout of BGCs requires bioinformatic analysis to prioritise the selection of the BGC. Natural products with structures from new compound families are favoured, therefore online tools that correlate biosynthetic gene clusters of NRPS and RiPPs with analytically acquired data are developed [182,183] as described in more detail in Section 5.2. BGCs are also being analysed for resistance genes located close to the biosynthetic genes that could provide information about the mode of action of the associated natural product. Predictions are available as an online tool PRISM3 [184,185] and recently developed the Antibiotic Resistant Target Seeker (ARTS) [186]. Cases of resistance genes located within the biosynthetic gene cluster are reported, for example, oxytetracycline with *otrA* gene encoding the elongation factors EF-Tu and EF-G that protect ribosomes of the producer strain [187] and griselimycin's *griR* resistance gene encoding an additional copy of the DNA polymerase III beta subunit [188]. Therefore, the prioritisation of BGCs according to the resistance genes found in the gene cluster shows great promise specifically to identify novel antibiotics. Beside these specialised self-resistance conferring genes, there are more general resistance genes located within BGCs; during the heterologous expression of bottromycins in *S. coelicolor*, the replacement of the native promoter by the strong *ermE** promoter in front of the gene encoding the respective efflux pump *botT*, showed a 20-fold increased production concentration compared to the natively expressed resistance gene [189].

4.2. Utilising the Complexity of the Biosynthetic Machinery for the Discovery of Novel Natural Products

Genetic modification of producer strains of interesting natural products is often challenging, as they are featured with several BGCs and manipulation often affects more than one metabolic pathway. In some cases, knocking out one biosynthetic gene of a known compound in a producer strain can therefore enhance the biosynthesis of other encoded natural products in the genome. The newly discovered ten pentangular polyphenols, namely amexanthomycins A–J [190], were only found after knocking out the *rifA* PKS gene responsible for rifampicin biosynthesis from *Amycolatopsis mediterranei* S699 (Figure 9).

In addition, transcription regulators encoded nearby the biosynthetic gene clusters of known natural products have a great influence on secondary metabolite expression [191–193]. Some act as activators and some as repressors, but exchange of their natural promoters with constitutive or inducible promoters can lead to overexpression or inhibition of the production of targeted BGC. For example the null mutation of the *tetR* repressor within two silent BGCs in *Streptomyces* sp. PGA64 and *S. ambofaciens* culminated in the isolation of the novel angucyclinone metabolite UVM6 [194] and the previously described kinamycins [195], whereas the production of gaburedin A was induced through the inactivation of the repressor *gbnR* in *S. venezuelae* [196]. In contrast, the discovery of stambomycins from *S. ambofaciens* was achieved by constitutive expression of a LAL family regulator, acting as activator of the stambomycin biosynthesis [197]. Exchange of the native promoter with an inducible promoter, upstream of the first biosynthetic gene in the operon, which can be tightly regulated, provides control over biosynthetic genes. For example, the polycyclic tetramate macrolactam 6-*epi*-alteramide A was obtained by introducing the strong *ermE** promoter in front of the hybrid type I PKS-NRPS operon [198] (Figure 9). Without induction no metabolite can be detected, comparable to a knockout mutant, but addition of inducer promotes metabolite overexpression [199]. For the above-mentioned promoter exchange strategies, decent understanding of operon structures is required in combination with comprehensive metabolic profiling with accurate detection systems such as high-resolution mass spectrometry to detect small concentrations and corresponding the target masses. Currently-used native, modified and synthetic promoters were summarised by Rebets et al. [191]. As promoters might be necessary for yield improvement, research groups are working on expanding promoter libraries and they already reported the successful generation of 56 synthetic promoters. Three of these promoters of different strengths are active in multiple actinobacterial strains [200] and 38 synthetic promoters ranging from weak, medium to strong promoters are active in several *Streptomyces* species [201]. This finding opens the potential for much better control of genetic regulation in the future. This is of importance as highest expression often does not go hand in hand with increase of product yield due to limitations in translational machinery or self-resistance, just to name a few.

Considering this observation, the generation of strains with minimalised genomes is also crucial to have a heterologous expression platform where gene functions can be studied in detail and metabolic changes can easily be correlated to specific manipulations in the strain. For actinomycetes several strains with minimalised genomes are available such as *S. albus* Del1, *S. albus* J1074 and *S. avermitilis* SUKA17 [202,203]. Unnecessary genomic parts, such as insertion sequence (IS) elements and secondary metabolite gene clusters, are removed to increase genetic stability and reduce the metabolic burden. *S. coelicolor* M1152 with four deleted native biosynthetic gene clusters and additional point mutations introduced into *rpoB* and *rpsL* genes enhanced transcription and translation, respectively; depending on the heterologous cluster type, the obtained strain produced 20–40 times more of the compounds under study when compared with parental *S. coelicolor* M145 [204]. After successful implementation of such platforms with minimalised genomes, they can be used for the expression of silent BGCs from various strains and heterologous expression of BGCs found with metagenomic platforms.

4.3. Silent BGC Activation by Chemical Elicitors, Ribosome Engineering and Chromatin Remodelling

Many BGCs found in genome sequence that do not have detectable natural product, due to small concentrations being produced under certain growth conditions, probably promoting negative regulation, are termed silent or cryptic biosynthetic gene clusters [205,206]. To access "silent" BGCs there are two strategies currently applied. Either they are targeted by empirical optimisation of growth conditions [206], addition of chemical elicitors [106,207], trace metal ions [208], provision of exogenous small molecules to producer strain [205], ribosome engineering [209–211] or using targeted approaches based on genome sequence [191] with some of them mainly focusing regulatory genes [193]. Empirical variation in growth conditions such as temperature, pH, co-cultivation, or addition of chemical elicitors can induce BGC expression. Natural products such as goadsporin promotes secondary metabolism and morphogenesis in *Streptomyces* [212], subinhibitory concentrations of trimethoprim activated the expression of the malleilactone BGC from *Burkholderia thailandensis*—silent under standard laboratory conditions [213]—siderophore desferrioxamine E produced by *S. griseus* stimulates growth and development of *S. tanashiensis* in co-cultivation experiments [214]. In 2016, a Canadian compound collection of 30,569 small molecules was screened for their ability to alter the pigmentation of *S. coelicolor* colonies, usually linked with production of different secondary metabolites, during growth on solid medium. Several compounds referred as ARC2 served as a potent general elicitor inducing production of cryptic metabolites [215]. Its derivative Cl-ARC was used on 50 different actinomycete strains; 10 μM Cl-ARC was added to five growth solid media per bacteria and bacteria were grown for seven days at 30 °C. *N*-butanol extracts of controls and Cl-ACR treated bacteria were subjected to comparative LC-MS analysis. At least 23% of BGCs got activated in addition to three rare secondary metabolites that showed activity against bacteria and/or against eukaryotes [216]. However, follow-up studies to show that indeed novel metabolites are produced at larger scale have yet to be reported.

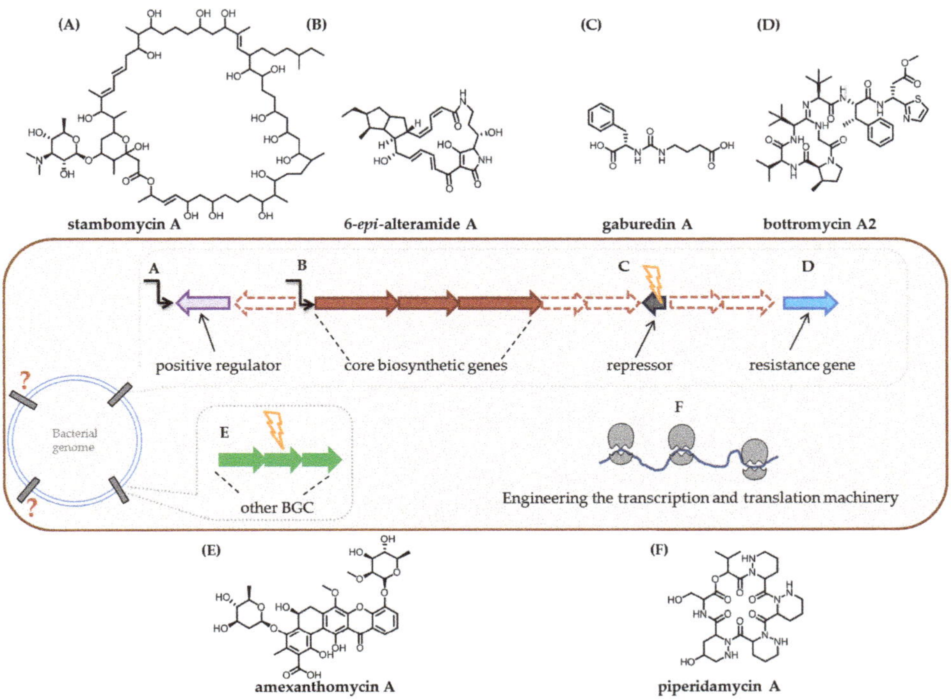

Figure 9. Generic representation of BGC consisting of core biosynthetic genes, genes for tailoring enzymes (dashed line), positive and negative regulators and self-resistance conferring genes. Strategies can be employed in the native or heterologous host. (**A**) A promoter is either inserted in front of a positive regulator or an additional copy of the positive regulator with a promoter is inserted in the genome; consequently, a higher concentration of natural product is expected. The macrolide stambomycin A was obtained through constitutive expression of an LAL family regulator [197]. (**B**) Promoter is inserted in front of the operon of biosynthetic genes enhancing transcription and production of natural product. The polycyclic tetramate macrolactam 6-*epi*-alteramide A was obtained by introducing the *ermE** promoter in front of the hybrid type I PKS-NRPS operon [198]. (**C**) Repressor gene is disrupted, production of natural product is enhanced. Inactivation of the repressor *gbn*R in *S. venezuelae* induced the production of gaburedin A [196]. (**D**) Resistance gene protects producer strain against its own natural product. During the heterologous expression of bottromycins in *S. coelicolor*, the replacement of the native promoter by the strong *ermE** promoter in front of the efflux pump *bot*T, showed a 20-fold increased production concentration compared to the natively expressed resistance gene [189]. (**E**) In some cases, knocking out the biosynthetic genes of known compounds in producer strains enhances biosynthesis of other encoded natural products in the genome. Amexanthomycin A was found after knocking out PKS *rifA* gene responsible for rifampicin biosynthesis from *Amycolatopsis mediterranei* S699 [190]. (**F**) Targeted induction of mutation in RNA polymerase and ribosomal proteins by antibiotics can cause upregulation of BGC expression. Mutations in *rpsL* and *rpoB* genes activated the silent BGC of piperidamycins from *S. mauvecolor*, culminating in the isolation of the antibacterial piperidamycin A [209].

Targeted mutation induction in RNA polymerase and ribosomal proteins by antibiotics rifamycin, streptomycin, and gentamicin can cause upregulation of BGC expression. Mutations in *rpsL* and *rpoB* genes activated silent BGC of piperidamycins from *S. mauvecolor* [209] and are shown to awaken cryptic BGCs [210] (Figure 9). Chromatin remodelling approaches were used on fungi, but interestingly these approaches were also applicable on *Streptomyces* species [106,205] as bacterial genomes are compacted presumably by nucleoid-associated proteins, RNAs and differential supercoiling leading to a comparable compaction for certain genes such as in eukaryotic organisms [217]. Cryptic BGCs could be located in tightly packed heterochromosome regions therefore chromatin remodelling or epigenetic modifications with addition of DNA methyltransferase (DNMT) and histone deacetylase (HDAC) can influence their expression [218]. Up to date, no new natural products were found applying these approaches, but differences in biosynthetic gene expression were observed even though in majority of cases expression was found reduced.

4.4. Conclusions

The majority of the above mentioned empirical methods or targeted approaches yielded examples for the identification of new natural products. Development of bioinformatic prediction tools improved genome mining and prioritisation of BGC, however there is definitely a lack of standardisation. New genomes are sequenced and prediction tools are being developed and provide useful platform although they lack the ability to perfectly predict compound structures. Even though biosynthetic gene cluster prediction algorithms are available, *in vivo* or *in vitro* confirmation are obviously the only way forward. Deciphering of biosynthetic genes is still not routine, so there remains much more to be learned about biosynthetic pathways as well as about BGC regulation in actinomycetes. However, the field now has a well-established set of genetic tools that greatly facilitate future work. Accurate analytical methods are of great importance, as comparative metabolic profiling is necessary for detection of differences between wild-type strains and their mutants.

5. Metabolomics for the Discovery of New Antibiotics Produced by Actinomycetes

In the pre-genomic era of antibiotic discovery most of the identified molecules were isolated by a classical "top-down" approach, which implies that they were produced in sufficient amount for their subsequent detection [219]. Since this conventional approach more and more failed because of frequent rediscovery of known metabolites, new approaches were required to systematically access novel antibiotics [220]. The increasing availability of genomic data for producers of natural products led to the recognition of the hidden potential of microbes to produce an even bigger variety of secondary metabolites as originally expected, and this notion changed the overall strategy for antibiotic discovery [211,221–224]. Genome mining and metagenomic analysis were added to the portfolio of methods for the prioritisation of newly isolated microbial strains, as well as for realising the previously unseen biosynthetic potential of already analysed strains [85,225]. As powerful as these *in silico* strategies are, they can only succeed when combined with analytical chemistry techniques, particularly high-resolution mass spectrometry and NMR (nuclear magnetic resonance spectroscopy) for structure elucidation of new molecules [226]. To access the full biosynthetic potential of microbes—in particular actinomycetes—as antibiotic producers, metabolomics is one of the most important tools due to its capability to systematically assess all primary and secondary metabolites in a biological sample [227,228]. The use of hyphenated techniques and the extension of the scope of already existing technologies in combination with advanced data evaluation set the stage to enlarge the window of observable metabolites. In this part of the review, we therefore highlight advances in the field of analytical chemistry including imaging mass spectrometry (IMS), liquid chromatography coupled to nuclear magnetic resonance spectroscopy (LC-NMR) and supercritical fluid chromatography (SFC). Molecules detected with IMS in the past 10 years are summarised and differences in ionisation methods compared. LC-NMR and SFC are briefly outlined, as they represent to date underexploited technologies in the field of actinomycetes compound discovery, showing already promising results in other natural product applications. Moreover, an overview is given about conceptual improvements for natural products dereplication using metabolomic data during the last 10 years.

5.1. Innovations in Analytical Instrumentation for Natural Product Discovery

5.1.1. Imaging Mass Spectrometry (IMS)

The introduction of ultra-high performance liquid chromatography (UHPLC) in combination with high-resolution MS detectors such as time of flight mass spectrometry (TOF-MS) remains one of the breakthroughs for natural product discovery [229]. This combination has since provided a rapid and robust analysis of the chemical composition of microbial extracts [16,230]. One of the main limitations of this approach comprises the use of a mixture of cells in different growth states for preparation of extracts for LC-MS measurement, wherefore chemotypes cannot be connected to phenotypes [231]. Among the Gram-positive bacteria, actinobacteria show the greatest morphological differentiation. They are able to form complex structures such as spore chain, sporangia or substrate mycelia with long-branching hyphae [232]. The impact of specific phenotypes on the antibiotics production was strikingly shown by Onaka et al. [233] for *Streptomyces endus* S-522: mycolic acid-containing bacteria such as *Tsukamurella pulmonis* stimulate the antibiotic production of other actinomycetes when grown together on agar-plates. Following this co-cultivation approach, Onaka et al. [233] were able to isolate alchivemycin A, an antibiotic unknown before. Modern mass spectrometry ionisation techniques such as nanospray desorption electrospray ionisation (NanoDESI), matrix-assisted laser desorption electrospray ionisation (MALDI-TOF) and secondary ion mass spectrometry (SIMS) imaging opened a way to further investigate the precise mechanism of these cell-to-cell contact interactions [234,235]. Even though IMS is nowadays mostly used for discovery of disease-related biomarkers, it has also proven useful for the detection of natural products. Figure 10 gives an overview of anti-infectives produced by actinomycetes detected by imaging mass spectrometry in the past 10 years.

Interestingly MALDI was the ionisation method of choice in most of the experiments, even though utilisation of a matrix for covering the sample and to adsorb laser energy comes with more time-consuming sample preparation [235]. A reason for this might be that MALDI is currently the most widely accessible IMS technique and provides soft ionisation. This, in turn enables high-resolution mapping of ions within a sample without destroying sensitive molecules [235,236]. NanoDESI belongs to the so-called soft ionisation methods as well, but is a rather new technique compared to MALDI. Desorption electrospray ionisation (DESI) is based on a two capillary system building a small bridge of charged solvent on the sample surface that absorbs the analytes and directs them into the atmospheric inlet of a mass spectrometer [236]. Compared to the quite robust MALDI, DESI is accompanied by great technical challenges since obtaining an ideal analyte signal relies on a big variety of geometrical and instrumental parameters [237]. Even though SIMS is the oldest IMS techniques out of the three methods compared here, it only plays a subordinate role for natural product discovery. This is very likely caused by the fact that a high degree of optimisation is needed to record useful data. Furthermore, measurements of intact natural products is often not possible because of the requirement for a direct ion beam that is focused on the sample surface [235].

How IMS can be successfully implemented for natural product discovery was shown by Kersten et al. [238] through their application of a "peptidogenomic" workflow. MALDI-TOF MS and subsequent MS^n sequence tagging was used to match the mass spectrometry data in an iterative approach to the genome-derived peptide structures generated with NRP prediction tools. Limiting the m/z range to 1500–5000 Da, they were able to detect nine novel lassopeptides from seven genome-sequenced *Streptomyces* strains, as well as the already known stendomycin produced by *S. hygroscopicus* ATCC 53653 [238].

IMS has proven its capability to detect various classes of known natural products produced by actinomycetes. It is an important tool for the direct observation of interspecies interactions and can be used for the detection of novel antibiotics. Nevertheless, most publications reporting the use of IMS focused on known metabolites and even though unassigned masses have been detected, the corresponding compounds still await their isolation and structure elucidation.

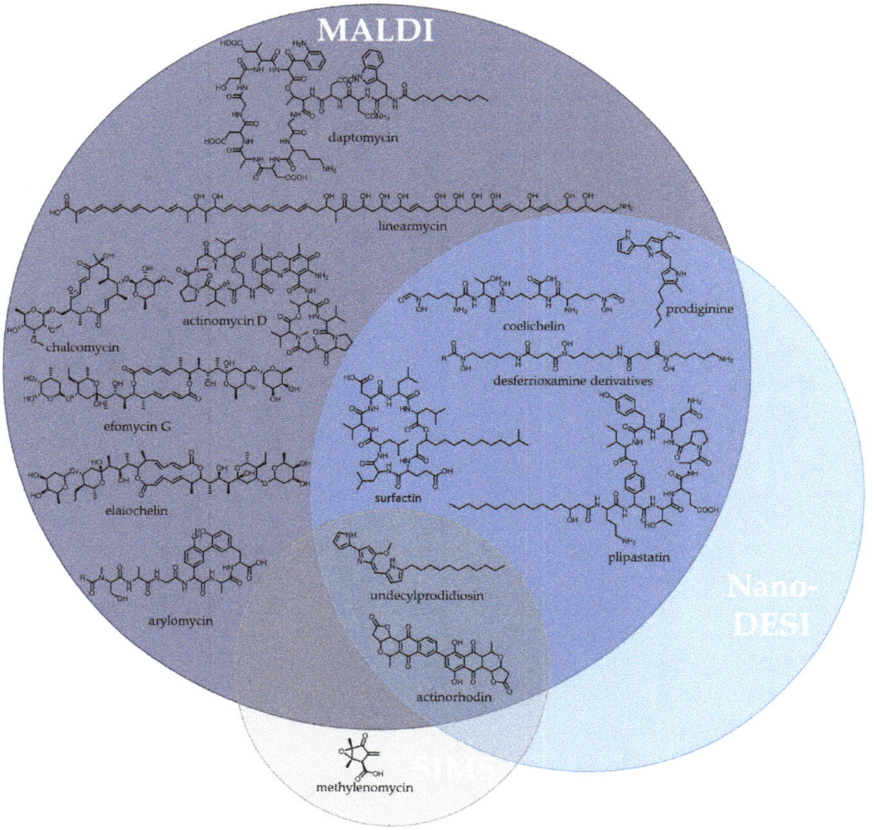

Figure 10. Natural products produced by actinomycetes detected and annotated by IMS. Fifteen known bioactive compounds were identified in different co-cultivation experiments. Only methylenomycin [239] exclusively had been detected using SIMS, whereas all molecules identified with Nano-DESI were also detectable by MALDI-TOF. Undecylprodigiosin [239,240] and actinorhodin [239,240] could be identified by all of the three methods. Prodiginine [240], coelichelin [240], different desferrioxamine derivatives [240], surfactin and plipastatin [241,242] could be assigned by NanoDESI and MALDI-TOF. Chalcomycin [243], daptomycin [243], actinomycin D [244], efomycin G [244], elaiochelin [244], linearmycin [243,245] and arylomycin [231] were only detected by MALDI-TOF.

5.1.2. Liquid Chromatography Coupled to Nuclear Magnetic Resonance Spectroscopy (LC-NMR)

MS and NMR are sharing the first place of most important technologies for identification and structure elucidation of new natural products [246]. Whilst NMR is commonly used for pre-purified samples, its hyphenation with liquid chromatography expands its application towards the screening stage in natural products discovery. Compared to the combination of liquid chromatography with mass spectrometry (LC-MS), NMR is limited by lower sensitivity but offers the possibility to provide important structural information on-line that are not accessible by MS. It was already proposed for many years, that a combination of NMR and LC (hence LC-NMR) could make structural information available earlier in the discovery workflow compared to classical procedures utilising various steps of fractionation, purification and subsequent structure elucidation using different spectroscopic methods (whereas NMR is usually the final step) [247]. Nevertheless, only the implementation of flow-through probes such as SPE-NMR or CapNMR and technical advantages in NMR spectrometer field strength in the last decades finally could pave the way into the natural products labs for LC-NMR technology [248]. LC-NMR has already proven its value for dereplication of plant extracts for various compound classes such as alkaloids [249,250], phenolic compounds [251,252], isoflavones [253], flavonoids [254] and lignans [255]. Johansen et al. [256] described in 2011 the characterisation of 15 compounds, of which four have not been previously identified, in the extract of the safflower *Carthamus oxyacantha*. They used a combined HPLC-PDA-HRMS-SPE-NMR approach allowing the preselection of peaks observed by PDA-HRMS on a SPE cartridge for subsequent NMR analysis. A following microfractionation in 96-well plates for bioactivity assays as described by Lang et al. [257] would complete a full automated screening process with maximum information about the composition of extracts. Consequently, Lang et al. classified LC-NMR as an evolving trend in the dereplication of fungal extracts.

For bacterial extracts, identification and structure elucidation of six linear peptides (macyranones A–F) from a myxobacterial extract was performed by Keller et al [258] using LC-SPE-NMR-MS techniques in 2015. Furthermore, Lin et al. [259] successfully used a LC-MS-NMR platform for the identification of four known compounds from a single LC injection of a cyanobacterial extract and additionally identified one new bioactive compound in the extract. With regards to actinobacteria LC-NMR has apparently been used only for structure elucidation of four new polyketides (granaticin C and metenaticin A–C) by Pham et al. [260] from a *Streptomyces violaceoruber* extract back in 2005. The finding that LC-NMR did not seem to be used frequently for the isolation and direct structure elucidation of natural products from microbial extracts in the past 10 years may be explained at least in part by the fact that most compounds produced in sufficient amount for identification by NMR—notably being much less sensitive than MS—are already known and for that matter not a target for structure elucidation. Nevertheless, for the investigation of newly isolated strains from new genera, species and families that are very likely to produce chemical diverse unknown metabolites, LC-NMR should be considered a feasible screening strategy [15].

5.1.3. Super Critical Fluid Chromatography (SFC)

Liquid and gas chromatography are irreplaceable instruments for the separation of highly complex bacterial crude extracts [261–263]. Another evolving analytical separation technique that has already proven its impact in phytochemistry is supercritical fluid chromatography (SFC) [264]. SFC uses supercritical fluids, most commonly CO_2, as mobile phase. In this stage, the fluids uniquely combine two of the most desirable features of a mobile phase: high dissolving capabilities but densities comparable to a liquid. As orthogonal method to liquid chromatography it offers the possibility to cope with metabolites not separable by normal phase or reversed phase liquid chromatography [264–267]. The use of co-solvents such as alcohols extends the spectrum of compound mixtures accessible by SFC. Almost all stationary phases used in HPLC are also feasible for SFC and diverse detectors such as diode array detector (DAD), evaporative light scattering detector (ELSD), MS, DC or charged aerosol detector (CAD) can be fitted to the instrument [268,269]. For plant extracts SFC has already been used for the isolation of nonpolar compounds such as terpenes [270] or flavonoids [271] but it also has proven to be a useful tool for the isolation of more polar natural products such as carsonic acid from rosemary extracts [272] and valerenic acid from *Valeriana officinalis* [273]. Notwithstanding, to our knowledge SFC has never been used for the discovery of new antibiotics from actinomycetes. In consideration of the wide field of natural products it has been used, it certainly will not take long until first research groups will successfully employ SFC for the separation of microbial natural products.

5.2. Dereplication Using Metabolomic Data

Metabolomics is the systematic acquisition of small molecules produced by a biological system at a certain point in time. Increasing resolution of the data acquired by state-of-the-art technologies produces an expanding volume of data hereby [274]. To evaluate these data in the best way, dereplication using bioinformatics tools is indispensable [275]. First introduced by Beutler et al. [276] back in 1990 as "a process of quickly identifying known chemotypes" used for the identification of compounds from bacterial extracts responsible for a biological activity in a simple phorboldibutyrate (PDBu) receptor binding assay, the term "dereplication" today covers various strategies and different fields of research from pharmacology, chemistry, plant sciences to biotechnology and food science [226]. One of the main issues in the field of natural product antibiotic discovery is the high rediscovery rate, which can only be limited by careful dereplication. Chanana et al. [277] made the statement, that it is more likely to identify a known metabolite from actinobacteria than finding a new one which is certainly true as many academic and industrial groups have worked on actinomycetes for decades. This group used comprehensive principal component analysis (PCA) for the prioritisation of most interesting bacterial strains and molecules and were therefore able to isolate two novel natural products from an *Actinomadura* sp. (strain WMMB-449) and patented another one [277]. For the prioritisation of bacterial extracts showing promising antibacterial activities it is crucial to sort out extracts containing known metabolites possibly responsible for the observed activity [278]. In-house databases—containing as many chromatographic data from known analytes as possible—provide the advantage of being perfectly matched to the conditions used for the sample of interest, but are difficult to keep up-to-date due to the requirement of continuous manual curation. Commercially available libraries such as Dictionary of Natural Products [279], AntiBase [280], MarinLit [281] and MS databases such as MassBank [282], Metlin [283], mzCloud [284] and ReSpect [285] certainly are useful sources, but are either not freely available, do not contain MS data or do not allow customised use of its reference library. In addition, connectivity between these databases is rather poor, leading to an unpredictable degree of overlap. Dereplication by database search is therefore a cumbersome and time-consuming procedure and last but not least an expensive one [286]. A better connection between already existing libraries, lowering the access barrier and a more intense data exchange between groups working in the field of natural products is strongly required. An important first step in the direction of better data exchange is Global Natural Products Social molecular networking (GNPS), an open-access knowledge platform for sharing tandem mass spectrometry data [287]. In the course of only a few years, GNPS has already proven its high impact for the natural product community. Crüsemann et al. [288] for example used GNPS for the analysis of 146 marine *Salinispora* and *Streptomyces* strains. They were able to identify 15 molecular families of diverse natural products and showed that GNPS is in principle ready for increased throughput screening thereby [288]. However, the simple comparison of MS/MS spectra with spectra deposited in the GNPS is not yet sufficient for a comprehensive analysis of secondary metabolome data. Algorithms such as DEREPLICATOR [289] or the just recently published VarQuest [290] were designed to expand the library search to identify variants of known peptidic natural products. The complete workflow from genome sequence to the putative biosynthetic gene cluster and its product with the associated spectra would ideally be done in a one-step procedure for high-throughput data analysis. Several tools have taken on this challenge and specifically address the connection of genes with secondary metabolites. Table 1 is a comparison of six different bioinformatic software packages comprising a structure prediction tool for chemical products derived from biosynthetic gene clusters: PRISM 3, SeMPI, antiSMASH 4.0, Pep2Path, RiPPquest and NRPquest. As some of them are designed for much broader analysis of genomic data, we would like to point out that they are henceforth just compared concerning structure prediction. Tools with more than one software release were summarised with functions available in the newest version to date.

Table 1. Comparison of six different bioinformatic tools for structure prediction of chemical products derived from biosynthetic gene clusters: PRISM 3, SeMPI, antiSMASH 4.0, Pep2Path, RiPPquest and NRPquest. Tools with more than one software release were summarised with functions available in the newest version to date.

Acronym	Input	Output	Type of BGC	Highlights & Limitations	Accessibility	Source
PRISM 3	Genome sequence	Gene clusters	NRPS, PKS	Tailoring reactions implemented [1]	Open-source web application	[184,185]
SeMPI	Genome sequence (*also raw DNA code*)	Domains and 10 best matching compounds	Type I PKS	Generation of the non-modified PKS products [1]	Open-source web application	[291]
antiSMASH 4.0	Genome sequence	NRPS/PKS domains, chemical structure prediction, Cluster Blast	NRPS, PKS, RiPPs terpenes	Terpene prediction Trans AT PKS domain alignments [1]	Open-source webserver	[121,122,292]
Pep2Path	Genome sequence mass shifts or amino acid-sequence	Tag-BGC alignment and scoring	NRPS, RiPPs	Automatic identification of BGC corresponding to mass shift sequence or amino acid-sequence tags [2]	Open-source application	[293]
RiPPquest	Genome sequence MS/MS dataset	Peptide-spectrum match (*p*-value)	RiPPs	Molecular network analysis using input from various MS/MS datasets possible [3]	Open-source application, (implemented in GNPS)	[294]
NRPquest	Genome sequence MS/MS dataset	Spectral network of matching peptides with *p*-values	NRPs	Molecular network analysis of various MS/MS datasets possible [4]	Open-source application	[183]

[1] No matching with experimental MS/MS data; [2] Limited to peptide natural products; [3] Limited to RiPPs; [4] Limited to NRPs.

AntiSMASH 4.0 presumably is one of the prediction tools that offers the broadest range of *in silico* analysis. Biosynthetic gene clusters are predicted using hidden Markov models and manually curated BLAST databases for identification of biosynthetic domains involved in secondary metabolite production. Subsequently the domains are arranged into clusters and chemical structures of the products are predicted. The hits are finally compared with a library of known natural products [122,295]. PRISM 3 follows a similar concept. However recent software releases of AntiSMASH were more focused on improving the genomic data evaluation and accessibility of prediction for additional cluster types such as terpenes [121,122,292]. Therefore, PRISM 3 remains the most promising bioinformatics tool for prediction of tailoring reactions on NRPS and PKS products [184,185]. Other bioinformatic tools such as Pep2Path, RiPPQuest and NRPQuest are limited to peptidic natural products [183,293,294]. Peptidic natural products (PNPs) however feature characteristic MS/MS-fragmentation patterns and usually reliable ionisation behaviour [293].

MS/MS metabolomic data can be used to complement the theoretical *in silico* prediction of chemical products produced by biosynthetic gene clusters. Thus Liu et al. [296] were able to identify the stenothricin biosynthetic gene cluster in *Streptomyces roseosporus* in 2014. Peptidogenomics in this manner is a powerful tool for full-automated high-throughput screening for bacterial extracts using both experimental and *in silico* data. Another elegant way to correlate natural products to their corresponding BGCs is using the tools mentioned above to predict specific physicochemical properties. The macrolactam salinilactam isolated from *Salinospora tropica* could be identified in a crude extract by searching for characteristic UV absorption of polyenes [297]. Since detection by UV absorption is not very specific when applied alone, rediscovery rate with such an approach is high. Therefore, MS-based methods using the identification of chlorine, bromine, fluorine and sulphur in a molecule by their characteristic isotopic pattern in combination with the exact molecular mass are more specific for the identification of secondary metabolites. In particular, this method has proven to be useful for the identification of related compounds as exemplified by the isolation of the thioholgamides from a crude extract of *Streptomyces malaysiense* MUSC 136 57. A combination of search for BGCs related to the one of thioviridamide and prioritising of potential products by their characteristic isotopic pattern revealed these natural products with promising cytotoxic activity [298]. Additionally, feeding experiments of isotopically labelled precursors are useful to narrow down potential products of the BGC present in the crude extract and provide further information about the biosynthesis of the secondary metabolite.

5.3. Conclusions

The widespread availability of high-resolution mass spectrometry in the past two decades has been a milestone for analytical chemistry not only for actinobacterial extracts. Especially in combination with HPLC this technology provides the opportunity for rapid and robust analysis of the bacterial metabolome. This advance increased the quality as well as the quantity of data acquired on a daily basis. The focus of metabolomics-driven microbial natural product research therefore was on extending the scope of molecules accessible to high resolution MS and subsequent data analysis including dereplication. IMS, LC-NMR and SFC are recent technical improvements useful for the characterisation of bacterial extracts. Although most of the modern analytical systems are commercially available, the application of most of the presented techniques was focused on one specific research field such as phytochemistry for SFC. Mass spectrometry and NMR still form the stable foundation for metabolomics of microorganisms. Their sphere of influence has already been extended by hyphenated techniques such as LC-NMR and will be certainly further increased by the combination with underrepresented techniques such as SFC in the coming years. The rising quality and quantity of metabolomic data obtained with hyphenated techniques in the past decades increased the importance of bioinformatics tools to organise and prioritise the data. There are various tools published, that either can be used for library matching or even for structure formula prediction. One big issue in natural product research is data exchange within the different research groups to hinder rediscovery of already known compounds. The establishment of the GNPS that allows sharing MS^2-data of natural products and other raw data is a good start towards a better exchange within the community. Nevertheless, as much as technical improvements can only unfold their full potential when combined with bioinformatic tools, they are more powerful, when genetic information is included in the search for new antibiotics from actinobacteria.

6. Summary and Conclusions

Actinomycetes are still a prolific source for natural products with intriguing bioactivities and inspiring chemical characteristics as exemplified by the new antimicrobial agents discovered in the past 10 years [164]. Considering that the majority of microbial secondary metabolite producers are not culturable under laboratory conditions yet, the potential for discovery of novel actinomycete metabolites is promising and far from exhausted [14]. However, as actinomycete research has been extensively performed in the past by academic and industrial groups, chances for the identification of novel chemistry may be higher in currently less studied microbial resources such as cyanobacteria or myxobacteria [299,300]. Natural product research is fundamentally dependent on the interplay of different disciplines of natural sciences. Particularly, microbiology for the isolation of new secondary metabolite producers, molecular biology for the characterisation of the biosynthetic machinery and analytical chemistry for identification and isolation of natural products form the basis for the characterisation of novel compounds from actinobacteria (Figure 11).

In the past years, focus on natural product discovery from actinomycetes shifted from the extensively investigated soil-dwelling isolates towards underexplored habitats of rare actinomycetes from unusual ecosystems. This strategy has been shown to give an impact on the discovery platform for novel compounds with promising bioactivities despite the fact that a large amount of the reservoir of habitats still awaits exploration [301]. The full potential of the natural product "treasure trove" of environmental soil samples from exotic or conventional habitats is still underexplored, since present isolation procedures fail to access the uncultured microbial majority. Nevertheless, the "genomic revolution" has revealed that actinobacteria carry more biosynthetic gene clusters in their genome than secondary metabolites found under standard laboratory conditions. This finding shows that harvesting novel strains just for few compounds understates their immense potential. The simulation of environmental conditions to mimic essential aspects of the environment extended the scope of microorganisms culturable under laboratory conditions [35]. Although technologies such as the iChip platform have shown great potential for increasing throughput and improving the isolation of previously uncultured bacteria, novel microbiology-based methods are still underrepresented in the natural product discovery pipeline. Genomics as a field that has seen tremendous development over the past 10 years, in contrast has rapidly triggered the establishment of several innovative methods to explore the diversity and distributions of BGCs encoding natural products on the genetic level [302]. To overcome obstacles of conventional isolation procedures and access the uncultured microbial majority of rare actinomycetes, metagenomic approaches have been developed. With the growing underlying in-depth knowledge of biosynthetic gene organisation, affordable whole genome sequencing technologies and prediction methods of the resulting natural product structure, "silent" BGCs can be leveraged to yield novel compounds. BGCs from genetically non-manipulable or uncultured actinomycetes can be accessed via heterologous expression through capturing the BGC and refactoring the gene arrangement. The combination of methods from synthetic biology, biotechnology and genetics is thus sets to expand Nature's chemical diversity through production of "unnatural" natural products and naturally expression of downregulated BGCs [303,304]. Despite the enormous impact of genomics, we must urgently remind ourselves that all sequence-based developments need to be supported by microbiology and analytical chemistry for production and isolation of compounds. In the field of analytical chemistry, advances have been made for the development of bioinformatic tools for an easier and faster dereplication. Furthermore, applied analytical chemistry in the field of natural product research is up to date focused on retro-biosynthetic logic to correlate secondary metabolites to their BGCs. Technologically, hyphenated techniques such as LC-NMR and separation techniques such as SFC have entered the stage more recently for the discovery of novel antibiotics but are not fully utilised for the actinobacterial natural product research yet [247]. Furthermore, imaging mass spectrometry has proven its ability to directly provide deep insights in the connection between phenotypic traits and secondary metabolism, but upscaling remains a formidable challenge [242]. On this occasion, advances in microbiological cultivation strategies are required to utilise the full potential of this method for the discovery of antibiotics. This is just one of many examples where improvements in one of the three key disciplines in natural product research will have synergistic effects when efficiently combined with other methods. Only the combination of technologies and exchange of knowledge between microbiology, molecular biology and analytical chemistry will help to not only scratch the surface of the untapped natural product reservoir but comprehensively exploit its full potential.

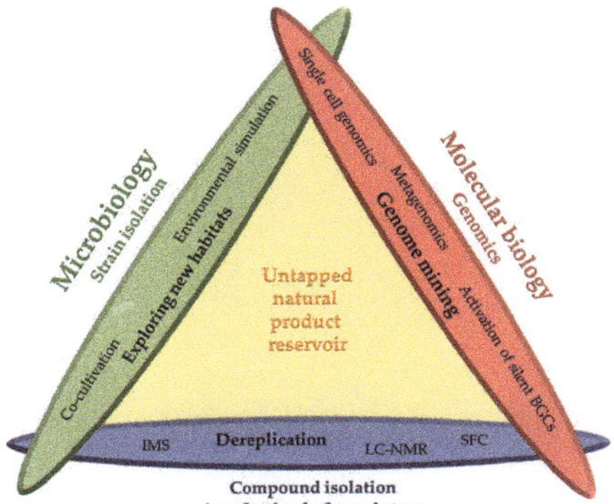

Figure 11. Interplay of disciplines for the exploitation of the untapped natural product reservoir with major developments in the past 10 years covered by this review.

Funding: This research received no external funding.

Acknowledgments: We are thankful to Daniel Krug, Chengzhang Fu and Tadeja Lukezic for their scientific advice. J. Hug was supported by a PhD fellowship of the Boehringer Ingelheim Fonds.

Conflicts of Interest: The authors declare no conflict of interest.

References

1. World Health Organization. Global Antimicrobial Resistance Surveillance System (GLASS) Report: Early Implementation. Available online: http://apps.who.int/iris/bitstream/handle/10665/259744/9789241513449-eng.pdf;jsessionid=97E045EBFE3B29F286CBA3EE8360F0D1?sequence=1 (accessed on 31 January 2018).
2. Aminov, R.I. A brief history of the antibiotic era: Lessons learned and challenges for the future. *Front. Microbiol.* **2010**, *1*. [CrossRef] [PubMed]
3. De Lima Procópio, R.E.; da Silva, I.R.; Martins, M.K.; de Azevedo, J.L.; de Araújo, J.M. Antibiotics produced by *Streptomyces*. *Braz. J. Infect. Dis.* **2012**, *16*, 466–471. [CrossRef] [PubMed]
4. Li, J.W.-H.; Vederas, J.C. Drug discovery and natural products: End of an era or an endless frontier? *Science* **2009**, *325*, 161–165. [CrossRef] [PubMed]
5. Luzhetskyy, A.; Pelzer, S.; Bechthold, A. The future of natural products as a source of new antibiotics. *Curr. Opin. Investig. Drugs* **2007**, *8*, 608–613. [PubMed]
6. Thumar, J.T.; Dhulia, K.; Singh, S.P. Isolation and partial purification of an antimicrobial agent from halotolerant alkaliphilic *Streptomyces aburaviensis* strain Kut-8. *World J. Microbiol. Biotechnol.* **2010**, *26*, 2081–2087. [CrossRef]
7. Hozzein, W.N.; Rabie, W.; Ali, M.I.A. Screening the Egyptian desert actinomycetes as candidates for new antimicrobial compounds and identification of a new desert *Streptomyces* strain. *Afr. J. Biotechnol.* **2011**, *10*, 2295–2301.
8. Tiwari, K.; Gupta, R.K. *Studies in Natural Products Chemistry. Bioactive Metabolites from Rare Actinomycetes*; Elsevier: New York, NY, USA, 2014; Chapter 14.
9. Bister, B.; Bischoff, D.; Strobele, M.; Riedlinger, J.; Reicke, A.; Wolter, F.; Bull, A.T.; Zahner, H.; Fiedler, H.P.; Sussmuth, R.D. Abyssomicin C—A polycyclic antibiotic from a marine *Verrucosispora* strain as an inhibitor of the *p*-aminobenzoic acid/tetrahydrofolate biosynthesis pathway. *Angew. Chem. Int. Ed. Engl.* **2004**, *43*, 2574–2576. [CrossRef] [PubMed]
10. Fiedler, H.-P.; Bruntner, C.; Riedlinger, J.; Bull, A.T.; Knutsen, G.; Goodfellow, M.; Jones, A.; Maldonado, L.; Pathom-Aree, W.; Beil, W.; et al. Proximicin A, B and C, novel aminofuran antibiotic and anticancer compounds isolated from marine strains of the actinomycete *Verrucosispora*. *J. Antibiot.* **2008**, *61*, 158–163. [CrossRef] [PubMed]
11. Genilloud, O.; Gonzalez, I.; Salazar, O.; Martin, J.; Tormo, J.R.; Vicente, F. Current approaches to exploit actinomycetes as a source of novel natural products. *J. Ind. Microbiol. Biotechnol.* **2011**, *38*, 375–389. [CrossRef] [PubMed]
12. Subramani, R.; Aalbersberg, W. Culturable rare Actinomycetes: Diversity, isolation and marine natural product discovery. *Appl. Microbiol. Biotechnol.* **2013**, *97*, 9291–9321. [CrossRef] [PubMed]
13. Bentley, S.D.; Chater, K.F.; Cerdeño-Tárraga, A.-M.; Challis, G.L.; Thomson, N.R.; James, K.D.; Harris, D.E.; Quail, M.A.; Kieser, H.; Harper, D.; et al. Complete genome sequence of the model actinomycete *Streptomyces coelicolor* A3(2). *Nature* **2002**, *417*, 141–147. [CrossRef] [PubMed]
14. Vartoukian, S.R.; Palmer, R.M.; Wade, W.G. Strategies for culture of "unculturable" bacteria. *FEMS Microbiol. Lett.* **2010**, *309*, 1–7. [CrossRef] [PubMed]
15. Hoffmann, T.; Krug, D.; Bozkurt, N.; Duddela, S.; Jansen, R.; Garcia, R.; Gerth, K.; Steinmetz, H.; Müller, R. Correlating chemical diversity with taxonomic distance for discovery of natural products in myxobacteria. *Nat. Commun.* **2018**, *9*. [CrossRef] [PubMed]
16. Krug, D.; Müller, R. Secondary metabolomics: The impact of mass spectrometry-based approaches on the discovery and characterization of microbial natural products. *Nat. Prod. Rep.* **2014**, *31*, 768–783. [CrossRef] [PubMed]
17. Wenzel, S.C.; Müller, R. The impact of genomics on the exploitation of the myxobacterial secondary metabolome. *Nat. Prod. Rep.* **2009**, *26*, 1385–1407. [CrossRef] [PubMed]

18. Bode, H.B.; Müller, R. The impact of bacterial genomics on natural product research. *Angew. Chem. Int. Ed. Engl.* **2005**, *44*, 6828–6846. [CrossRef] [PubMed]
19. Challinor, V.L.; Bode, H.B. Bioactive natural products from novel microbial sources. *Ann. N. Y. Acad. Sci.* **2015**, *1354*, 82–97. [CrossRef] [PubMed]
20. Lam, K.S. Discovery of novel metabolites from marine actinomycetes. *Curr. Opin. Microbiol.* **2006**, *9*, 245–251. [CrossRef] [PubMed]
21. Mahajan, G.B.; Balachandran, L. Sources of antibiotics: Hot springs. *Biochem. Pharmacol.* **2017**, *134*, 35–41. [CrossRef] [PubMed]
22. Passari, A.K.; Mishra, V.K.; Saikia, R.; Gupta, V.K.; Singh, B.P. Isolation, abundance and phylogenetic affiliation of endophytic actinomycetes associated with medicinal plants and screening for their *in vitro* antimicrobial biosynthetic potential. *Front. Microbiol.* **2015**, *6*. [CrossRef] [PubMed]
23. Blunt, J.W.; Carroll, A.R.; Copp, B.R.; Davis, R.A.; Keyzers, R.A.; Prinsep, M.R. Marine natural products. *Nat. Prod. Rep.* **2018**, *35*, 8–53. [CrossRef] [PubMed]
24. Tiwari, K.; Gupta, R.K. Diversity and isolation of rare actinomycetes: An overview. *Crit. Rev. Microbiol.* **2013**, *39*, 256–294. [CrossRef] [PubMed]
25. Kurtboke, D.I.; French, J.R.J. Use of phage battery to investigate the actinofloral layers of termite gut microflora. *J. Appl. Microbiol.* **2007**, *103*, 722–734. [CrossRef] [PubMed]
26. Rateb, M.E.; Houssen, W.E.; Harrison, W.T.A.; Deng, H.; Okoro, C.K.; Asenjo, J.A.; Andrews, B.A.; Bull, A.T.; Goodfellow, M.; Ebel, R.; et al. Diverse metabolic profiles of a *Streptomyces* strain isolated from a hyper-arid environment. *J. Nat. Prod.* **2011**, *74*, 1965–1971. [CrossRef] [PubMed]
27. Okoro, C.K.; Brown, R.; Jones, A.L.; Andrews, B.A.; Asenjo, J.A.; Goodfellow, M.; Bull, A.T. Diversity of culturable actinomycetes in hyper-arid soils of the Atacama Desert, Chile. *Anton. Leeuwenhoek* **2009**, *95*, 121–133. [CrossRef] [PubMed]
28. Hadj Rabia-Boukhalfa, Y.; Eveno, Y.; Karama, S.; Selama, O.; Lauga, B.; Duran, R.; Hacène, H.; Eparvier, V. Isolation, purification and chemical characterization of a new angucyclinone compound produced by a new halotolerant *Nocardiopsis* sp. HR-4 strain. *World J. Microbiol. Biotechnol.* **2017**, *33*. [CrossRef] [PubMed]
29. Duraipandiyan, V.; Al-Dhabi, N.A.; Ignacimuthu, S. New antimicrobial anthraquinone $6,6^1$-bis (1,5,7-trihydroxy-3-hydroxymethylanthraquinone) isolated from *Streptomyces* sp. isolate ERI-26. *Saudi J. Biol. Sci.* **2016**, *23*, 731–735. [CrossRef] [PubMed]
30. Arasu, M.V.; Duraipandiyan, V.; Agastian, P.; Ignacimuthu, S. Antimicrobial activity of *Streptomyces* spp. ERI-26 recovered from Western Ghats of Tamil Nadu. *J. Mycol. Med.* **2008**, *18*, 147–153. [CrossRef]
31. Stewart, E.J. Growing Unculturable Bacteria. *J. Bacteriol.* **2012**, *194*, 4151–4160. [CrossRef] [PubMed]
32. Bollmann, A.; Lewis, K.; Epstein, S.S. Incubation of environmental samples in a diffusion chamber increases the diversity of recovered isolates. *Appl. Environ. Microbiol.* **2007**, *73*, 6386–6390. [CrossRef] [PubMed]
33. Nichols, D.; Cahoon, N.; Trakhtenberg, E.M.; Pham, L.; Mehta, A.; Belanger, A.; Kanigan, T.; Lewis, K.; Epstein, S.S. Use of iChip for high-throughput *in situ* cultivation of "uncultivable" microbial species. *Appl. Environ. Microbiol.* **2010**, *76*, 2445–2450. [CrossRef] [PubMed]
34. Peoples, A.J.; Zhang, Q.; Millett, W.P.; Rothfeder, M.T.; Pescatore, B.C.; Madden, A.A.; Ling, L.L.; Moore, C.M. Neocitreamicins I and II, novel antibiotics with activity against methicillin resistant *Staphylococcus aureus* and vancomycin-resistant *Enterococci*. *J. Antibiot.* **2008**, *61*, 457–463. [CrossRef] [PubMed]
35. Kaeberlein, T.; Lewis, K.; Epstein, S.S. Isolating "uncultivable" microorganisms in pure culture in a simulated natural environment. *Science* **2002**, *296*, 1127–1129. [CrossRef] [PubMed]
36. Peoples, A.; Zhang, Q.; Moore, C.; Ling, L.; Rothfeder, M.; Lewis, K. Macrolactam Compounds. Patent WO2009026527A1, 17 January 2012.
37. Gavrish, E.; Sit, C.S.; Cao, S.; Kandror, O.; Spoering, A.; Peoples, A.; Ling, L.; Fetterman, A.; Hughes, D.; Bissell, A.; et al. Lassomycin, a ribosomally synthesized cyclic peptide, kills *Mycobacterium tuberculosis* by targeting the ATP-dependent protease ClpC1P1P2. *Chem. Biol.* **2014**, *21*, 509–518. [CrossRef] [PubMed]
38. Ling, L.L.; Schneider, T.; Peoples, A.J.; Spoering, A.L.; Engels, I.; Conlon, B.P.; Mueller, A.; Schäberle, T.F.; Hughes, D.E.; Epstein, S.; et al. A new antibiotic kills pathogens without detectable resistance. *Nature* **2015**, *517*, 455–459. [CrossRef] [PubMed]
39. Goers, L.; Freemont, P.; Polizzi, K.M. Co-culture systems and technologies: Taking synthetic biology to the next level. *J. R. Soc. Interface* **2014**, *11*. [CrossRef] [PubMed]

40. Dashti, Y.; Grkovic, T.; Abdelmohsen, U.R.; Hentschel, U.; Quinn, R.J. Production of induced secondary metabolites by a co-culture of sponge-associated actinomycetes, *Actinokineospora* sp. EG49 and *Nocardiopsis* sp. RV163. *Mar. Drugs* **2014**, *12*, 3046–3059. [CrossRef] [PubMed]
41. Zengler, K.; Toledo, G.; Rappe, M.; Elkins, J.; Mathur, E.J.; Short, J.M.; Keller, M. Cultivating the uncultured. *Proc. Natl. Acad. Sci. USA* **2002**, *99*, 15681–15686. [CrossRef] [PubMed]
42. Zengler, K.; Walcher, M.; Clark, G.; Haller, I.; Toledo, G.; Holland, T.; Mathur, E.J.; Woodnutt, G.; Short, J.M.; Keller, M. High-throughput cultivation of microorganisms using microcapsules. In *Environmental Microbiology*; Leadbetter, J.R., Ed.; Elsevier Academic Press: Amsterdam, The Netherlands; Boston, MA, USA, 2005; pp. 124–130.
43. Ben-Dov, E.; Kramarsky-Winter, E.; Kushmaro, A. An *in situ* method for cultivating microorganisms using a double encapsulation technique. *FEMS Microbiol. Ecol.* **2009**, *68*, 363–371. [CrossRef] [PubMed]
44. Bomar, L.; Maltz, M.; Colston, S.; Graf, J. Directed culturing of microorganisms using metatranscriptomics. *mBio* **2011**, *2*. [CrossRef] [PubMed]
45. Morris, J.J.; Kirkegaard, R.; Szul, M.J.; Johnson, Z.I.; Zinser, E.R. Facilitation of robust growth of prochlorococcus colonies and dilute liquid cultures by "helper" heterotrophic bacteria. *Appl. Environ. Microbiol.* **2008**, *74*, 4530–4534. [CrossRef] [PubMed]
46. Tanaka, Y.; Hanada, S.; Manome, A.; Tsuchida, T.; Kurane, R.; Nakamura, K.; Kamagata, Y. Catellibacterium nectariphilum gen. nov., sp. nov., which requires a diffusible compound from a strain related to the genus *Sphingomonas* for vigorous growth. *Int. J. Syst. Evol. Microbiol.* **2004**, *54*, 955–959. [CrossRef] [PubMed]
47. Kim, K.; Kim, J.-J.; Masui, R.; Kuramitsu, S.; Sung, M.-H. A commensal symbiotic interrelationship for the growth of *Symbiobacterium toebii* with its partner bacterium, *Geobacillus toebii*. *BMC Res. Notes* **2011**, *4*. [CrossRef] [PubMed]
48. Morris, J.J.; Lenski, R.E.; Zinser, E.R. The Black Queen Hypothesis: Evolution of dependencies through adaptive gene loss. *mBio* **2012**, *3*, e00036-12. [CrossRef] [PubMed]
49. Schofield, M.M.; Jain, S.; Porat, D.; Dick, G.J.; Sherman, D.H. Identification and analysis of the bacterial endosymbiont specialized for production of the chemotherapeutic natural product ET-743. *Environ. Microbiol.* **2015**, *17*, 3964–3975. [CrossRef] [PubMed]
50. Rashad, F.M.; Fathy, H.M.; El-Zayat, A.S.; Elghonaimy, A.M. Isolation and characterization of multifunctional *Streptomyces* species with antimicrobial, nematicidal and phytohormone activities from marine environments in Egypt. *Microbiol. Res.* **2015**, *175*, 34–47. [CrossRef] [PubMed]
51. Bode, H.B.; Bethe, B.; Höfs, R.; Zeeck, A. Big effects from small changes: Possible ways to explore nature's chemical diversity. *ChemBioChem* **2002**, *3*, 619–627. [CrossRef]
52. Engelhardt, K.; Degnes, K.F.; Kemmler, M.; Bredholt, H.; Fjaervik, E.; Klinkenberg, G.; Sletta, H.; Ellingsen, T.E.; Zotchev, S.B. Production of a new thiopeptide antibiotic, TP-1161, by a marine *Nocardiopsis* species. *Appl. Environ. Microbiol.* **2010**, *76*, 4969–4976. [CrossRef] [PubMed]
53. Dufour, C.; Wink, J.; Kurz, M.; Kogler, H.; Olivan, H.; Sablé, S.; Heyse, W.; Gerlitz, M.; Toti, L.; Nußer, A.; et al. Isolation and structural elucidation of armeniaspirols A–C: Potent antibiotics against Gram-positive pathogens. *Chem. Eur. J.* **2012**, *18*, 16123–16128. [CrossRef] [PubMed]
54. Ma, J.; Huang, H.; Xie, Y.; Liu, Z.; Zhao, J.; Zhang, C.; Jia, Y.; Zhang, Y.; Zhang, H.; Zhang, T.; et al. Biosynthesis of ilamycins featuring unusual building blocks and engineered production of enhanced anti-tuberculosis agents. *Nat. Commun.* **2017**, *8*. [CrossRef] [PubMed]
55. Boubetra, D.; Sabaou, N.; Zitouni, A.; Bijani, C.; Lebrihi, A.; Mathieu, F. Taxonomy and chemical characterization of new antibiotics produced by *Saccharothrix* SA198 isolated from a Saharan soil. *Microbiol. Res.* **2013**, *168*, 223–230. [CrossRef] [PubMed]
56. Lee, L.-H.; Cheah, Y.-K.; Mohd Sidik, S.; Ab Mutalib, N.-S.; Tang, Y.-L.; Lin, H.-P.; Hong, K. Molecular characterization of Antarctic actinobacteria and screening for antimicrobial metabolite production. *World J. Microbiol. Biotechnol.* **2012**, *28*, 2125–2137. [CrossRef] [PubMed]
57. Liu, L.; Salam, N.; Jiao, J.-Y.; Jiang, H.-C.; Zhou, E.-M.; Yin, Y.-R.; Ming, H.; Li, W.-J. Diversity of culturable thermophilic Actinobacteria in hot springs in Tengchong, china and studies of their biosynthetic gene profiles. *Microb. Ecol.* **2016**, *72*, 150–162. [CrossRef] [PubMed]
58. Silva-Lacerda, G.R.; Santana, R.C.F.; Vicalvi-Costa, M.C.V.; Solidônio, E.G.; Sena, K.X.F.R.; Lima, G.M.S.; Araújo, J.M. Antimicrobial potential of actinobacteria isolated from the rhizosphere of the Caatinga biome plant *Caesalpinia pyramidalis* Tul. *Genet. Mol. Res.* **2016**, *15*. [CrossRef] [PubMed]

59. Hamm, P.S.; Caimi, N.A.; Northup, D.E.; Valdez, E.W.; Buecher, D.C.; Dunlap, C.A.; Labeda, D.P.; Lueschow, S.; Porras-Alfaro, A. Western bats as a reservoir of novel *Streptomyces* species with antifungal activity. *Appl. Environ. Microbiol.* **2017**, *83*. [CrossRef] [PubMed]
60. Braña, A.F.; Sarmiento-Vizcaíno, A.; Pérez-Victoria, I.; Otero, L.; Fernández, J.; Palacios, J.J.; Martín, J.; De La Cruz, M.; Díaz, C.; Vicente, F.; et al. Branimycins B and C, antibiotics produced by the abyssal actinobacterium *Pseudonocardia carboxydivorans* M-227. *J. Nat. Prod.* **2017**, *80*, 569–573. [CrossRef] [PubMed]
61. Xu, J.-H.; Gu, K.-B.; Zhang, D.-J.; Li, Y.-G.; Tian, L. Ghanamycins A and B, two novel γ-butyrolactones from marine-derived *Streptomyces ghanaensis* TXC6-16. *J. Antibiot.* **2017**, *70*, 733–736. [CrossRef] [PubMed]
62. Niu, S.; Li, S.; Chen, Y.; Tian, X.; Zhang, H.; Zhang, G.; Zhang, W.; Yang, X.; Zhang, S.; Ju, J.; et al. Lobophorins E and F, new spirotetronate antibiotics from a South China Sea-derived *Streptomyces* sp. SCSIO 01127. *J. Antibiot.* **2011**, *64*, 711–716. [CrossRef] [PubMed]
63. Fu, P.; MacMillan, J.B. Spithioneines A and B, two new bohemamine derivatives possessing ergothioneine moiety from a marine-derived *Streptomyces spinoverrucosus*. *Org. Lett.* **2015**, *17*, 3046–3049. [CrossRef] [PubMed]
64. Keller, S.; Nicholson, G.; Drahl, C.; Sorensen, E.; Fiedler, H.-P.; Sussmuth, R.D. Abyssomicins G and H and atrop-abyssomicin C from the marine *Verrucosispora* strain AB-18-032. *J. Antibiot.* **2007**, *60*, 391–394. [CrossRef] [PubMed]
65. Igarashi, Y.; Yu, L.; Miyanaga, S.; Fukuda, T.; Saitoh, N.; Sakurai, H.; Saiki, I.; Alonso-Vega, P.; Trujillo, M.E. Abyssomicin I, a modified polycyclic polyketide from *Streptomyces* sp. CHI39. *J. Nat. Prod.* **2010**, *73*, 1943–1946. [CrossRef] [PubMed]
66. Niu, X.-M.; Li, S.-H.; Gorls, H.; Schollmeyer, D.; Hilliger, M.; Grabley, S.; Sattler, I. Abyssomicin E, a highly functionalized polycyclic metabolite from *Streptomyces* species. *Org. Lett.* **2007**, *9*, 2437–2440. [CrossRef] [PubMed]
67. Feling, R.H.; Buchanan, G.O.; Mincer, T.J.; Kauffman, C.A.; Jensen, P.R.; Fenical, W. Salinosporamide A: A highly cytotoxic proteasome inhibitor from a novel microbial source, a marine bacterium of the new genus *Salinospora*. *Angew. Chem. Int. Ed. Engl.* **2003**, *42*, 355–357. [CrossRef] [PubMed]
68. Stadler, M.; Bitzer, J.; Mayer-Bartschmid, A.; Muller, H.; et-Buchholz, J.; Gantner, F.; Tichy, H.V.; Reinemer, P.; Bacon, K.B. Cinnabaramides A–G: Analogues of lactacystin and salinosporamide from a terrestrial streptomycete. *J. Nat. Prod.* **2007**, *70*, 246–252. [CrossRef] [PubMed]
69. Messaoudi, O.; Bendahou, M.; Benamar, I.; Abdelwouhid, D.-E. Identification and preliminary characterization of non-polyene antibiotics secreted by new strain of actinomycete isolated from sebkha of Kenadsa, Algeria. *Asian Pac. J. Trop. Biomed.* **2015**, *5*, 438–445. [CrossRef]
70. Shin, B.; Kim, B.-Y.; Cho, E.; Oh, K.-B.; Shin, J.; Goodfellow, M.; Oh, D.-C. Actinomadurol, an antibacterial norditerpenoid from a rare Actinomycete, *Actinomadura* sp. KC 191. *J. Nat. Prod.* **2016**, *79*, 1886–1890. [CrossRef] [PubMed]
71. Al-Dhabi, N.A.; Esmail, G.A.; Duraipandiyan, V.; Valan Arasu, M.; Salem-Bekhit, M.M. Isolation, identification and screening of antimicrobial thermophilic *Streptomyces* sp. Al-Dhabi-1 isolated from Tharban hot spring, Saudi Arabia. *Extremophiles* **2016**, *20*, 79–90. [CrossRef] [PubMed]
72. Strobel, G.; Daisy, B. Bioprospecting for microbial endophytes and their natural products. *Microbiol. Mol. Biol. Rev.* **2003**, *67*, 491–502. [CrossRef] [PubMed]
73. Matsumoto, A.; Takahashi, Y. Endophytic actinomycetes: Promising source of novel bioactive compounds. *J. Antibiot.* **2017**, *70*, 514–519. [CrossRef] [PubMed]
74. Chagas, F.O.; Ruzzini, A.C.; Bacha, L.V.; Samborskyy, M.; Conti, R.; Pessotti, R.C.; de Oliveira, L.G.; Clardy, J.; Pupo, M.T. Genome sequence of *Streptomyces* sp. strain RTd22, an endophyte of the Mexican sunflower. *Genome Announc.* **2016**, *4*. [CrossRef] [PubMed]
75. Barke, J.; Seipke, R.F.; Grüschow, S.; Heavens, D.; Drou, N.; Bibb, M.J.; Goss, R.J.M.; Yu, D.W.; Hutchings, M.I. A mixed community of actinomycetes produce multiple antibiotics for the fungus farming ant *Acromyrmex octospinosus*. *BMC Biol.* **2010**, *8*. [CrossRef] [PubMed]
76. Oh, D.-C.; Poulsen, M.; Currie, C.R.; Clardy, J. Sceliphrolactam, a polyene macrocyclic lactam from a wasp-associated *Streptomyces* sp. *Org. Lett.* **2011**, *13*, 752–755. [CrossRef] [PubMed]
77. Beemelmanns, C.; Ramadhar, T.R.; Kim, K.H.; Klassen, J.L.; Cao, S.; Wyche, T.P.; Hou, Y.; Poulsen, M.; Bugni, T.S.; Currie, C.R.; et al. Macrotermycins A–D, glycosylated macrolactams from a termite-associated *Amycolatopsis* sp. M39. *Org. Lett.* **2017**, *19*, 1000–1003. [CrossRef] [PubMed]

78. Kim, K.H.; Ramadhar, T.R.; Beemelmanns, C.; Cao, S.; Poulsen, M.; Currie, C.R.; Clardy, J. Natalamycin A, an ansamycin from a termite-associated *Streptomyces* sp. *Chem. Sci.* **2014**, *5*, 4333–4338. [CrossRef] [PubMed]
79. Sit, C.S.; Ruzzini, A.C.; van Arnam, E.B.; Ramadhar, T.R.; Currie, C.R.; Clardy, J. Variable genetic architectures produce virtually identical molecules in bacterial symbionts of fungus-growing ants. *Proc. Natl. Acad. Sci. USA* **2015**, *112*, 13150–13154. [CrossRef] [PubMed]
80. Nettleton, D.E.; Doyle, T.W.; Krishnan, B.; Matsumoto, G.K.; Clardy, J. Isolation and structure of rebeccamycin—A new antitumor antibiotic from *Nocardia aerocoligenes*. *Tetrahedron Lett.* **1985**, *26*, 4011–4014. [CrossRef]
81. Van Arnam, E.B.; Ruzzini, A.C.; Sit, C.S.; Currie, C.R.; Clardy, J. A rebeccamycin analog provides plasmid-encoded niche defense. *J. Am. Chem. Soc.* **2015**, *137*, 14272–14274. [CrossRef] [PubMed]
82. Wyche, T.P.; Ruzzini, A.C.; Schwab, L.; Currie, C.R.; Clardy, J. Tryptorubin A: A polycyclic peptide from a fungus-derived streptomycete. *J. Am. Chem. Soc.* **2017**, *139*, 12899–12902. [CrossRef] [PubMed]
83. Pettit, R.K. Culturability and secondary metabolite diversity of extreme microbes: Expanding contribution of deep sea and deep-sea vent microbes to natural product discovery. *Mar. Biotechnol.* **2011**, *13*, 1–11. [CrossRef] [PubMed]
84. Wilson, Z.E.; Brimble, M.A. Molecules derived from the extremes of life. *Nat. Prod. Rep.* **2009**, *26*, 44–71. [CrossRef] [PubMed]
85. Brady, S.F.; Simmons, L.; Kim, J.H.; Schmidt, E.W. Metagenomic approaches to natural products from free-living and symbiotic organisms. *Nat. Prod. Rep.* **2009**, *26*, 1488–1503. [CrossRef] [PubMed]
86. Techtmann, S.M.; Hazen, T.C. Metagenomic applications in environmental monitoring and bioremediation. *J. Ind. Microbiol. Biotechnol.* **2016**, *43*, 1345–1354. [CrossRef] [PubMed]
87. Kavitha, S.; Selvakumar, R.; Sathishkumar, M.; Swaminathan, K.; Lakshmanaperumalsamy, P.; Singh, A.; Jain, S.K. Nitrate removal using *Brevundimonas diminuta* MTCC 8486 from ground water. *Water Sci. Technol.* **2009**, *60*, 517–524. [CrossRef] [PubMed]
88. Green, S.J.; Prakash, O.; Jasrotia, P.; Overholt, W.A.; Cardenas, E.; Hubbard, D.; Tiedje, J.M.; Watson, D.B.; Schadt, C.W.; Brooks, S.C.; et al. Denitrifying bacteria from the genus *Rhodanobacter* dominate bacterial communities in the highly contaminated subsurface of a nuclear legacy waste site. *Appl. Environ. Microbiol.* **2012**, *78*, 1039–1047. [CrossRef] [PubMed]
89. Dubinsky, E.A.; Conrad, M.E.; Chakraborty, R.; Bill, M.; Borglin, S.E.; Hollibaugh, J.T.; Mason, O.U.; Piceno, Y.M.; Reid, F.C.; Stringfellow, W.T.; et al. Succession of hydrocarbon-degrading bacteria in the aftermath of the deepwater horizon oil spill in the gulf of Mexico. *Environ. Sci. Technol.* **2013**, *47*, 10860–10867. [CrossRef] [PubMed]
90. Lok, C. Mining the microbial dark matter. *Nature* **2015**, *522*, 270–273. [CrossRef] [PubMed]
91. Brady, S.F.; Chao, C.J.; Handelsman, J.; Clardy, J. Cloning and heterologous expression of a natural product biosynthetic gene cluster from eDNA. *Org. Lett.* **2001**, *3*, 1981–1984. [CrossRef] [PubMed]
92. Owen, J.G.; Charlop-Powers, Z.; Smith, A.G.; Ternei, M.A.; Calle, P.Y.; Reddy, B.V.B.; Montiel, D.; Brady, S.F. Multiplexed metagenome mining using short DNA sequence tags facilitates targeted discovery of epoxyketone proteasome inhibitors. *Proc. Natl. Acad. Sci. USA* **2015**, *112*, 4221–4226. [CrossRef] [PubMed]
93. Katz, M.; Hover, B.M.; Brady, S.F. Culture-independent discovery of natural products from soil metagenomes. *J. Ind. Microbiol. Biotechnol.* **2016**, *43*, 129–141. [CrossRef] [PubMed]
94. Altaee, N.; Jawad, M.; Hameed, I. Characterization of metabolites produced by *E. coli* and analysis of its chemical compounds using GC-MS. *IJCPR* **2017**, *7*, 393–399.
95. Wexler, M.; Johnston, A.W.B. Wide host-range cloning for functional metagenomics. In *Metagenomics: Methods and Protocols*; Streit, W.R., Daniel, R., Eds.; Humana: London, UK; Springer: New York, NY, USA, 2010; pp. 77–96.
96. Gabor, E.M.; Alkema, W.B.; Janssen, D.B. Quantifying the accessibility of the metagenome by random expression cloning techniques. *Environ. Microbiol.* **2004**, *6*, 879–886. [CrossRef] [PubMed]
97. Stevens, D.C.; Conway, K.R.; Pearce, N.; Villegas-Penaranda, L.R.; Garza, A.G.; Boddy, C.N. Alternative sigma factor over-expression enables heterologous expression of a type II polyketide biosynthetic pathway in *Escherichia coli*. *PLoS ONE* **2013**, *8*, e64858. [CrossRef] [PubMed]
98. Pfeifer, B.A.; Admiraal, S.J.; Gramajo, H.; Cane, D.E.; Khosla, C. Biosynthesis of complex polyketides in a metabolically engineered strain of *E. coli*. *Science* **2001**, *291*, 1790–1792. [CrossRef] [PubMed]

99. Boddy, C.N.; Hotta, K.; Tse, M.L.; Watts, R.E.; Khosla, C. Precursor-directed biosynthesis of epothilone in *Escherichia coli*. *J. Am. Chem. Soc.* **2004**, *126*, 7436–7437. [CrossRef] [PubMed]
100. Watanabe, K.; Hotta, K.; Praseuth, A.P.; Koketsu, K.; Migita, A.; Boddy, C.N.; Wang, C.C.; Oguri, H.; Oikawa, H. Total biosynthesis of antitumor nonribosomal peptides in *Escherichia coli*. *Nat. Chem. Biol.* **2006**, *2*, 423–428. [CrossRef] [PubMed]
101. Ortega, M.A.; Van Der Donk, W.A. New insights into the biosynthetic logic of ribosomally synthesized and post-translationally modified peptide natural products. *Cell Chem. Biol.* **2016**, *23*, 31–44. [CrossRef] [PubMed]
102. Ongley, S.; Bian, X.; Neilan, B.A.; Müller, R. Recent advances in the heterologous expression of microbial natural product biosynthetic pathways. *Nat. Prod. Rep.* **2013**, *30*, 1121–1138. [CrossRef] [PubMed]
103. Wenzel, S.C. Biosynthesis and Heterologous Production of Myxobacterial Secondary Metabolites. Ph.D. Thesis, Fachbereich Biowissenschaftern und Psychologie, Technische Carolo-Wilhemina-Universität, Braunschweig, Germany, 2005.
104. Ferrari, B.C.; Binnerup, S.J.; Gillings, M. Microcolony cultivation on a soil substrate membrane system selects for previously uncultured soil bacteria. *Appl. Environ. Microbiol.* **2005**, *71*, 8714–8720. [CrossRef] [PubMed]
105. Cueto, M.; Jensen, P.R.; Kauffman, C.; Fenical, W.; Lobkovsky, E.; Clardy, J. Pestalone, a new antibiotic produced by a marine fungus in response to bacterial challenge. *J. Nat. Prod.* **2001**, *64*, 1444–1446. [CrossRef] [PubMed]
106. Moore, J.M.; Bradshaw, E.; Seipke, R.F.; Hutchings, M.I.; McArthur, M. Use and discovery of chemical elicitors that stimulate biosynthetic gene clusters in *Streptomyces* bacteria. *Methods Enzymol.* **2012**, *517*, 367–385. [PubMed]
107. Montiel, D.; Kang, H.-S.; Chang, F.-Y.; Charlop-Powers, Z.; Brady, S.F. Yeast homologous recombination-based promoter engineering for the activation of silent natural product biosynthetic gene clusters. *Proc. Natl. Acad. Sci. USA* **2015**, *112*, 8953–8958. [CrossRef] [PubMed]
108. Chang, F.-Y.; Ternei, M.A.; Calle, P.Y.; Brady, S.F. Discovery and synthetic refactoring of tryptophan dimer gene clusters from the environment. *J. Am. Chem. Soc.* **2013**, *135*, 17906–17912. [CrossRef] [PubMed]
109. Trindade, M.; van Zyl, L.J.; Navarro-Fernández, J.; Abd Elrazak, A. Targeted metagenomics as a tool to tap into marine natural product diversity for the discovery and production of drug candidates. *Front. Microbiol.* **2015**, *6*. [CrossRef] [PubMed]
110. Williamson, L.L.; Borlee, B.R.; Schloss, P.D.; Guan, C.; Allen, H.K.; Handelsman, J. Intracellular screen to identify metagenomic clones that induce or inhibit a quorum-sensing biosensor. *Appl. Environ. Microbiol.* **2005**, *71*, 6335–6344. [CrossRef] [PubMed]
111. Uchiyama, T.; Abe, T.; Ikemura, T.; Watanabe, K. Substrate-induced gene-expression screening of environmental metagenome libraries for isolation of catabolic genes. *Nat. Biotechnol.* **2005**, *23*, 88–93. [CrossRef] [PubMed]
112. Lim, H.K.; Chung, E.J.; Kim, J.-C.; Choi, G.J.; Jang, K.S.; Chung, Y.R.; Cho, K.Y.; Lee, S.-W. Characterization of a forest soil metagenome clone that confers indirubin and indigo production on *Escherichia coli*. *Appl. Environ. Microbiol.* **2005**, *71*, 7768–7777. [CrossRef] [PubMed]
113. Brady, S.F.; Chao, C.J.; Clardy, J. New natural product families from an environmental DNA (eDNA) gene cluster. *J. Am. Chem. Soc.* **2002**, *124*, 9968–9969. [CrossRef] [PubMed]
114. Gillespie, D.E.; Brady, S.F.; Bettermann, A.D.; Cianciotto, N.P.; Liles, M.R.; Rondon, M.R.; Clardy, J.; Goodman, R.M.; Handelsman, J. Isolation of antibiotics turbomycin A and B from a metagenomic library of soil microbial DNA. *Appl. Environ. Microbiol.* **2002**, *68*, 4301–4306. [CrossRef] [PubMed]
115. Zhang, M.M.; Qiao, Y.; Ang, E.L.; Zhao, H. Using natural products for drug discovery: The impact of the genomics era. *Expert Opin. Drug Discov.* **2017**, *12*, 475–487. [CrossRef] [PubMed]
116. Ferrer, M.; Beloqui, A.; Vieites, J.M.; Guazzaroni, M.E.; Berger, I.; Aharoni, A. Interplay of metagenomics and *in vitro* compartmentalization. *Microb. Biotechnol.* **2009**, *2*, 31–39. [CrossRef] [PubMed]
117. Owen, J.G.; Robins, K.J.; Parachin, N.S.; Ackerley, D.F. A functional screen for recovery of 4′-phosphopantetheinyl transferase and associated natural product biosynthesis genes from metagenome libraries. *Environ. Microbiol.* **2012**, *14*, 1198–1209. [CrossRef] [PubMed]
118. Beld, J.; Sonnenschein, E.C.; Vickery, C.R.; Noel, J.P.; Burkart, M.D. The phosphopantetheinyl transferases: Catalysis of a post-translational modification crucial for life. *Nat. Prod. Rep.* **2013**, *31*, 61–108. [CrossRef] [PubMed]

119. Bitok, J.K.; Lemetre, C.; Ternei, M.A.; Brady, S.F. Identification of biosynthetic gene clusters from metagenomic libraries using PPTase complementation in a *Streptomyces* host. *FEMS Microbiol. Lett.* **2017**, *364*. [CrossRef] [PubMed]
120. Gaitatzis, N.; Kunze, B.; Müller, R. *In vitro* reconstitution of the myxochelin biosynthetic machinery of *Stigmatella aurantiaca* Sg a15: Biochemical characterization of a reductive release mechanism from nonribosomal peptide synthetases. *Proc. Natl. Acad. Sci. USA* **2001**, *98*, 11136–11141. [CrossRef] [PubMed]
121. Weber, T.; Blin, K.; Duddela, S.; Krug, D.; Kim, H.U.; Bruccoleri, R.; Lee, S.Y.; Fischbach, M.A.; Müller, R.; Wohlleben, W.; et al. antiSMASH 3.0—A comprehensive resource for the genome mining of biosynthetic gene clusters. *Nucleic Acids Res.* **2015**, *43*, W237–W243. [CrossRef] [PubMed]
122. Blin, K.; Wolf, T.; Chevrette, M.G.; Lu, X.; Schwalen, C.J.; Kautsar, S.A.; Suarez Duran, H.G.; de Los Santos, E.L.C.; Kim, H.U.; Nave, M.; et al. antiSMASH 4.0—Improvements in chemistry prediction and gene cluster boundary identification. *Nucleic Acids Res* **2017**, *45*, W36–W41. [CrossRef] [PubMed]
123. Reddy, B.V.B.; Milshteyn, A.; Charlop-Powers, Z.; Brady, S.F. eSNaPD: A versatile, web-based bioinformatics platform for surveying and mining natural product biosynthetic diversity from metagenomes. *Chem. Biol.* **2014**, *21*, 1023–1033. [CrossRef] [PubMed]
124. Ziemert, N.; Podell, S.; Penn, K.; Badger, J.H.; Allen, E.; Jensen, P.R. The natural product domain seeker NaPDoS: A phylogeny based bioinformatic tool to classify secondary metabolite gene diversity. *PLoS ONE* **2012**, *7*, e34064. [CrossRef] [PubMed]
125. Owen, J.G.; Reddy, B.V.; Ternei, M.A.; Charlop-Powers, Z.; Calle, P.Y.; Kim, J.H.; Brady, S.F. Mapping gene clusters within arrayed metagenomic libraries to expand the structural diversity of biomedically relevant natural products. *Proc. Natl. Acad. Sci. USA* **2013**, *110*, 11797–11802. [CrossRef] [PubMed]
126. Charlop-Powers, Z.; Milshteyn, A.; Brady, S.F. Metagenomic small molecule discovery methods. *Curr. Opin. Microbiol.* **2014**, *0*, 70–75. [CrossRef] [PubMed]
127. Woodhouse, J.N.; Fan, L.; Brown, M.V.; Thomas, T.; Neilan, B.A. Deep sequencing of non-ribosomal peptide synthetases and polyketide synthases from the microbiomes of Australian marine sponges. *ISME J.* **2013**, *7*, 1842–1851. [CrossRef] [PubMed]
128. Niu, G.; Tan, H. Nucleoside antibiotics: Biosynthesis, regulation, and biotechnology. *Trends Microbiol.* **2015**, *23*, 110–119. [CrossRef] [PubMed]
129. Feng, Z.Y.; Kallifidas, D.; Brady, S.F. Functional analysis of environmental DNA-derived type II polyketide synthases reveals structurally diverse secondary metabolites. *Proc. Natl. Acad. Sci. USA* **2011**, *108*, 12629–12634. [CrossRef] [PubMed]
130. Banik, J.J.; Craig, J.W.; Calle, P.Y.; Brady, S.F. Tailoring enzyme-rich environmental DNA clones: A source of enzymes for generating libraries of unnatural natural products. *J. Am. Chem. Soc.* **2010**, *132*, 15661–15670. [CrossRef] [PubMed]
131. Yao, J.; Fan, X.J.; Lu, Y.; Liu, Y.H. Isolation and characterization of a novel tannase from a metagenomic library. *J. Agric. Food Chem.* **2011**, *59*, 3812–3818. [CrossRef] [PubMed]
132. Wang, Q.; Wu, H.; Wang, A.; Du, P.; Pei, X.; Li, H.; Yin, X.; Huang, L.; Xiong, X. Prospecting metagenomic enzyme subfamily genes for DNA family shuffling by a novel PCR-based approach. *J. Biol. Chem.* **2010**, *285*, 41509–41516. [CrossRef] [PubMed]
133. Wang, K.; Li, G.; Yu, S.Q.; Zhang, C.T.; Liu, Y.H. A novel metagenome-derived β-galactosidase: Gene cloning, overexpression, purification and characterization. *Appl. Microbiol. Biotechnol.* **2010**, *88*, 155–165. [CrossRef] [PubMed]
134. Brady, S.F.; Bauer, J.D.; Clarke-Pearson, M.F.; Daniels, R. Natural products from *isnA*-containing biosynthetic gene clusters recovered from the genomes of cultured and uncultured bacteria. *J. Am. Chem. Soc.* **2007**, *129*, 12102–12103. [CrossRef] [PubMed]
135. Chang, F.-Y.; Ternei, M.A.; Calle, P.Y.; Brady, S.F. Targeted metagenomics: Finding rare tryptophan dimer natural products in the environment. *J. Am. Chem. Soc.* **2015**, *137*, 6044–6052. [CrossRef] [PubMed]
136. Kang, H.-S.; Brady, S.F. Arimetamycin A: Improving clinically relevant families of natural products through sequence-guided screening of soil metagenomes. *Angew. Chem. Int. Ed.* **2013**, *52*, 11063–11067. [CrossRef] [PubMed]
137. Kallifidas, D.; Kang, H.-S.; Brady, S.F. Tetarimycin A, an MRSA-active antibiotic identified through induced expression of environmental DNA gene clusters. *J. Am. Chem. Soc.* **2012**, *134*, 19552–19555. [CrossRef] [PubMed]

138. Hover, B.M.; Kim, S.-H.; Katz, M.; Charlop-Powers, Z.; Owen, J.G.; Ternei, M.A.; Maniko, J.; Estrela, A.B.; Molina, H.; Park, S.; et al. Culture-independent discovery of the malacidins as calcium-dependent antibiotics with activity against multidrug-resistant Gram-positive pathogens. *Nat. Microbiol.* **2018**, *3*, 415–422. [CrossRef] [PubMed]
139. Kallifidas, D.; Jiang, G.; Ding, Y.; Luesch, H. Rational engineering of *Streptomyces albus* J1074 for the overexpression of secondary metabolite gene clusters. *Microb. Cell Fact.* **2018**, *17*. [CrossRef] [PubMed]
140. Leis, B.; Angelov, A.; Liebl, W. Screening and expression of genes from metagenomes. *Adv. Appl. Microbiol.* **2013**, *83*, 1–68. [PubMed]
141. Montaser, R.; Luesch, H. Marine natural products: A new wave of drugs? *Future Med. Chem.* **2011**, *3*, 1475–1489. [CrossRef] [PubMed]
142. Towbin, H.; Bair, K.W.; DeCaprio, J.A.; Eck, M.J.; Kim, S.; Kinder, F.R.; Morollo, A.; Mueller, D.R.; Schindler, P.; Song, H.K.; et al. Proteomics-based target identification—Bengamides as a new class of methionine aminopeptidase inhibitors. *J. Biol. Chem.* **2003**, *278*, 52964–52971. [CrossRef] [PubMed]
143. Wenzel, S.C.; Hoffmann, H.; Zhang, J.; Debussche, L.; Haag-Richter, S.; Kurz, M.; Nardi, F.; Lukat, P.; Kochems, I.; Tietgen, H.; et al. Production of the bengamide class of marine natural products in myxobacteria: Biosynthesis and structure-activity relationships. *Angew. Chem. Int. Ed. Engl.* **2015**, *54*, 15560–15564. [CrossRef] [PubMed]
144. Quinoa, E.; Adamczeski, M.; Crews, P.; Bakus, G.J. Bengamides, heterocyclic anthelmintics from a Jaspidae marine sponge. *J. Org. Chem.* **1986**, *51*, 4494–4497. [CrossRef]
145. Piel, J. Metabolites from symbiotic bacteria. *Nat. Prod. Rep.* **2009**, *26*, 338–362. [CrossRef] [PubMed]
146. Bewley, C.A.; Debitus, C.; Faulkner, D.J. Microsclerodermins A and B. Antifungal cyclic peptides from the lithistid sponge *Microscleroderma* sp. *J. Am. Chem. Soc.* **1994**, *116*, 7631–7636. [CrossRef]
147. Schmidt, E.W.; Faulkner, D.J. Microsclerodermins C–E, antifungal cyclic peptides from the lithistid marine sponges *Theonella* sp. and *Microscleroderma* sp. *Tetrahedron* **1998**, *54*, 3043–3056. [CrossRef]
148. Hoffmann, T.; Müller, S.; Nadmid, S.; Garcia, R.; Müller, R. Microsclerodermins from terrestrial myxobacteria: An intriguing biosynthesis likely connected to a sponge symbiont. *J. Am. Chem. Soc.* **2013**, *45*, 16904–16911. [CrossRef] [PubMed]
149. Simister, R.L.; Deines, P.; Botté, E.S.; Webster, N.S.; Taylor, M.W. Sponge-specific clusters revisited: A comprehensive phylogeny of sponge-associated microorganisms. *Environ. Microbiol.* **2012**, *14*, 517–524. [CrossRef] [PubMed]
150. Amann, R.I.; Ludwig, W.; Schleifer, K.H. Phylogenetic identification and *in situ* detection of individual microbial cells without cultivation. *Microbiol. Rev.* **1995**, *59*, 143–169. [PubMed]
151. Hamada, T.; Matsunaga, S.; Yano, G.; Fusetani, N. Polytheonamides A and B, highly cytotoxic, linear polypeptides with unprecedented structural features, from the marine sponge, *Theonella swinhoei*. *J. Am. Chem. Soc.* **2005**, *127*, 110–118. [CrossRef] [PubMed]
152. Inoue, M.; Shinohara, N.; Tanabe, S.; Takahashi, T.; Okura, K.; Itoh, H.; Mizoguchi, Y.; Iida, M.; Lee, N.; Matsuoka, S. Total synthesis of the large non-ribosomal peptide polytheonamide B. *Nat. Chem.* **2010**, *2*, 280–285. [CrossRef] [PubMed]
153. Freeman, M.F.; Gurgui, C.; Helf, M.J.; Morinaka, B.I.; Uria, A.R.; Oldham, N.J.; Sahl, H.-G.; Matsunaga, S.; Piel, J. Metagenome mining reveals polytheonamides as posttranslationally modified ribosomal peptides. *Science* **2012**, *338*, 387–390. [CrossRef] [PubMed]
154. Freeman, M.F.; Helf, M.J.; Bhushan, A.; Morinaka, B.I.; Piel, J. Seven enzymes create extraordinary molecular complexity in an uncultivated bacterium. *Nat. Chem.* **2017**, *9*, 387–395. [CrossRef] [PubMed]
155. Wilson, M.C.; Mori, T.; Rückert, C.; Uria, A.R.; Helf, M.J.; Takada, K.; Gernert, C.; Steffens, U.A.E.; Heycke, N.; Schmitt, S.; et al. An environmental bacterial taxon with a large and distinct metabolic repertoire. *Nature* **2014**, *506*. [CrossRef] [PubMed]
156. Kato, Y.; Fusetani, N.; Matsunaga, S.; Hashimoto, K.; Fujita, S.; Furuya, T. Bioactive marine metabolites. Part 16. Calyculin A. A novel antitumor metabolite from the marine sponge *Discodermia calyx*. *J. Am. Chem. Soc.* **1986**, *108*, 2780–2781. [CrossRef]
157. Edrada, R.A.; Ebel, R.; Supriyono, A.; Wray, V.; Schupp, P.; Steube, K.; van Soest, R.; Proksch, P. Swinhoeiamide A, a new highly active calyculin derivative from the marine sponge *Theonella swinhoei*. *J. Nat. Prod.* **2002**, *65*, 1168–1172. [CrossRef] [PubMed]

158. Hrvatin, S.; Piel, J. Rapid isolation of rare clones from highly complex DNA libraries by PCR analysis of liquid gel pools. *J. Microbiol. Methods* **2007**, *68*, 434–436. [CrossRef] [PubMed]
159. Wakimoto, T.; Egami, Y.; Nakashima, Y.; Wakimoto, Y.; Mori, T.; Awakawa, T.; Ito, T.; Kenmoku, H.; Asakawa, Y.; Piel, J.; et al. Calyculin biogenesis from a pyrophosphate protoxin produced by a sponge symbiont. *Nat. Chem. Biol.* **2014**, *10*. [CrossRef] [PubMed]
160. Repka, L.M.; Chekan, J.R.; Nair, S.K.; Van Der Donk, W.A. Mechanistic understanding of lanthipeptide biosynthetic enzymes. *Chem. Rev.* **2017**, *117*, 5457–5520. [CrossRef] [PubMed]
161. Smith, T.E.; Pond, C.D.; Pierce, E.; Harmer, Z.P.; Kwan, J.; Zachariah, M.M.; Harper, M.K.; Wyche, T.P.; Matainaho, T.K.; Bugni, T.S.; et al. Accessing chemical diversity from the uncultivated symbionts of small marine animals. *Nat. Chem. Biol.* **2018**, *14*. [CrossRef] [PubMed]
162. Minowa, Y.; Araki, M.; Kanehisa, M. Comprehensive analysis of distinctive polyketide and nonribosomal peptide structural motifs encoded in microbial genomes. *J. Mol. Biol.* **2007**, *368*, 1500–1517. [CrossRef] [PubMed]
163. Gawad, C.; Koh, W.; Quake, S.R. Single-cell genome sequencing: Current state of the science. *Nat. Rev. Genet.* **2016**, *17*. [CrossRef] [PubMed]
164. Genilloud, O. Actinomycetes: Still a source of novel antibiotics. *Nat. Prod. Rep.* **2017**, *34*, 1203–1232. [CrossRef] [PubMed]
165. Herrmann, J.; Fayad, A.A.; Müller, R. Natural products from myxobacteria: Novel metabolites and bioactivities. *Nat. Prod. Rep.* **2017**, *34*, 135–160. [CrossRef] [PubMed]
166. Gomez-Escribano, J.P.; Alt, S.; Bibb, M.J. Next generation sequencing of Actinobacteria for the discovery of novel natural products. *Mar. Drugs* **2016**, *14*. [CrossRef] [PubMed]
167. Grindberg, R.V.; Ishoey, T.; Brinza, D.; Esquenazi, E.; Coates, R.C.; Liu, W.-T.; Gerwick, L.; Dorrestein, P.C.; Pevzner, P.; Lasken, R.; et al. Single cell genome amplification accelerates identification of the apratoxin biosynthetic pathway from a complex microbial assemblage. *PLoS ONE* **2011**, *6*, e18565. [CrossRef] [PubMed]
168. Mason, O.U.; Hazen, T.C.; Borglin, S.; Chain, P.S.G.; Dubinsky, E.A.; Fortney, J.L.; Han, J.; Holman, H.-Y.N.; Hultman, J.; Lamendella, R.; et al. Metagenome, metatranscriptome and single-cell sequencing reveal microbial response to Deepwater Horizon oil spill. *ISME J.* **2012**, *6*, 1715–1727. [CrossRef] [PubMed]
169. Engel, K.; Pinnell, L.; Cheng, J.; Charles, T.C.; Neufeld, J.D. Nonlinear electrophoresis for purification of soil DNA for metagenomics. *J. Microbiol. Methods* **2012**, *88*, 35–40. [CrossRef] [PubMed]
170. Pel, J.; Broemeling, D.; Mai, L.; Poon, H.-L.; Tropini, G.; Warren, R.L.; Holt, R.A.; Marziali, A. Nonlinear electrophoretic response yields a unique parameter for separation of biomolecules. *Proc. Natl. Acad. Sci. USA* **2009**, *106*, 14796–14801. [CrossRef] [PubMed]
171. Liles, M.R.; Williamson, L.L.; Rodbumrer, J.; Torsvik, V.; Goodman, R.M.; Handelsman, J. Recovery, purification, and cloning of high-molecular-weight DNA from soil microorganisms. *Appl. Environ. Microbiol.* **2008**, *74*, 3302–3305. [CrossRef] [PubMed]
172. Teeling, H.; Glöckner, F.O. Current opportunities and challenges in microbial metagenome analysis—A bioinformatic perspective. *Brief. Bioinform.* **2012**, *13*, 728–742. [CrossRef] [PubMed]
173. Lakey, J.H.; Lea, E.J.; Rudd, B.A.; Wright, H.M.; Hopwood, D.A. A new channel-forming antibiotic from *Streptomyces coelicolor* A3(2) which requires calcium for its activity. *J. Gen. Microbiol.* **1983**, *129*, 3565–3573. [CrossRef] [PubMed]
174. Rudd, B.A.; Hopwood, D.A. A pigmented mycelial antibiotic in *Streptomyces coelicolor*: Control by a chromosomal gene cluster. *J. Gen. Microbiol.* **1980**, *119*, 333–340. [CrossRef] [PubMed]
175. Tsao, S.-W.; Rudd, B.A.M.; He, X.-G.; Chang, C.-J.; Floss, H.G. Identification of a red pigment from *Streptomyces coelicolor* A3(2) as a mixture of prodigiosin derivatives. *J. Antibiot.* **1985**, *38*, 128–131. [CrossRef] [PubMed]
176. Wright, L.F.; Hopwood, D.A. Actinorhodin is a chromosomally-determined antibiotic in *Streptomyces coelicolor* A3(2). *J. Gen. Microbiol.* **1976**, *96*, 289–297. [CrossRef] [PubMed]
177. Medema, M.H.; Blin, K.; Cimermancic, P.; de Jager, V.; Zakrzewski, P.; Fischbach, M.A.; Weber, T.; Takano, E.; Breitling, R. antiSMASH: Rapid identification, annotation and analysis of secondary metabolite biosynthesis gene clusters in bacterial and fungal genome sequences. *Nucleic Acids Res.* **2011**, *39*, W339–W346. [CrossRef] [PubMed]
178. Blin, K.; Kazempour, D.; Wohlleben, W.; Weber, T. Improved lanthipeptide detection and prediction for antiSMASH. *PLoS ONE* **2014**, *9*, e89420. [CrossRef] [PubMed]

179. Röttig, M.; Medema, M.H.; Blin, K.; Weber, T.; Rausch, C.; Kohlbacher, O. NRPSpredictor2—A web server for predicting NRPS adenylation domain specificity. *Nucleic Acids Res.* **2011**, *39*, W362–W367. [CrossRef] [PubMed]
180. Stachelhaus, T.; Mootz, H.D.; Marahiel, M.A. The specificity-conferring code of adenylation domains in nonribosomal peptide synthetases. *Chem. Biol.* **1999**, *6*, 493–505. [CrossRef]
181. Yadav, G.; Gokhale, R.S.; Mohanty, D. Computational approach for prediction of domain organization and substrate specificity of modular polyketide synthases. *J. Mol. Biol.* **2003**, *328*, 335–363. [CrossRef]
182. Mohimani, H.; Pevzner, P.A. Dereplication, sequencing and identification of peptidic natural products: From genome mining to peptidogenomics to spectral networks. *Nat. Prod. Rep.* **2016**, *33*, 73–86. [CrossRef] [PubMed]
183. Mohimani, H.; Liu, W.-T.; Kersten, R.D.; Moore, B.S.; Dorrestein, P.C.; Pevzner, P.A. NRPquest: Coupling mass spectrometry and genome mining for nonribosomal peptide discovery. *J. Nat. Prod.* **2014**, *77*, 1902–1909. [CrossRef] [PubMed]
184. Skinnider, M.A.; Merwin, N.J.; Johnston, C.W.; Magarvey, N.A. PRISM 3: Expanded prediction of natural product chemical structures from microbial genomes. *Nucleic Acids Res.* **2017**, *45*, W49–W54. [CrossRef] [PubMed]
185. Skinnider, M.A.; Dejong, C.A.; Rees, P.N.; Johnston, C.W.; Li, H.; Webster, A.L.H.; Wyatt, M.A.; Magarvey, N.A. Genomes to natural products PRediction Informatics for Secondary Metabolomes (PRISM). *Nucleic Acids Res.* **2015**, *43*, 9645–9662. [CrossRef] [PubMed]
186. Alanjary, M.; Kronmiller, B.; Adamek, M.; Blin, K.; Weber, T.; Huson, D.; Philmus, B.; Ziemert, N. The Antibiotic Resistant Target Seeker (ARTS), an exploration engine for antibiotic cluster prioritization and novel drug target discovery. *Nucleic Acids Res.* **2017**, *45*, W42–W48. [CrossRef] [PubMed]
187. Doyle, D.; McDowall, K.J.; Butler, M.J.; Hunter, I.S. Characterization of an oxytetracycline-resistance gene, *otrA*, of *Streptomyces rimosus*. *Mol. Microbiol.* **1991**, *5*, 2923–2933. [CrossRef] [PubMed]
188. Kling, A.; Lukat, P.; Almeida, D.V.; Bauer, A.; Fontaine, E.; Sordello, S.; Zaburannyi, N.; Herrmann, J.; Wenzel, S.C.; König, C.; et al. Targeting DnaN for tuberculosis therapy using novel griselimycins. *Science* **2015**, *348*, 1106–1112. [CrossRef] [PubMed]
189. Huo, L. Synthetic Biotechnology to Study and Engineer Natural Product Biosynthesis in Actinomycetes. Ph.D. Thesis, Saarland University, Saarbrücken, Germany, 2014.
190. Li, X.; Wu, X.; Zhu, J.; Shen, Y. Amexanthomycins A–J, pentangular polyphenols produced by *Amycolatopsis mediterranei* S699Δ*rifA*. *Appl. Microbiol. Biotechnol.* **2018**, *102*, 689–702. [CrossRef] [PubMed]
191. Rebets, Y.; Brötz, E.; Tokovenko, B.; Luzhetskyy, A. Actinomycetes biosynthetic potential: How to bridge *in silico* and *in vivo*? *J. Ind. Microbiol. Biotechnol.* **2014**, *41*, 387–402. [CrossRef] [PubMed]
192. Zerikly, M.; Challis, G.L. Strategies for the discovery of new natural products by genome mining. *ChemBioChem* **2009**, *10*, 625–632. [CrossRef] [PubMed]
193. Rigali, S.; Anderssen, S.; Naômé, A.; van Wezel, G.P. Cracking the regulatory code of biosynthetic gene clusters as a strategy for natural product discovery. *Biochem. Pharmacol.* **2018**. [CrossRef] [PubMed]
194. Metsä-Ketelä, M.; Ylihonko, K.; Mäntsälä, P. Partial activation of a silent angucycline-type gene cluster from a rubromycin β producing *Streptomyces* sp. PGA64. *J. Antibiot.* **2004**, *57*, 502–510. [CrossRef] [PubMed]
195. Bunet, R.; Song, L.; Mendes, M.V.; Corre, C.; Hotel, L.; Rouhier, N.; Framboisier, X.; Leblond, P.; Challis, G.L.; Aigle, B. Characterization and manipulation of the pathway-specific late regulator AlpW reveals *Streptomyces ambofaciens* as a new producer of kinamycins. *J. Bacteriol.* **2011**, *193*, 1142–1153. [CrossRef] [PubMed]
196. Sidda, J.D.; Song, L.; Poon, V.; Al-Bassam, M.; Lazos, O.; Buttner, M.J.; Challis, G.L.; Corre, C. Discovery of a family of γ-aminobutyrate ureas via rational derepression of a silent bacterial gene cluster. *Chem. Sci.* **2013**, *5*, 86–89. [CrossRef]
197. Laureti, L.; Song, L.J.; Huang, S.; Corre, C.; Leblond, P.; Challis, G.L.; Aigle, B. Identification of a bioactive 51-membered macrolide complex by activation of a silent polyketide synthase in *Streptomyces ambofaciens*. *Proc. Natl. Acad. Sci. USA* **2011**, *108*, 6258–6263. [CrossRef] [PubMed]
198. Olano, C.; García, I.; González, A.; Rodriguez, M.; Rozas, D.; Rubio, J.; Sánchez-Hidalgo, M.; Braña, A.F.; Méndez, C.; Salas, J.A. Activation and identification of five clusters for secondary metabolites in *Streptomyces albus* J1074. *Microb. Biotechnol.* **2014**, *7*, 242–256. [CrossRef] [PubMed]

199. Bode, E.; Brachmann, A.O.; Kegler, C.; Simsek, R.; Dauth, C.; Zhou, Q.; Kaiser, M.; Klemmt, P.; Bode, H.B. Simple "on-demand" production of bioactive natural products. *ChemBioChem* **2015**, *16*, 1115–1119. [CrossRef] [PubMed]
200. Siegl, T.; Tokovenko, B.; Myronovskyi, M.; Luzhetskyy, A. Design, construction and characterisation of a synthetic promoter library for fine-tuned gene expression in actinomycetes. *Metab. Eng.* **2013**, *19*, 98–106. [CrossRef] [PubMed]
201. Seghezzi, N.; Amar, P.; Koebmann, B.; Jensen, P.R.; Virolle, M.-J. The construction of a library of synthetic promoters revealed some specific features of strong *Streptomyces* promoters. *Appl. Microbiol. Biotechnol.* **2011**, *90*, 615–623. [CrossRef] [PubMed]
202. Baltz, R.H. Genetic manipulation of secondary metabolite biosynthesis for improved production in *Streptomyces* and other actinomycetes. *J. Ind. Microbiol. Biotechnol.* **2016**, *43*, 343–370. [CrossRef] [PubMed]
203. Zaburannyi, N.; Rabyk, M.; Ostash, B.; Fedorenko, V.; Luzhetskyy, A. Insights into naturally minimised *Streptomyces albus* J1074 genome. *BMC Genom.* **2014**, *15*. [CrossRef] [PubMed]
204. Gomez-Escribano, J.P.; Bibb, M.J. Engineering Streptomyces coelicolor for heterologous expression of secondary metabolite gene clusters. *Microb. Biotechnol.* **2011**, *4*, 207–215. [CrossRef] [PubMed]
205. Okada, B.K.; Seyedsayamdost, M.R. Antibiotic dialogues: Induction of silent biosynthetic gene clusters by exogenous small molecules. *FEMS Microbiol. Rev.* **2017**, *41*, 19–33. [CrossRef] [PubMed]
206. Rutledge, P.J.; Challis, G.L. Discovery of microbial natural products by activation of silent biosynthetic gene clusters. *Nat. Rev. Microbiol.* **2015**, *13*, 509–523. [CrossRef] [PubMed]
207. Seyedsayamdost, M.R. High-throughput platform for the discovery of elicitors of silent bacterial gene clusters. *Proc. Natl. Acad. Sci. USA* **2014**, *111*, 7266–7271. [CrossRef] [PubMed]
208. Locatelli, F.M.; Goo, K.-S.; Ulanova, D. Effects of trace metal ions on secondary metabolism and the morphological development of streptomycetes. *Metallomics* **2016**, *8*, 469–480. [CrossRef] [PubMed]
209. Hosaka, T.; Ohnishi-Kameyama, M.; Muramatsu, H.; Murakami, K.; Tsurumi, Y.; Kodani, S.; Yoshida, M.; Fujie, A.; Ochi, K. Antibacterial discovery in actinomycetes strains with mutations in RNA polymerase or ribosomal protein S12. *Nat. Biotechnol.* **2009**, *27*, 462–464. [CrossRef] [PubMed]
210. Tanaka, Y.; Kasahara, K.; Hirose, Y.; Murakami, K.; Kugimiya, R.; Ochi, K. Activation and products of the cryptic secondary metabolite biosynthetic gene clusters by rifampin resistance (*rpoB*) mutations in actinomycetes. *J. Bacteriol.* **2013**, *195*, 2959–2970. [CrossRef] [PubMed]
211. Ochi, K.; Hosaka, T. New strategies for drug discovery: Activation of silent or weakly expressed microbial gene clusters. *Appl. Microbiol. Biotechnol.* **2013**, *97*, 87–98. [CrossRef] [PubMed]
212. Onaka, H.; Tabata, H.; Igarashi, Y.; Sato, Y.; Furumai, T. Goadsporin, a chemical substance which promotes secondary metabolism and morphogenesis in streptomycetes. I. Purification and Characterization. *J. Antibiot.* **2001**, *54*, 1036–1044. [CrossRef] [PubMed]
213. Truong, T.T.; Seyedsayamdost, M.; Greenberg, E.P.; Chandler, J.R. A Burkholderia thailandensis acyl-homoserine lactone-independent orphan luxr homolog that activates production of the cytotoxin malleilactone. *J. Bacteriol.* **2015**, *197*, 3456–3462. [CrossRef] [PubMed]
214. Yamanaka, K.; Oikawa, H.; Ogawa, H.O.; Hosono, K.; Shinmachi, F.; Takano, H.; Sakuda, S.; Beppu, T.; Ueda, K. Desferrioxamine E produced by *Streptomyces griseus* stimulates growth and development of *Streptomyces tanashiensis*. *Microbiology* **2005**, *151*, 2899–2905. [CrossRef] [PubMed]
215. Craney, A.; Ozimok, C.; Pimentel-Elardo, S.M.; Capretta, A.; Nodwell, J.R. Chemical perturbation of secondary metabolism demonstrates important links to primary metabolism. *Chem. Biol.* **2012**, *19*, 1020–1027. [CrossRef] [PubMed]
216. Pimentel-Elardo, S.M.; Sørensen, D.; Ho, L.; Ziko, M.; Bueler, S.A.; Lu, S.; Tao, J.; Moser, A.; Lee, R.; Agard, D.; et al. Activity-independent discovery of secondary metabolites using chemical elicitation and cheminformatic inference. *ACS Chem. Biol.* **2015**, *10*, 2616–2623. [CrossRef] [PubMed]
217. Lombardi, P.M.; Angell, H.D.; Whittington, D.A.; Flynn, E.F.; Rajashankar, K.R.; Christianson, D.W. Structure of prokaryotic polyamine deacetylase reveals evolutionary functional relationships with eukaryotic histone deacetylases. *Biochemistry* **2011**, *50*, 1808–1817. [CrossRef] [PubMed]
218. Gacek, A.; Strauss, J. The chromatin code of fungal secondary metabolite gene clusters. *Appl. Microbiol. Biotechnol.* **2012**, *95*, 1389–1404. [CrossRef] [PubMed]
219. Katz, L.; Baltz, R.H. Natural product discovery: Past, present, and future. *J. Ind. Microbiol. Biotechnol.* **2016**, *43*, 155–176. [CrossRef] [PubMed]

220. Penesyan, A.; Kjelleberg, S.; Egan, S. Development of novel drugs from marine surface associated microorganisms. *Mar. Drugs* **2010**, *8*, 438–459. [CrossRef] [PubMed]
221. Chen, R.; Zhang, Q.; Tan, B.; Zheng, L.; Li, H.; Zhu, Y.; Zhang, C. Genome mining and activation of a silent PKS/NRPS gene cluster direct the production of totopotensamides. *Org. Lett.* **2017**, *19*, 5697–5700. [CrossRef] [PubMed]
222. Reen, F.J.; Romano, S.; Dobson, A.D.W.; O'Gara, F. The Sound of silence: Activating silent biosynthetic gene clusters in marine microorganisms. *Mar. Drugs* **2015**, *13*, 4754–4783. [CrossRef] [PubMed]
223. Thanapipatsiri, A.; Gomez-Escribano, J.P.; Song, L.; Bibb, M.J.; Al-Bassam, M.; Chandra, G.; Thamchaipenet, A.; Challis, G.L.; Bibb, M.J. Discovery of unusual biaryl polyketides by activation of a silent *Streptomyces venezuelae* biosynthetic gene cluster. *ChemBioChem* **2016**, *17*, 2189–2198. [CrossRef] [PubMed]
224. Zhang, M.M.; Wong, F.T.; Wang, Y.; Luo, S.; Lim, Y.H.; Heng, E.; Yeo, W.L.; Cobb, R.E.; Enghiad, B.; Ang, E.L.; et al. CRISPR-Cas9 strategy for activation of silent *Streptomyces* biosynthetic gene clusters. *Nat. Chem. Biol.* **2017**. [CrossRef] [PubMed]
225. Challis, G.L. Genome mining for novel natural product discovery. *J. Med. Chem.* **2008**, *51*, 2618–2628. [CrossRef] [PubMed]
226. Hubert, J.; Nuzillard, J.-M.; Renault, J.-H. Dereplication strategies in natural product research: How many tools and methodologies behind the same concept? *Phytochem. Rev.* **2017**, *16*, 55–95. [CrossRef]
227. Breitling, R.; Ceniceros, A.; Jankevics, A.; Takano, E. Metabolomics for secondary metabolite research. *Metabolites* **2013**, *2013*, 1076–1083. [CrossRef] [PubMed]
228. Nguyen, Q.T.; Merlo, M.E.; Medema, M.H.; Jankevics, A.; Breitling, R.; Takano, E. Metabolomics methods for the synthetic biology of secondary metabolism. *FEBS Lett.* **2012**, *586*, 2177–2183. [CrossRef] [PubMed]
229. Wolfender, J.-L.; Marti, G.; Ferreira Queiroz, E. Advances in techniques for profiling crude extracts and for the rapid identification of natural products: Dereplication, quality control and metabolomics. *Curr. Org. Chem.* **2010**, *14*, 1808–1832. [CrossRef]
230. Hou, Y.; Braun, D.R.; Michel, C.R.; Klassen, J.L.; Adnani, N.; Wyche, T.P.; Bugni, T.S. Microbial strain prioritization using metabolomics tools for the discovery of natural products. *Anal. Chem.* **2012**, *84*, 4277–4283. [CrossRef] [PubMed]
231. Liu, W.-T.; Kersten, R.D.; Yang, Y.-L.; Moore, B.S.; Dorrestein, P.C. Imaging mass spectrometry and genome mining via short sequence tagging identified the anti-infective agent arylomycin in *Streptomyces roseosporus*. *J. Am. Chem. Soc.* **2011**, *133*, 18010–18013. [CrossRef] [PubMed]
232. Li, Q.; Chen, X.; Jiang, Y.; Jiang, C. Morphological identification of actinobacteria. In *Actinobacteria—Basics and Biotechnological Applications*; Dhanasekaran, D., Jiang, Y., Eds.; InTech: Rijeka, Croatia, 2016.
233. Onaka, H.; Mori, Y.; Igarashi, Y.; Furumai, T. Mycolic acid-containing bacteria induce natural-product biosynthesis in *Streptomyces* species. *Appl. Environ. Microbiol.* **2011**, *77*, 400–406. [CrossRef] [PubMed]
234. Yang, Y.L.; Xu, Y.; Straight, P.; Dorrestein, P.C. Translating metabolic exchange with imaging mass spectrometry. *Nat. Chem. Biol.* **2009**, *5*, 885–887. [CrossRef] [PubMed]
235. Esquenazi, E.; Yang, Y.L.; Watrous, J.; Gerwick, W.H.; Dorrestein, P.C. Imaging mass spectrometry of natural products. *Nat. Prod. Rep.* **2009**, *26*, 1521–1534. [CrossRef] [PubMed]
236. Watrous, J.D.; Dorrestein, P.C. Imaging mass spectrometry in microbiology. *Nat. Rev. Microbiol.* **2011**, *9*, 683–694. [CrossRef] [PubMed]
237. Costa, A.B.; Cooks, R.G. Simulation of atmospheric transport and droplet-thin film collisions in desorption electrospray ionization. *Chem. Commun.* **2007**, 3915–3917. [CrossRef] [PubMed]
238. Kersten, R.D.; Yang, Y.L.; Xu, Y.; Cimermancic, P.; Nam, S.J.; Fenical, W.; Fischbach, M.A.; Moore, B.S.; Dorrestein, P.C. A mass spectrometry-guided genome mining approach for natural product peptidogenomics. *Nat. Chem. Biol.* **2011**, *7*, 794–802. [CrossRef] [PubMed]
239. Vaidyanathan, S.; Fletcher, J.S.; Goodacre, R.; Lockyer, N.P.; Micklefield, J.; Vickerman, J.C. Subsurface biomolecular imaging of *Streptomyces coelicolor* using secondary ion mass spectrometry. *Anal. Chem.* **2008**, *80*, 1942–1951. [CrossRef] [PubMed]
240. Traxler, M.F.; Watrous, J.D.; Alexandrov, T.; Dorrestein, P.C.; Kolter, R. Interspecies interactions stimulate diversification of the *Streptomyces coelicolor* secreted metabolome. *mBio* **2013**, *4*. [CrossRef] [PubMed]

241. Nguyen, S.N.; Liyu, A.V.; Chu, R.K.; Anderton, C.R.; Laskin, J. Constant-distance mode nanospray desorption electrospray ionization mass spectrometry imaging of biological samples with complex topography. *Anal. Chem.* **2017**, *89*, 1131–1137. [CrossRef] [PubMed]
242. Gonzalez, D.J.; Haste, N.M.; Hollands, A.; Fleming, T.C.; Hamby, M.; Pogliano, K.; Nizet, V.; Dorrestein, P.C. Microbial competition between *Bacillus subtilis* and *Staphylococcus aureus* monitored by imaging mass spectrometry. *Microbiology* **2011**, *157*, 2485–2492. [CrossRef] [PubMed]
243. Barger, S.R.; Hoefler, B.C.; Cubillos-Ruiz, A.; Russell, W.K.; Russell, D.H.; Straight, P.D. Imaging secondary metabolism of *Streptomyces* sp. Mg1 during cellular lysis and colony degradation of competing *Bacillus subtilis*. *Anton. Leeuwenhoek* **2012**, *102*, 435–445. [CrossRef] [PubMed]
244. Boya, P.C.A.; Fernández-Marín, H.; Mejía, L.C.; Spadafora, C.; Dorrestein, P.C.; Gutiérrez, M. Imaging mass spectrometry and MS/MS molecular networking reveals chemical interactions among cuticular bacteria and pathogenic fungi associated with fungus-growing ants. *Sci. Rep.* **2017**, *7*. [CrossRef] [PubMed]
245. Hoefler, B.C.; Stubbendieck, R.M.; Josyula, N.K.; Moisan, S.M.; Schulze, E.M.; Straight, P.D. A link between linearmycin biosynthesis and extracellular vesicle genesis connects specialized metabolism and bacterial membrane physiology. *Cell Chem. Biol.* **2017**, *24*, 1238–1249. [CrossRef] [PubMed]
246. Schmitt, E.K.; Moore, C.M.; Krastel, P.; Petersen, F. Natural products as catalysts for innovation: A pharmaceutical industry perspective. *Curr. Opin. Chem. Biol.* **2011**, *15*, 497–504. [CrossRef] [PubMed]
247. Exarchou, V.; Krucker, M.; van Beek, T.A.; Vervoort, J.; Gerothanassis, I.P.; Albert, K. LC-NMR coupling technology: Recent advancements and applications in natural products analysis. *Magn. Reson. Chem.* **2005**, *43*, 681–687. [CrossRef] [PubMed]
248. Schroeder, F.C.; Gronquist, M. Extending the scope of NMR spectroscopy with microcoil probes. *Angew. Chem. Int. Ed. Engl.* **2006**, *45*, 7122–7131. [CrossRef] [PubMed]
249. Iwasa, K.; Takahashi, T.; Nishiyama, Y.; Moriyasu, M.; Sugiura, M.; Takeuchi, A.; Tode, C.; Tokuda, H.; Takeda, K. Online structural elucidation of alkaloids and other constituents in crude extracts and cultured cells of *Nandina domestica* by combination of LC-MS/MS, LC-NMR, and LC-CD analyses. *J. Nat. Prod.* **2008**, *71*, 1376–1385. [CrossRef] [PubMed]
250. Iwasa, K.; Cui, W.; Takahashi, T.; Nishiyama, Y.; Kamigauchi, M.; Koyama, J.; Takeuchi, A.; Moriyasu, M.; Takeda, K. Biotransformation of phenolic tetrahydroprotoberberines in plant cell cultures followed by LC-NMR, LC-MS, and LC-CD. *J. Nat. Prod.* **2010**, *73*, 115–122. [CrossRef] [PubMed]
251. Cui, W.; Iwasa, K.; Sugiura, M.; Takeuchi, A.; Tode, C.; Nishiyama, Y.; Moriyasu, M.; Tokuda, H.; Takeda, K. Biotransformation of phenolic 1-benzyl-*N*-methyltetrahydroisoquinolines in plant cell cultures followed by LC/NMR, LC/MS, and LC/CD. *J. Nat. Prod.* **2007**, *70*, 1771–1778. [CrossRef] [PubMed]
252. Exarchou, V.; Fiamegos, Y.C.; van Beek, T.A.; Nanos, C.; Vervoort, J. Hyphenated chromatographic techniques for the rapid screening and identification of antioxidants in methanolic extracts of pharmaceutically used plants. *J. Chromatogr. A* **2006**, *1112*, 293–302. [CrossRef] [PubMed]
253. Krucker, M.; Lienau, A.; Putzbach, K.; Grynbaum, M.D.; Schuler, P.; Albert, K. Hyphenation of capillary HPLC to microcoil ^{1}H NMR spectroscopy for the determination of tocopherol homologues. *Anal. Chem.* **2004**, *76*, 2623–2628. [CrossRef] [PubMed]
254. Zehl, M.; Braunberger, C.; Conrad, J.; Crnogorac, M.; Krasteva, S.; Vogler, B.; Beifuss, U.; Krenn, L. Identification and quantification of flavonoids and ellagic acid derivatives in therapeutically important *Drosera* species by LC-DAD, LC-NMR, NMR, and LC-MS. *Anal. Bioanal. Chem.* **2011**, *400*, 2565–2576. [CrossRef] [PubMed]
255. Hu, J.-F.; Garo, E.; Yoo, H.-D.; Cremin, P.A.; Goering, M.G.; O'Neil-Johnson, M.; Eldridge, G.R. Cyclolignans from *Scyphocephalium ochocoa* via high-throughput natural product chemistry methods. *Phytochemistry* **2005**, *66*, 1077–1082. [CrossRef] [PubMed]
256. Johansen, K.T.; Wubshet, S.G.; Nyberg, N.T.; Jaroszewski, J.W. From retrospective assessment to prospective decisions in natural product isolation: HPLC-SPE-NMR analysis of *Carthamus oxyacantha*. *J. Nat. Prod.* **2011**, *74*, 2454–2461. [CrossRef] [PubMed]
257. Lang, G.; Mayhudin, N.A.; Mitova, M.I.; Sun, L.; van der, S.S.; Blunt, J.W.; Cole, A.L.; Ellis, G.; Laatsch, H.; Munro, M.H. Evolving trends in the dereplication of natural product extracts: New methodology for rapid, small-scale investigation of natural product extracts. *J. Nat. Prod.* **2008**, *71*, 1595–1599. [CrossRef] [PubMed]

258. Keller, L.; Plaza, A.; Dubiella, C.; Groll, M.; Kaiser, M.; Muller, R. Macyranones: Structure, biosynthesis, and binding mode of an unprecedented epoxyketone that targets the 20S proteasome. *J. Am. Chem. Soc.* **2015**, *137*, 8121–8130. [CrossRef] [PubMed]
259. Lin, Y.; Schiavo, S.; Orjala, J.; Vouros, P.; Kautz, R. Microscale LC-MS-NMR platform applied to the identification of active cyanobacterial metabolites. *Anal. Chem.* **2008**, *80*, 8045–8054. [CrossRef] [PubMed]
260. Pham, L.H.; Vater, J.; Rotard, W.; Mügge, C. Identification of secondary metabolites from *Streptomyces violaceoruber* TU22 by means of *on-flow* LC-NMR and LC-DAD-MS. *Magn. Reson. Chem.* **2005**, *43*, 710–723. [CrossRef] [PubMed]
261. Alshaibani, M.; Zin, N.; Jalil, J.; Sidik, N.; Ahmad, S.J.; Kamal, N.; Edrada-Ebel, R. Isolation, purification, and characterization of five active diketopiperazine derivatives from endophytic *Streptomyces* SUK 25 with antimicrobial and cytotoxic activities. *J. Microbiol. Biotechnol.* **2017**, *27*, 1249–1256. [CrossRef] [PubMed]
262. Shah, A.; Rather, M.A.; Hassan, Q.P.; Aga, M.A.; Mushtaq, S.; Shah, A.M.; Hussain, A.; Baba, S.A.; Ahmad, Z. Discovery of anti-microbial and anti-tubercular molecules from *Fusarium solani*: An endophyte of *Glycyrrhiza glabra*. *J. Appl. Microbiol.* **2017**, *122*, 1168–1176. [CrossRef] [PubMed]
263. Greule, A.; Zhang, S.; Paululat, T.; Bechthold, A. From a natural product to its biosynthetic gene cluster: a demonstration using polyketomycin from *Streptomyces diastatochromogenes* Tü6028. *J. Vis. Exp.* **2017**. [CrossRef] [PubMed]
264. Gibitz Eisath, N.; Sturm, S.; Stuppner, H. Supercritical fluid chromatography in natural product analysis—An update. *Planta Med.* **2017**. [CrossRef] [PubMed]
265. Khater, S.; Lozac'h, M.-A.; Adam, I.; Francotte, E.; West, C. Comparison of liquid and supercritical fluid chromatography mobile phases for enantioselective separations on polysaccharide stationary phases. *J. Chromatogr. A* **2016**, *1467*, 463–472. [CrossRef] [PubMed]
266. De Klerck, K.; Mangelings, D.; Vander Heyden, Y. Supercritical fluid chromatography for the enantioseparation of pharmaceuticals. *J. Pharm. Biomed. Anal.* **2012**, *69*, 77–92. [CrossRef] [PubMed]
267. Novakova, L.; Perrenoud, A.G.-G.; Francois, I.; West, C.; Lesellier, E.; Guillarme, D. Modern analytical supercritical fluid chromatography using columns packed with sub-2 µm particles: A tutorial. *Anal. Chim. Acta* **2014**, *824*, 18–35. [CrossRef] [PubMed]
268. Lesellier, E.; West, C. The many faces of packed column supercritical fluid chromatography—A critical review. *J. Chromatogr. A* **2015**, *1382*, 2–46. [CrossRef] [PubMed]
269. West, C.; Lemasson, E.; Bertin, S.; Hennig, P.; Lesellier, E. An improved classification of stationary phases for ultra-high performance supercritical fluid chromatography. *J. Chromatogr. A* **2016**, *1440*, 212–228. [CrossRef] [PubMed]
270. Wang, M.; Carrell, E.J.; Chittiboyina, A.G.; Avula, B.; Wang, Y.-H.; Zhao, J.; Parcher, J.F.; Khan, I.A. Concurrent supercritical fluid chromatographic analysis of terpene lactones and ginkgolic acids in *Ginkgo biloba* extracts and dietary supplements. *Anal. Bioanal. Chem.* **2016**, *408*, 4649–4660. [CrossRef] [PubMed]
271. Huang, Y.; Feng, Y.; Tang, G.; Li, M.; Zhang, T.; Fillet, M.; Crommen, J.; Jiang, Z. Development and validation of a fast SFC method for the analysis of flavonoids in plant extracts. *J. Pharm. Biomed. Anal.* **2017**, *140*, 384–391. [CrossRef] [PubMed]
272. Vicente, G.; García-Risco, M.R.; Fornari, T.; Reglero, G. Isolation of carnosic acid from rosemary extracts using semi-preparative supercritical fluid chromatography. *J. Chromatogr. A* **2013**, *1286*, 208–215. [CrossRef] [PubMed]
273. Salimi, A.; Fatemi, S.; Zakizadeh Nei Nei, H.; Safaralie, A. Mathematical modeling of supercritical extraction of valerenic acid from *Valeriana officinalis* L. *Chem. Eng. Technol.* **2008**, *31*, 1470–1480. [CrossRef]
274. Ramsden, J.J. Metabolomics and metabonomics. In *Bioinformatics: An Introduction*, 2nd ed.; Ramsden, J., Ed.; Springer: New York, NY, USA, 2009; pp. 1–6.
275. Mohamed, A.; Nguyen, C.H.; Mamitsuka, H. Current status and prospects of computational resources for natural product dereplication: A review. *Brief. Bioinform.* **2016**, *17*, 309–321. [CrossRef] [PubMed]
276. Beutler, J.A.; Alvarado, A.B.; Schaufelberger, D.E.; Andrews, P.; McCloud, T.G. Dereplication of phorbol bioactives: *Lyngbya majuscula* and *Croton cuneatus*. *J. Nat. Prod.* **1990**, *53*, 867–874. [CrossRef] [PubMed]
277. Chanana, S.; Thomas, C.S.; Braun, D.R.; Hou, Y.; Wyche, T.P.; Bugni, T.S. Natural product discovery using planes of principal component analysis in R (PoPCAR). *Metabolites* **2017**, *7*. [CrossRef] [PubMed]
278. Bitzer, J.; Köpcke, B.; Stadler, M.; Hellwig, V.; Ju, Y.-M.; Seip, S.; Henkel, T. Accelerated dereplication of natural products, supported by reference libraries. *CHIMIA* **2007**, *61*, 332–338. [CrossRef]

279. Dictionary of Natural Products. Available online: http://dnp.chemnetbase.com/faces/chemical/ChemicalSearch.xhtml (accessed on 31 January 2018).
280. Laatsch, H. *AntiBase 2014: The Natural Compound Identifier*; Wiley-VCH: Weinheim, Germany, 2014.
281. Blunt, J.W.; Copp, B.R.; Keyzers, R.A.; Munro, M.H.G.; Prinsep, M.R. Marine natural products. *Nat. Prod. Rep.* **2017**, *34*, 235–294. [CrossRef] [PubMed]
282. Horai, H.; Arita, M.; Kanaya, S.; Nihei, Y.; Ikeda, T.; Suwa, K.; Ojima, Y.; Tanaka, K.; Tanaka, S.; Aoshima, K.; et al. MassBank: A public repository for sharing mass spectral data for life sciences. *J. Mass Spectrom.* **2010**, *45*, 703–714. [CrossRef] [PubMed]
283. Smith, C.A.; O'Maille, G.; Want, E.J.; Qin, C.; Trauger, S.A.; Brandon, T.R.; Custodio, D.E.; Abagyan, R.; Siuzdak, G. METLIN: A metabolite mass spectral database. *Ther. Drug Monit.* **2005**, *27*, 747–751. [CrossRef] [PubMed]
284. mzCloud: Advanced Mass Spectral Database. Available online: https://www.mzcloud.org/ (accessed on 10 May 2018).
285. Sawada, Y.; Nakabayashi, R.; Yamada, Y.; Suzuki, M.; Sato, M.; Sakata, A.; Akiyama, K.; Sakurai, T.; Matsuda, F.; Aoki, T.; et al. RIKEN tandem mass spectral database (ReSpect) for phytochemicals: A plant-specific MS/MS-based data resource and database. *Phytochemistry* **2012**, *82*, 38–45. [CrossRef] [PubMed]
286. Johnston, C.W.; Skinnider, M.A.; Wyatt, M.A.; Li, X.; Ranieri, M.R.M.; Yang, L.; Zechel, D.L.; Ma, B.; Magarvey, N.A. An automated Genomes-to-Natural Products platform (GNP) for the discovery of modular natural products. *Nat. Commun.* **2015**, *6*. [CrossRef] [PubMed]
287. Wang, M.; Carver, J.J.; Phelan, V.V.; Sanchez, L.M.; Garg, N.; Peng, Y.; Nguyen, D.D.; Watrous, J.; Kapono, C.A.; Luzzatto-Knaan, T.; et al. Sharing and community curation of mass spectrometry data with Global Natural Products Social Molecular Networking. *Nat. Biotechnol.* **2016**, *34*, 828–837. [CrossRef] [PubMed]
288. Crüsemann, M.; O'Neill, E.C.; Larson, C.B.; Melnik, A.V.; Floros, D.J.; da Silva, R.R.; Jensen, P.R.; Dorrestein, P.C.; Moore, B.S. Prioritizing natural product diversity in a collection of 146 bacterial strains based on growth and extraction protocols. *J. Nat. Prod.* **2016**. [CrossRef] [PubMed]
289. Mohimani, H.; Gurevich, A.; Mikheenko, A.; Garg, N.; Nothias, L.-F.; Ninomiya, A.; Takada, K.; Dorrestein, P.C.; Pevzner, P.A. Dereplication of peptidic natural products through database search of mass spectra. *Nat. Chem. Biol.* **2017**, *13*, 30–37. [CrossRef] [PubMed]
290. Gurevich, A.; Mikheenko, A.; Shlemov, A.; Korobeynikov, A.; Mohimani, H.; Pevzner, P.A. Increased diversity of peptidic natural products revealed by modification-tolerant database search of mass spectra. *Nat. Microbiol.* **2018**. [CrossRef] [PubMed]
291. Zierep, P.F.; Padilla, N.; Yonchev, D.G.; Telukunta, K.K.; Klementz, D.; Günther, S. SeMPI: A genome-based secondary metabolite prediction and identification web server. *Nucleic Acids Res.* **2017**, *45*, W64–W71. [CrossRef] [PubMed]
292. Blin, K.; Medema, M.H.; Kazempour, D.; Fischbach, M.A.; Breitling, R.; Takano, E.; Weber, T. antiSMASH 2.0—A versatile platform for genome mining of secondary metabolite producers. *Nucleic Acids Res.* **2013**, *41*, W204–W212. [CrossRef] [PubMed]
293. Medema, M.H.; Paalvast, Y.; Nguyen, D.D.; Melnik, A.; Dorrestein, P.C.; Takano, E.; Breitling, R. Pep2Path: Automated mass spectrometry-guided genome mining of peptidic natural products. *PLoS Comput. Biol.* **2014**, *10*, e1003822. [CrossRef] [PubMed]
294. Mohimani, H.; Kersten, R.D.; Liu, W.-T.; Wang, M.; Purvine, S.O.; Wu, S.; Brewer, H.M.; Pasa-Tolic, L.; Bandeira, N.; Moore, B.S.; et al. Automated genome mining of ribosomal peptide natural products. *ACS Chem. Biol.* **2014**, *9*, 1545–1551. [CrossRef] [PubMed]
295. Blin, K.; Medema, M.H.; Kottmann, R.; Lee, S.Y.; Weber, T. The antiSMASH database, a comprehensive database of microbial secondary metabolite biosynthetic gene clusters. *Nucleic Acids Res.* **2017**, *45*, D555–D559. [CrossRef] [PubMed]
296. Liu, W.T.; Lamsa, A.; Wong, W.R.; Boudreau, P.D.; Kersten, R.; Peng, Y.; Moree, W.J.; Duggan, B.M.; Moore, B.S.; Gerwick, W.H.; et al. MS/MS-based networking and peptidogenomics guided genome mining revealed the stenothricin gene cluster in *Streptomyces roseosporus*. *J. Antibiot.* **2014**, *67*, 99–104. [CrossRef] [PubMed]

297. Udwary, D.W.; Zeigler, L.; Asolkar, R.N.; Singan, V.; Lapidus, A.; Fenical, W.; Jensen, P.R.; Moore, B.S. Genome sequencing reveals complex secondary metabolome in the marine actinomycete *Salinispora tropica*. *Proc. Natl. Acad. Sci. USA* **2007**, *104*, 10376–10381. [CrossRef] [PubMed]
298. Kjaerulff, L.; Sikandar, A.; Zaburannyi, N.; Adam, S.; Herrmann, J.; Koehnke, J.; Müller, R. Thioholgamides: Thioamide-containing cytotoxic RiPPs natural products. *ACS Chem. Biol.* **2017**, *12*, 2837–2841. [CrossRef] [PubMed]
299. Salvador-Reyes, L.A.; Luesch, H. Biological targets and mechanisms of action of natural products from marine cyanobacteria. *Nat. Prod. Rep.* **2015**, *32*, 478–503. [CrossRef] [PubMed]
300. Wenzel, S.C.; Müller, R. Myxobacteria—Unique microbial secondary metabolite factories. In *Comprehensive Natural Products Chemistry II, Vol 2: Structural Diversity II—Secondary Metabolite Sources, Evolution and Selected Molecular Structures*; Moore, B.S., Ed.; Elsevier: Oxford, UK, 2010; pp. 189–222.
301. Tiwari, K.; Gupta, R.K. Rare actinomycetes: A potential storehouse for novel antibiotics. *Crit. Rev. Biotechnol.* **2012**, *32*, 108–132. [CrossRef] [PubMed]
302. Jensen, P.R. Natural products and the gene cluster revolution. *Trends Microbiol.* **2016**, *24*, 968–977. [CrossRef] [PubMed]
303. Goss, R.J.M.; Shankar, S.; Fayad, A.A. The generation of "unnatural" products: Synthetic biology meets synthetic chemistry. *Nat. Prod. Rep.* **2012**, *29*, 870–889.
304. Smanski, M.J.; Zhou, H.; Claesen, J.; Shen, B.; Fischbach, M.A.; Voigt, C.A. Synthetic biology to access and expand nature's chemical diversity. *Nat. Rev. Microbiol.* **2016**, *14*, 135–149. [CrossRef] [PubMed]

antibiotics

MDPI

Review

Streptomyces Differentiation in Liquid Cultures as a Trigger of Secondary Metabolism

Ángel Manteca and Paula Yagüe *

Área de Microbiología, Departamento de Biología Funcional IUOPA, Facultad de Medicina, Universidad de Oviedo, 33006 Oviedo, Spain; mantecaangel@uniovi.es

* Correspondence: paula.yague@gmail.com; Tel.: +34-9-8510-3000 (ext. 5289)

Received: 26 February 2018; Accepted: 9 May 2018; Published: 14 May 2018

check for updates

Abstract: *Streptomyces* is a diverse group of gram-positive microorganisms characterised by a complex developmental cycle. *Streptomycetes* produce a number of antibiotics and other bioactive compounds used in the clinic. Most screening campaigns looking for new bioactive molecules from actinomycetes have been performed empirically, e.g., without considering whether the bacteria are growing under the best developmental conditions for secondary metabolite production. These screening campaigns were extremely productive and discovered a number of new bioactive compounds during the so-called "golden age of antibiotics" (until the 1980s). However, at present, there is a worrying bottleneck in drug discovery, and new experimental approaches are needed to improve the screening of natural actinomycetes. *Streptomycetes* are still the most important natural source of antibiotics and other bioactive compounds. They harbour many cryptic secondary metabolite pathways not expressed under classical laboratory cultures. Here, we review the new strategies that are being explored to overcome current challenges in drug discovery. In particular, we focus on those aimed at improving the differentiation of the antibiotic-producing mycelium stage in the laboratory.

Keywords: *streptomyces*; screening; antibiotics; secondary metabolism; differentiation; elicitors; morphology; liquid cultures

1. Introduction

The *Streptomyces* genus includes an important group of biotechnological bacteria. They produce two-thirds of the antibiotics of medical and agricultural interest, several antitumor agents, antifungals, and a great number of eukaryotic cell differentiation effectors, such as apoptosis inducers and inhibitors [1]. Drug discovery from *streptomycetes* fell considerably after initial screenings where the most common compounds were discovered. Antibiotic resistance is increasing dramatically, and new antibiotics are urgently required in the clinic. Alternative methods, such as the exploration of chemical libraries and combinatorial chemistry, have provided limited yields. Screening from nature has resumed through methods such as exploring new environments, looking for elicitors, accessing the metagenome, etc.

One of the most important characteristics of *Streptomyces* is its complex life cycle, which is closely related to secondary metabolite production [2] (outlined in Figure 1). In solid sporulating cultures, development starts with spore germination and the rapid development of compartmentalised hyphae into the medium (early substrate mycelium or MI) [3]. After that, programmed cell death (PCD) occurs (red cellular segments in Figure 1) which triggers the differentiation of the multinucleated (MII) antibiotic-producing hyphae (late substrate mycelium, early MII) [3,4]. Then, the mycelium starts to grow into the air forming the aerial mycelium (late MII). At the end of the cycle, there is a second round of PCD, and most of the remaining viable hyphae undergo a process of compartmentalisation that culminates in the formation of unigenomic spores [5].

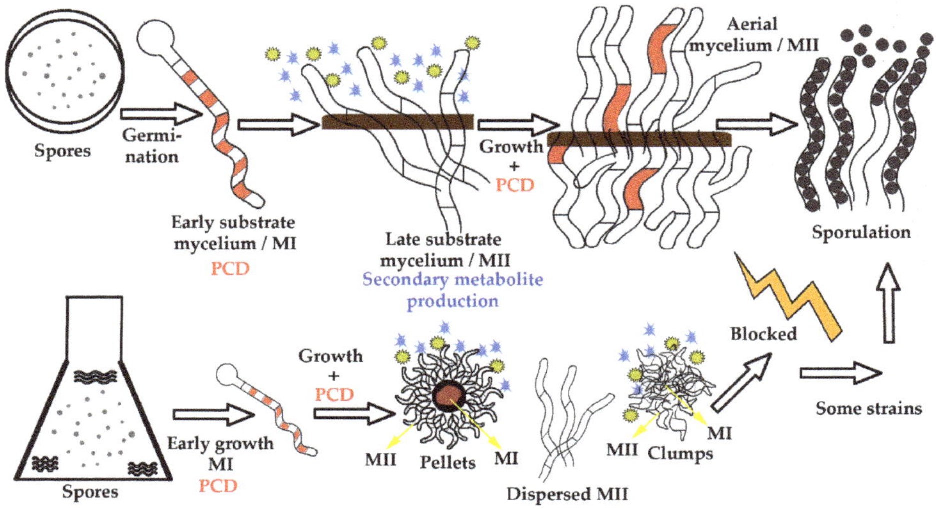

Figure 1. *Streptomyces* growth in solid cultures (**upper panels**) and liquid cultures (**lower panels**). In solid cultures (petri plates), spores germinate developing a compartmentalised mycelium (early substrate mycelium, MI) with 1 µm average cross-membrane spacing [6]. Some of the MI cells suffer a first round of programmed cell death PCD (red segments). The remaining viable segments start to grow as a multinucleated mycelium with sporadic septa (early MII, late substrate mycelium) [6]. The mycelium substrate suffers a second round of PCD (red segments) and differentiates into a mycelium that starts to grow into the air (the medium/agar border is indicated by a brown line) (late MII, aerial mycelium). Part of the aerial hyphae form spore chains (black circles). In liquid cultures, there is germination, MI development, PCD (in the centre of the mycelial pellets) and MII differentiation (in the periphery of the pellets). In most species, there is no aerial mycelium formation or sporulation, and hyphae form pellets and clumps [2]. Secondary metabolites (outlined as yellow circles and blue starts) are produced by the MII hyphae.

Most *streptomycetes* do not sporulate in liquid cultures. Therefore, it was previously assumed that under these conditions, there was no differentiation. However, industrial antibiotic production is mostly performed in liquid cultures (flasks and bioreactors). Currently, it is known that in liquid cultures, differentiation is comparable to that observed in solid cultures (Figure 1). In liquid cultures, there is a first mycelium stage (MI), PCD and the differentiation of a secondary metabolite, producing mycelium (MII). However, in most *Streptomyces* strains, aerial mycelium formation and sporulation are blocked [6] (Figure 1). *S. coelicolor* proteomic and transcriptomic studies have shown that physiological differentiation in liquid and solid cultures is comparable [6,7]. MII expresses/translates the genes/proteins involved in secondary metabolism in both solid and liquid cultures [6,7].

Surprisingly, *Streptomyces* differentiation as a trigger for antibiotic production remains almost unexplored. The absence of a developmental model to describe differentiation in liquid cultures has inhibited the understanding of the relationship between macroscopic morphology (pellet and clump formation) and differentiation. Pellet and clump formation has been classically correlated with secondary metabolite production, but the relationship between both processes remains obscure. Most authors have affirmed that pellets and clumps are fundamental for secondary metabolite production (e.g., retamycin in *S. olindensis* [8], nikkomycins in *S. tendae* [9], hybrid antibiotics in *S. lividans* [10]), while some authors have affirmed that pellet and clump formation reduces antibiotic production (e.g., nystatin in *S. noursei* [11], tylosin in *S. fradiae* [12]). More recently, our group demonstrated that one of the key events in the activation of secondary metabolite production in

Streptomyces liquid cultures is the differentiation of MII (e.g., actinorhodin/undecylprodigiosin production in *S. coelicolor* [2,13], microbial transglutaminase production in *S. mobarensis* [14], apigenin and luteolin production in *S. albus* [15]). The differentiation of this mycelium is conditioned by PCD of the vegetative hyphae (MI) [2], which, in liquid cultures, depends on the growth rate of the strain and hypha aggregation (pellet/clump formation) [2,7,14–16]. However, secondary metabolism has additional regulations (elicitors activate specific biosynthetic pathways) [17], and most *Streptomyces* strains do not display all their potential secondary metabolites under standard developmental laboratory conditions, even if they are differentiated at the MII stage [7].

Each *Streptomyces* strain can harbour up to 30 secondary metabolite pathways, but only a few of these are active in usual screening processes [18]. Activating these pathways in the lab will be crucial in the process of screening for new secondary metabolites from actinomycetes. Here, we review the most important strategies that are being explored to activate cryptic pathways and/or those that are being explored to enhance secondary metabolites production.

2. Screening for New Secondary Metabolites from *Streptomycetes*

The search for new actinomycetes in unexplored niches or from the screening of strains that have not been previously cultivated is useful, but usually leads to the rediscovery of already known compounds [19]. New screening strategies are necessary to overcome the current challenges of discovering new bioactive compounds [19]. In 2013, Arryn Craney et al. [20] summarised the new strategies that are being used to enhance secondary metabolite production and activate cryptic pathways, dividing them into unselective and selective methods [20]. Unselective methods are non-specific methods that are used to screen for new activities, whereas selective methods are biosynthetic cluster-specific methods that are used to improve the production of already known molecules [20].

Non-specific methods were largely used during "the golden age of antibiotics", and they are still useful. These methods include classical strategies, such as changing media components, increasing general precursors (metabolic engineering), inducing stress responses (with heat/ethanol/salt/acid shock, nutrient limitations) [21], and obtaining strains that overproduce secondary metabolites by random mutagenesis [22–24]. More novel non-specific methods include ribosomal engineering (the alteration of ribosomal proteins to activate cryptic secondary metabolites in *streptomycetes*) [20,25] and the use of small molecules as elicitors of secondary metabolism [20,26] (Table 1). Differentiation of the antibiotic producer mycelium (MII) as a non-specific method to activate antibiotic production remains almost unexplored. There has been no previous analysis of the frequency of *Streptomyces* strains that do not produce secondary metabolites because they are not differentiated at the MII stage in the laboratory.

Biosynthetic cluster-specific methods include self-resistance engineering (upregulation of self-resistance genes), regulatory engineering (overexpression of activators or elimination of repressors) and genome mining to search for new biosynthetic pathways [20] (Table 2). One of the most important biosynthetic cluster-specific methods is heterologous expression. Heterologous expression has been used to express *Streptomyces* industrial enzymes, such as laccases, in microorganisms with simpler developmental cycles than *Streptomyces*, such as *E. coli* [27]. However, the complex biosynthetic pathways of *Streptomyces* rarely can be expressed in simple expression hosts, such as *E. coli* or *Bacillus*. Thus, other *streptomycetes*, such as *S. lividans*, *S. albus*, *S. coelicolor* or *S. avermitilis*, are commonly used as expression hosts [28]. The activation of cryptic metabolites through the expression of the *Streptomyces coelicolor* pleiotropic regulator, *Afs*Q, in other *streptomycetes* [29] has been successfully achieved. Combinatorial biosynthesis, chemical modification of existing molecules, has been largely developed over the last 20 years, in particular, progress has been made in the last few years thanks to genome mining and synthetic biology [30–32]. Differentiation of *Streptomyces* MII was successfully used to enhance the production of various products [2,13–15] through its role as a trigger for antibiotic production (described in Section 2.3).

2.1. *Streptomyces Differentiation Strategies Based on Elicitors*

In the last few years, effort has been made to elucidate the mechanism by which some small molecules (elicitors) affect differentiation and secondary metabolite production in *Streptomyces* strains. Elicitors can be defined as diffusible signals that are able to induce cryptic pathways and/or differentiation in *Streptomyces* cultures [17]. Some elicitors act as signals for interspecies interaction [33]. Thus, subinhibitory concentrations of certain antibiotics produced by a given *Streptomyces* strain accelerate differentiation and antibiotic production in other *Streptomyces* strains through "pseudo" gamma-butyrolactone receptors [33]. Another good strategy is the use of random chemical probes (natural or synthetic) as elicitors (reviewed in [21]).

One of the most common strategies used to activate secondary metabolism and differentiation is mimicking the ecological environment through co-cultures of different microbes [17,34]. This methodology typically uses species that have symbiotic relationships with *Streptomyces* in nature [35,36] or pathogen partners that activate the production of antimicrobial compounds [37–39]. For instance, fungal elicitors (complex mix of cell walls and filtered cultures) positively affect the production of natamycin [40], bacterial and yeast elicitors improve valinomycin production [41], nutrients such as glucose and xylose repress the production of actinorhodin [42,43], and small molecules, such as GlcNAc or phosphate, can trigger differentiation and antibiotic production in *S. coelicolor* through the activation of *act*II-ORF4/*redZ* genes [44].

Pimentel-Elardo et al. [45] developed an activity-independent screening method based on the use of elicitors, to prevent the rediscovery of the most active/abundant compounds. In addition, cheminformatics techniques are used to identify the putative biological activities of identified compounds [45]. The use of elicitors increases the production of low-abundant compounds which were undetected in the classical activity dependent screening. The chemical elicitor "CI-ARC" has been identified as being responsible for triggering several cryptic biosynthetic genes [45].

2.2. *Differentiation Strategies Based on Macroscopic Morphology*

2.2.1. The Genetic Control of Aggregation and Macroscopic Morphology in Liquid Cultures

Large-scale antibiotic production is mostly performed in liquid cultures. It is almost unanimously accepted that the macroscopic morphology of the mycelium (pellets and clump formation) is correlated with the production of secondary metabolites. However, it was not until recently that the genes controlling pellet and clump formation have been characterised. The *S. coelicolor mat* gene cluster [46] and the *cslA*, *glxA*, *dtpA* genes [47–49] are responsible for mycelial aggregation and pellet formation. These genes could be a great tool for controlling the morphology in industrial fermentation.

The *Streptomyces* life cycle in liquid cultures starts with the germination of spores. Awakening from the dormant spore state depends on the level of AMPc in the cultures [50] and involves the small hydrophobic protein NepA, [51]. The expression of several sigma factors involved in osmotic and oxidative stress (SigH, SigB, SigI, SigJ) undergoes remarkable changes during germination, indicating that germination evokes stress-like cell responses [52]. Several genes encoding proteins involved in lipid metabolism and membrane transport are overexpressed during germination [52]. The conservation of D-alanyl-D-alanine carboxypeptidase (SCO4439) contributes to the swelling phase of germination [53]. Cell wall hydrolases participate in germination [54]. SsgA protein marks the germinative tube emission points [55]. Recently it was described that during germination, spores aggregate due to extracellular glycans synthesized by the MatA, MatB [46,56] and the CslA/GlxA/DtpA proteins [56]. These aggregates determine the macroscopic morphology (pellets and clumps) of the culture [56] which triggers PCD and the physiological differentiation of the antibiotic producer, mycelium MII [2].

Another issue that influence secondary metabolite production is sporulation. Several *streptomycetes* are able to sporulate in liquid cultures [57] and some strains, that normally do not sporulate are also able to sporulate in bioreactors due to the stress generated in the fermenter [13]. Sporulation stops metabolism, including secondary metabolite production. Consequently, in industrial fermentations and during screening for new secondary metabolites, it is important to avoid sporulation to increase and maintain secondary metabolism for as long as possible [13].

2.2.2. Monitoring of *Streptomyces* Macroscopic Morphology and Differentiation in Liquid Cultures

Pellet and clump formation led to differentiation and secondary metabolism [2]. Consequently, new methodologies to monitor macroscopic morphology have been developed. Laser diffraction has been used to measure pellet size [58]. Flow cytometry has been used to establish pellet size distribution of culture populations [59,60]. Recently, a useful algorithm was developed as a plug-in for the open-source software, ImageJ, to characterize the morphology of filamentous microorganisms in liquid cultures [61]. Mathematical models have been performed to predict the behaviour of *Streptomyces* liquid cultures based on pellet/clump morphology [62,63].

Biophysical parameters (e.g., pH, viscosity, agitation, dissolved oxygen levels and surface tension, among others) directly affect morphology and differentiation [13,64]. These parameters must be considered when scaling up production to industrial conditions [65]. Interestingly, a recent study downscaled liquid cultures to the 100 μL scale in microtiter plates [66], reproducing the same range of production and morphology as large-scale bioreactors, making screening easier and facilitating further upscaling.

2.2.3. Macroscopic Morphology Conditions, Programmed Cell Death and Second Mycelium Differentiation in Liquid Cultures

PCD is the key event that triggers the differentiation of the antibiotic producer, mycelium (MII), in liquid and solid cultures [2]. However, the specific signals derived from cell death are not yet known. The production of *N*-acetylglucosamine from peptidoglycan dismantling accelerates development and antibiotic production [67,68] and might be one of the signals released during PCD.

A simple methodology based on fluorometric measures of cultures stained with SYTO9 and propidium iodide was designed to quantify PCD in liquid cultures [69]. This method allows the efficiency of antibiotic production to be predicted based on the level of PCD [69].

Strains showing dispersed growth take a long time to suffer PCD, and sometimes, PCD does not occur. Modify the developmental conditions to enhance PCD and MII differentiation, leads to an improvement in secondary metabolite production. This approach was recently applied to enhance flavonoid production in a strain of *Streptomyces albus* [15] and to enhance microbial transglutaminase production from *Streptomyces mobaraensis* [14]. The "PCD-MII" approach complements other approaches well; there is no secondary metabolite production without differentiation of MII, but there are biosynthetic pathways that in addition to MII differentiation, need specific elicitors to become active [70].

2.3. L-Forms

An interesting alternative that would avoid the problems of mycelial growth in industry, is the use of L-forms, which are individual cells without cell walls [71]. However, until now, the antibiotic levels reached by *Streptomyces* L-forms have been quite minor compared to those reached by the regular form. Therefore, future research should explore whether L-forms could offer an industrial alternative.

2.4. Other Strategies

A big challenge in screening for new secondary metabolites is exploring non-cultivated bacteria. The scientific community is aware of the huge quantity of microorganisms that are not cultivated under laboratory conditions. Next Generation Sequencing revealed the big pharmacological potential of uncultured bacteria. Innovative culturing techniques, such as the isolation chip (iChip), are being used successfully in combination with co-cultures to grow previously uncultured bacteria [72]. The study of unexplored niches to look for new Actinomycetes is another strategy that enables the discovery of new species and compounds [73–75]. The combination of these two methods is a promising strategy to identify new compounds.

One of the newest strategies focusses on primary metabolism and vegetative growth. Very recently, work by Schniete et al. [76] showed how genetic redundancy within actinobacterial genomes allows functional specialization of two pyruvate kinases in *Streptomyces* under different life cycle stages and environmental conditions. Genetic redundancy within actinobacteria genomes as being a key to understanding how the plasticity of this microorganism enhances the production of clinically useful molecules. Furthermore, Cihak et al. [77] recently described the production of secondary metabolites during germination in *Streptomyces coelicolor*. The germination stage was ignored in most secondary metabolite screening campaigns and constitutes a potential source of bioactive compounds to be explored [77].

Table 1. Non-specific methods and some successful examples of their enforcement. "Enhance" means an improvement in production; "cryptic" means activation of the expression of cryptic pathways.

Methods	Microorganism	Product	Effect	Ref.
Media manipulation	*S. roseosporus*	Daptomycin	Enhance	[78]
Stress Response	*S. venezuelae*	Jadomycin B	Enhance	[79]
	S. hygroscopicus	Validamycin A	Enhance	[21]
	S. parvulus	Manumycin family	Cryptic	[80]
	S. coelicolor	Ectoine, 5-hydroxyectoine	Enhance	[81]
	S. coelicolor	Methylenomycin	Enhance	[82]
One Strain Many Compounds (OSMAC)	*S. parvulus*	20 cryptic compounds	Cryptic	[80]
Random Mutagenesis	*S. clavuligerus*	Clavulanic acid	Enhance	[83]
	S. hygroscopicus	Rapamycin	Enhance	[84]
	S. coelicolor	Actinorhodin, Undecylprodigiosin	Enhance	[22]
Ribosomal Engineering	*S. coelicolor*	Actinorhodin	Enhance	[85]
Engineering Global Regulation	*S. coelicolor*	Actinorhodin, Prodigiosin, Calcium-Dependent Antibiotic	Enhance	[86]
	S. griseus	Streptomycin	Enhance	[86]
	S. griseochromogenes	Blasticidin S	Enhance	[86]
Elicitors	*S. coelicolor*	Actinorhodin	Enhance	[87]
	S. pristinaespiralis	Desferrioxamine B/E	Enhance	[20]
	S. peucetius	Doxorubicin, Baumycin	Enhance	[20]
	S. coelicolor	Actinorhodin, Undecylprodigiosin	Enhance	[68]
	S. lividans	Prodiginine	Enhance	[88]
	S. griseus	Streptomycin	Enhance	[21]
	S. natalensis	Pimaricin	Enhance	[89]
	29 strains	Cryptic compounds	Cryptic	[45]
Metabolic Engineering	*S. clavuligerus*	FK606	Enhance	[90]
	S. coelicolor	Actinorhodin	Enhance	[90]
	S. rimosus	Oxytetracycline	Enhance	[91]
Co-cultures	*S. rimosus* MY02	Antifungal activity	Enhance	[36]
	S. coelicolor	Actinorhodin	Enhance	[37]
	S. fradiae 007	Phenolic polyketides	Enhance	[38]
	Marine *streptomycetes*	See tables in reference	Cryptic	[39]
Conditioning Morphology (PCD + MII)	*S. cattleya*	Tienamycin	Enhance	[92]
	S. cinereoruber	Rodomycin	Enhance	
	Saccharopolyspora erythraea	Erithromycin	Enhance	
	S. coelicolor	Actinorhodin	Enhance	

Table 2. Biosynthetic cluster specific methods and some successful examples of their enforcement.

Methods	Microorganism	Product	Effect	Ref.
Engineering Self-Resistance	*S. peucetius*	Doxorubicin, Daunorubicin	Enhance	[93]
	S. avermitilis	Avermectin,	Enhance	[94]
	S. coelicolor	Actinorhodin	Enhance	[95]
Regulatory Engineering				
Delete repressor AbsA2~P	*S. coelicolor*	Actinorhodin, Undecylprodigiosin, Calcium-dependent antibiotic	Enhance	[96]
Overexpress AverR/StrR	*S. avermitilis*	Avermectin	Enhance	[97]
Overexpress AverR/StrR	*S. griseous*	Streptomycin	Enhance	[98]
Overexpress SamR0484	*S. ambofaciens*	Stambomicin A-D	Cryptic	[99]
Delete repressor *cmmRII*	*S. griseus*	Chromomycin	Enhance	[100]
Delete repressor AlpW	*S. ambofaciens*	Alpomycin	Enhance	[101]
Heterologous Expression	*S. avermitilis*	Streptomycin	Enhance	[102]
	S. coelicolor	Chloramphenicol	Enhance	[103]
	S. coelicolor	Congocidine	Enhance	[103]
	S. cyaneus	CECT 3335 laccase	Enhance	[27]
	S. lividans TK24	Mithramycin A	Enhance	[104]
	Streptomyces sp.	Neothioviridamide	Cryptic	[105]
	Several wild-type	Siamycin-I	Cryptic	[29]
Combinatorial Biosynthesis	*S. albus J1074*	Novel paulomycin	Cryptic	[31]
	See table 1 in ref.			[30]
Conditioning Morphology (PCD + MII)	*S. albus*	Apigenin, Luteolin	Enhance	[15]
	S. mobarensis	Microbial transglutaminase	Enhance	[14]

3. Conclusions

We generally face the great challenge of fighting antibiotic resistance, which is growing much faster than our capacity to find new antimicrobials and new strategies to face this problem. The *Streptomyces* genus is still a huge source of natural bioactive compounds, but we need to form new strategies to avoid rediscovering compounds. There is not a single methodology to trigger differentiation, activate cryptic secondary metabolism pathways and improve the discovery of new bioactive compounds. However, the multidisciplinary biosynthetic cluster specific and non-specific approaches discussed in this manuscript, will be key to improving the screening for new secondary metabolites from *streptomycetes*.

Author Contributions: P.Y. planned the topic of the review and searched the information; A.M. and P.Y. wrote the manuscript.

Acknowledgments: We thank the Spanish "Ministerio de Economía y Competitividad" (MINECO; BIO2015-65709-R) for financial support.

Conflicts of Interest: The authors declare no conflict of interest.

References

1. Worrall, J.A.; Vijgenboom, E. Copper mining in *streptomyces*: Enzymes, natural products and development. *Nat. Prod. Rep.* **2010**, *27*, 742–756. [CrossRef] [PubMed]
2. Manteca, A.; Alvarez, R.; Salazar, N.; Yague, P.; Sanchez, J. Mycelium differentiation and antibiotic production in submerged cultures of *Streptomyces coelicolor*. *Appl. Environ. Microbiol.* **2008**, *74*, 3877–3886. [CrossRef] [PubMed]
3. Yague, P.; Willemse, J.; Koning, R.I.; Rioseras, B.; Lopez-Garcia, M.T.; Gonzalez-Quinonez, N.; Lopez-Iglesias, C.; Shliaha, P.V.; Rogowska-Wrzesinska, A.; Koster, A.J.; et al. Subcompartmentalization by cross-membranes during early growth of *streptomyces* hyphae. *Nat. Commun.* **2016**, *7*, 12467. [CrossRef] [PubMed]

4. Manteca, A.; Fernandez, M.; Sanchez, J. A death round affecting a young compartmentalized mycelium precedes aerial mycelium dismantling in confluent surface cultures of *streptomyces* antibioticus. *Microbiology* **2005**, *151*, 3689–3697. [CrossRef] [PubMed]
5. Flardh, K. Growth polarity and cell division in *streptomyces*. *Curr. Opin. Microbiol.* **2003**, *6*, 564–571. [CrossRef] [PubMed]
6. Manteca, A.; Jung, H.R.; Schwammle, V.; Jensen, O.N.; Sanchez, J. Quantitative proteome analysis of *Streptomyces coelicolor* nonsporulating liquid cultures demonstrates a complex differentiation process comparable to that occurring in sporulating solid cultures. *J. Proteome Res.* **2010**, *9*, 4801–4811. [CrossRef] [PubMed]
7. Yague, P.; Rodriguez-Garcia, A.; Lopez-Garcia, M.T.; Rioseras, B.; Martin, J.F.; Sanchez, J.; Manteca, A. Transcriptomic analysis of liquid non-sporulating *Streptomyces coelicolor* cultures demonstrates the existence of a complex differentiation comparable to that occurring in solid sporulating cultures. *PLoS ONE* **2014**, *9*, e86296. [CrossRef] [PubMed]
8. Giudici, R.; Pamboukian, C.R.; Facciotti, M.C. Morphologically structured model for antitumoral retamycin production during batch and fed-batch cultivations of streptomyces olindensis. *Biotechnol. Bioeng.* **2004**, *86*, 414–424. [CrossRef] [PubMed]
9. Vecht-Lifshitz, S.E.; Sasson, Y.; Braun, S. Nikkomycin production in pellets of streptomyces tendae. *J. Appl. Bacteriol.* **1992**, *72*, 195–200. [CrossRef] [PubMed]
10. Sarra, M.; Casas, C.; Godia, F. Continuous production of a hybrid antibiotic by *Streptomyces lividans* TK21 pellets in a three-phase fluidized-bed bioreactor. *Biotechnol. Bioeng.* **1997**, *53*, 601–610. [CrossRef]
11. Jonsbu, E.; McIntyre, M.; Nielsen, J. The influence of carbon sources and morphology on nystatin production by streptomyces noursei. *J. Biotechnol.* **2002**, *95*, 133–144. [CrossRef]
12. Park, Y.; Tamura, S.; Koike, Y.; Toriyama, M.; Okabe, M. Mycelial pellet intrastructure visualization and viability prediction in a culture of streptomyces fradiae using confocal scanning laser microscopy. *J. Ferment. Bioeng.* **1997**, *84*, 483–486. [CrossRef]
13. Rioseras, B.; Lopez-Garcia, M.T.; Yague, P.; Sanchez, J.; Manteca, A. Mycelium differentiation and development of *Streptomyces coelicolor* in lab-scale bioreactors: Programmed cell death, differentiation, and lysis are closely linked to undecylprodigiosin and actinorhodin production. *Bioresour. Technol.* **2014**, *151*, 191–198. [CrossRef] [PubMed]
14. Treppiccione, L.; Ottombrino, A.; Luongo, D.; Maurano, F.; Manteca, A.; Lombó, F.; Rossi, M. Development of gluten with immunomodulatory properties using mTG-active food grade supernatants from *Streptomyces mobaraensis* cultures. *J. Funct. Foods* **2017**, *34*, 390–397. [CrossRef]
15. Marin, L.; Gutierrez-Del-Rio, I.; Yague, P.; Manteca, A.; Villar, C.J.; Lombo, F. De novo biosynthesis of apigenin, luteolin, and eriodictyol in the actinomycete *Streptomyces albus* and production improvement by feeding and spore conditioning. *Front. Microbiol.* **2017**, *8*, 921. [CrossRef] [PubMed]
16. Zhang, L.; Zhang, L.; Han, X.; Du, M.; Zhang, Y.; Feng, Z.; Yi, H.; Zhang, Y. Enhancement of transglutaminase production in *Streptomyces mobaraensis* as achieved by treatment with excessive $MgCl_2$. *Appl. Microbiol. Biotechnol.* **2012**, *93*, 2335–2343. [CrossRef] [PubMed]
17. Onaka, H. Novel antibiotic screening methods to awaken silent or cryptic secondary metabolic pathways in actinomycetes. *J. Antibiot.* **2017**, *70*, 865–870. [CrossRef] [PubMed]
18. Genilloud, O. The re-emerging role of microbial natural products in antibiotic discovery. *Antonie Van Leeuwenhoek* **2014**, *106*, 173–188. [CrossRef] [PubMed]
19. Genilloud, O. Actinomycetes: Still a source of novel antibiotics. *Nat. Prod. Rep.* **2017**, *34*, 1203–1232. [CrossRef] [PubMed]
20. Craney, A.; Ahmed, S.; Nodwell, J. Towards a new science of secondary metabolism. *J. Antibiot.* **2013**, *66*, 387–400. [CrossRef] [PubMed]
21. Yoon, V.; Nodwell, J.R. Activating secondary metabolism with stress and chemicals. *J. Ind. Microbiol. Biotechnol.* **2014**, *41*, 415–424. [CrossRef] [PubMed]
22. Xu, Z.; Wang, Y.; Chater, K.F.; Ou, H.Y.; Xu, H.H.; Deng, Z.; Tao, M. Large-scale transposition mutagenesis of *Streptomyces coelicolor* identifies hundreds of genes influencing antibiotic biosynthesis. *Appl. Environ. Microbiol.* **2017**, *83*, e02889-16. [CrossRef] [PubMed]

23. Khaliq, S.; Akhtar, K.; Afzal Ghauri, M.; Iqbal, R.; Mukhtar Khalid, A.; Muddassar, M. Change in colony morphology and kinetics of tylosin production after UV and gamma irradiation mutagenesis of streptomyces fradiae NRRL-2702. *Microbiol. Res.* **2009**, *164*, 469–477. [CrossRef] [PubMed]
24. Korbekandi, H.; Darkhal, P.; Hojati, Z.; Abedi, D.; Hamedi, J.; Pourhosein, M. Overproduction of clavulanic acid by UV mutagenesis of *Streptomyces clavuligerus*. *Iran. J. Pharm. Res.* **2010**, *9*, 177–181. [CrossRef] [PubMed]
25. Hosaka, T.; Ohnishi-Kameyama, M.; Muramatsu, H.; Murakami, K.; Tsurumi, Y.; Kodani, S.; Yoshida, M.; Fujie, A.; Ochi, K. Antibacterial discovery in actinomycetes strains with mutations in RNA polymerase or ribosomal protein S12. *Nat. Biotechnol.* **2009**, *27*, 462–464. [CrossRef] [PubMed]
26. Ahmed, S.; Craney, A.; Pimentel-Elardo, S.M.; Nodwell, J.R. A synthetic, speciesspecific activator of secondary metabolism and sporulation in *Streptomyces coelicolor*. *Chembiochem* **2013**, *14*, 83–91. [CrossRef] [PubMed]
27. Ece, S.; Lambertz, C.; Fischer, R.; Commandeur, U. Heterologous expression of a *Streptomyces cyaneus* laccase for biomass modification applications. *AMB Express* **2017**, *7*, 86. [CrossRef] [PubMed]
28. Baltz, R.H. Streptomyces and saccharopolyspora hosts for heterologous expression of secondary metabolite gene clusters. *J. Ind. Microbiol. Biotechnol.* **2010**, *37*, 759–772. [CrossRef] [PubMed]
29. Daniel-Ivad, M.; Hameed, N.; Tan, S.; Dhanjal, R.; Socko, D.; Pak, P.; Gverzdys, T.; Elliot, M.A.; Nodwell, J.R. An engineered allele of afsQ1 facilitates the discovery and investigation of cryptic natural products. *ACS Chem. Biol.* **2017**, *12*, 628–634. [CrossRef] [PubMed]
30. Olano, C.; Mendez, C.; Salas, J.A. Post-PKS tailoring steps in natural product-producing actinomycetes from the perspective of combinatorial biosynthesis. *Nat. Prod. Rep.* **2010**, *27*, 571–616. [CrossRef] [PubMed]
31. González, A.; Rodríguez, M.; Braña, A.F.; Méndez, C.; Salas, J.A.; Olano, C. New insights into paulomycin biosynthesis pathway in *Streptomyces albus* j1074 and generation of novel derivatives by combinatorial biosynthesis. *Microb. Cell Fact.* **2016**, *15*. [CrossRef] [PubMed]
32. Baltz, R.H. Synthetic biology, genome mining, and combinatorial biosynthesis of NRPS-derived antibiotics: A perspective. *J. Ind. Microbiol. Biotechnol.* **2017**. [CrossRef] [PubMed]
33. Wang, W.; Ji, J.; Li, X.; Wang, J.; Li, S.; Pan, G.; Fan, K.; Yang, K. Angucyclines as signals modulate the behaviors of *Streptomyces coelicolor*. *Proc. Natl. Acad. Sci. USA* **2014**, *111*, 5688–5693. [CrossRef] [PubMed]
34. Marmann, A.; Aly, A.H.; Lin, W.; Wang, B.; Proksch, P. Co-cultivation—A powerful emerging tool for enhancing the chemical diversity of microorganisms. *Mar. Drugs* **2014**, *12*, 1043–1065. [CrossRef] [PubMed]
35. Piel, J. Metabolites from symbiotic bacteria. *Nat. Prod. Rep.* **2004**, *21*, 519–538. [CrossRef] [PubMed]
36. Yu, J.; Liu, Q.; Chen, C.; Qi, X. Antifungal activity change of streptomyces rimosus MY02 mediated by confront culture with other microorganism. *J. Basic Microbiol.* **2017**, *57*, 276–282. [CrossRef] [PubMed]
37. Perez, J.; Munoz-Dorado, J.; Brana, A.F.; Shimkets, L.J.; Sevillano, L.; Santamaria, R.I. *Myxococcus Xanthus* induces actinorhodin overproduction and aerial mycelium formation by *Streptomyces coelicolor*. *Microb. Biotechnol.* **2011**, *4*, 175–183. [CrossRef] [PubMed]
38. Wang, Y.; Wang, L.; Zhuang, Y.; Kong, F.; Zhang, C.; Zhu, W. Phenolic polyketides from the co-cultivation of marine-derived *Penicillium* sp. Wc-29-5 and *Streptomyces fradiae* 007. *Mar. Drugs* **2014**, *12*, 2079–2088. [CrossRef] [PubMed]
39. Sung, A.A.; Gromek, S.M.; Balunas, M.J. Upregulation and identification of antibiotic activity of a marine-derived *Streptomyces* sp. Via co-cultures with human pathogens. *Mar. Drugs* **2017**, *15*, 250. [CrossRef] [PubMed]
40. Wang, D.; Yuan, J.; Gu, S.; Shi, Q. Influence of fungal elicitors on biosynthesis of natamycin by streptomyces natalensis HW-2. *Appl. Microbiol. Biotechnol.* **2013**, *97*, 5527–5534. [CrossRef] [PubMed]
41. Sharma, R.; Jamwal, V.; Singh, V.P.; Wazir, P.; Awasthi, P.; Singh, D.; Vishwakarma, R.A.; Gandhi, S.G.; Chaubey, A. Revelation and cloning of valinomycin synthetase genes in streptomyces lavendulae acr-da1 and their expression analysis under different fermentation and elicitation conditions. *J. Biotechnol.* **2017**, *253*, 40–47. [CrossRef] [PubMed]
42. Gubbens, J.; Janus, M.M.; Florea, B.I.; Overkleeft, H.S.; van Wezel, G.P. Identification of glucose kinase-dependent and -independent pathways for carbon control of primary metabolism, development and antibiotic production in *Streptomyces coelicolor* by quantitative proteomics. *Mol. Microbiol.* **2017**, *105*, 175. [CrossRef] [PubMed]

43. Park, S.S.; Yang, Y.H.; Song, E.; Kim, E.J.; Kim, W.S.; Sohng, J.K.; Lee, H.C.; Liou, K.K.; Kim, B.G. Mass spectrometric screening of transcriptional regulators involved in antibiotic biosynthesis in *Streptomyces coelicolor* A3(2). *J. Ind. Microbiol. Biotechnol.* **2009**, *36*, 1073–1083. [CrossRef] [PubMed]
44. Liu, G.; Chater, K.F.; Chandra, G.; Niu, G.; Tan, H. Molecular regulation of antibiotic biosynthesis in streptomyces. *Microbiol. Mol. Biol. Rev.* **2013**, *77*, 112–143. [CrossRef] [PubMed]
45. Pimentel-Elardo, S.M.; Sorensen, D.; Ho, L.; Ziko, M.; Bueler, S.A.; Lu, S.; Tao, J.; Moser, A.; Lee, R.; Agard, D.; et al. Activity-independent discovery of secondary metabolites using chemical elicitation and cheminformatic inference. *ACS Chem. Biol.* **2015**, *10*, 2616–2623. [CrossRef] [PubMed]
46. Van Dissel, D.; Claessen, D.; Roth, M.; van Wezel, G.P. A novel locus for mycelial aggregation forms a gateway to improved streptomyces cell factories. *Microb. Cell Fact.* **2015**, *14*, 44. [CrossRef] [PubMed]
47. Chaplin, A.K.; Petrus, M.L.; Mangiameli, G.; Hough, M.A.; Svistunenko, D.A.; Nicholls, P.; Claessen, D.; Vijgenboom, E.; Worrall, J.A. Glxa is a new structural member of the radical copper oxidase family and is required for glycan deposition at hyphal tips and morphogenesis of *Streptomyces lividans*. *Biochem. J.* **2015**, *469*, 433–444. [CrossRef] [PubMed]
48. Petrus, M.L.; Vijgenboom, E.; Chaplin, A.K.; Worrall, J.A.; van Wezel, G.P.; Claessen, D. The dyp-type peroxidase dtpa is a tat-substrate required for glxa maturation and morphogenesis in streptomyces. *Open Biol.* **2016**, *6*, 150149. [CrossRef] [PubMed]
49. Petrus, M.L.; Claessen, D. Pivotal roles for streptomyces cell surface polymers in morphological differentiation, attachment and mycelial architecture. *Antonie Van Leeuwenhoek* **2014**, *106*, 127–139. [CrossRef] [PubMed]
50. Susstrunk, U.; Pidoux, J.; Taubert, S.; Ullmann, A.; Thompson, C.J. Pleiotropic effects of camp on germination, antibiotic biosynthesis and morphological development in *Streptomyces coelicolor*. *Mol. Microbiol.* **1998**, *30*, 33–46. [CrossRef] [PubMed]
51. De Jong, W.; Manteca, A.; Sanchez, J.; Bucca, G.; Smith, C.P.; Dijkhuizen, L.; Claessen, D.; Wosten, H.A. Nepa is a structural cell wall protein involved in maintenance of spore dormancy in *Streptomyces coelicolor*. *Mol. Microbiol.* **2009**, *71*, 1591–1603. [CrossRef] [PubMed]
52. Bobek, J.; Strakova, E.; Zikova, A.; Vohradsky, J. Changes in activity of metabolic and regulatory pathways during germination of *S. coelicolor*. *BMC Genomics* **2014**, *15*, 1173. [CrossRef] [PubMed]
53. Rioseras, B.; Yague, P.; Lopez-Garcia, M.T.; Gonzalez-Quinonez, N.; Binda, E.; Marinelli, F.; Manteca, A. Characterization of SCO4439, a D-alanyl-D-alanine carboxypeptidase involved in spore cell wall maturation, resistance, and germination in *Streptomyces coelicolor*. *Sci. Rep.* **2016**, *6*, 21659. [CrossRef] [PubMed]
54. Sexton, D.L.; St-Onge, R.J.; Haiser, H.J.; Yousef, M.R.; Brady, L.; Gao, C.; Leonard, J.; Elliot, M.A. Resuscitation-promoting factors are Cell wall-lytic enzymes with important roles in the germination and growth of *Streptomyces coelicolor*. *J. Bacteriol.* **2015**, *197*, 848–860. [CrossRef] [PubMed]
55. Noens, E.E.; Mersinias, V.; Willemse, J.; Traag, B.A.; Laing, E.; Chater, K.F.; Smith, C.P.; Koerten, H.K.; van Wezel, G.P. Loss of the controlled localization of growth stage-specific cell-wall synthesis pleiotropically affects developmental gene expression in an ssga mutant of *Streptomyces coelicolor*. *Mol. Microbiol.* **2007**, *64*, 1244–1259. [CrossRef] [PubMed]
56. Zacchetti, B.; Willemse, J.; Recter, B.; van Dissel, D.; van Wezel, G.P.; Wosten, H.A.; Claessen, D. Aggregation of germlings is a major contributing factor towards mycelial heterogeneity of *streptomyces*. *Sci. Rep.* **2016**, *6*, 27045. [CrossRef] [PubMed]
57. Girard, G.; Traag, B.A.; Sangal, V.; Mascini, N.; Hoskisson, P.A.; Goodfellow, M.; van Wezel, G.P. A novel taxonomic marker that discriminates between morphologically complex actinomycetes. *Open Biol.* **2013**, *3*, 130073. [CrossRef] [PubMed]
58. Ronnest, N.P.; Stocks, S.M.; Lantz, A.E.; Gernaey, K.V. Comparison of laser diffraction and image analysis for measurement of *Streptomyces coelicolor* cell clumps and pellets. *Biotechnol. Lett.* **2012**, *34*, 1465–1473. [CrossRef] [PubMed]
59. Van Veluw, G.J.; Petrus, M.L.; Gubbens, J.; de Graaf, R.; de Jong, I.P.; van Wezel, G.P.; Wosten, H.A.; Claessen, D. Analysis of two distinct mycelial populations in liquid-grown streptomyces cultures using a flow cytometry-based proteomics approach. *Appl. Microbiol. Biotechnol.* **2012**, *96*, 1301–1312. [CrossRef] [PubMed]
60. Petrus, M.L.; van Veluw, G.J.; Wosten, H.A.; Claessen, D. Sorting of streptomyces cell pellets using a complex object parametric analyzer and sorter. *J. Vis. Exp.* **2014**, e51178. [CrossRef] [PubMed]

61. Willemse, J.; Buke, F.; van Dissel, D.; Grevink, S.; Claessen, D.; van Wezel, G.P. Sparticle, an algorithm for the analysis of filamentous microorganisms in submerged cultures. *Antonie Van Leeuwenhoek* **2017**, 171–182. [CrossRef]
62. Celler, K.; Cristian, P.; van Loosdrecht, M.C.; van Wezel, G.P. Structured morphological modeling as a framework for rational strain design of streptomyces species. *Antonie Van Leeuwenhoek* **2012**, *102*, 409–423. [CrossRef] [PubMed]
63. Nieminen, L.; Webb, S.; Smith, M.C.; Hoskisson, P.A. A flexible mathematical model platform for studying branching networks: Experimentally validated using the model actinomycete, *Streptomyces coelicolor*. *PLoS ONE* **2013**, *8*, e54316. [CrossRef] [PubMed]
64. Van Dissel, D.; Claessen, D.; van Wezel, G.P. Morphogenesis of streptomyces in submerged cultures. *Adv. Appl. Microbiol.* **2014**, *89*, 1–45. [PubMed]
65. Ha, S.; Lee, K.J.; Lee, S.I.; Gwak, H.J.; Lee, J.H.; Kim, T.W.; Choi, H.J.; Jang, J.Y.; Choi, J.S.; Kim, C.J.; et al. Optimization of herbicidin a production in submerged culture of streptomyces scopuliridis m40. *J. Microbiol. Biotechnol.* **2017**, *27*, 947–955. [CrossRef] [PubMed]
66. Van Dissel, D.; van Wezel, G.P. Morphology-driven downscaling of streptomyces lividans to micro-cultivation. *Antonie Van Leeuwenhoek* **2017**, 457–469. [CrossRef]
67. Rigali, S.; Nothaft, H.; Noens, E.E.; Schlicht, M.; Colson, S.; Müller, M.; Joris, B.; Koerten, H.K.; Hopwood, D.A.; Titgemeyer, F.; van Wezel, G.P. The sugar phosphotransferase system of *Streptomyces coelicolor* is regulated by the gntr-family regulator dasr and links n-acetylglucosamine metabolism to the control of development. *Mol. Microbiol.* **2006**, *61*, 1237–1251. [CrossRef] [PubMed]
68. Rigali, S.; Titgemeyer, F.; Barends, S.; Mulder, S.; Thomae, A.W.; Hopwood, D.A.; van Wezel, G.P. Feast or famine: The global regulator dasr links nutrient stress to antibiotic production by streptomyces. *EMBO Rep.* **2008**, *9*, 670–675. [CrossRef] [PubMed]
69. Yague, P.; Manteca, A.; Simon, A.; Diaz-Garcia, M.E.; Sanchez, J. New method for monitoring programmed cell death and differentiation in submerged streptomyces cultures. *Appl. Environ. Microbiol.* **2010**, *76*, 3401–3404. [CrossRef] [PubMed]
70. Yague, P.; Lopez-Garcia, M.T.; Rioseras, B.; Sanchez, J.; Manteca, A. Pre-sporulation stages of streptomyces differentiation: State-of-the-art and future perspectives. *FEMS Microbiol. Lett.* **2013**, *342*, 79–88. [CrossRef] [PubMed]
71. Innes, C.M.; Allan, E.J. Induction, growth and antibiotic production of streptomyces viridifaciens l-form bacteria. *J. Appl. Microbiol.* **2001**, *90*, 301–308. [CrossRef] [PubMed]
72. Kealey, C.; Creaven, C.A.; Murphy, C.D.; Brady, C.B. New approaches to antibiotic discovery. *Biotechnol. Lett.* **2017**, *39*, 805–817. [CrossRef] [PubMed]
73. Bai, L.; Liu, C.; Guo, L.; Piao, C.; Li, Z.; Li, J.; Jia, F.; Wang, X.; Xiang, W. *Streptomyces formicae* sp. Nov., a novel actinomycete isolated from the head of *Camponotus japonicus* mayr. *Antonie Van Leeuwenhoek* **2016**, *109*, 253–261. [CrossRef] [PubMed]
74. Sarmiento-Vizcaino, A.; Brana, A.F.; Gonzalez, V.; Nava, H.; Molina, A.; Llera, E.; Fiedler, H.P.; Rico, J.M.; Garcia-Florez, L.; Acuna, J.L.; et al. Atmospheric dispersal of bioactive *Streptomyces albidoflavus* strains among terrestrial and marine environments. *Microb. Ecol.* **2016**, *71*, 375–386. [CrossRef] [PubMed]
75. Sarmiento-Vizcaino, A.; Gonzalez, V.; Brana, A.F.; Palacios, J.J.; Otero, L.; Fernandez, J.; Molina, A.; Kulik, A.; Vazquez, F.; Acuna, J.L.; et al. Pharmacological potential of phylogenetically diverse actinobacteria isolated from deep-sea coral ecosystems of the submarine aviles canyon in the cantabrian sea. *Microb. Ecol.* **2017**, *73*, 338–352. [CrossRef] [PubMed]
76. Schniete, J.K.; Cruz-Morales, P.; Selem-Mojica, N.; Fernandez-Martinez, L.T.; Hunter, I.S.; Barona-Gomez, F.; Hoskisson, P.A. Expanding primary metabolism helps generate the metabolic robustness to facilitate antibiotic biosynthesis in streptomyces. *MBio* **2018**, *9*. [CrossRef] [PubMed]
77. Cihak, M.; Kamenik, Z.; Smidova, K.; Bergman, N.; Benada, O.; Kofronova, O.; Petrickova, K.; Bobek, J. Secondary metabolites produced during the germination of *Streptomyces coelicolor*. *Front. Microbiol.* **2017**, *8*, 2495. [CrossRef] [PubMed]
78. Yu, G.; Jia, X.; Wen, J.; Lu, W.; Wang, G.; Caiyin, Q.; Chen, Y. Strain improvement of streptomyces roseosporus for daptomycin production by rational screening of he-ne laser and NTG induced mutants and kinetic modeling. *Appl. Biochem. Biotechnol.* **2011**, *163*, 729–743. [CrossRef] [PubMed]

79. Jakeman, D.L.; Graham, C.L.; Young, W.; Vining, L.C. Culture conditions improving the production of jadomycin b. *J. Ind. Microbiol. Biotechnol.* **2006**, *33*, 767–772. [CrossRef] [PubMed]
80. Bode, H.B.; Bethe, B.; Hofs, R.; Zeeck, A. Big effects from small changes: Possible ways to explore nature's chemical diversity. *Chembiochem* **2002**, *3*, 619–627. [CrossRef]
81. Bursy, J.; Kuhlmann, A.U.; Pittelkow, M.; Hartmann, H.; Jebbar, M.; Pierik, A.J.; Bremer, E. Synthesis and uptake of the compatible solutes ectoine and 5-hydroxyectoine by *Streptomyces coelicolor* A3(2) in response to salt and heat stresses. *Appl. Environ. Microbiol.* **2008**, *74*, 7286–7296. [CrossRef] [PubMed]
82. Hayes, A.; Hobbs, G.; Smith, C.P.; Oliver, S.G.; Butler, P.R. Environmental signals triggering methylenomycin production by *Streptomyces coelicolor* A3(2). *J. Bacteriol.* **1997**, *179*, 5511–5515. [CrossRef] [PubMed]
83. Medema, M.H.; Alam, M.; Breitling, R.; Takano, E. The future of industrial antibiotic production: From random mutagenesis to synthetic biology. *Bioeng. Bugs.* **2011**, 230–233. [CrossRef]
84. Cheng, Y.R.; Huang, J.; Qiang, H.; Lin, W.L.; Demain, A.L. Mutagenesis of the rapamycin producer streptomyces hygroscopicus FC904. *J. Antibiot.* **2001**, *54*, 967–972. [CrossRef] [PubMed]
85. Wang, G.; Hosaka, T.; Ochi, K. Dramatic activation of antibiotic production in *Streptomyces coelicolor* by cumulative drug resistance mutations. *Appl. Environ. Microbiol.* **2008**, *74*, 2834–2840. [CrossRef] [PubMed]
86. McKenzie, N.L.; Thaker, M.; Koteva, K.; Hughes, D.W.; Wright, G.D.; Nodwell, J.R. Induction of antimicrobial activities in heterologous streptomycetes using alleles of the *Streptomyces coelicolor* gene absa1. *J. Antibiot.* **2010**, *63*, 177–182. [CrossRef] [PubMed]
87. Foley, T.L.; Young, B.S.; Burkart, M.D. Phosphopantetheinyl transferase inhibition and secondary metabolism. *FEBS J.* **2009**, *276*, 7134–7145. [CrossRef] [PubMed]
88. Onaka, H. Biosynthesis of indolocarbazole and goadsporin, two different heterocyclic antibiotics produced by actinomycetes. *Biosci. Biotechnol. Biochem.* **2009**, *73*, 2149–2155. [CrossRef] [PubMed]
89. Recio, E.; Colinas, A.; Rumbero, A.; Aparicio, J.F.; Martin, J.F. Pi factor, a novel type quorum-sensing inducer elicits pimaricin production in streptomyces natalensis. *J. Biol. Chem.* **2004**, *279*, 41586–41593. [CrossRef] [PubMed]
90. Mo, S.; Ban, Y.H.; Park, J.W.; Yoo, Y.J.; Yoon, Y.J. Enhanced FK506 production in *Streptomyces clavuligerus* ckd1119 by engineering the supply of methylmalonyl-coa precursor. *J. Ind. Microbiol. Biotechnol.* **2009**, *36*, 1473–1482. [CrossRef] [PubMed]
91. Liu, Z.; Guo, M.; Qian, J.; Zhuang, Y.; Zhang, S. Disruption of ZWF2 gene to improve oxytetraclyline biosynthesis in streptomyces rimosus M4018. *Wei Sheng Wu Xue Bao* **2008**, *48*, 21–25. [PubMed]
92. Yagüe, P.; Manteca, A. Optimization of the antibiotic producction in different Streptomyces strains by conditioning the MII differentiation. **2018**. Unpublished results.
93. Malla, S.; Niraula, N.P.; Liou, K.; Sohng, J.K. Self-resistance mechanism in streptomyces peucetius: Overexpression of drra, drrb and drrc for doxorubicin enhancement. *Microbiol. Res.* **2010**, *165*, 259–267. [CrossRef] [PubMed]
94. Qiu, J.; Zhuo, Y.; Zhu, D.; Zhou, X.; Zhang, L.; Bai, L.; Deng, Z. Overexpression of the abc transporter avtab increases avermectin production in streptomyces avermitilis. *Appl. Microbiol. Biotechnol.* **2011**, *92*, 337–345. [CrossRef] [PubMed]
95. Xu, Y.; Willems, A.; Au-Yeung, C.; Tahlan, K.; Nodwell, J.R. A two-step mechanism for the activation of actinorhodin export and resistance in *Streptomyces coelicolor*. *MBio* **2012**, *3*, e00191-12. [CrossRef] [PubMed]
96. McKenzie, N.L.; Nodwell, J.R. Phosphorylated absa2 negatively regulates antibiotic production in *Streptomyces coelicolor* through interactions with pathway-specific regulatory gene promoters. *J. Bacteriol.* **2007**, *189*, 5284–5292. [CrossRef] [PubMed]
97. Guo, J.; Zhao, J.; Li, L.; Chen, Z.; Wen, Y. Li, J. The pathway-specific regulator aver from *Streptomyces avermitilis* positively regulates avermectin production while it negatively affects oligomycin biosynthesis. *Mol. Genet. Genomics* **2010**, *283*, 123–133. [CrossRef] [PubMed]
98. Retzlaff, L.; Distler, J. The regulator of streptomycin gene expression, strr, of streptomyces griseus is a DNA binding activator protein with multiple recognition sites. *Mol. Microbiol.* **1995**, *18*, 151–162. [CrossRef] [PubMed]
99. Laureti, L.; Song, L.; Huang, S.; Corre, C.; Leblond, P.; Challis, G.L.; Aigle, B. Identification of a bioactive 51-membered macrolide complex by activation of a silent polyketide synthase in *Streptomyces ambofaciens*. *Proc. Natl. Acad. Sci. USA* **2011**, *108*, 6258–6626. [CrossRef] [PubMed]

100. Menendez, N.; Brana, A.F.; Salas, J.A.; Mendez, C. Involvement of a chromomycin abc transporter system in secretion of a deacetylated precursor during chromomycin biosynthesis. *Microbiology* **2007**, *153*, 3061–3070. [CrossRef] [PubMed]
101. Bunet, R.; Song, L.; Mendes, M.V.; Corre, C.; Hotel, L.; Rouhier, N.; Framboisier, X.; Leblond, P.; Challis, G.L.; Aigle, B. Characterization and manipulation of the pathway-specific late regulator alpw reveals *Streptomyces ambofaciens* as a new producer of kinamycins. *J. Bacteriol.* **2010**. [CrossRef] [PubMed]
102. Gomez-Escribano, J.P.; Bibb, M.J. Engineering *Streptomyces coelicolor* for heterologous expression of secondary metabolite gene clusters. *Microb. Biotechnol.* **2011**, *4*, 207–215. [CrossRef] [PubMed]
103. Komatsu, M.; Uchiyama, T.; Omura, S.; Cane, D.E.; Ikeda, H. Genome-minimized streptomyces host for the heterologous expression of secondary metabolism. *Proc. Natl. Acad. Sci. USA* **2010**, *107*, 2646–2651. [CrossRef] [PubMed]
104. Novakova, R.; Nunez, L.E.; Homerova, D.; Knirschova, R.; Feckova, L.; Rezuchova, B.; Sevcikova, B.; Menendez, N.; Moris, F.; Cortes, J.; et al. Increased heterologous production of the antitumoral polyketide mithramycin a by engineered streptomyces lividans TK24 strains. *Appl. Microbiol. Biotechnol.* **2018**, *102*, 857–869. [CrossRef] [PubMed]
105. Kawahara, T.; Izumikawa, M.; Kozone, I.; Hashimoto, J.; Kagaya, N.; Koiwai, H.; Komatsu, M.; Fujie, M.; Sato, N.; Ikeda, H.; et al. Neothioviridamide, a polythioamide compound produced by heterologous expression of a *Streptomyces* sp. Cryptic ripp biosynthetic gene cluster. *J. Nat. Prod.* **2018**, 264–269. [CrossRef] [PubMed]

antibiotics

MDPI

Review

Unraveling Nutritional Regulation of Tacrolimus Biosynthesis in *Streptomyces tsukubaensis* through *omic* Approaches

María Ordóñez-Robles [1,2], Fernando Santos-Beneit [2,3] and Juan F. Martín [1,*]

1 Área de Microbiología, Departamento de Biología Molecular, Universidad de León, León 24071, Spain; mordr@unileon.es
2 Instituto de Biotecnología de León, INBIOTEC, Avda. Real no. 1, León 24006, Spain; fersanben3@yahoo.es
3 Departamento de Biología Funcional, Universidad de Oviedo, Oviedo 33006, Spain
* Correspondence: jf.martin@unileon.es; Tel.: +34-619-114-487

Received: 27 February 2018; Accepted: 26 April 2018; Published: 1 May 2018

check for updates

Abstract: *Streptomyces tsukubaensis* stands out among actinomycetes by its ability to produce the immunosuppressant tacrolimus. Discovered about 30 years ago, this macrolide is widely used as immunosuppressant in current clinics. Other potential applications for the treatment of cancer and as neuroprotective agent have been proposed in the last years. In this review we introduce the discovery of *S. tsukubaensis* and tacrolimus, its biosynthetic pathway and gene cluster (*fkb*) regulation. We have focused this work on the *omic* studies performed in this species in order to understand tacrolimus production. Transcriptomics, proteomics and metabolomics have improved our knowledge about the *fkb* transcriptional regulation and have given important clues about nutritional regulation of tacrolimus production that can be applied to improve production yields. Finally, we address some points of *S. tsukubaensis* biology that deserve more attention.

Keywords: *Streptomyces tsukubaensis*; tacrolimus; FK506; *omics*

1. Discovery of *S. tsukubaensis* and Tacrolimus Use in Current Clinics

Streptomyces tsukubaensis and its secondary metabolite tacrolimus were discovered in 1984, during a screening performed by the Fujisawa Pharmaceutical Co. (since 2005 merged to Yamanouchi Pharmaceutical Co. to form Astellas Pharma). *S. tsukubaensis* was isolated from a soil sample in the Tsukuba region (Japan) and tacrolimus was identified in its culture broths, becoming the first immunosuppressant discovered with macrolide structure [1,2]. The strain, patented as *S. tsukubaensis* No. 9993, is currently known as *S. tsukubaensis* NRRL 18488 and is the parental strain of most of the strains used for the industrial production of tacrolimus.

Macrolides such as erythromycin are composed of 14–16 C-membered macrolactone rings to which one or more deoxysugars are attached. Tacrolimus, a 23-carbon macrolide (822 Da), was initially named as compound FR900506 but, later on, it received other names such as FK506 or fujimycin. The name of tacrolimus was established as an acronym of "Tsukuba Macrolide Immunosuppressant" [3]. The first reference to tacrolimus was made at the 11th International Congress of the Transplantation Society, held in Helsinki in 1986, one year before the first publications by Kino and coworkers. The first clinical assays, focused on hepatic transplantation, were developed at the University of Pittsburgh in 1989. Two years later the first international congress on tacrolimus was celebrated in that city [3]. Tacrolimus acts as a calcineurin inhibitor, showing a mechanism of action very similar to that of cyclosporine (Figure 1) [4]. When tacrolimus interacts with its cytosolic receptors, mainly FKBP12- [5], the calmodulin-dependent serine/threonine phosphatase activity of calcineurin is inhibited, resulting in the arrest of T cell proliferation [6]. The mechanism of action is conserved in human T cells and yeast and thus, tacrolimus also has antifungal activity [7,8]. This activity is useful for the qualitative detection of tacrolimus by bioassay against susceptible strains such as *Saccharomyces cerevisiae* TB23 [9].

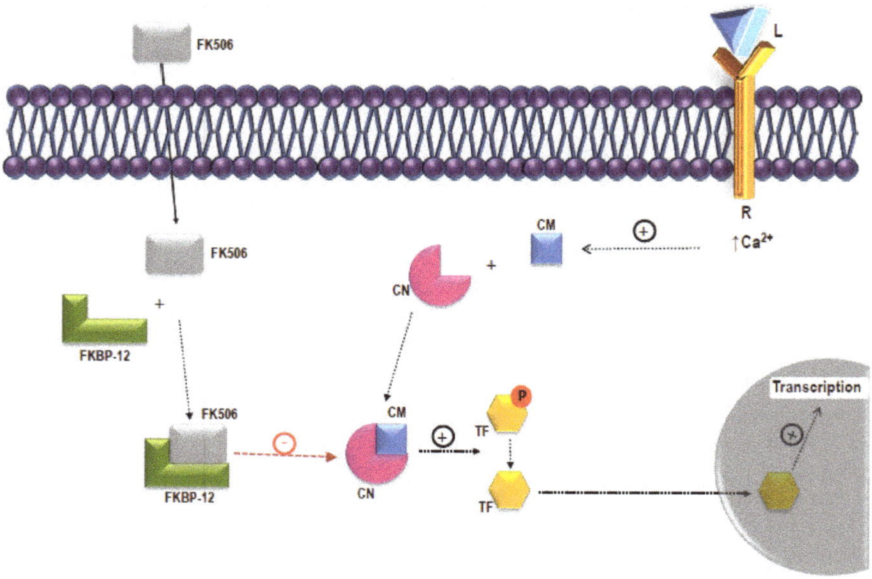

Figure 1. Mechanism of action of tacrolimus (FK506). Tacrolimus interacts with cytosolic receptors such as FKBP12. The complex FKBP12-FK506 inhibits the calmodulin-dependent serine/threonine phosphatase activity of calcineurin. In this situation, calcineurin can no longer dephosphorylate transcriptional factors (e.g., NFAT). The dephosphorylated TFs are required for governing T cell proliferation. L: ligand; R: receptor; CM: calmodulin; CN: calcineurin; TF: transcription factor; P: phosphate group; FKBP-12: FK506 binding protein 12.

Since its approval by the FDA for the treatment of hepatic transplantation in 1994, tacrolimus has been also applied to medulla, kidney and heart transplantation [10–12]. This macrolide is also used for the treatment of other diseases such as atopic dermatitis [13,14] and is applied to the stents implanted in coronary arteries [15]. Several works have been published about its use in immune diseases such as rheumatoid arthritis and intestinal inflammatory diseases [16,17]. Tacrolimus has shown antiviral activity against orthopoxvirus, HIV and feline immunodeficiency virus (FIV) [18–21] and has properties such as a hair growth stimulator [22]. Neuroprotective and neuroregenerative activities have been also reported [23–25] as well as its potential application in the treatment of cancer [26]. More recently, the efficacy of tacrolimus ointment in the treatment of allergic ocular diseases has been reported [27].

The efficacy of tacrolimus in the treatment of organ transplantation is the basis of its industrial importance. Tacrolimus is between 10 and 100 times more potent than cyclosporine and has been shown to be more effective in several clinical trials [28,29]. Tacrolimus generates important benefits for the pharmaceutical market; for example, the sales of tacrolimus under the commercial names "Prograf" and "Protopic" yielded a total of $1727 million to Astellas Pharma in 2016 (data from http://www.pharmacompass.com).

2. Biosynthetic Pathway and Gene Cluster

The first studies on the tacrolimus biosynthetic pathway were performed by researchers from the pharmaceutical company Merck (USA) during the 90's [30–33]. Tacrolimus is a polyketide synthesized by a hybrid polyketide I synthase-non-ribosomal peptide synthase (PKSI-NRPS) system encoded by the *fkb* cluster, which encompasses a minimum of 19 genes (Figure 2). Until now, more than 15 tacrolimus-producing species have been reported [34], the last being *S. tsukubaensis* F601 [35]. There are two types of *fkb* clusters in the tacrolimus producing strains [36]: (i) A short version comprising the genes *fkbQ*, *fkbN*, *fkbM*, *fkbD*, *fkbA*, *fkbP*, *fkbO*, *fkbB*, *fkbC*, *fkbL*, *fkbK*, *fkbJ*, *fkbI*, *fkbH*, *fkbG*, *allD*, *allR*, *allK* and *allA* (found in *Streptomyces tacrolimicus* and *Streptomyces kanamyceticus* KCTC 9225) and; (ii) An extended version found in *S. tsukubaensis* NRRL 18488, *S. tsukubaensis* L19 and *Streptomyces* sp. KCTC 11604BP that includes 5 additional genes in the 5′ region of the *fkbG gene* (*allMNPOS*/*tcs12345*) and one or two extra genes (depending on the species) in the 3′ region (*tcs6*-fkbR/*tcs67*). Deletion of *allMNPOS* genes in *Streptomyces* sp. KCTC 11604BP does not significantly affect tacrolimus production; thus, it is dubious that they are involved in tacrolimus biosynthesis [37]. Actually, their transcription levels are low, which supports this assumption [38,39].

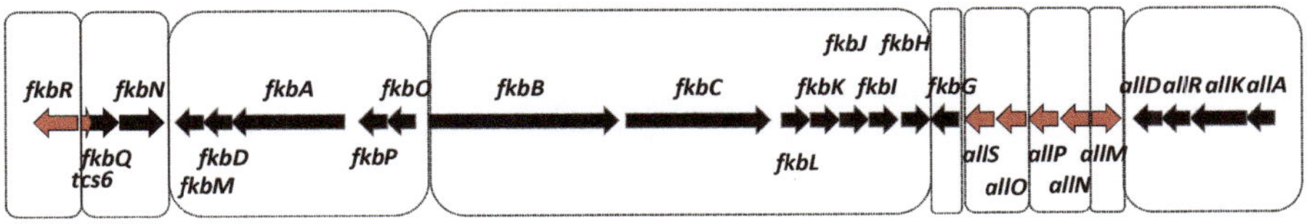

Figure 2. Tacrolimus biosynthesis cluster (*fkb*). Genes present in both the short and extended version of the *fkb* cluster are depicted in black. Genes present only in the extended version are depicted in red. These groups also correspond to their FkbN transcriptional dependence (**Black**) or independence (**Red**). The transcriptional units identified to date are indicated by boxes.

The first step in tacrolimus biosynthesis is the formation of (4R, 5R)-4,5-dihydroxycyclohex-1-enecarboxylic acid (abbreviated DHCHC) from chorismate through the so-called chorismatase activity of FkbO (Figure 3) [40]. DHCHC acts as starter unit for the subsequent formation of the carbon skeleton and corresponds to the cyclohexane ring in the final structure of tacrolimus. This ring is the most tolerant target for structural modifications that do not eliminate the immunosuppressant activity [41]. The polyketide synthases FkbA, FkbB and FkbC catalyze 10 elongation steps from DHCHC using as extender units malonyl-CoA (2 molecules), methylmalonyl-CoA (5 molecules), methoxymalonyl-ACP (2 molecules) and allylmalonyl-CoA (one molecule). The latter two extender units are unusual in the formation of polyketides and result in the methoxyl group of C13 and C15 and the allyl radical of C21, respectively [36,37,42,43]. The biosynthesis of methoxymalonyl-ACP from 1,3-biphosphoglycerate depends on the enzymes encoded by the *fkbGHIJK* subcluster [42–45]. The incorporation of allylmalonyl-CoA is the sole difference between tacrolimus and ascomycin (FK520), in which biosynthesis ethylmalonyl-CoA is used instead. The *all* subcluster is involved in the formation of allylmalonyl-CoA and encodes a polyketide synthase of unusual structure [46]. Nevertheless, ketoreductase and dehydratase activities encoded outside the *fkb* cluster might be involved in some steps of allylmalonyl-CoA formation and these activities could be shared with fatty acid synthases [37]. The tacrolimus cluster does not encode an ACP-CoA transacylase necessary for the final reaction leading to allylmalonyl-CoA [37], but the acyltransferase domain of the fourth module in FkbB (AT4FkbB) is able to transfer an allylmalonyl unit to the ACP domain [47].

For the cyclation of the macrolide, FkbL generates L-pipecolate from L-lysine [48], which is then incorporated into the carbon skeleton by NRPS FkbP [30,49,50]. Finally, two modification steps are necessary to achieve the final molecule with biological activity: a methylation of the hydroxyl group located at C31 and an oxidation at C9. Both groups are important for the binding of tacrolimus to FKBP12 [51,52]. The methylation is catalyzed by the S-adenosylmethione dependent O-methyltransferase FkbM and the oxidation by the cytochrome P450-oxidoreductase FkbD [31,53]. Both activities are encoded in the same operon and can occur in any order [31,45,54]. Interestingly, the reaction catalyzed by FkbD (a double step oxidation involving 4 electron transfers and the formation of the alcoholic intermediate 9-hydroxy-FK506) is known for terpenoid biosynthesis but was first described for polyketide biosynthesis [45].

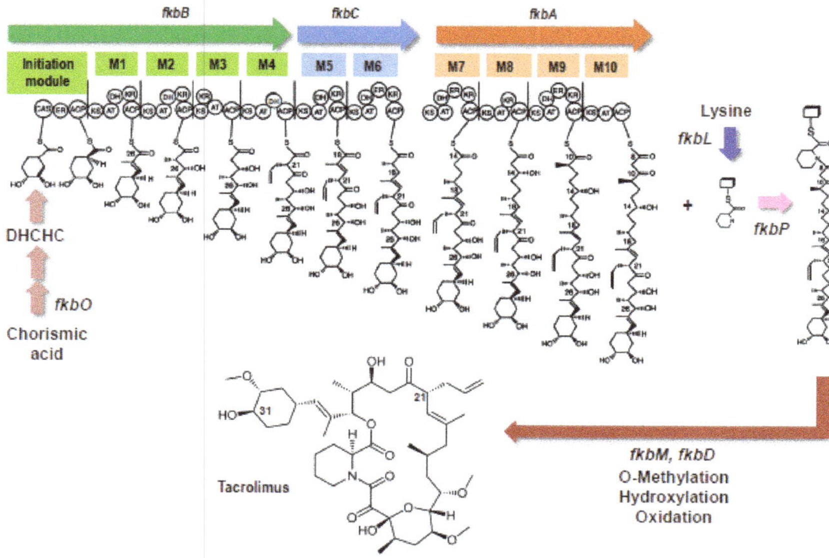

Figure 3. Scheme representing the assembly of the tacrolimus polyketide and the early and late biosynthetic steps. In the upper part the arrows represent the three PKS genes (*fkbA*, *fkbB*, *fkbC*) of the cluster. Note that the *fkbA* gene is physically separated from *fkbB* and *fkbC* genes in the *fkb* cluster (see Figure 2). The modules of the PKSs are boxed and indicated as M1 to M10. Domains in the modules are indicated by circles: ACP, acyl carrier protein; AT, acyltransferase; ER, enoyl reductase; CAS, CoA synthetase; KR, 3-oxoacyl (ACP) reductase; DH, 3-oxoacyl thioester dehydratase; KS, 3-oxoacyl (ACP) synthase. DHCHC: (4R, 5R)-4,5-dihydroxycyclohex-1-enecarboxylic acid. Biosynthetic and late modification steps, and the encoding genes for the starter (*fkbO*), elongation units (*fkbL*, *fkbP*) and late modification reactions (*fkbM*, *fkbD*). Based on data from Motamedi and Shafiee [30].

3. Transcriptional Regulators and Recent Insights through Transcriptomic and RNAseq Studies

The first sequence analyses of the *fkb* cluster revealed three potential regulators: *fkbN*, *fkbR* and *allN* (belonging to the LAL, LysR and AsnC families, respectively). FkbN is a large regulatory protein of the LAL family (Large ATP binding regulators of the LuxR family). The LAL regulators are large proteins (872–1159 amino acids) that contain a LuxR-type HTH DNA binding region near the C-terminal end of the protein and an ATP binding motif in the N-terminal end [55,56]. Similar FkbN-like genes have been found in several other macrolide gene clusters including RapH of the rapamycin producer *Streptomyces hygroscopicus* [57], PikD of the pikromycin producer *Streptomyces venezuelae* [58], GdmR1 and GdmR2 of the geldanamycin producer *Streptomyces hygroscopicus* [59], FkbN of the ascomycin producer *S. hygroscopicus* var. *ascomyceticus* [44], FscRI in the candicidin producer *Streptomyces griseus* [60,61], PimM of the pimaricin producer *Streptomyces natalensis* [62,63], NysR from the nystatin producer *Streptomyces noursei* [64], AmphRIV in the amphotericin B producer *Streptomyces nodosus* [65] and PteF in the filipin producer *Streptomyces avermitilis* [66,67].

The second regulatory protein FkbR belongs to the family of the LysR-type transcriptional regulators, also named LTTR, which are very common autoregulatory genes in bacteria [68]. In fact, they are widely distributed in *Streptomyces*: genome sequencing revealed about 40 LTTRs in *S. coelicolor* [69]. FkbR, as occurs with other members of the LTTR family, is a relatively small protein of less than

325 amino acids that is characterized by an HTH DNA binding motif in the C-terminal and by a ligand (co-inducer) binding sequence in the N-terminal region [70,71]. Other LTTRs acting as pathway-specific regulators include SCLAV_p1262 of *S. clavuligerus* (77% identity), ThnI from *Streptomyces cattleya* (39% identity), AbaB from *Streptomyces antibioticus* or ClaR from *S. avermitilis* [72,73].

The third putative regulatory gene of the tacrolimus gene cluster is *allN*. This gene is located in the 5′ end of the extended version of the tacrolimus gene cluster and encodes a protein that has similarity with regulatory proteins involved in nitrogen metabolism, particularly with regulators of AsnC family [74]. This gene is included in a region that is involved in the formation of the precursor allylmalonyl-CoA (all genes) [37,46].

Functional analysis of the role of FkbN, FkbR and AllN in *S. tsukubaensis* was performed by gene disruption and complementation studies. Whilst the inactivation of *fkbN* resulted in the lack of tacrolimus production, disruption of *fkbR* reduced tacrolimus yields to 20% of that of the parental strain and the inactivation of *allN* did not affect tacrolimus production [36]. Thus, it was concluded that both *fkbN* and *fkbR* encode positive regulators whilst *allN* has no influence on tacrolimus production [36]. In addition, AllN (also named Tcs2) seems to be not involved in tacrolimus production in other strains such as *S. tsukubaensis* L19 [75]. Overexpression of *fkbN* or *fkbR* in the wild type strain using the *ermE** promoter produced an increase of the final yield of tacrolimus of 55% and 30%, respectively, using a culture medium optimized for tacrolimus production. These results agree with the observations published by Mo and coworkers on the effect of FkbN in *Streptomyces* sp. KCTC 11604BP [76].

There are important differences between FkbN and FkbR that we summarize here as follows: (1) *fkbN* is present in both the extended and the short version of the *fkb* cluster but *fkbR* is only present in the extended cluster version [37]; (2) FkbN always shows a positive effect on tacrolimus production whilst FkbR can have positive or negative effects [36,76,77]; (3) A complete lack of tacrolimus production is only produced by inactivation of *fkbN* (but not with that of *fkbR*) [36,38]; (4) transcription of *fkbR* is constant and low throughout the culture whilst that of *fkbN* increases before the onset of tacrolimus production and is maintained during the production phase (Figure 4) [36,38,75].

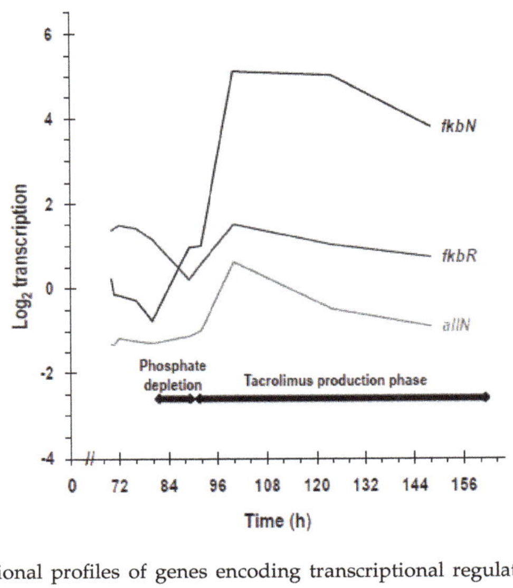

Figure 4. Transcriptional profiles of genes encoding transcriptional regulators of the *fkb* cluster. Transcription of *fkbN*, *fkbR* and *allN* in *S. tsukubaensis* NRRL 18488 grown in MGm-2.5 production media. As indicated in the graph, phosphate depletion occurs between 80 h and 89 h and tacrolimus is detected from 89 h. The cultures were performed in duplicated flasks. Error bars have been omitted to facilitate the visualization of the results.

3.1. Characterization of *fkb* Cluster Transcriptional Subunits

Early studies using the *rppA* chalcone synthase reporter systems and qRT-PCR showed that the inactivation of *fkbR* or *fkbN* prevents transcription of certain genes in the *S. tsukubaensis fkb* cluster such as *fkbG* or *fkbB*, implying that some *fkb* genes are regulated by FkbN while others are not [36]. However, more recent transcriptomic studies with the same *fkbN* inactivated mutant have confirmed that FkbN controls the expression of most of the genes of the *fkb* cluster [38]. Two types of gene expression were observed in response to *fkbN* inactivation: (a) Genes clearly induced by FkbN coinciding with the onset of tacrolimus biosynthesis (in the so called "induction phase") and whose expression is significantly reduced in the *fkbN* mutant (i.e., *fkbABC*, *fkbGHIJK*, *fkbL*, *allAKRD*, *fkbO*, *fkbP*, *fkbD* and *fkbM*) and (b) Genes poorly expressed through the culture time and not affected by *fkbN* inactivation (i.e., *allMNPOS* and *fkbR*) (Figure 2). Thus, the complete transcriptional dependency of the *fkb* genes on FkbN, with the exception of *allMNPOS* and *fkbR* (only present in the extended versions of the *fkb* cluster), which are FkbN-independent, was demonstrated.

The use of tiling probes covering the *fkb* cluster allowed the identification of 6 transcriptional units: *fkbR*, *tcs6-fkbQ-fkbN*, *fkbOPADM*, *fkbBCLKJIH*, *fkbG* and *allAKRD*. It was concluded that *fkbR* is transcribed as a leaderless mRNA and that *fkbN* forms an operon along with *tcs6* and *fkbQ* whose transcription depends on two different promoters, one FkbN-dependent and the other FkbN-independent [38]. These results are supported by the EMSAs performed with the FkbN-DNA binding domain in *S. tsukubaensis* L19 by Zhang and coworkers [75], who reported FkbN binding to the promoter regions of the same six transcriptional units and identified two new ones corresponding to *allNPOS* and *allM*. More recently, differential RNA-seq (dRNA-seq) transcriptional profiling has been performed in *S. tsukubaensis* by Bauer and coworkers [39], who identified 9 transcriptional units that are in good agreement with previous studies (Figure 2). The main finding is that *allOS* and *allNP* are transcribed as independent mRNAs [39].

fkbR seems to be transcribed as a leaderless mRNA and is not directly regulated by FkbN [38]. In fact, it is likely that FkbR regulates its own expression, although detailed information is not available. Recently, the binding of FkbR to the promoter regions of *tcs6-fkbQ-fkbN* and *fkbR* in *S. tsukubaensis* L19 has been reported [75].

3.2. Genes Located Outside of the Tacrolimus Gene Cluster Regulated by FkbN

It has been reported that cluster-situated regulators (CSR) can regulate genes located outside their own cluster [78,79] and, therefore, the utilization of transcriptomic studies is a good tool to identify them. The transcriptomic analysis performed with the *fkbN* mutant by Ordóñez-Robles and coworkers [38] revealed potential genes located outside the *fkb* cluster that might be targets of FkbN such as *ppt1*, encoding a 4′-phosphopantetheinyl transferase that is known to be involved in CDA formation in *S. coelicolor* [80]. This gene showed an FkbN-dependent profile and a putative FkbN binding sequence [38]. In agreement with these results it was reported that the orthologue of *ppt1* is involved in tacrolimus production in *S. tsukubaensis* L19 [81] and later, it was observed that *ppt1* and *fkbN* share a common transcriptional response to glucose, glycerol and *N*-acetylglucosamine additions (see below). The study identified acyl-CoA dehydrogenase and methoxymalonate biosynthesis coding genes that were negatively affected by the *fkbN* inactivation and thus, might be involved in tacrolimus biosynthesis. On the contrary, some PKS coding genes located in a chromosomal region that has been predicted to encode a cluster for the production of a bafilomycin-like compound [82] were upregulated after *fkbN* inactivation, which might reflect competition for precursors between these two clusters for the biosynthesis of secondary metabolites.

Using the information-theory of Schneider [83], a putative FkbN binding sequence would be composed by two 7 nt inverted repeats [38]. This sequence would be similar to that identified for binding of PimM in the genome of *S. natalensis* [63].

In-depth knowledge of the *fkb* cluster regulation is necessary to achieve higher tacrolimus production yields. In this sense, the identification of transcriptional start sites (TSS) is useful for the introduction of artificial promoters without affecting the structure of mRNAs. Bauer and coworkers [39] reported that 22% of the transcripts identified by dRNAseq are predicted to present long leader mRNAs (greater than 150 nt), which points out the importance of post-transcriptional regulation of the *fkb* cluster through the formation of RNA secondary structures [84]. In fact, the *allAKRD* operon was reported to be transcribed with a rather long untranslated 5′ region (5′-UTR; 247 bp) that is predicted to form a secondary structure.

4. Classical Strategies to Increase Tacrolimus Production

Despite the efficacy of tacrolimus in the treatment of organ transplantation, its use in clinical therapy is expensive. This is mainly due to the low production yields of the producer strains used but also to the formation of byproducts such as ascomycin (FK520) or FK525, which are structurally similar to tacrolimus but differ in the nature of some radical groups [85]. The presence of byproducts in the culture broths hampers extraction and purification of tacrolimus; thus, different approaches involving the use of organic solvents and/or chromatography have been developed to increase tacrolimus purity [86]. As an example, ascomycin production can represent 20% of tacrolimus production in *S. tsukubaensis* NRRL 18488 and 8% in *Streptomyces clavuligerus* KCTC 10561BP [86,87]. The chemical synthesis of tacrolimus was described in the 90′s but it is not applied in practice due to its low efficacy and high costs [88,89].

In the last decades, the research on tacrolimus production enhancement has been mainly focused on culture media optimization and genetic engineering of the strains. For a recent review on the improvement of tacrolimus biosynthesis through synthetic biology approaches see [90,91]. The optimization of culture media encompasses formulation of defined compositions, precursor supply and the addition of stressing agents. Defined media are highly necessary to perform nutritional studies in which the stimulating or inhibitory effect of a particular nutrient on growth and antibiotic production is tested. The first defined media for the growth of *Streptomyces* sp. MA6858 (ATCC 55098) was formulated by Yoon and Choi [92]; later, Martínez-Castro and coworkers [93] developed two additional media, MGm-2.5 and ISPz. MGm-2.5, which contain starch as the main carbon source and glutamate as carbon and nitrogen sources whilst ISPz, an optimization of ISP4 medium, contains glucose and corn dextrin as the main carbon source. MGm-2.5 has been further used to perform transcriptomic analyses on the carbon and phosphate control of *S. tsukubaensis* [94,95]. This medium supports dispersed growth and high tacrolimus production yields. Moreover, this medium permits an estimate of the onset of tacrolimus production since this process has been shown to take place when phosphate is depleted from this medium [93].

Considering that the availability of precursors is a limiting factor in the biosynthesis of secondary metabolites, precursor supply is a straightforward strategy to increase antibiotic yields [96]. A summary of the compounds that have been applied to increase tacrolimus production is shown in Table 1. At this point of the review and as a conclusion of all the mentioned work, it is interesting to note that (1) The effect of a precursor depends on its concentration; (2) The combination of positive additions does not always have an additive positive effect and (3) The positive effect can be exerted through growth promotion, production stimulation or both.

Table 1. Common precursors used for tacrolimus production enhancement in different *S. tsukubaensis* strains. The precursor, *S. tsukubaensis* strain used and bibliographic reference are indicated.

Precursor	Strain	Reference
Soybean oil	*Streptomyces* sp. MA6858 B3178 *S. tsukubaensis* TJ-04 *S. tsukubaensis* D852	[97–101]
L-lysine	*Streptomyces* sp. MA6858 B3178 *S. tsukubaensis* D852 *S. tsukubaensis* NRRL18488	[93,97,98,100–103]

Table 1. *Cont.*

Precursor	Strain	Reference
Methyl-oleate	*S. clavuligerus* CKD1119	[98,104]
Pipecolic acid	*S. tsukubaensis* NRRL18488 *S. tsukubaensis* D852	[100,101,105]
Picolinic acid	*S. tsukubaensis* NRRL18488	[105]
Nicotinamide	*S. tsukubaensis* NRRL18488	[105]
Nicotinic acid	*S. tsukubaensis* NRRL18488	[105]
Chorismate	*S. tsukubaensis* D852	[100,101]
Shikimate	*S. tsukubaensis* D852 *S. tsukubaensis* TJ-04 *S. tsukubaensis* NRRL18488	[99–103]
Lactate	*S. tsukubaensis* D852 *Streptomyces* sp. MA6858	[92,100–102]
Succinate	*S. tsukubaensis* D852 *S. tsukubaensis* TJ-04 *S. tsukubaensis* NRRL18488	[99–101,103]
Isoleucine	*S. tsukubaensis* D852	[100,101]
Valine	*S. tsukubaensis* D852 *S. tsukubaensis* TJ-04	[99–101]
Proline	*S. tsukubaensis* TJ-04	[99]
Leucine	*S. tsukubaensis* TJ-04	[99]
Threonine	*S. tsukubaensis* TJ-04	[99]
Propilenglycol	*S. tsukubaensis* FERM BP-927	[106]
Propanol	*S. tsukubaensis* FERM BP-927	[106]
Propionic acid	*S. tsukubaensis* FERM BP-927	[106]
Malonate	*S. tsukubaensis* D852 *S. tsukubaensis* NRRL18488	[102,103]
Citrate	*S. tsukubaensis* D852 *S. tsukubaensis* NRRL18488	[102,103]

Nevertheless, the addition of precursors in industrial fermentations can be a non-efficient strategy from an economical point of view (i.e., shikimate, chorismate and pipecolate are expensive; [107]); thus, an alternative strategy is to increase the copy number of tacrolimus biosynthetic genes by genetic engineering. In this manner, the overexpression of genes coding for the synthesis of methylmalonyl-CoA, methoxymalonyl-ACP and allylmalonyl-CoA has been shown to have a positive impact on the tacrolimus production yields [104,108].

Finally, the addition of stressing agents, such as dimethylsulfoxide (DMSO) or sodium thiosulfate, has been shown to stimulate polyketide production in different bacteria [109,110] as well as tacrolimus production in *S. tsukubaensis* NRRL 18488 [90].

5. Omic Approaches in *S. Tsukubaensis* and Their Application in Tacrolimus Production

5.1. Metabolomic and Proteomic Studies

The inactivation or overexpression of a particular gene involved in a certain biosynthetic pathway can affect other metabolic pathways and also the growth of the microorganism. For this reason, global studies covering the whole transcriptome, proteome or metabolome are usually preferred. In *S. tsukubaensis*, several metabolomic studies have been performed in the last decade. Huang and coworkers [100,111] developed a genome-scale metabolic model (GSMM) for *S. tsukubaensis* D852 including 865 chemical reactions and 621 metabolites to predict targets for genetic manipulation. These models reconstruct the organism metabolism from the genome annotation, taking into account genes encoding enzymes and transporters. By this means it was predicted that some of those modifications in the primary metabolism pathways leading to the accumulation of erythrose-4-phosphate, α-ketoglutarate, fumarate, succinate, pyruvate, phosphoenolpyruvate, NADPH, chorismate and malonyl-CoA have a positive effect on tacrolimus production. This implies that both the pentose phosphate pathway and the TCA cycle are positively correlated with tacrolimus production. Regarding the biosynthetic cluster, the overexpression of genes involved in the formation of the starter unit DHCHC, pipecolate and in different modification reactions (*fkbO*, *fkbL*, *fkbP*, *fkbM* and *fkbD*; see Table 2) also has a positive effect. Interestingly, as mentioned before, the combination of positive mutations does not always have an additive effect, i.e., the combined overexpression of *fkbL* and *fkbP* reduced biomass formation due to the use of lysine for tacrolimus production. More recently, a metabolomic approach has been reported in which lysine, shikimate, malonate, and citrate (the last three ones in the form of sodium salts) were supplied to the culture media of *S. tsukubaensis* D852 [102]. In this study, the addition of compounds targeting different precursor pathways facilitates the comprehension of the metabolic switches that are positive for tacrolimus production, and the application of weighted correlation network analysis (WGCNA; [112]) allowed the identification of hub modules and key metabolites depending on the culture stage. For example, 48 h after the feeding, pyruvate, phosphoenolpyruvate and methylmalonate show a high degree of connectivity whilst 72 h after the feeding, shikimate and aspartate control tacrolimus production. Supporting previous results, it was reported that the pentose phosphate, shikimate and aspartate pathways are crucial for the biosynthesis of the immunosuppressant. Overexpression of *aroC* and *dapA* (involved in shikimate pathway and lysine biosynthesis, respectively) increased production of the macrolide by 40% and 23%, respectively. See a summary of the distinct gene modifications that produce a positive impact on tacrolimus production in Table 2.

Table 2. Genetic modifications predicted through metabolic modelling in *S. tsukubaensis* to improve tacrolimus production. The target gene, type of modification, strain and bibliographic reference are indicated.

Gene/Modification	Strain	Reference
fkbO/overexpression	*S. tsukubaensis* D852	[100]
fkbL/overexpression	*S. tsukubaensis* D852	[100]
fkbM/overexpression	*S. tsukubaensis* D852	[100]
fkbP/overexpression	*S. tsukubaensis* D852	[100]
fkbD/overexpression	*S. tsukubaensis* D852	[100]
gdhA/inactivation	*S. tsukubaensis* D852	[111]
ppc/inactivation	*S. tsukubaensis* D852	[111]
dahp/overexpression	*S. tsukubaensis* D852	[111]
pntAB/overexpression	*S. tsukubaensis* D852	[111]
accA2/overexpression	*S. tsukubaensis* D852	[111]
zwf2/overexpression	*S. tsukubaensis* D852	[111]
fkbD/overexpression	*S. tsukubaensis* D852	[111]
aroC/overexpression	*S. tsukubaensis* D852	[102]
dapA/overexpression	*S. tsukubaensis* D852	[102]
gcdh/inactivation	*S. tsukubaensis* NRRL 18488	[113]
tktB/overexpression	*S. tsukubaensis* NRRL 18488	[113]
msdh/overexpression	*S. tsukubaensis* NRRL 18488	[113]
ask/overexpression	*S. tsukubaensis* NRRL 18488	[113]

The GSMM developed by Huang and coworkers [111] is a pseudo-steady metabolic model, that is to say, it assumes that there is no depletion or accumulation of intracellular metabolites. Dynamic flux balance analysis (DFBA) takes into consideration the fluctuations in metabolite concentrations and thus allows the study of the interaction between metabolism and environmental changes [114]. Wang C. and coworkers [113] developed a genome-scale DFBA (GS-DFBA) model for *S. tsukubaensis* NRRL 18488 which uncovered new targets for genetic manipulation (see Table 2) that resulted in increased tacrolimus production; i.e., inactivation of *gcdh* (glutaryl-CoA dehydrogenase) and overexpression of *tktB* (transketolase), *msdh* (methylmalonate semialdehyde dehydrogenase) and *ask* (aspartate kinase).

The approached used by Xia and coworkers [99] consisted of the growth of *S. tsukubaensis* TJ-04 in two media of similar composition but resulting in different tacrolimus productivity. They analyzed the concentration of a wide range of metabolites and compared them between the two media to identify key metabolites that correlate positively with tacrolimus production. In good agreement with the results of Huang and coworkers [100,111], intermediates of the TCA cycle such as oxaloacetate, citrate, α-ketoglutarate and, especially, succinyl-CoA and acetyl-CoA, showed a positive correlation with tacrolimus production. In addition, the intracellular levels of pentose phosphate pathway intermediates were lower in the high production media, supporting the assumption that this pathway is positively correlated with tacrolimus production. Regarding metabolites from the tacrolimus biosynthetic pathway, methylmalonyl-CoA showed the best correlation.

More recently, Wang and coworkers [103] performed a comparative proteomic and metabolomic approach in *S. tsukubaensis* NRRL 18488 grown under soybean oil feeding. The positive effect of this carbon source on growth and on tacrolimus production has been already reported in other producing strains [97–101] and, as expected, increased tacrolimus production by 89%. This work has unraveled the effect of soybean oil on tacrolimus production, which mainly affects primary metabolism proteins (42%), redox proteins (12.5%), transcriptional regulators, signal transduction components and translation proteins (11%). The key metabolites associated with tacrolimus production correlate well with those identified previously by Xia and coworkers [99] and include malic acid, gluconic acid, citric acid, α-ketoglutarate, hexadecanoic acid, threonine, fumaric acid, succinic acid, proline, valine, oleic acid, trehalose, pyruvate, ornithine, 10-undecenoic acid, shikimic acid, mannose, and lactate. Several enzymes involved in the lower glycolytic pathway and the TCA cycle (i.e., triosephosphate isomerase, phosphoglycerate mutase, pyruvate kinase or citrate synthase) were overproduced under the soybean oil condition, and the rate-limiting enzyme of the pentose phosphate pathway glucose-6-phosphate dehydrogenase showed higher amounts in the fed condition, which supports the above-mentioned positive correlation of the pentose phosphate and TCA cycle pathways with the tacrolimus production process. Finally, enzymes related to fatty acid, shikimic acid, valine and isoleucine metabolisms (which can be transformed in the extender units methylmalonyl-CoA and propionyl-CoA) were also upregulated (valine and isoleucine can be transformed in the extender units methylmalonyl-CoA and propionyl-CoA). Interestingly, higher amounts of the transcriptional regulators Crp and AfsQ1 were detected under the soybean oil feeding condition, pointing to their possible involvement in tacrolimus production regulation.

5.2. Transcriptomic Studies on Phosphate Regulation of the *fkb* Cluster

Understanding how a biosynthetic cluster is regulated is important to develop strategies to improve secondary metabolite production. Our group has studied the phosphate regulation of antibiotic production in different *Streptomyces* species in the last two decades, including *S. tsukubaensis* [94,115,116]. It is well known that high phosphate concentrations in the culture media downregulate antibiotic production [117]. This regulatory phenomenon is exerted, at least in part, through the two-component system PhoR-PhoP, which is formed by a sensor kinase and a response regulator, respectively [115,118]. When phosphate is depleted from the culture media, PhoR phosphorylates PhoP. The binding of phosphorylated PhoP (PhoP-P) to its target sequences (known as PHO boxes) can have a positive or negative transcriptional effect depending on the location of the PhoP-P binding site [118–120]. In *S. tsukubaensis*, the negative regulation of tacrolimus biosynthesis by phosphate was reported in 2013 [93] and later the PhoR-PhoP system was studied in detail [94]. In the work, transcriptomics were applied to identify genes that are transcriptionally activated after phosphate depletion. The study allowed the identification of not only common Pho members but also of potential new species-specific members, like, for example, three overlapping genes encoding a two component system and a small hydrophilic protein. In addition, a bioinformatic search for PHO boxes was developed [121]. Putative PHO boxes were identified in most of the genes responding to phosphate starvation, supporting the transcriptional results. A putative PHO box was identified in the promoter region of *fkbN* and also in primary metabolism genes that might be involved in tacrolimus precursor supply such as STSU_30046, encoding an acetoacetate-CoA ligase [94].

5.2.1. Transcriptomics of Carbon Catabolite Regulation of Tacrolimus Biosynthesis

A second regulatory mechanism governing secondary metabolite production is carbon repression. Similar to phosphate, the presence of ready-to-use carbon sources in the media reduces or blocks antibiotic production and this can happen at the transcriptional or at the posttranslational level [122,123]. The mechanisms involved in this nutritional regulation are not completely understood in streptomycetes and, as it can be deduced, its unveiling is very interesting in order to use easily assimilated carbon sources that allow faster growth in the culture broths without hampering tacrolimus biosynthesis. Regarding this subject, our group observed that glucose and glycerol, when added as carbon sources at a concentration of 0.22 M at the first growth phase (and before phosphate depletion), arrest tacrolimus production in *S. tsukubaensis*; the glucose effect being stronger than that of glycerol [95]. Both glucose and glycerol additions resulted in a lack of transcriptional activation of the *fkb* cluster; thus, it was concluded that transcriptional repression plays a role in this regulatory mechanism. In addition, the effect of these carbon sources can be exerted at the intermediary metabolism level: glucose addition increased transcription of genes involved in glycolysis, pyruvate and oxaloacetate formation but downregulated genes involved in the TCA cycle. These results are coherent with the previous assumption that the TCA cycle is positively correlated with tacrolimus production whilst glycolytic metabolites show a negative correlation [99].

In the MGm-2.5 medium used in the work, transcription of *fkbN* increases in a two-step fashion before tacrolimus is detected in the broths [38]: a slight increase in mRNA levels occurs between 80 h and 89 h and then it is followed by a higher increase from 92 h to 100 h (Figure 4). The first step coincides with phosphate depletion, supporting the proposal that *fkbN* is under phosphate control [95] (Figure 4). Taking into account that *fkbN* transcription is not strongly self-regulated [38], it seems that a key transcriptional regulator, co-activator molecule or sigma factor might be absent in the presence of glucose or glycerol. Therefore, the identification of this additional factor would be useful to trigger tacrolimus production under carbon repressing conditions. Actually, key sigma factors (i.e., *hrdA* or *bldN*) and transcriptional regulators (i.e., *eshA*, *atrA*, *afsR*) were downregulated under glucose or glycerol addition conditions [95]. HrdA might control secondary metabolism genes [124], and EshA and AtrA are both involved in antibiotic production in *S. coelicolor* and *S. griseus* [125–128]; thus, it seems interesting to analyze the effect of their inactivation and overexpression on tacrolimus production. Finally, AfsR is a very interesting candidate for these studies since it is overexpressed in an *S. tsukubaensis* strain that overproduces tacrolimus [101].

5.2.2. Transcriptomics of *N*-acetylglucosamine Addition in Tacrolimus Biosynthesis

A third example of the nutritional regulation of secondary metabolite production is that exerted by *N*-acetylglucosamine, the monomer of chitin. This compound shows a dual regulatory role, accelerating differentiation and antibiotic production under poor nutritional conditions and arresting them under rich nutritional conditions, which have been traditionally named as "famine" and "feast" conditions, respectively [129,130]. We observed a negative effect of *N*-acetylglucosamine addition on tacrolimus production when *S. tsukubaensis* was grown in MGm-2.5 medium, which might be due, at least in part, to the transcriptional repression of *fkbN*, since we observed a significant decrease in its transcription soon after *N*-acetylglucosamine addition (Ordóñez-Robles et al., unpublished data). The transcriptional response to *N*-acetylglucosamine addition is very similar to that exerted by glucose, which is not surprising since both carbon sources share a common catabolic pathway from fructose-6-phosphate.

Overall, the application of transcriptomics to nutritional studies in *S. tsukubaensis* unveils potential candidates for the rational engineering of industrial strains. It has also improved our knowledge about other aspects of its physiology such as the possible members of the PHO regulon in this species or the mechanisms operating in the presence of repressing carbon sources. These findings are worthy to detect potential targets for the bypass of nutritional repression of secondary metabolism in *Streptomyces*.

6. Conclusions and Future Prospective

It has been more than 30 years since *S. tsukubaensis* and its secondary metabolite tacrolimus were discovered. Despite the importance of this immunosuppressant macrolide in current clinics, there are still many aspects to be elucidated about the transcriptional and nutritional regulation of tacrolimus biosynthesis, and further studies are necessary to improve the yield and reduce the costs of its industrial production. In this sense, the *omic* approaches constitute an important basis to understand the producer microorganism physiology from a genome- [131], proteome- and metabolome-wide point of view. Initial *omic* studies performed in *S. tsukubaensis* have given important clues such as the positive correlation of the pentose phosphate pathway and TCA cycle with tacrolimus production or the identification of targets for genetic manipulation. These types of studies can be applied not only to the overproduction of tacrolimus but also to the awakening of cryptic clusters [132]. In fact, similar to most streptomycetes, *S. tsukubaensis'* genome contains several clusters for the production of secondary metabolites which might encode useful compounds. One of the potential products encoded is predicted to be similar to bafilomycin [133] and two other clusters show homology to those for biosynthesis of nigericin and enduracidin [134,135]. Nevertheless, we must keep in mind the interpretation of the *omic* results in the framework of the strain and culture media used since there are important physiological differences depending on the strain and the culture conditions. Therefore, the comparison of different models can broaden our perspective of tacrolimus production and *S. tsukubaensis'* physiology.

There are still some interesting points to address in the study of the *fkb* cluster such as the role of the *allMNPOS* subcluster in the strains that contain it. Although not strictly required for tacrolimus production, the *all* subcluster might be involved in the generation of macrolide variants with useful properties. Thus, the overexpression of these genes under promoters regulated by FkbN seems an interesting study. In addition, the *ppt1* and *scoT* genes, which are affected by the inactivation of *fkbN*, might be potential targets for tacrolimus biosynthesis improvement. Considering the transcriptional regulation of *eshA* and *atrA* under tacrolimus producing and repressing conditions, both genes seem good candidates for genetic engineering of the strains.

The transcriptional regulation of *fkbN* is also interesting given that it is the main transcriptional activator of the *fkb* cluster. The identification of transcriptional regulators that bind to its promoter region is a good approach to identify new targets for genetic engineering of the strains that overexpress *fkbN* and therefore, to increase tacrolimus production. Finally, the post-transcriptional regulation of the *fkb* cluster deserves further attention. As reported by Bauer and coworkers [39], a high percentage of genes are transcribed with long leader sequences in *S. tsukubaensis* (i.e., *allAKRD*). Long 5'-UTRs might be involved in the formation of secondary structures that regulate transcription of the cistrons and might be potential targets for manipulation.

Author Contributions: Juan F. Martín wrote the sections on biosynthesis of tacrolimus and regulatory genes, corrected the text and supervised the final version. María Ordóñez-Robles wrote the other sections and Fernando Santos-Beneit corrected and improved the text.

Acknowledgments: We acknowledge Paloma Liras for helpful scientific discussion.

Conflicts of Interest: The authors declare no conflict of interest.

References

1. Kino, T.; Hatanaka, H.; Hashimoto, M.; Nishiyama, M.; Goto, T.; Okuhara, M.; Kohsaka, M.; Aoki, H.; Imanaka, H. FK-506, a novel immunosuppressant isolated from a *Streptomyces*. I. Fermentation, isolation, and physico-chemical and biological characteristics. *J. Antibiot.* **1987**, *40*, 1249–1255. [CrossRef] [PubMed]
2. Kino, T.; Hatanaka, H.; Miyata, S.; Inamura, N.; Nishiyama, M.; Yajima, T.; Goto, T.; Okuhara, M.; Kohsaka, M.; Aoki, H. FK-506, a novel immunosuppressant isolated from a *Streptomyces*. II. Immunosuppressive effect of FK-506 in vitro. *J. Antibiot.* **1987**, *40*, 1256–1265. [CrossRef] [PubMed]
3. Wallemacq, P.E.; Reding, R. FK506 (tacrolimus), a novel immunosuppressant in organ transplantation: Clinical, biomedical, and analytical aspects. *Clin. Chem.* **1993**, *39*, 2219–2228. [PubMed]
4. Liu, J.; Farmer, J., Jr.; Lane, W.S.; Friedman, J.; Weissman, I.; Schreiber, S.L. Calcineurin is a common target of cyclophilin-cyclosporin A and FKBP-FK506 complexes. *Cell* **1991**, *66*, 807–815. [CrossRef]
5. Harding, M.W.; Galat, A.; Uehling, D.E.; Schreiber, S.L. A receptor for the immunosuppressant FK506 is a cis-trans peptidyl-prolyl isomerase. *Nature* **1989**, *341*, 758–760. [CrossRef] [PubMed]
6. Tocci, M.J.; Matkovich, D.A.; Collier, K.A.; Kwok, P.; Dumont, F.; Lin, S.; Degudicibus, S.; Siekierka, J.J.; Chin, J.; Hutchinson, N.I. The immunosuppressant FK506 selectively inhibits expression of early T cell activation genes. *J. Immunol.* **1989**, *143*, 718–726. [PubMed]
7. Foor, F.; Parent, S.A.; Morin, N.; Dahl, A.M.; Ramadan, N.; Chrebet, G.; Bostian, K.A.; Nielsen, J.B. Calcineurin mediates inhibition by FK506 and cyclosporin of recovery from alpha-factor arrest in yeast. *Nature* **1992**, *360*, 682–684. [PubMed]
8. Kunz, J.; Hall, M.N.; Cyclosporin, A. FK506 and rapamycin: More than just immunosuppression. *Trends Biochem. Sci.* **1993**, *18*, 334–338. [CrossRef]
9. Breuder, T.; Hemenway, C.S.; Movva, N.R.; Cardenas, M.E.; Heitman, J. Calcineurin is essential in cyclosporin A- and FK506-sensitive yeast strains. *Proc. Natl. Acad. Sci. USA* **1994**, *91*, 5372–5376. [CrossRef] [PubMed]
10. Trede, N.S.; Warwick, A.B.; Rosoff, P.M.; Rohrer, R.; Bierer, B.E.; Guinan, E. Tacrolimus (FK506) in allogeneic bone marrow transplantation for severe aplastic anemia following orthotopic liver transplantation. *Bone Marrow Transplant.* **1997**, *20*, 257–260. [CrossRef] [PubMed]
11. Meier-Kriesche, H.-U.; Li, S.; Gruessner, R.W.G.; Fung, J.J.; Bustami, R.T.; Barr, M.L.; Leichtman, A.B. Immunosuppression: Evolution in practice and trends, 1994–2004. *Am. J. Transplant.* **2006**, *6*, 1111–1131. [CrossRef] [PubMed]
12. McCormack, P.L.; Keating, G.M. Tacrolimus: In heart transplant recipients. *Drugs* **2006**, *66*, 2269–2279. [CrossRef] [PubMed]
13. Ingram, J.R.; Martin, J.A.; Finlay, A.Y. Impact of topical calcineurin inhibitors on quality of life in patients with atopic dermatitis. *Am. J. Clin. Dermatol.* **2009**, *10*, 229–237. [CrossRef] [PubMed]
14. Remitz, A.; Reitamo, S. Long-term safety of tacrolimus ointment in atopic dermatitis. *Expert Opin. Drug Saf.* **2009**, *8*, 501–506. [CrossRef] [PubMed]
15. Romano, A.; Jensen, M.R.; McAlpine, J. Toward the optimization of stent-based treatment for coronary artery disease. *Curr. Opin. Drug Discov. Devel.* **2010**, *13*, 157–158. [PubMed]
16. Akimoto, K.; Kusunoki, Y.; Nishio, S.; Takagi, K.; Kawai, S. Safety profile of tacrolimus in patients with rheumatoid arthritis. *Clin. Rheumatol.* **2008**, *27*, 1393–1397. [CrossRef] [PubMed]
17. Benson, A.; Barrett, T.; Sparberg, M.; Buchman, A.L. Efficacy and safety of tacrolimus in refractory ulcerative colitis and Crohn's disease: A single-center experience. *Inflamm. Bowel Dis.* **2008**, *14*, 7–12. [CrossRef] [PubMed]
18. Reis, S.A.; Moussatché, N.; Damaso, C.R.A. FK506, a secondary metabolite produced by *Streptomyces*, presents a novel antiviral activity against Orthopoxvirus infection in cell culture. *J. Appl. Microbiol.* **2006**, *100*, 1373–1380. [CrossRef] [PubMed]
19. Karpas, A.; Lowdell, M.; Jacobson, S.K.; Hill, F. Inhibition of human immunodeficiency virus and growth of infected T cells by the immunosuppressive drugs cyclosporin A and FK 506. *Proc. Natl. Acad. Sci. USA* **1992**, *89*, 8351–8355. [CrossRef] [PubMed]
20. Briggs, C.J.; Ott, D.E.; Coren, L.V.; Oroszlan, S.; Tözsér, J. Comparison of the effect of FK506 and cyclosporin A on virus production in H9 cells chronically and newly infected by HIV-1. *Arch. Virol.* **1999**, *144*, 2151–2160. [CrossRef] [PubMed]

21. Mortola, E.; Endo, Y.; Ohno, K.; Watari, T.; Tsujimoto, H.; Hasegawa, A. The use of two immunosuppressive drugs, cyclosporin A and tacrolimus, to inhibit virus replication and apoptosis in cells infected with feline immunodeficiency virus. *Vet. Res. Commun.* **1998**, *22*, 553–563. [CrossRef] [PubMed]
22. Yamamoto, S.; Jiang, H.; Kato, R. Stimulation of hair growth by topical application of FK506, a potent immunosuppressive agent. *J. Investig. Dermatol.* **1994**, *102*, 160–164. [CrossRef] [PubMed]
23. Klettner, A.; Herdegen, T. FK506 and its analogs—Therapeutic potential for neurological disorders. *Curr. Drug Targets CNS Neurol. Disord.* **2003**, *2*, 153–162. [CrossRef] [PubMed]
24. Sierra-Paredes, G.; Sierra-Marcuño, G. Ascomycin and FK506: Pharmacology and therapeutic potential as anticonvulsants and neuroprotectants. *CNS Neurosci. Ther.* **2008**, *14*, 36–46. [CrossRef] [PubMed]
25. Konofaos, P.; Terzis, J.K. FK506 and nerve regeneration: Past, present, and future. *J. Reconstr. Microsurg.* **2013**, *29*, 141–148. [CrossRef] [PubMed]
26. Periyasamy, S.; Warrier, M.; Tillekeratne, M.P.M.; Shou, W.; Sanchez, E.R. The immunophilin ligands cyclosporin A and FK506 suppress prostate cancer cell growth by androgen receptor-dependent and -independent mechanisms. *Endocrinology* **2007**, *148*, 4716–4726. [CrossRef] [PubMed]
27. Barot, R.K.; Shitole, S.C.; Bhagat, N.; Patil, D.; Sawant, P.; Patil, K. Therapeutic effect of 0.1% Tacrolimus Eye Ointment in Allergic Ocular Diseases. *JCDR* **2016**, *10*, NC05-9. [CrossRef] [PubMed]
28. Pirsch, J.D.; Miller, J.; Deierhoi, M.H.; Vincenti, F.; Filo, R.S. A comparison of tacrolimus (FK506) and cyclosporine for immunosuppression after cadaveric renal transplantation. FK506 Kidney Transplant Study Group. *Transplantation* **1997**, *63*, 977–983. [CrossRef] [PubMed]
29. Jiang, H.; Kobayashi, M. Differences between cyclosporin A and tacrolimus in organ transplantation. *Transplant. Proc.* **1999**, *31*, 1978–1980. [CrossRef]
30. Motamedi, H.; Shafiee, A. The biosynthetic gene cluster for the macrolactone ring of the immunosuppressant FK506. *Eur. J. Biochem.* **1998**, *256*, 528–534. [CrossRef] [PubMed]
31. Motamedi, H.; Shafiee, A.; Cai, S.J.; Streicher, S.L.; Arison, B.H.; Miller, R.R. Characterization of methyltransferase and hydroxylase genes involved in the biosynthesis of the immunosuppressants FK506 and FK520. *J. Bacteriol.* **1996**, *178*, 5243–5248. [CrossRef] [PubMed]
32. Motamedi, H.; Cai, S.J.; Shafiee, A.; Elliston, K.O. Structural organization of a multifunctional polyketide synthase involved in the biosynthesis of the macrolide immunosuppressant FK506. *Eur. J. Biochem.* **1997**, *244*, 74–80. [CrossRef] [PubMed]
33. Shafiee, A.; Motamedi, H.; Chen, T. Enzymology of FK-506 biosynthesis. Purification and characterization of 31-O-desmethylFK-506 O:methyltransferase from *Streptomyces* sp. MA6858. *Eur. J. Biochem.* **1994**, *225*, 755–764. [CrossRef] [PubMed]
34. Barreiro, C.; Martínez-Castro, M. Trends in the biosynthesis and production of the immunosuppressant tacrolimus (FK506). *Appl. Microbiol. Biotechnol.* **2014**, *98*, 497–507. [CrossRef] [PubMed]
35. Zong, G.; Zhong, C.; Fu, J.; Qin, R.; Cao, G. Draft genome sequence of the tacrolimus-producing bacterium *Streptomyces tsukubaensis* F601. *Genome Announc.* **2017**, *5*, e00385-17. [CrossRef] [PubMed]
36. Goranovič, D.A.; Blažič, I.M.; Magdevska, V.; Horvat, J.; Kuščer, E.; Polak, T.; Santos-Aberturas, J.; Martínez-Castro, M.; Barreiro, C.; Mrak, P.; et al. FK506 biosynthesis is regulated by two positive regulatory elements in *Streptomyces tsukubaensis*. *BMC Microbiol.* **2012**, *12*, 238. [CrossRef] [PubMed]
37. Mo, S.; Kim, D.H.; Lee, J.H.; Park, J.W.; Basnet, D.B.; Ban, Y.H.; Yoo, Y.J.; Chen, S.-W.; Park, S.R.; Choi, E.A.; et al. Biosynthesis of the allylmalonyl-CoA extender unit for the FK506 polyketide synthase proceeds through a dedicated polyketide synthase and facilitates the mutasynthesis of analogues. *J. Am. Chem. Soc.* **2011**, *133*, 976–985. [CrossRef] [PubMed]
38. Ordóñez-Robles, M.; Rodríguez-García, A.; Martín, J.F. Target genes of the *Streptomyces tsukubaensis* FkbN regulator include most of the tacrolimus biosynthesis genes, a phosphopantetheinyl transferase and other PKS genes. *Appl. Microbiol. Biotechnol.* **2016**, *100*, 8091–8103. [CrossRef] [PubMed]
39. Bauer, J.S.; Fillinger, S.; Förstner, K.; Herbig, A.; Jones, A.C.; Flinspach, K.; Sharma, C.; Gross, H.; Nieselt, K.; Apel, A.K. dRNA-seq transcriptional profiling of the FK506 biosynthetic gene cluster in *Streptomyces tsukubaensis* NRRL18488 and general analysis of the transcriptome. *RNA Biol.* **2017**, *14*, 1617–1626. [CrossRef] [PubMed]
40. Andexer, J.N.; Kendrew, S.G.; Nur-e-Alam, M.; Lazos, O.; Foster, T.A.; Zimmermann, A.-S.; Warneck, T.D.; Suthar, D.; Coates, N.J.; Koehn, F.E.; et al. Biosynthesis of the immunosuppressants FK506, FK520, and rapamycin involves a previously undescribed family of enzymes acting on chorismate. *Proc. Natl. Acad. Sci. USA* **2011**, *108*, 4776–4781. [CrossRef] [PubMed]

41. Goulet, M.T.; Rupprecht, K.M.; Sinclair, P.J.; Wyvratt, M.J.; Parsons, W.H. The medicinal chemistry of FK-506. *Perspect. Drug Discov. Des.* **1994**, *2*, 145–162. [CrossRef]
42. Carroll, B.J.; Moss, S.J.; Bai, L.; Kato, Y.; Toelzer, S.; Yu, T.-W.; Floss, H.G. Identification of a set of genes involved in the formation of the substrate for the incorporation of the unusual "glycolate" chain extension unit in ansamitocin biosynthesis. *J. Am. Chem. Soc.* **2002**, *124*, 4176–4177. [CrossRef] [PubMed]
43. Kato, Y.; Bai, L.; Xue, Q.; Revill, W.P.; Yu, T.-W.; Floss, H.G. Functional expression of genes involved in the biosynthesis of the novel polyketide chain extension unit, methoxymalonyl-acyl carrier protein, and engineered biosynthesis of 2-desmethyl-2-methoxy-6-deoxyerythronolide B. *J. Am. Chem. Soc.* **2002**, *124*, 5268–5269. [CrossRef] [PubMed]
44. Wu, K.; Chung, L.; Revill, W.P.; Katz, L.; Reeves, C.D. The FK520 gene cluster of *Streptomyces hygroscopicus* var. *ascomyceticus* (ATCC 14891) contains genes for biosynthesis of unusual polyketide extender units. *Gene* **2000**, *251*, 81–90. [PubMed]
45. Chen, D.; Zhang, L.; Pang, B.; Chen, J.; Xu, Z.; Abe, I.; Liu, W. FK506 maturation involves a cytochrome p450 protein-catalyzed four-electron C-9 oxidation in parallel with a C-31 o-methylation. *J. Bacteriol.* **2013**, *195*, 1931–1939. [CrossRef] [PubMed]
46. Goranovič, D.; Kosec, G.; Mrak, P.; Fujs, S.; Horvat, J.; Kuščer, E.; Kopitar, G.; Petković, H. Origin of the allyl group in FK506 biosynthesis. *J. Biol. Chem.* **2010**, *285*, 14292–14300. [CrossRef] [PubMed]
47. Jiang, H.; Wang, Y.Y.; Guo, Y.Y.; Shen, J.J.; Zhang, X.S.; Luo, H.D.; Ren, N.N.; Jiang, X.H.; Li, Y.Q. An acyltransferase domain of FK506 polyketide synthase recognizing both an acyl carrier protein and coenzyme A as acyl donors to transfer allylmalonyl and ethylmalonyl units. *FEBS J.* **2015**, *282*, 2527–2539. [CrossRef] [PubMed]
48. Byrne, K.; Shafiee, A.; Nielsen, J.; Arison, B.; Monaghan, R.; Kaplan, L. The biosynthesis and enzymology of an immunosuppressant, immunomycin, produced by *Streptomyces hygroscopicus* var. *ascomyceticus*. *Dev. Ind. Microbiol.* **1993**, *32*, 29–45.
49. Gatto, G.J., Jr.; McLoughlin, S.M.; Kelleher, N.L.; Walsh, C.T. Elucidating the substrate specificity and condensation domain activity of FkbP, the FK520 pipecolate-incorporating enzyme. *Biochemistry* **2005**, *44*, 5993–6002. [CrossRef] [PubMed]
50. Gatto, G.J., Jr.; Boyne, M.T., 2nd; Kelleher, N.L.; Walsh, C.T. Biosynthesis of pipecolic acid by RapL, a lysine cyclodeaminase encoded in the rapamycin gene cluster. *J. Am. Chem. Soc.* **2006**, *128*, 3838–3847. [CrossRef] [PubMed]
51. Van Duyne, G.D.; Standaert, R.F.; Karplus, P.A.; Schreiber, S.L.; Clardy, J. Atomic structure of FKBP-FK506, an immunophilin-immunosuppressant complex. *Science* **1991**, *252*, 839–842. [CrossRef] [PubMed]
52. Becker, J.W.; Rotonda, J.; McKeever, B.M.; Chan, H.K.; Marcy, A.I.; Wiederrecht, G.; Hermes, J.D.; Springer, J.P. FK-506-binding protein: Three-dimensional structure of the complex with the antagonist L-685,818. *J. Biol. Chem.* **1993**, *268*, 11335–11339. [PubMed]
53. Shafiee, A.; Motamedi, H.; Dumont, F.J.; Arison, B.H.; Miller, R.R. Chemical and biological characterization of two FK506 analogs produced by targeted gene disruption in *Streptomyces* sp. MA6548. *J. Antibiot.* **1997**, *50*, 418–423. [CrossRef] [PubMed]
54. Ban, Y.H.; Shinde, P.B.; Hwang, J.-Y.; Song, M.-C.; Kim, D.H.; Lim, S.-K.; Sohng, J.K.; Yoon, Y.J. Characterization of FK506 biosynthetic intermediates involved in post-PKS elaboration. *J. Nat. Prod.* **2013**, *76*, 1091–1098. [CrossRef] [PubMed]
55. Schrijver, A.D.; Mot, R.D. A subfamily of MalT-related ATP-dependent regulators in the LuxR family. *Microbiology* **1999**, *145*, 1287–1288. [CrossRef] [PubMed]
56. Bibb, M.J. Regulation of secondary metabolism in *Streptomycetes*. *Curr. Opin. Microbiol.* **2005**, *8*, 208–215. [CrossRef] [PubMed]
57. Molnár, I.; Aparicio, J.F.; Haydock, S.F.; Khaw, L.E.; Schwecke, T.; König, A.; Staunton, J.; Leadlay, P.F. Organisation of the biosynthetic gene cluster for rapamycin in *Streptomyces hygroscopicus*: Analysis of genes flanking the polyketide synthase. *Gene* **1996**, *169*, 1–7. [CrossRef]
58. Wilson, D.J.; Xue, Y.; Reynolds, K.A.; Sherman, D.H. Characterization and analysis of the PikD regulatory factor in the pikromycin biosynthetic pathway of *Streptomyces venezuelae*. *J. Bacteriol.* **2001**, *183*, 3468–3475. [CrossRef] [PubMed]

59. He, W.; Lei, J.; Liu, Y.; Wang, Y. The LuxR family members GdmRI and GdmRII are positive regulators of geldanamycin biosynthesis in *Streptomyces hygroscopicus* 17997. *Arch. Microbiol.* **2008**, *189*, 501–510. [CrossRef] [PubMed]
60. Chen, S.; Huang, X.; Zhou, X.; Bai, L.; He, J.; Jeong, K.J.; Lee, S.Y.; Deng, Z. Organizational and Mutational Analysis of a Complete FR-008/Candicidin Gene Cluster Encoding a Structurally Related Polyene Complex. *Chem. Biol.* **2003**, *10*, 1065–1076. [CrossRef] [PubMed]
61. Martín, J.F.; Aparicio, J.F. Biosynthesis Enzymology of the Polyenes Pimaricin and Candicidin. *Methods Enzymol.* **2009**, *459*, 215–242. [PubMed]
62. Santos-Aberturas, J.; Payero, T.D.; Vicente, C.M.; Guerra, S.M.; Cañibano, C.; Martín, J.F.; Aparicio, J.F. Functional conservation of PAS-LuxR transcriptional regulators in polyene macrolide biosynthesis. *Metab. Eng.* **2011**, *13*, 756–757. [CrossRef] [PubMed]
63. Santos-Aberturas, J.; Vicente, C.M.; Guerra, S.M.; Payero, T.D.; Martín, J.F.; Aparicio, J.F. Molecular control of polyene macrolide biosynthesis: Direct binding of the regulator PimM to eight promoters of pimaricin genes and identification of binding boxes. *J. Biol. Chem.* **2011**, *286*, 9150–9161. [CrossRef] [PubMed]
64. Brautaset, T.; Sekurova, O.N.; Sletta, H.; Ellingsen, T.E.; Strøm, A.R.; Valla, S.; Zotchev, S.B. Biosynthesis of the polyene antifungal antibiotic nystatin in *Streptomyces noursei* ATCC 11455: Analysis of the gene cluster and deduction of the biosynthetic pathway. *Chem. Biol.* **2000**, *7*, 395–403. [CrossRef]
65. Carmody, M.; Byrne, B.; Murphy, B.; Breen, C.; Lynch, S.; Flood, E.; Finnan, S.; Caffrey, P. Analysis andmanipulation of amphotericin biosynthetic genes by means of modified phage KC515 transduction techniques. *Gene* **2004**, *343*, 107–115. [CrossRef] [PubMed]
66. Omura, S.; Ikeda, H.; Ishikawa, J.; Hanamoto, A.; Takahashi, C.; Shinose, M.; Hattori, M. Genome Sequence of an Industrial Microorganism *Streptomyces avermitilis*: Deducing the Ability of Producing Secondary Metabolites. *Proc. Natl. Acad. Sci. USA* **2001**, *98*, 12215–12220. [CrossRef] [PubMed]
67. Ikeda, H.; Ishikawa, J.; Hanamoto, A.; Shinose, M.; Kikuchi, H.; Shiba, T.; Sakaki, Y.; Hattori, M.; Omura, S. Complete genome sequence and comparative analysis of the industrial microorganism *Streptomyces avermitilis*. *Nat. Biotechnol.* **2003**, *21*, 526–531. [CrossRef] [PubMed]
68. Song, K.; Wei, L.; Liu, J.; Wang, J.; Qi, H.; Wen, J. Engineering of the LysR family transcriptional regulator FkbR1 and its target gene to improve ascomycin production. *Appl. Microbiol. Biotechnol.* **2017**, *101*, 4581–4592. [CrossRef] [PubMed]
69. Bentley, S.D.; Chater, K.F.; Cerdeño-Tárraga, A.-M.; Challis, G.L.; Thomson, N.R.; James, K.D.; Harris, D.E.; Quail, M.A.; Kieser, H.; Harper, D.; et al. Complete genome sequence of the model actinomycete *Streptomyces coelicolor* A3. *Nature* **2002**, *417*, 141–147. [CrossRef] [PubMed]
70. Schell, M.A. Molecular biology of the LysR family of transcriptional regulators. *Annu. Rev. Microbiol.* **1993**, *47*, 597–626. [CrossRef] [PubMed]
71. Maddocks, S.E.; Oyston, P.C. Structure and function of the LysR-type transcriptional regulator (LTTR) family proteins. *Microbiology* **2008**, *154 (Pt 12)*, 3609–3623. [CrossRef] [PubMed]
72. Rodríguez, M.; Nuñez, L.E.; Braña, A.F.; Mendez, C.; Salas, J.A.; Blanco, G. Identification of transcriptional activators for thienamycin and cephamycin C biosynthetic genes within the thienamycin gene cluster from *Streptomyces cattleya*. *Mol. Microbiol.* **2008**, *69*, 633–645. [CrossRef] [PubMed]
73. Pérez-Redondo, R.; Rodríguez-García, A.; Martín, J.F.; Liras, P. The *claR* gene of *Streptomyces clavuligerus*, encoding a LysR-type regulatory protein controlling clavulanic acid biosynthesis, is linked to the clavulanate-9-aldehyde reductase (car) gene. *Gene* **1998**, *211*, 311–321. [CrossRef]
74. Kölling, R.; Lother, H. AsnC: An autogenlously regulated activator of asparragine syntetase A transcription in *Escherícha coli*. *J. Bacteriol.* **1985**, *164*, 310–315. [PubMed]
75. Zhang, X.S.; Luo, H.D.; Tao, Y.; Wang, Y.Y.; Jiang, X.H.; Jiang, H.; Li, Y.Q. FkbN and Tcs7 are pathway-specific regulators of the FK506 biosynthetic gene cluster in *Streptomyces tsukubaensis* L19′. *J. Ind. Microbiol. Biotechnol.* **2016**, *43*, 1693–1703. [CrossRef] [PubMed]
76. Mo, S.; Yoo, Y.J.; Ban, Y.H.; Lee, S.-K.; Kim, E.; Suh, J.-W.; Yoon, Y.J. Roles of *fkbN* in positive regulation and *tcs7* in negative regulation of FK506 biosynthesis in *Streptomyces* sp. strain KCTC 11604BP. *Appl. Environ. Microbiol.* **2012**, *78*, 2249–2255. [CrossRef] [PubMed]
77. Jones, A.C.; Gust, B.; Kulik, A.; Heide, L.; Buttner, M.J.; Bibb, M.J. Phage p1-derived artificial chromosomes facilitate heterologous expression of the FK506 gene cluster. *PLoS ONE* **2013**, *8*, e69319. [CrossRef] [PubMed]

78. Liu, G.; Chater, K.F.; Chandra, G.; Niu, G.; Tan, H. Molecular regulation of antibiotic biosynthesis in *Streptomyces*. *Microbiol. Mol. Biol. Rev.* **2013**, *77*, 112–143. [CrossRef] [PubMed]
79. Martínez-Burgo, Y.; Álvarez-Álvarez, R.; Rodríguez-García, A.; Liras, P. The Pathway-Specific Regulator *ClaR* of *Streptomyces clavuligerus* has a Global Effect on the Expression of Genes for Secondary Metabolism and Differentiation. *Appl. Environ. Microbiol.* **2015**, *81*, 6637–6648. [CrossRef] [PubMed]
80. Lu, Y.W.; San Roman, A.K.; Gehring, A.M. Role of Phosphopantetheinyl Transferase Genes in Antibiotic Production by *Streptomyces coelicolor*. *J. Bacteriol.* **2008**, *190*, 6903–6908. [CrossRef] [PubMed]
81. Wang, Y.Y.; Zhang, X.S.; Luo, H.D.; Ren, N.N.; Jiang, X.H.; Jiang, H.; Li, Y.Q. Characterization of discrete phosphopantetheinyl transferases in *Streptomyces tsukubaensis* L19 unveils a complicate phosphopantetheinylation network. *Sci. Rep.* **2016**, *6*, 24255. [CrossRef] [PubMed]
82. Blazic, M.; Starcevic, A.; Lisfi, M.; Baranasic, D.; Goranovic, D.; Fujs, S.; Kuščer, E.; Kosec, G.; Petkovic, H.; Cullum, J.; et al. Annotation of the modular polyketide synthase and nonribosomal peptide synthetase gene clusters in the genome of *Streptomyces tsukubaensis* NRRL18488. *Appl. Environ. Microbiol.* **2012**, *78*, 8183–8190. [CrossRef] [PubMed]
83. Schneider, T.D. Information content of individual genetic sequences. *J. Theor. Biol.* **1997**, *189*, 427–441. [CrossRef] [PubMed]
84. Vockenhuber, M.P.; Sharma, C.M.; Statt, M.G.; Schmidt, D.; Xu, Z.; Dietrich, S.; Liesegang, H.; Mathews, D.H.; Suess, B. Deep sequencing based identification of small non-coding RNAs in *Streptomyces coelicolor*. *RNA Biol.* **2011**, *8*, 468–477. [CrossRef] [PubMed]
85. Hatanaka, H.; Kino, T.; Asano, M.; Goto, T.; Tanaka, H.; Okuhara, M. FK-506 related compounds produced by *Streptomyces tsukubaensis* No. 9993. *J. Antibiot.* **1989**, *42*, 620–622. [CrossRef] [PubMed]
86. Kosec, G.; Goranovič, D.; Mrak, P.; Fujs, S.; Kuščer, E.; Horvat, J.; Kopitar, G.; Petković, H. Novel chemobiosynthetic approach for exclusive production of FK506. *Metab. Eng.* **2012**, *14*, 39–46. [CrossRef] [PubMed]
87. Park, J.W.; Mo, S.-J.; Park, S.R.; Ban, Y.-H.; Yoo, Y.J.; Yoon, Y.J. Liquid chromatography-mass spectrometry characterization of FK506 biosynthetic intermediates in *Streptomyces clavuligerus* KCTC 10561BP. *Anal. Biochem.* **2009**, *393*, 1–7. [CrossRef] [PubMed]
88. Nakatsuka, M.; Ragan, J.A.; Sammakia, T.; Smith, D.B.; Uehling, D.E.; Schreiber, S.L. Total synthesis of FK506 and an FKBP probe reagent, [C,C-13C2]-FK506. *J. Am. Chem. Soc.* **1990**, *112*, 5583–5601. [CrossRef]
89. Ireland, R.E.; Gleason, J.L.; Gegnas, L.D.; Highsmith, T.K. A Total Synthesis of FK-506. *J. Org. Chem.* **1996**, *61*, 6856–6872. [CrossRef] [PubMed]
90. Ban, Y.H.; Park, S.R.; Yoon, Y.J. The biosynthetic pathway of FK506 and its engineering: From past achievements to future prospects. *J. Ind. Microbiol. Biotechnol.* **2016**, *43*, 389–400. [CrossRef] [PubMed]
91. Fu, L.F.; Tao, Y.; Jin, M.Y.; Jiang, H. Improvement of FK506 production by synthetic biology approaches. *Biotechnol. Lett.* **2016**, *38*, 2015–2021. [CrossRef] [PubMed]
92. Yoon, Y.J.; Choi, C.Y. Nutrient Effects on FK-506, a New Immunosuppressant, Production by *Streptomyces* sp. in a Defined Medium. *J. Ferment. Bioeng.* **1997**, *83*, 599–603. [CrossRef]
93. Martínez-Castro, M.; Salehi-Najafabadi, Z.; Romero, F.; Pérez-Sanchiz, R.; Fernández-Chimeno, R.I.; Martín, J.F.; Barreiro, C. Taxonomy and chemically semi-defined media for the analysis of the tacrolimus producer '*Streptomyces tsukubaensis*'. *Appl. Microbiol. Biotechnol.* **2013**, *97*, 2139–2152. [CrossRef] [PubMed]
94. Ordóñez-Robles, M.; Santos-Beneit, F.; Rodríguez-García, A.; Martín, J.F. Analysis of the PHO regulon in *Streptomyces tsukubaensis*. *Microbiol. Res.* **2017**, *205*, 80–87. [CrossRef] [PubMed]
95. Ordóñez-Robles, M.; Santos-Beneit, F.; Albillos, S.M.; Liras, P.; Martín, J.F.; Rodríguez-García, A. *Streptomyces tsukubaensis* as a new model for carbon repression: Transcriptomic response to tacrolimus repressing carbon sources. *Appl. Microbiol. Biotechnol.* **2017**, *101*, 8181–8195. [CrossRef] [PubMed]
96. Reeves, A.R.; Cernota, W.H.; Brikun, I.A.; Wesley, R.K.; Weber, J.M. Engineering precursor flow for increased erythromycin production in *Aeromicrobium erythreum*. *Metab. Eng.* **2004**, *6*, 300–312. [CrossRef] [PubMed]
97. Singh, B.P.; Behera, B.K. Regulation of tacrolimus production by altering primary source of carbons and amino acids. *Lett. Appl. Microbiol.* **2009**, *49*, 254–259. [CrossRef] [PubMed]
98. Mishra, A.; Verma, S. Optimization of process parameters for tacrolimus (FK 506) production by new isolate of *Streptomyces* sp. using response surface methodology. *J. Biochem. Technol.* **2012**, *3*, 419–425.
99. Xia, M.; Huang, D.; Li, S.; Wen, J.; Jia, X.; Chen, Y. Enhanced FK506 production in *Streptomyces tsukubaensis* by rational feeding strategies based on comparative metabolic profiling analysis. *Biotechnol. Bioeng.* **2013**, *110*, 2717–2730. [CrossRef] [PubMed]

100. Huang, D.; Xia, M.; Li, S.; Wen, J.; Jia, X. Enhancement of FK506 production by engineering secondary pathways of *Streptomyces tsukubaensis* and exogenous feeding strategies. *J. Ind. Microbiol. Biotechnol.* **2013**, *40*, 1023–1037. [CrossRef] [PubMed]
101. Du, W.; Huang, D.; Xia, M.; Wen, J.; Huang, M. Improved FK506 production by the precursors and product-tolerant mutant of *Streptomyces tsukubaensis* based on genome shuffling and dynamic fed-batch strategies. *J. Ind. Microbiol. Biotechnol.* **2014**, *41*, 1131–1143. [CrossRef] [PubMed]
102. Wang, C.; Liu, J.; Liu, H.; Liang, S.; Wen, J. Combining metabolomics and network analysis to improve tacrolimus productionin *Streptomyces tsukubaensis* using different exogenous feedings. *J. Ind. Microbiol. Biotechnol.* **2017**, *44*, 1527–1540. [CrossRef] [PubMed]
103. Wang, J.; Liu, H.; Huang, D.; Jin, L.; Wang, C.; Wen, J. Comparative proteomic and metabolomic analysis of *Streptomyces tsukubaensis* reveals the metabolic mechanism of FK506 overproduction by feeding soybean oil. *Appl. Microbiol. Biotechnol.* **2017**, *101*, 2447–2465. [CrossRef] [PubMed]
104. Mo, S.; Ban, Y.-H.; Park, J.W.; Yoo, Y.J.; Yoon, Y.J. Enhanced FK506 production in *Streptomyces clavuligerus* CKD1119 by engineering the supply of methylmalonyl-CoA precursor. *J. Ind. Microbiol. Biotechnol.* **2009**, *36*, 1473–1482. [CrossRef] [PubMed]
105. Turlo, J.; Gajzlerska, W.; Klimaszewska, M.; Król, M.; Dawidowski, M.; Gutkowska, B. Enhancement of tacrolimus productivity in *Streptomyces tsukubaensis* by the use of novel precursors for biosynthesis. *Enzyme Microb. Technol.* **2012**, *51*, 388–395. [CrossRef] [PubMed]
106. Gajzlerska, W.; Kurkowiak, J.; Turlo, J. Use of three-carbon chain compounds as biosynthesis precursors to enhance tacrolimus production in *Streptomyces tsukubaensis*. *New Biotechnol.* **2014**, *32*, 32–39. [CrossRef] [PubMed]
107. Zhu, X.; Zhang, W.; Chen, X.; Wu, H.; Duan, Y.; Xu, Z. Generation of high rapamycin producing strain via rational metabolic pathway-based mutagenesis and further titer improvement with fed-batch bioprocess optimization. *Biotechnol. Bioeng.* **2010**, *107*, 506–515. [CrossRef] [PubMed]
108. Chen, D.; Zhang, Q.; Zhang, Q.; Cen, P.; Xu, Z.; Liu, W. Improvement of FK506 production in *Streptomyces tsukubaensis* by genetic enhancement of the supply of unusual polyketide extender units via utilization of two distinct site-specific recombination systems. *Appl. Environ. Microbiol.* **2012**, *78*, 5093–5103. [CrossRef] [PubMed]
109. Chen, G.; Wang, G.Y.; Li, X.; Waters, B.; Davies, J. Enhanced production of microbial metabolites in the presence of dimethyl sulfoxide. *J. Antibiot.* **2000**, *53*, 1145–1153. [CrossRef] [PubMed]
110. Butler, A.R.; Cundliffe, E. Influence of dimethylsulfoxide on tylosin production in *Streptomyces fradiae*. *J. Ind. Microbiol. Biotechnol.* **2001**, *27*, 46–51. [CrossRef] [PubMed]
111. Huang, D.; Li, S.; Xia, M.; Wen, J.; Jia, X. Genome-scale metabolic network guided engineering of *Streptomyces tsukubaensis* for FK506 production improvement. *Microb. Cell. Fact.* **2013**, *12*, 52. [CrossRef] [PubMed]
112. Pei, G.; Chen, L.; Zhang, W. Chapter nine-WGCNA application to proteomic and metabolomic data analysis. *Methods Enzymol.* **2017**, *585*, 135–158. [PubMed]
113. Wang, C.; Liu, J.; Liu, H.; Wang, J.; Wen, J. A genome-scale dynamic flux balance analysis model of *Streptomyces tsukubaensis* NRRL18488 to predict the targets for increasing FK506 production. *Biochem. Eng. J.* **2017**, *123*, 45–56. [CrossRef]
114. Höffner, K.; Harwood, S.; Barton, P. A reliable simulator for dynamic flux balance analysis. *Biotechnol. Bioeng* **2013**, *110*, 792–802. [CrossRef] [PubMed]
115. Sola-Landa, A.; Moura, R.S.; Martín, J.F. The two-component PhoR-PhoP system controls both primary metabolism and secondary metabolite biosynthesis in *Streptomyces lividans*. *Proc. Natl. Acad. Sci. USA* **2003**, *100*, 6133–6138. [CrossRef] [PubMed]
116. Sola-Landa, A.; Rodríguez-García, A.; Franco-Domínguez, E.; Martín, J.F. Binding of PhoP to promoters of phosphate-regulated genes in *Streptomyces coelicolor*: Identification of PHO boxes. *Mol. Microbiol.* **2005**, *56*, 1373–1385. [CrossRef] [PubMed]
117. Martín, J.F. Phosphate control of the biosynthesis of antibiotics and other secondary metabolites is mediated by the PhoR-PhoP system: An unfinished story. *J. Bacteriol.* **2004**, *186*, 5197–5201. [CrossRef] [PubMed]
118. Martín, J.F.; Santos-Beneit, F.; Rodríguez-García, A.; Sola-Landa, A.; Smith, M.C.M.; Ellingsen, T.E.; Nieselt, K.; Burroughs, N.J.; Wellington, E.M.H. Transcriptomic studies of phosphate control of primary and secondary metabolism in *Streptomyces coelicolor*. *Appl. Microbiol. Biotechnol.* **2012**, *95*, 61–75. [CrossRef] [PubMed]
119. Santos-Beneit, F. The Pho regulon: A huge regulatory network in bacteria. *Front. Microbiol.* **2015**, *6*, 402. [CrossRef] [PubMed]

120. Martín, J.F.; Rodríguez-García, A.; Liras, P. The master regulator PhoP coordinates phosphate and nitrogen metabolism, respiration, cell differentiation and antibiotic biosynthesis: Comparison in *Streptomyces coelicolor* and *Streptomyces avermitilis*. *J. Antibiot.* **2017**, *70*, 534–541. [CrossRef] [PubMed]
121. Makino, K.; Shinagawa, H.; Amemura, M.; Kimura, S.; Nakata, A.; Ishihama, A. Regulation of the phosphate regulon of *Escherichia coli* Activation of *pstS* transcription by PhoB protein in vitro. *J. Mol. Biol.* **1988**, *203*, 85–95. [CrossRef]
122. Ruiz, B.; Chávez, A.; Forero, A.; García-Huante, Y.; Romero, A.; Sánchez, M.; Rocha, D.; Sánchez, B.; Rodríguez-Sanoja, R.; Sánchez, S.; et al. Production of microbial secondary metabolites: Regulation by the carbon source. *Crit. Rev. Microbiol.* **2010**, *36*, 146–167. [CrossRef] [PubMed]
123. Sánchez, S.; Chávez, A.; Forero, A.; García-Huante, Y.; Romero, A.; Sánchez, M.; Rocha, D.; Sánchez, B.; Avalos, M.; Guzmán-Trampe, S.; et al. Carbon source regulation of antibiotic production. *J. Antibiot.* **2010**, *63*, 442–459. [CrossRef] [PubMed]
124. Strakova, E.; Zikova, A.; Vohradsky, J. Inference of sigma factor controlled networks by using numerical modeling applied to microarray time series data of the germinating prokaryote. *Nucl. Acids Res.* **2014**, *42*, 748–763. [CrossRef] [PubMed]
125. Kawamoto, S.; Watanabe, M.; Saito, N.; Hesketh, A.; Vachalova, K.; Matsubara, K.; Ochi, K. Molecular and functional analyses of the gene (*eshA*) encoding the 52-kilodalton protein of *Streptomyces coelicolor* A3 required for antibiotic production. *J. Bacteriol.* **2001**, *183*, 6009–6016. [CrossRef] [PubMed]
126. Saito, N.; Xu, J.; Hosaka, T.; Okamoto, S.; Aoki, H.; Bibb, M.J.; Ochi, K. EshA accentuates ppGpp accumulation and is conditionally required for antibiotic production in *Streptomyces coelicolor* A3. *J. Bacteriol.* **2006**, *188*, 4952–4961. [CrossRef] [PubMed]
127. Uguru, G.C.; Stephens, K.E.; Stead, J.A.; Towle, J.E.; Baumberg, S.; McDowall, K.J. Transcriptional activation of the pathway-specific regulator of the actinorhodin biosynthetic genes in *Streptomyces coelicolor*. *Mol. Microbiol.* **2005**, *58*, 131–150. [CrossRef] [PubMed]
128. Vujaklija, D.; Horinouchi, S.; Beppu, T. Detection of an A-factor-responsive protein that binds to the upstream activation sequence of *strR*, a regulatory gene for streptomycin biosynthesis in *Streptomyces griseus*. *J. Bacteriol.* **1993**, *175*, 2652–2661. [CrossRef] [PubMed]
129. Rigali, S.; Nothaft, H.; Noens, E.E.E.; Schlicht, M.; Colson, S.; Müller, M.; Joris, B.; Koerten, H.K.; Hopwood, D.A.; Titgemeyer, F.; et al. The sugar phosphotransferase system of *Streptomyces coelicolor* is regulated by the GntR-family regulator DasR and links *N*-acetylglucosamine metabolism to the control of development. *Mol. Microbiol.* **2006**, *61*, 1237–1251. [CrossRef] [PubMed]
130. Rigali, S.; Titgemeyer, F.; Barends, S.; Mulder, S.; Thomae, A.W.; Hopwood, D.A.; van Wezel, G.P. Feast or famine: The global regulator DasR links nutrient stress to antibiotic production by *Streptomyces*. *EMBO Rep.* **2008**, *9*, 670–675. [CrossRef] [PubMed]
131. Barreiro, C.; Prieto, C.; Sola-Landa, A.; Solera, E.; Martínez-Castro, M.; Pérez-Redondo, R.; García-Estrada, C.; Aparicio, J.F.; Fernández-Martínez, L.T.; Santos-Aberturas, J.; et al. Draft genome of *Streptomyces tsukubaensis* NRRL 18488, the producer of the clinically important immunosuppressant tacrolimus (FK506). *J. Bacteriol.* **2012**, *194*, 3756–3757. [CrossRef] [PubMed]
132. Martín, J.F.; Liras, P. Novel Antimicrobial and other Bioactive Metabolites obtained from Silent Gene Clusters. In *Antibiotics: Current Innovations and Future Trends*; Demain, A.L., Sánchez, S., Eds.; Horizon Scientific Press and Caister Academic Press: Norkfolk, UK, 2015; pp. 275–292. ISBN 978-1-908230-54-6.
133. Werner, G.; Hagenmaier, H.; Drautz, H.; Baumgartner, A.; Zähner, H. Metabolic products of microorganisms. 224. Bafilomycins, a new group of macrolide antibiotics. Production, isolation, chemical structure and biological activity. *J. Antibiot.* **1984**, *37*, 110–117. [CrossRef] [PubMed]
134. Harvey, B.M.; Mironenko, T.; Sun, Y.; Hong, H.; Deng, Z.; Leadlay, P.F.; Weissman, K.J.; Haydock, S.F. Insights into polyether biosynthesis from analysis of the nigericin biosynthetic gene cluster in *Streptomyces* sp. DSM4137. *Chem. Biol.* **2007**, *14*, 703–714. [CrossRef] [PubMed]
135. Yin, X.; Zabriskie, T.M. The enduracidin biosynthetic gene cluster from *Streptomyces fungicidicus*. *Microbiology* **2006**, *152 Pt 10*, 2969–2983. [CrossRef] [PubMed]

antibiotics

MDPI

Review

The Cellular Mechanisms that Ensure an Efficient Secretion in *Streptomyces*

Sonia Gullón and Rafael P. Mellado *

Departamento de Biotecnología Microbiana, Centro Nacional de Biotecnología (CNB-CSIC), c/Darwin 3, 28049 Madrid, Spain; sgullon@cnb.csic.es

* Correspondence: rpmellado@cnb.csic.es; Tel.: +34-915-854-547

Received: 27 February 2018; Accepted: 11 April 2018; Published: 14 April 2018

check for updates

Abstract: Gram-positive soil bacteria included in the genus *Streptomyces* produce a large variety of secondary metabolites in addition to extracellular hydrolytic enzymes. From the industrial and commercial viewpoints, the *S. lividans* strain has generated greater interest as a host bacterium for the overproduction of homologous and heterologous hydrolytic enzymes as an industrial application, which has considerably increased scientific interest in the characterization of secretion routes in this bacterium. This review will focus on the secretion machinery in *S. lividans*.

Keywords: *Streptomyces lividans*; secretion pathways; secretory proteins; signal peptides

---

1. Introduction

In their natural soil environment, the streptomycetes bacteria are characterized by the formation of an aerial mycelium during their life cycle before the sporulation. Along their life cycle, the streptomycetes produce and secrete a variety of hydrolytic enzymes [1,2] and enzyme inhibitors as well as signaling molecules and antibiotics in order to ensure the continuation of their existence in competitive habitats.

Their capacity to produce extracellular homologous and heterologous proteins of industrial application has remarkably increased the interest and targeted research for increasing the knowledge about the functioning of the streptomycetes secretion machinery. Therefore, this information could be used to engineer efficient streptomycetes strains for the overproduction of commercially valuable proteins needed for biotechnological application.

Streptomyces lividans possess a relaxed restriction-modification system that facilitates its transformation with exogenous DNA without degrading it. Moreover, it has been a satisfactory potential host for extracellular protein overproduction because of its ability to secrete a variety of hydrolytic enzymes. Additionally, its genome sequence is known [3,4]. The use of *S. lividans* as a host to engineer the production of secretory proteins is well documented [5–7].

There are two pathways commonly used to secrete extracellular proteins in *S. lividans*. The Sec route is the major system used to release extracellular proteins through the cellular membrane [8]. A secondary system, also found in *S. lividans*, is the Tat route. The Tat-secreted proteins have the property of appearing fully folded in the culture supernatant [9]. Different types of proteins with the predicted Sec or Tat signal peptides together with the antibiotic undecylprodigiosin and other proteins lacking signal peptides have detected in vesicles present inside droplets found in *Streptomyces* grown on top agar plates. This produces harmful effects in fungi and is a new way to deliver proteins with potentially different applications, which deserves to be explored further [10]. The function of the secretion system known as Esx or Type VII described for the secretion of small size proteins in Gram-positive bacteria [11] is not found in streptomycetes [12].

The characterization of the *S. lividans* cellular response when challenged to overproduce vast amounts of particular secretory model proteins has also been researched in the recent past. This research has provided in-depth information and increased our knowledge of the bacterial factors that may affect the efficiency of secretory protein production by causing different types of stress in the bacterial cell.

2. The Sec Pathway

A characteristic signal peptide present in the pre-secretory amino end is the signal recognized by intracellular chaperones to target these proteins to the membrane. A typical signal peptide consists of three regions including a short stretch of positively charged amino acids close to the N terminus (n-region), a much longer stretch of hydrophobic amino acids (h-region), and the region at the end of the signal peptide, which contains the stretch of amino acids that the signal peptidase will cleave (c-region). *Escherichia coli* SecB chaperon prevents pre-secretory protein folding by targeting it to the translocase complex in the membrane by cooperating with SecA [13]. This post-translational secretory protein transport mechanism is widely used for bacterial protein secretion via the Sec pathway. Although genes homologous to SecB are found in Gram-positive bacteria, they are not present in the *Bacillus subtilis* genome [14] or in the genomes of *S. coelicolor* [15] or the genomes of *S. lividans* 66 [3] strains. This fact has triggered more research to find out alternative means that these Gram-positive bacteria use to transport secretory proteins to the membrane.

The signal recognition particle (SRP) seems to be involved in a co-translational protein targeting mechanism conserved from bacteria to mammals [16–18]. In mammals, SRP interacts with the ribosome nascent chain complex (RNC) to translocate proteins across the endoplasmic reticulum membrane. The entire complex links with an SRP-receptor complex to be attached to the membrane [19].

The *E. coli* SRP mechanism comprises a 4.5S RNA (scRNA), which is the fifty-four homolog protein (Ffh) that is homologous to the mammalian SRP54 protein. It is also homologous to the FtsY protein, which has a C-terminal domain 300-residues long that is similar to the alpha subunit of the mammalian SRP receptor [20]. The interaction with the nascent pre-protein chain takes place via the M domain located at the Ffh carboxyl end. The minimal functional SRP conserved consists of Ffh and the scRNA [21]. The signal for targeting FtsY to the membrane is contained in the A domain located at the FtsY amino end [22–24]. SRP is responsible for targeting *E. coli* integral membrane proteins [25,26] and the export of *E. coli* SecB independent secretory proteins takes place without the postulated intervention of SRP [27]. The *E. coli* SRP based and the SecB-based targeting systems are substrate specific [28] and all the *E. coli* SRP components are essential for cell growth.

The *B. subtilis* SRP contains the scRNA, Ffh, and the histone-like protein HbsU [29]. The role of HbsU is still unknown. Although some studies indicated that the SRP could target *B. subtillis* secretory proteins via the Sec pathway [30], other authors have suggested that SecA could have a dual function in which it targets the newly made secretory protein to the membrane and pushes it through the translocase [30,31]. Therefore, there is a lack of fully convincing evidence showing whether, in the absence of SecB, the *B. subtilis* SRP could target both membrane and secretory proteins. The *S. lividans* SRP system consists of Ffh and an 82 nt long small size RNA (scRNA) [32]. The receptor protein FtsY forms part of the system and the three components are apparently expressed throughout cellular growth. No viable mutants in any of these three components have been obtained, and a possible essential role has been assigned to each of them. Experimental evidence obtained by co-immunoprecipitation studies has shown that *S. lividans* SRP is involved in targeting model secretory proteins [32,33]. The level of the SRP components seems to not be a limiting factor for the overproduction of secretory proteins in *S. lividans*, according to the results obtained with overproduced model proteins [34]. The *S. lividans* FtsY hydrophobic N-terminal segment identifies the protein as a homolog to other integral FtsY-like membrane proteins from other Actinobacteria with similar hydrophobic profiles [35]. The *S. lividans* strain is defective in the major type I signal peptidase (SipY; [34]) and has provided a useful tool for temporarily blocking translocation favoring the accumulation of pre-protein linked to SRP at the membrane. Therefore, it is helping detect the SRP ability to escort the pre-protein to the translocase complex [33]. This SipY deficiency also favored detecting an in vivo interaction of the SRP with a soluble form of FtsY, which acts as a functional cytoplasmic SRP receptor [36].

In *E. coli*, SecY, SecE, and SecG form a heterotrimeric stable complex that gives rise to the channel that conducts the protein to be translocated [8] and the translocation of the *Streptomyces* secretory proteins is thought to take place in *E. coli*. Streptomycetes appear to have only one gene for each of these Sec proteins in their genomes including SecA [15]. In vitro experiments showed that *S. lividans* SecA was required for secretion of the Sec-dependent alpha-amylase [37], which suggests a comparable functioning of the *E. coli* equivalents. *S. coelicolor*, *S. lividans* [38], and *B. subtilis* [39] encoded SecG proteins have a functional homology to that of *E. coli*. Deletion of the *E. coli* or *B. subtilis* SecG causes a cold-sensitive phenotype while this seems not to occur in *S. lividans* [39,40]. Protein secretion is impaired in the *S. lividans* SecG deficiency, which delays the extracellular appearance of model secretory enzymes [38,41]. In *E. coli*, the heterotrimeric membrane complex formed by SecD, SecF, and YajC associates with the SecYEG channel, although YajC does not seem to be required for the complex to function [8]. The *S. coelicolor* genome contains two sets of SecD and SecF homologs required for an efficient secretion of some proteins [42]. Genes homologous to *yajC* have been identified in eleven *Streptomyces* genomes [43] even though a homolog to this does not seem to be present in the *S. lividans* 66 genome [3]. The functionality of these genes has not been determined experimentally. The YidC protein is essential in *E. coli* since it helps insert membrane proteins in a Sec-dependent [44] or a Sec-independent manner [45,46]. Two genes encoding potentially equivalent YidC proteins seem to form part of the *S. lividans* genome [3], where the YidC protein acts in *Streptomyces* in a Sec-dependent manner [47]. Figure 1 shows a scheme of how the Sec pathway works.

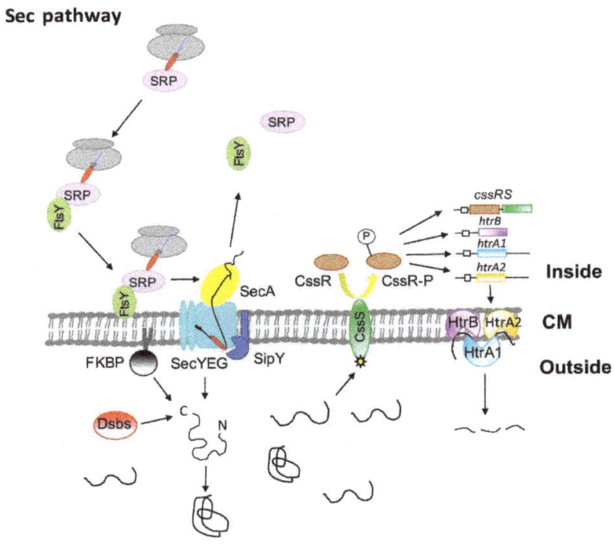

Figure 1. *Streptomyces lividans* major secretory route (Sec pathway). The secretory protein precursor is targeted to the membrane translocation complex by SRP, which is a GTP-dependent process, that requires the help of the FtsY receptor. When the signal peptidases (exemplified by the major signal peptidase, SipY) act, the SRP components are released to be used again. The extracellular presence of incorrectly folded mature proteins triggers the expression of the CssRS two-component system, which, in turn, induces the synthesis of the three HtrA-like proteins that are working cooperatively and degrades the incorrectly folded extracellular proteins. The acquisition of an active, correctly folded structure may need the action of enzymes involved in the formation of disulfide bonds (Dsbs) and/or peptidyl-prolyl cis-trans isomerases (FKBP-like).

3. The Twin-Arginine (Tat) Pathway

The Tat route is an export pathway found in the thylakoid membranes of plant chloroplasts and described in several bacteria [48]. Proteins using this route appear to be secreted across the cytoplasmic membrane in a folded state [48]. This feature confers to the Tat pathway a particular interest to be exploited for the overproduction of secretory proteins in streptomycetes. The Tat pathway seems to consist of four proteins in *E. coli* including TatA, TatB, TatC, and TatE. The TatB-TatC complex binds to the signal peptide in an energy-independent manner to target the pre-protein to the membrane where Tat A polymerizes to form the channel for the extracellular protein to exit the cell. TatE has been shown in vivo and in vitro to interact with the translocase complex and is considered to be a regular element of the *E. coli* translocase [49]. In *B. subtilis* the TatB component seems to be absent and the Tat machinery is formed by three TatA subunits (TatAc, TatAd and TatAy) and two TatC subunits (TatCd and TatCy). In addition, two independent acting complexes with different substrate specifity are formed (TatAd-TatCd and TatAy-TatCy) and remain ambiguous to the TatAc function [50–53].

Streptomyces lividans contains three Tat components (TatA, TatB, and TatC) with two different *tatA* genes [54]. Recognition of the Tat pre-secretory proteins in *S. lividans* is carried out by the heterodimeric complex formed by TatA and TatB, which theoretically target the pre-proteins to the membrane [55]. This heterodimer complex interacts with TatC seated in the membrane for the pre-proteins insertion. Then TatA oligomerises form the pore used by the folded secretory protein to leave the cell [55]. The eubacteria signal peptide of Tat peptides has some characteristic features that differentiate them from those of the Sec proteins. The n-region of the Tat signal peptide contains a conserved motif R-R-x-U-U where R-R is the twin-arginine motif and x and U are usually polar and hydrophobic amino acids, respectively [56–58]. Their h region is less hydrophobic and the c-region may contain basic residues [59]. The algorithms designed to predict the existence of this type of signal peptides identified a large number of potential Tat proteins (145–189) in *S. coelicolor* [60,61]. When membrane proteins from TatC deficient strains were compared to those of the isogenic wild type strain, only 27 proteins were unequivocally defined as Tat-dependent [58]. The algorithms predicted up to 127 Tat proteins in *S. lividans* [62]. However, the final number has not yet been experimentally determined even though the DypA peroxidase has been found to be a *S. lividans* Tat-dependent protein in this bacterium [63]. The annotated sequence of the *S. coelicolor* genome predicts more than 800 secretory proteins [15]. Therefore, in comparison, the assigned 27 Tat proteins involve a small number that reflects the minor role played by the Tat route in secretion. However, these 27 Tat proteins still represent a significant number when compared to those found in other bacteria. This reinforces the interest in exploring the Tat route for the secretion of correctly folded proteins in streptomycetes. The short number of predicted *E. coli* Tat proteins seems to bind redox cofactors, which are important for respiratory metabolism [64] while only three of the 27 experimentally confirmed *S. coelicolor* Tat-dependent proteins may bind cofactors. This reveals that the streptomycetes Tat pathway can export a broader spectrum of proteins than other bacteria [58]. Figure 2 illustrates the *S. lividans* Tat secretion pathway.

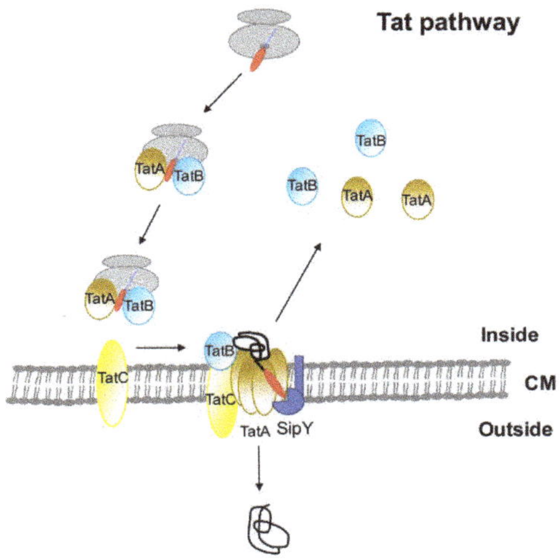

Figure 2. *Streptomyces lividans* minor secretory route (Tat pathway). The nascent secretory pre-protein using the Tat route is thought to be targeted to the membrane by the heterodimeric TatA-TatB complex. The signal peptide is cleaved by the signal peptidases (exemplified by the major signal peptidase, SipY) before releasing the mature protein to the cell wall.

4. Stresses Caused by Overproduction of Secretory Proteins in Streptomycetes

The engineering of streptomycetes strains to overproduce homologous or heterologous secretory proteins require more in-depth knowledge of the potential detrimental effects that this overproduction may cause to the cell. Therefore, acquiring the best possible understanding of the physiological limiting steps that could be encountered when engineering bacterial secretory protein factories is optimal.

4.1. Temporal Translocase Blockage

Mutations in the *S. lividans* translocase component SecG or in the major signal peptidase SipY lead to a temporal translocation blockage in which the extracellular presence of several secretory factors are considerably reduced in both mutants, which is determined by the corresponding proteomics analyses and is consistent with the transcriptional profiles in each case [65]. The accumulation of unprocessed pre-proteins in the translocase complex generates a membrane stress in the SipY and SecG deficient mutants. A large number of genes encoding secretory proteins were down-regulated in the SipY deficient strain than in the SecG one, which was determined by transcriptional analyses and is in agreement with the proteomics results. This suggests a more dramatic effect of the SipY mutation than that of the SecG one. The formation of the aerial mycelium involves the action of extracellular proteins [12,66]. Temporal blockage of the pre-protein processing method would likely result in a severe limitation in efficiency protein secretion, which may lead to malfunction of the aerial mycelium formation and a deficient sporulation (the so-called bald phenotype). Comparison of the wild type and the mutant strains transcriptomic profiles identified a different set of bald-related genes down-regulated in each mutant strain in which a morphological analysis of the mycelial filaments under the scanning electron microscope showed that the effect was more dramatic in the SipY that in the SecG one. The differences in the expression of these bald-related genes together with the differences detected on the mutant strains extracellular proteins when compared to those of the wild type revealed by tandem mass spectrometry analysis constitute a characteristic profile of the "translocation stress" caused in the bacterial cell when the natural translocation process is impaired [65].

4.2. Induction of the Stringent Response

Bacteria frequently use a variety of sensor kinase genes contained in their genomes to respond to changes in the environment. The two-component environmental sensor systems are well represented in the streptomycetes genomes. A two-gene operon has been identified among the *S. coelicolor* sixty-seven two-component systems [67] that shares homology with the *B. subtilis degS-degU* operon, which regulates synthesis of some secretory proteins and the level of antibiotic production. This temporarily modulates the secondary metabolism of the bacterium [68,69]. Transcriptomic and tandem mass spectrometry analyses of the two-component operon defective strain as well as the strain harboring the regulator gene in high copy number allowed us to conclude that, when present in a high copy number, the regulator gene enhances antibiotic production and sporulation in early times of growth and considerably increases the expression of genes encoding the secretory protein to such an extent that the cellular amino acid pool suffers from depletion. This precursors deficiency triggers a stringent response in the bacterium, which mainly down regulates ribosomal genes [67]. Propagation of the regulator gene in high copy number could be useful if associated with overproduction of homologous or heterologous extracellular proteins in *Streptomyces*, but, at the same time, clearly points out the danger of provoking amino acid depletion, which may affect cell viability.

Lipoproteins are membrane-translocated proteins that remain associated with the cytoplasmic membrane. Bacterial lipoproteins are known to be involved in a number of essential cellular processes since most of the solute binding proteins are lipoproteins. They also take part in the cell envelope biogenesis process and in signal transduction pathways. In Gram-positive bacteria, a fair number of peptidyl-prolyl isomerases are lipoproteins responsible for the correct folding of extracellular proteins [70]. The pre-lipoproteins have a type II signal peptide that is specifically cleaved by a type II signal peptidase (Lsp) for which the presence of an essential sequence motif is necessary for the correct processing of the protein. After the translocation has taken place, a diacylgycerol transferase lipoprotein (Lgt) links a lipid molecule at the conserved cysteine residue where Lsp will cleave the signal peptide. Proteins homologous to Lgt have been found in the *S. coelicolor* and *S. lividans* genomes [3,71]. Proteins homologous to the N-acyl transferase Lnt add more lipids to the conserved cysteine residue in Gram negative bacteria and have also been described in streptomycetes strains. However, their deletion seems not to cause any effect to the *S. scabies* bacterium [72]. Most of the bacterial lipoproteins are Sec-dependent but some use the Tat system instead [71].

The transcriptomic study of Lsp-deficient *S. lividans* strain revealed that translocase blockage takes place, which the consequent negative effect on extracellular protein appearance causes sporulation delay and triggers the bacterial translocation stress response since the translocation is blocked by the accumulation of unprocessed lipoproteins [73]. This phenotype is observed when the cells were defective in the major type I signal peptidase, SipY, or the translocase SecG protein. Additionally, Lsp deficiency caused depletion of solute binding lipoproteins, which caused the depletion of nutrients inside the cell and triggered a stringent response [73]. This mimicked the effect caused by overproduction of secretory proteins when the regulator gene of the two-component operon described above was propagated in a high copy number [67]. These results are in accord with the key roles played by bacterial lipoproteins and they will have to take into account in the optimization of *Streptomyces* strains for the overproduction of secretory proteins. A Venn diagram [74] illustrates the relative degree of coincidence among the genes transcriptionally affected in the SipY, SecG, or Lsp deficient strains (see Figure 3). Therefore, 52.45% of the total genes regulated in the SecG deficient cells coincide with 42.10% of the total genes regulated in the SipY deficient cells. Lsp deficient cells show a coincidence on the 22.0% and the 14.0% of its genes with the 14.47% and the 11.47% of the genes regulated by the SipY and SecG deficiencies, respectively. Most of the affected genes were involved in nitrogen/amino acids metabolism, morphological differentiation and genes encoding a possible secreted protein that were downregulated in most cases (Table S1). Therefore, a selected subset of these genes could eventually be used to monitor an efficient secretory protein production in streptomycetes.

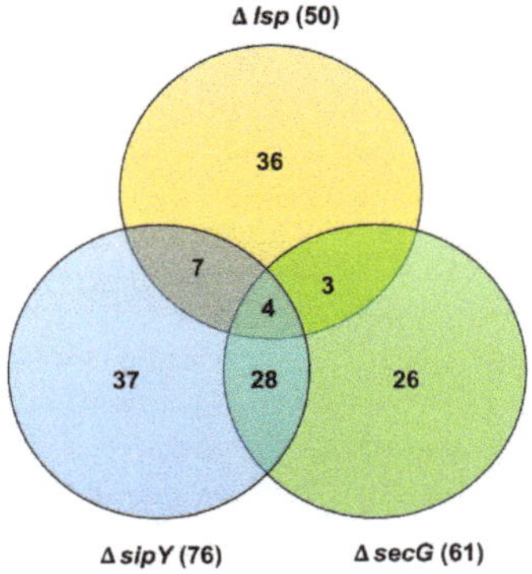

Figure 3. Venn diagram summarizing the relative degree of coincidence among the genes transcriptionally affected in the *Streptomyces lividans* SipY, SecG, or Lsp deficient strains. For comparative purposes, the set of genes involved in the stringent response in the Lsp deficient strains have been subtracted.

5. Overproduction of Model Secretory Proteins in *S. lividans*

The comparison of transcriptomic analyses of *S. lividans* strains overproducing secretory proteins via the Sec (alpha-amylase) or the Tat route (agarase) in *S. lividans* resulted in different cellular responses [75]. Basically, overproduction of the Tat model protein elicited the characteristic down regulation of genes involved in the stringent response while the same set of genes was upregulated when the Sec model protein was overproduced. The stringent response may cause cellular death and its potential induction has to be taken into account when engineering extracellular protein production in streptomycetes. Therefore, despite the potential attractiveness of using the Tat secretory route to obtain a potentially correctly folded extracellular protein, the Sec pathway could be the preferred route for secretory protein production in *Streptomyces* strains.

Additionally, some proteins appear to be able to use both routes. Therefore, when the Tat model protein agarase is overproduced in *S lividans*, a small amount of it was apparently secreted at the early phase of growth at a time when the Tat pathway is not functional, which was determined by co-immunoprecipitation studies using antibodies against agarase and the SRP components. This suggests that the protein could eventually be secreted via the Sec route [32]. A similar account has also been described for *E. coli* in which Tat signal peptides could trigger proteins to the Sec pathway [76] and the first two transmembrane domains of a Tat-membrane protein could be recognized by the Sec pathway. Sec and Tat routes seem to cooperate for the integration of a particular membrane protein [47]. Some membrane proteins with extracytoplasmic domains apparently need the cooperation of the Tat and Sec pathways [77].

The construction of chimeric precursor proteins interchanging the Sec-dependent alpha-amylase and the Tat-dependent agarase signal peptides showed that agarase was able to use the Sec route when led by the alpha-amylase signal peptide. However, the alpha-amylase did not succeed in using the Tat route led by the agarase signal peptide [41]. The inability of alpha-amylase to use the Tat signal peptide argues in favor of other parts of the protein apart from the signal peptide contributing to the streptomycetes secretion as described in *E. coli* [78]. The structure of the proteins may have evolved to select a particular secretory route. Most of the *S. coelicolor* unequivocally assessed Tat proteins [58] do not have predictable disulphide bonds while the major Sec proteins have at least one disulphide bond [65]. Structurally less complex secretory proteins may be engineered to use either secretory route while structurally complex proteins would have to be processed via the Sec pathway. Therefore, the Sec route may be the one to choose to overproduce extracellular proteins in streptomycetes.

6. Extracellular Stability of Secretory Proteins

Overproduction of secretory proteins in streptomycetes, as in other bacteria, could likely result in the extracellular accumulation of misfolded secreted proteins. In practical terms, they are inactive contaminants of the correctly folded, functionally active secreted proteins present in the total population of the overproduced ones. Therefore, previous identification and characterization of these quality factors are required in order to optimize extracellular protein production via the Sec route.

The *S. lividans* CssRS two-component system regulates the degradation of accumulated incorrectly-folded extracellular proteins outside the cell known as the secretion stress response. This two-component system is known to induce the synthesis of specific proteases to degrade the incorrectly-folded proteins and it has been described in *B. subtilis* [79] and *S. lividans* [80].

Overproduction of the *B. amyloliquefaciens* alpha-amylase in *B. subtilis* results in phosphorylation of the CssR regulator, which activates the expression of the genes *htrA* and *htrB* encoding two HtrA-like proteases [81,82]. An equivalent CssRS two-component system has been described in *S. lividans* [81]. When activated by the presence of incorrectly-folded alpha-amylase outside the bacterial cell, the regulator induced the expression of three different HtrA-like proteases encoding genes, *htrA1*, *htrA2*, and *htrB* ([83]; Figure 1). The CssR operon is a main element of the bacterial secretion stress response [80,84].

Genes encoding HtrA-like proteins are present in bacterial genomes and a significant number of bacteria have more than one HtrA-like protease [85]. HtrA proteins normally assemble into complex oligomers mainly via their PDZ domains in the carboxyl terminal part of the protein involved in substrate recognition, regulation of protease activity, and protein-protein interaction [86].

The *S. lividans* HtrA-like proteases seem to act in a cooperative manner since the alpha-amylase activity detected in each of the individual deficient strains was severely reduced when compared to the wild type one. A functional model for the combined action of the *S. lividans*' three proteases has been proposed ([83]; Figure 1).

Different ways of action have been described for HtrA-like proteins in different bacteria and it is clear that, whichever they work, the final outcome is the degradation of the incorrectly-folded proteins present outside the cell that result from Sec-dependent secretion. Theoretically, secretory proteins using the Tat pathway should not need to activate the synthesis of HtrA-like proteins since these are secreted in an active conformation (see Figure 2).

Degradation of the extracellular accumulated incorrectly-folded secretory proteins becomes necessary to avoid their possible interference with essential bacterial cell processes and it is an effective way to correct that potentially harmful situation. An additional efficient way to deal with this accumulation is to procure other means to intervene in the correct folding of Sec-dependent proteins such as the action of the peptidyl-prolyl-isomerases (FKBP in Figure 1) and/or or thiol-disulfide oxidoreductases (Dsbs in Figure 1) equivalent to those described in many bacteria. A lipoprotein with peptidyl-prolyl isomerase activity likely involved in the folding of secretory proteins has been identified in *S. coelicolor* [71]. The mode of action of all these streptomycetes folding proteins still remain to be fully elucidated.

Exploring the *Streptomyces* response to different stresses that could potentially take place when a protein is overproduced has helped identify potential drawbacks in the secretory protein production as those described above, which facilitates a means to avoid them. Despite the potential existence of these stress responses, the actual amount of accumulated knowledge on how the streptomycetes protein secretion takes place and the potential use of the Table S1 listed genes to monitor the scale-up of secretory production points to *S. lividans* as a sufficient candidate for engineering the overproduction of extracellular proteins at the industrial level.

Supplementary Materials: The following are available online at http://www.mdpi.com/2079-6382/7/2/33/s1, Table S1: Genes commonly modulated by the translocase blockage.

Acknowledgments: This work was funded by the Spanish Ministry of Economy, Industry, and Competitivity/European Regional Development Fund grant BIO2015-71504-R (AEI/FEDER, UE).

Author Contributions: Sonia Gullón and Rafael P. Mellado conceived and wrote the manuscript.

Conflicts of Interest: The authors declare no conflict of interest.

References

1. Gilbert, M.; Morosoli, R.; Shareck, F.; Kluepfel, D. Production and secretion of proteins by streptomycetes. *Crit. Rev. Biotechnol.* **1995**, *15*, 13–19. [CrossRef] [PubMed]
2. Chater, K.F. Taking a genetic scalpel to the *Streptomyces* colony. *Microbiology* **1998**, *144*, 1465–1478. [CrossRef]
3. Cruz-Morales, P.; Vijgenboom, E.; Iruegas-Bocargo, F.; Girard, G.; Yáñez-Guerra, L.A.; Ramos-Aboites, H.E.; Pernodet, J.L.; Anné, J.; van Wezel, G.P.; Barona-Gómez, F. The genome sequence of *Streptomyces lividans* 66 reveals a novel tRNA-dependent peptide biosynthetic system within a metal-related genomic island. *Genome Biol. Evol.* **2013**, *5*, 1165–1175. [CrossRef] [PubMed]
4. Rückert, C.; Albersmeier, A.; Busche, T.; Jaenicke, S.; Winkler, A.; Friðjónsson, Ó.H.; Hreggviðsson, G.Ó.; Lambert, C.; Badcock, D.; Bernaerts, K.; et al. Complete genome sequence of *Streptomyces lividans* TK24. *J. Biotechnol.* **2015**, *199*, 21–22. [CrossRef] [PubMed]
5. Parro, V.; Mellado, R.P. Effect of glucose on agarase overproduction by *Streptomyces*. *Gene* **1994**, *145*, 49–55. [CrossRef]
6. Isiegas, C.; Parro, V.; Mellado, R.P. *Streptomyces lividans* as a host to produce and secrete *Escherichia coli* TEM β-lactamase. *Lett. Appl. Microbiol.* **1999**, *28*, 321–326. [CrossRef] [PubMed]
7. Anné, J.; Maldonado, B.; Van Impe, J.; Van Mellaert, L.; Bernaerts, K. Recombinant protein production and streptomycetes. *J. Biotechnol.* **2012**, *158*, 159–167. [CrossRef] [PubMed]
8. Driessen, A.J.; Nouwen, N. Protein translocation across the bacterial cytoplasmic membrane. *Annu. Rev. Biochem.* **2008**, *77*, 643–667. [CrossRef] [PubMed]
9. Berks, B.C.; Sargent, F.; Palmer, T. The Tat protein export pathway. *Mol. Microbiol.* **2000**, *35*, 260–274. [CrossRef] [PubMed]
10. Schrempf, H.; Merling, P. Extracellular *Streptomyces lividans* vesicles: Composition, biogenesis and antimicrobial activity. 2015. *Microb. Biotechnol.* **2015**, *8*, 644–658. [CrossRef] [PubMed]
11. Pallen, M.J. The ESAT-6/WXG100 superfamily- and a new Gram-positive secretion system? *Trends Microbiol.* **2002**, *10*, 209–212. [CrossRef]
12. Chater, K.F.; Biró, S.; Lee, K.; Palmer, T.; Schrempf, H. The complex extracelular biology of *Streptomyces*. *FEMS Microbiol. Rev.* **2010**, *34*, 171–198. [CrossRef] [PubMed]
13. Sala, A.; Bordes, P.; Genevaux, P. Multitasking SecB chaperones in bacteria. *Front. Microbiol.* **2014**, *5*, 666. [CrossRef] [PubMed]
14. Kunst, F.; Ogasawara, N.; Moszer, I.; Albertini, A.M.; Alloni, G.; Azevedo, V.; Bertero, M.G.; Bessières, P.; Bolotin, A.; Borchert, S.; et al. The complete genome sequence of the Gram-positive bacterium *Bacillus subtilis*. *Nature* **1997**, *390*, 249–256. [CrossRef] [PubMed]
15. Bentley, S.D.; Chater, K.F.; Cerdeño-Tárraga, A.M.; Challis, G.L.; Thomson, N.R.; James, K.D.; Harris, D.E.; Quail, M.A.; Kieser, H.; Harper, D.; et al. Complete genome sequence of the model actinomycete *Streptomyces coelicolor* A3. *Nature* **2002**, *417*, 141–147. [CrossRef] [PubMed]
16. Keenan, R.J.; Freymann, D.M.; Stroud, R.M.; Walter, P. The signal recognition particle. *Annu. Rev. Biochem.* **2001**, *70*, 755–775. [CrossRef] [PubMed]
17. Cao, T.; Saier, M. The general protein secretory pathway: Phylogenetic analyses leading to evolutionary conclusions. *Biochim. Biophys. Acta* **2003**, *1609*, 115–125. [CrossRef]
18. Pohlschöder, M.; Dilks, K.; Hand, N.; Wesley Rose, R. Translocation of proteins across archaeal cytoplasmic membranes. *FEMS Microbiol. Rev.* **2004**, *28*, 3–24. [CrossRef] [PubMed]
19. Walter, P.; Johnson, A.E. Signal recognition and protein targeting to the endoplasmic reticulum membrane. *Annu. Rev. Cell Biol.* **1994**, *10*, 87–119. [CrossRef] [PubMed]
20. Wollin, S.L. From the elephant to *E. coli*: SRP-dependent protein targeting. *Cell* **1994**, *77*, 787–790. [CrossRef]
21. Miller, J.; Bernstein, H.D.; Walter, P. Interaction of *E. coli* Ffh/4.5 S ribonucleoprotein and FtsY mimics that of mammalian signal recognition particle and its receptor. *Nature* **1994**, *366*, 351–354. [CrossRef] [PubMed]
22. Powers, T.; Walter, P. Co-translational protein targeting catalised by the *Escherichia coli* signal recognition particle and its receptor. *EMBO J.* **1997**, *16*, 4880–4886. [CrossRef] [PubMed]

23. Millman, J.S.; Andrews, D.W. A site-specific, membrane-dependent cleavage event defines the membrane binding domain of FtsY. *J. Biol. Chem.* **1999**, *274*, 33227–33234. [CrossRef] [PubMed]
24. Herscovits, A.A.; Bochkareva, E.; Bibi, E. New prospects in studying the bacterial signal recoginiton particle pathway. *Mol. Microbiol.* **2000**, *38*, 927–939. [CrossRef]
25. De Gier, J.W.; Mansournia, P.; Valent, Q.A.; Philips, G.J.; Luirink, J.; von Heijne, G. Assembly of a cytoplasmic membrane protein in *Escherichia coli* is dependent on the signal recognition particle. *FEBS Lett.* **1996**, *399*, 307–309. [CrossRef]
26. Ulbrandt, N.D.; Newitt, J.A.; Bernstein, H.D. The *E. coli* signal recognition particle is required for the insertion of a subset of inner membrane proteins. *Cell* **1997**, *88*, 187–196. [CrossRef]
27. Beha, D.; Deitermann, S.; Müller, M.; Koch, H.G. Export of beta-lactamase is independent of the signal recognition particle. *J. Biol. Chem.* **2003**, *278*, 22161–22167. [CrossRef] [PubMed]
28. Beck, K.; Wu, L.F.; Brunner, J.; Muller, M. Discrimination between SRP-and SecA/SecB-dependent substrates involves selective recognition of nascent chains by SRP and trigger factor. *EMBO J.* **2000**, *19*, 134–143. [CrossRef] [PubMed]
29. Nakamura, D.; Yahagi, S.; Yamazaki, T.; Yamane, K. *Bacillus subtilis* histone-like protein HBsu, is an integral component of a SRP-like particle that can bind the *Alu* domain of small cytoplasmic RNA. *J. Biol. Chem.* **1999**, *274*, 13569–13576. [CrossRef] [PubMed]
30. Zanen, G.; Antelmann, H.; Meima, R.; Jongbloed, J.D.; Kolkman, M.; Hecker, M.; van Dijl, J.M.; Quax, W.J. Proteomic dissection of potential signal recognition particle dependence in protein secretion by *Bacillus subtilis*. *Proteomics* **2006**, *6*, 3636–3648. [CrossRef] [PubMed]
31. Zanen, G.; Houben, E.N.; Meima, R.; Tjalsma, J.; Jongbloed, J.D.; Westers, H.; Oudega, B.; Luirink, J.; van Dijl, J.M.; Quax, W.J. Signal peptide hydrophobibity is a critical for early stages in protein export by *Bacillus subtilis*. *FEBS J.* **2005**, *272*, 4617–4630. [CrossRef] [PubMed]
32. Palacín, A.; de la Fuente, R.; Valle, I.; Rivas, L.A.; Mellado, R.P. *Streptomyces lividans* contains a minimal functional signal recognition particle that is involved in protein secretion. *Microbiology* **2003**, *149*, 2435–2442. [CrossRef] [PubMed]
33. Palomino, C. Role of the Signal Recoginition Particle (SRP) in the Extracelular Protein Secretion in *Streptomyces lividans*. Ph.D. Thesis, Universidad Autónoma de Madrid, Madrid, Spain, 2006.
34. Palacín, A.; Parro, V.; Geukens, N.; Anné, J.; Mellado, R.P. SipY is the *Streptomyces lividans* type I signal peptidase exerting a major effect on protein secretion. *J. Bacteriol.* **2002**, *184*, 4875–4880. [CrossRef] [PubMed]
35. Bibi, E.; Herskovits, A.A.; Bochkareva, E.; Zelazny, A. Putative integral membrane SRP receptors. *Trends Biochem. Sci.* **2001**, *26*, 15–16. [CrossRef]
36. Palomino, C.; Mellado, R.P. The *Streptomyces lividans* cytoplasmic signal recognition particle receptor FtsY is involved in protein secretion. *J. Mol. Microbiol. Biotechnol.* **2005**, *9*, 57–62. [CrossRef] [PubMed]
37. Blanco, J.; Driessen, A.J.; Coque, J.J.; Martín, J.F. Biochemical characterization of the SecA protein of *Streptomyces lividans* interaction with nucleotides, binding to membrane vesicles and in vitro translocation of proAmy protein. *Eur. J. Biochem.* **1998**, *257*, 472–478. [CrossRef] [PubMed]
38. Palomino, C.; Mellado, R.P. Influence of a *Streptomyces lividans* SecG functional analogue on protein secretion. *Int. Microbiol.* **2008**, *11*, 25–31. [PubMed]
39. Van Welly, K.H.M.; Swaving, J.; Broekhuizen, C.P.; Rose, M.; Quax, W.J.; Driessen, A.J.M. Functional identification of the product of *Bacillus subtilis yvaL* gene as a SecG homologue. *J. Bacteriol.* **1999**, *181*, 1786–1792.
40. Nishiyama, K.; Hanada, M.; Tokuda, H. Disruption of the gene encoding p12 (SecG) reveals the direct involvement and important function of SecG in protein translocation of *Escherichia coli* at low temperature. *EMBO J.* **1994**, *13*, 3272–3277. [PubMed]
41. Gullón, S.; Vicente, R.L.; Valverde, J.R.; Marín, S.; Mellado, R.P. Exploring the feasibility of the Sec route to secrete proteins using the Tat route in *Streptomyces lividans*. *Mol. Biotechnol.* **2015**, *57*, 931–938. [CrossRef] [PubMed]
42. Zhou, Z.; Li, Y.; Sun, N.; Sun, Z.; Lv, L.; Wang, Y.; Shen, L.; Li, Y.Q. Function and evolution of two forms of SecDF homologs in *Streptomyces coelicolor*. *PLoS ONE* **2014**, *9*, e105237. [CrossRef] [PubMed]
43. Zhou, Z.; Sun, N.; Wu, S.; Li, Y.Q.; Wang, Y. Genomic data mining reveals a rich repertoire of transport proteins in *Streptomyces*. *BMC Genom.* **2016**, *17* (Suppl. 7), 510. [CrossRef] [PubMed]

44. Scotti, P.A.; Urbanus, M.L.; Brunner, J.; de Gier, J.W.; von Heijne, G.; van der Does, C.; Driessen, A.J.; Oudega, B.; Luirink, J.; Yid, C. The *Escherichia coli* homologue of mitochondrial Oxa1p, is a component of the Sec translocase. *EMBO J.* **2000**, *19*, 542–549. [CrossRef] [PubMed]
45. Van der Laan, M.; Bechtluft, P.; Kol, S.; Nouwen, N.; Driessen, A.J. F1F0 ATP synthase subunit c is a substrate of the novel YidC pathway for membrane protein biogenesis. *J. Cell Biol.* **2004**, *165*, 213–222. [CrossRef] [PubMed]
46. Serek, J.; Bauer-Manz, G.; Struhalla, G.; van der Berg, L.; Kiefer, D.; Dalbey, R.; Kuhn, A. *Escherichia coli* YidC is a membrane insertase for Sec-independent proteins. *EMBO J.* **2004**, *23*, 294–301. [CrossRef] [PubMed]
47. Keller, R.; de Keyzer, J.; Driessen, A.J.; Palmer, T. Co-operation between different targeting pathways during integration of a membrane protein. *J. Cell Biol.* **2012**, *199*, 303–315. [CrossRef] [PubMed]
48. Berks, B.C.; Palmer, T.; Sargent, F. The Tat protein translocation pathway and its role in microbial physiology. *Adv. Microb. Physiol.* **2003**, *47*, 187–254. [PubMed]
49. Eimer, E.; Fröbel, J.; Blümmel, A.S.; Müller, M. TatE as a regular constituent of bacterial twin-arginine protein translocases. *J. Biol. Chem.* **2015**, *290*, 29281–29289. [CrossRef] [PubMed]
50. Dilks, K.; Giménez, M.I.; Pohlschröder, M. Genetic and biochemical analysis of the twin-arginine translocation pathway in halophilic archarea. *J. Bacteriol.* **2005**, *187*, 8013–8104. [CrossRef] [PubMed]
51. Jongbloed, J.D.; Grieger, U.; Antelmann, H.; Hecker, M.; Nijland, R.; Bron, S.; van Dijl, J.M. Two minimal Tat translocases in *Bacillus*. *Mol. Microbiol.* **2004**, *545*, 1319–1325. [CrossRef] [PubMed]
52. Pop, O.; Martín, U.; Abel, C.; Muller, J.P. The twin-arginine signal peptide of PhoD and the TatAd/Cd proteins of *Bacillus subtilis* form an autonomous Tat translocation system. *J. Biol. Chem.* **2002**, *277*, 3268–3273. [CrossRef] [PubMed]
53. Goosens, V.; Monteferrante, C.G.; van Dijl, J.M. The tat system of Gram positive bacteria. *Biochim. Biophys. Acta* **2014**, *1843*, 1698–1706. [CrossRef] [PubMed]
54. Tsolis, K.C.; Tsare, E.P.; Orfanoudaki, G.; Busche, T.; Kanaki, K.; Ramakrishnan, R.; Rousseau, F.; Schymkowitz, J.; Rückert, C.; Kalinowski, J.; et al. Comprehensive subcellular topologies of polypeptides in Streptomyces. *Microb. Cell Fact.* **2018**, *17*, 43. [CrossRef] [PubMed]
55. De Keersmaeker, S.; Vrancken, K.; Van Mellaert, L.; Anné, J.; Geukens, N. The Tat pathway in *Streptomyces lividans*: Interaction of Tat subunits and their role in translocation. *Microbiology* **2007**, *153*, 1087–1094. [CrossRef] [PubMed]
56. Berks, B.C. A common export pathway for proteins binding complex redox cofactors? *Mol. Microbiol.* **1996**, *22*, 393–404. [CrossRef] [PubMed]
57. Stanley, N.T.; Palmer, T.; Berks, B.C. The twin-arginine consensus motif of Tat signal peptides is involved in Sec-independent protein targeting in *Escherichia coli*. *J. Biol. Chem.* **2000**, *275*, 11591–11596. [CrossRef] [PubMed]
58. Widdick, D.A.; Dilks, K.; Chandra, G.; Bottrill, A.; Naldrett, M.; Pohlschroder, M.; Palmer, T. The twin-arginine translocation pathway is a major route of protein export in *Streptomyces coelicolor*. *Proc. Natl. Acad. Sci. USA* **2006**, *103*, 17927–17932. [CrossRef] [PubMed]
59. Palmer, T.; Berks, B.C. The twin-arginine translocation (Tat) protein export pathway. *Nat. Rev. Microbiol.* **2012**, *10*, 483–496. [CrossRef] [PubMed]
60. Bendtsen, J.D.; Nielsen, H.; Widdick, D.; Palmer, T.; Brunak, S. Prediction of twin-arginine signal peptides. *BMC Bioinf.* **2005**, *6*, 167–175. [CrossRef] [PubMed]
61. Dilks, K.; Rose, R.W.; Hartmann, E.; Pohschröder, M. Prokaryotic utilization of the twin-arginine translocation pathway: A genomic survey. *J. Bacteriol.* **2003**, *185*, 1478–1483. [CrossRef] [PubMed]
62. Schaerlaekens, K.; van Mellaert, L.; Lammertyn, E.; Geukens, N.; Anné, J. The importance of the Tat-dependent protein secretion pathway in *Streptomyces* as revealed by phenotypic changes in tat deletion mutants and genome analysis. *Microbiology* **2004**, *150*, 21–31. [CrossRef] [PubMed]
63. Petrus, M.L.C.; Vijgenboom, E.; Chaplin, A.K.; Worrall, J.A.R.; van Wezel, G.P.; Claessen, D. The Dyp-type peroxidase DtpA is a Tat-substrate required for GlxA maturation and morphogenesis in Streptomyces. *Open Biol.* **2015**, *6*, 150149. [CrossRef] [PubMed]
64. Palmer, T.; Sargent, F.; Berks, B.C. Export of complex cofactor-containing proteins y the bacterial Tat pathway. *Trends Microbiol.* **2005**, *13*, 175–180. [CrossRef] [PubMed]

65. Gullón, S.; Palomino, C.; Navajas, R.; Paradela, A.; Mellado, R.P. Translocase and major signal peptidase malfunctions affect aerial mycelium formation in *Streptomyces lividans*. *J. Biotecnol.* **2012**, *160*, 112–122. [CrossRef] [PubMed]
66. Kim, D.W.; Hesketh, A.; Kim, E.S.; Song, J.Y.; Lee, D.H.; Kin, I.S.; Chater, K.F.; Lee, K.J. Complex extracellular interactions of proteases and a protease inhibitor influence multicellular development of *Streptomyces coelicolor*. *Mol. Microbiol.* **2008**, *70*, 1180–1193. [CrossRef] [PubMed]
67. Rozas, D.; Gullón, S.; Mellado, R.P. A novel two-component system involved in the transition to secondary metabolism in *Streptomyces coelicolor*. *PLoS ONE* **2012**, *7*, e31760. [CrossRef] [PubMed]
68. Ogura, M.; Yamaguchi, H.; Yoshida, K.; Fujita, Y.; Tanaka, T. DNA microarray analysis of *Bacillus subtilis* DegU, ComA and PhoP regulons: An approach to comprehensive analysis of *B. subtilis* two-component regulatory systems. *Nucl. Acids Res.* **2001**, *29*, 3804–3813. [CrossRef] [PubMed]
69. Mäder, U.; Antelmann, H.; Buder, T.; Dahl, M.K.; Hecker, M.; Homuth, G. *Bacillus subtilis* functional genomics: Genome-widw analysis of the DegS-DegU regulon by transcriptomics and proteomics. *Mol. Genet. Genom.* **2002**, *268*, 455–467. [CrossRef] [PubMed]
70. Hutchings, M.I.; Palmer, T.; Harrington, D.J.; Sutcliffe, I.C. Lipoprotein biogenesis in Gram-positive bacteria: Knowing when to hold 'em, knowing when to fold 'em. *Trends Microbiol.* **2009**, *17*, 13–21. [CrossRef] [PubMed]
71. Thompson, B.J.; Widdick, D.A.; Hicks, M.G.; Chandra, G.; Sutcliffe, I.C.; Palmer, T.; Hutchings, M.I. Investigating lipoprotein biogenesis and function in the model Gram-positive bacterium *Streptomyces coelicolor*. *Mol. Microbiol.* **2010**, *77*, 943–957. [CrossRef] [PubMed]
72. Widdick, D.A.; Hicks, M.G.; Thompson, B.J.; Tschumi, A.; Chandra, G.; Sutcliffe, I.C.; Brülle, J.K.; Sander, P.; Palmer, T.; Hutchings, M.I. Dissecting the complete lipoprotein biogenesis pathway in *Streptomyces scabies*. *Mol. Microbiol.* **2011**, *80*, 1395–1412. [CrossRef] [PubMed]
73. Gullón, S.; Arranz, E.I.G.; Mellado, R.P. Transcriptional characterisation of the negative effect exerted by a deficiency in type II signal peptidase on extracelular protein secretion in *Streptomyces lividans*. *Appl. Microbiol. Biotechnol.* **2013**, *97*, 10069–10080. [CrossRef] [PubMed]
74. Oliveros, J.C.; VENNY. An Interactive Tool for Comparing Lists with Venn's Diagrams. 2007. Available online: http://bioinfogp.cnb.csic.es/tools/venny/index.html (accessed on 27 February 2018).
75. Gullón, S.; Marín, S.; Mellado, R.P. Overproduction of a model Sec-and Tat-Dependent secretory protein elicits different cellular responses in *Streptomyces lividans*. *PLoS ONE* **2015**, *10*, e0133645. [CrossRef] [PubMed]
76. Tullman-Ercek, D.; DeLisa, M.P.; Kawarasaki, Y.; Iranpour, P.; Ribnicky, B.; Palmer, T.; Georgiou, G. Export pathway selectivity of *Escherichia coli* twin arginine translocation signal peptides. *J. Biol. Chem.* **2007**, *282*, 8309–8319. [CrossRef] [PubMed]
77. Tooke, F.J.; Babot, M.; Chandra, G.; Buchanan, G.; Palmer, T. A unifying mechanism for biogenesis of membrane proteins co-operatively integrated by the Sec and Tat pathways. *eLife* **2017**, *6*, e26577. [CrossRef] [PubMed]
78. Cristóbal, S.; de Gier, J.W.; Nielsen, H.; von Heijne, G. Competition between Sec-and TAT-dependent protein translocation in *Escherichia coli*. *EMBO J.* **1999**, *18*, 2982–2990. [CrossRef] [PubMed]
79. Hyyriläinen, H.L.; Bolhuis, A.; Darmon, E.; Muukkonen, L.; Koski, P.; Vitikainen, M.; Sarvas, M.; Prágai, Z.; Bron, S.; van Dijl, M.; et al. A novel two-component regulatory system in *Bacillus subtilis* for the survival of severe secretion stress. *Mol. Microbiol.* **2001**, *41*, 1159–1172. [CrossRef]
80. Gullón, S.; Vicente, R.L.; Mellado, R.-P. A novel two-component system involved in secretion stress response in *Streptomyces lividans*. *PLoS ONE* **2012**, *7*, e48987. [CrossRef] [PubMed]
81. Darmon, E.; Noone, D.; Masson, A.; Bron, S.; Kuipers, O.P.; Devine, K.M.; van Dijl, J.M. A novel class of heat and secretion stress-responsive genes is controlled by the autoregulated CssRS two-component system of *Bacillus subtilis*. *J. Bacteriol.* **2002**, *184*, 5661–5671. [CrossRef] [PubMed]
82. Lulko, A.T.; Veening, J.W.; Buist, G.; Smits, W.K.; Blom, E.J.; Beekman, A.C.; Bron, S.; Kuipers, O.P. Production and secretion stress caused by overexpression of heterologous α-amylase leads to inhibition of sporulation and a prolonged motile phase in *Bacillus subtilis*. *Appl. Environ. Microbiol.* **2007**, *73*, 5354–5362. [CrossRef] [PubMed]
83. Vicente, R.L.; Gullón, S.; Marín, S.; Mellado, R.P. The three *Streptomyces lividans* HtrA-like proteases involved in the secretion stress response act in a cooperative manner. *PLoS ONE* **2016**, *11*, e0168112. [CrossRef] [PubMed]

84. Westers, H.; Westers, L.; Darmon, E.; van Dijl, J.M.; Quax, W.J.; Zanen, G. The CssRS two-component regulatory system controls a general secretion stress response in *B. subtilis*. *FEBS J.* **2006**, *273*, 3816–3827. [CrossRef] [PubMed]
85. Noone, D.; Howell, A.; Collery, R.; Devine, K.M. YkdA and YvtA, HtrA-Like serine proteases in *Bacillus subtilis*, engage in negative autoregulation and reciprocal cross-regulation of *ykdA* and *yvtA* gene expression. *J. Bacteriol.* **2001**, *183*, 654–663. [CrossRef] [PubMed]
86. Clausen, T.; Kaiser, M.; Huber, R.; Ehrmann, M. HTRA proteases: Regulated proteolysis in protein quality control. *Nat. Rev. Mol. Cell Biol.* **2011**, *12*, 152–162. [CrossRef] [PubMed]

antibiotics

MDPI

Review

Complex Regulatory Networks Governing Production of the Glycopeptide A40926

Rosa Alduina [1,*], Margherita Sosio [2,3] and Stefano Donadio [2,3]

1 Department of Biological, Chemical and Pharmaceutical Sciences and Technologies (STEBICEF), Università degli Studi di Palermo, Viale delle Scienze Bd.16, 90128 Palermo, Italy
2 Via Ortles 22/4, Naicons Srl, 20139 Milano, Italy; msosio@naicons.com (M.S.); sdonadio@naicons.com (S.D.)
3 Via Ortles 22/4, Ktedogen Srl, 20139 Milano, Italy
* Correspondence: valeria.alduina@unipa.it; Tel.: +39-091-23897306

Received: 9 March 2018; Accepted: 3 April 2018; Published: 5 April 2018

check for updates

Abstract: Glycopeptides (GPAs) are an important class of antibiotics, with vancomycin and teicoplanin being used in the last 40 years as drugs of last resort to treat infections caused by Gram-positive pathogens, including methicillin-resistant *Staphylococcus aureus*. A few new GPAs have since reached the market. One of them is dalbavancin, a derivative of A40926 produced by the actinomycete *Nonomuraea* sp. ATCC 39727, recently classified as *N. gerenzanensis*. This review summarizes what we currently know on the multilevel regulatory processes governing production of the glycopeptide A40926 and the different approaches used to increase antibiotic yields. Some nutrients, e.g., valine, L-glutamine and maltodextrin, and some endogenous proteins, e.g., Dbv3, Dbv4 and RpoB^R, have a positive role on A40926 biosynthesis, while other factors, e.g., phosphate, ammonium and Dbv23, have a negative effect. Overall, the results available so far point to a complex regulatory network controlling A40926 in the native producing strain.

Keywords: glycopeptide antibiotics; *dbv* cluster; regulatory genes; StrR; LAL; LuxR solo; dalbavancin; A40926

---

1. The Glycopeptides

The glycopeptides are a class of antibiotics with a complex chemical structure and relatively high molecular weight. Since 1953, about 50 glycopeptide antibiotics (GPA) have been isolated [1], and several of these have been approved for clinical use. These include vancomycin, produced by *Amycolatopsis orientalis* and marketed in 1958, and teicoplanin, produced by *Actinoplanes teichomyceticus* and marketed in 1987. The second-generation glycopeptides telavancin, derived from vancomycin, dalbavancin, derived from A40926, and oritavancin, derived from choloroeremomycin, were introduced onto the market in 2009, 2014 and 2015, respectively. All glycopeptides are used to treat persistent infections by Gram-positive multi-resistant pathogens [2]. The second-generation glycopeptides are nearly 4- to 8-fold more effective than vancomycin against Gram-positive pathogens, and are also active against vancomycin-intermediate or vancomycin-resistant strains of *Staphylococcus* and *Enterococcus* spp. [3]. While dalbavancin impedes the late steps of cell wall biosynthesis principally by blocking transglycosylase activity, oritavancin and telavancin bind to the bacterial membrane by the lipophilic side chain linked to their disaccharide moiety, disturbing membrane integrity and leading to bacteriolysis [3].

Chemically, glycopeptides are a class of molecules constituted by a heptapeptide core consisting of both proteinogenic and non-proteinogenic amino acids, such as 3,5-dihydroxyphenylglycine (Dpg) and 4-hydroxyphenylglycine (Hpg). A heptapeptide is produced by a non-ribosomal peptide synthetase (NRPS) and, while tethered to the large multi-functional enzyme, the peptide scaffold is made rigid through oxidative cross-linking of the electron-rich aromatic side chains by P450s and chlorinated [4–6]. Further tailoring steps may include one or more glycosylations, methylation, sulfation and modification of the added sugar(s) by acylation and acetylation.

Glycopeptide producers are widespread among distantly related genera of actinomycetes [1]: vancomycin, balhymicin and ristocetin were isolated from distinct species of the genus *Amycolatopsis*, belonging to the family *Pseudonocardiaceae*; teicoplanin and UK-68597 are produced by members of the genus *Actinoplanes*, family *Micromonosporaceae*; A40926 is from the genus *Nonomuraea*, family *Streptosporangiaceae*; and pekiskomycin and A47394 are from the genus *Streptomyces*, family *Streptomycetaceae* [7]. Thus, production of GPAs is widespread among actinomycetes, as shown by the relatively high frequency at which glycopeptide producers can be detected in environmental samples after applying appropriate selection procedures [8].

The medical interest and importance of these molecules has prompted the analysis of the genes required for their synthesis. Different glycopeptide biosynthetic gene clusters have been reported [9]; combining the information obtained from these clusters, a function has been assigned to most genes involved in glycopeptide formation by in vivo gene disruption in the producing strain(s) and by biochemical studies of the overproduced enzymes. While these results have analyzed different pathways, the emerging overall picture has contributed to deciphering most of the biosynthetic steps and the timing of the events in the biosynthesis of all GPAs [4–6,8,10].

2. Development of Dalbavancin

Dalbavancin is a second-generation glycopeptide derived from A40926 with an improved antibacterial activity over teicoplanin, the most closely correlated marketed GPA. The enhanced pharmaco-dynamic properties of the molecule and lipophilic anchoring to the bacterial cell membrane confer more potent in vitro and in vivo activity than teicoplanin. The most prominent peculiarity of dalbavancin is a significantly extended half-life in plasma, which allows once-a-week dosing by intravenous injection. The drug has been approved for treating complicated acute bacterial skin and skin structure infections. Its synthesis involves the deacetylation of the final biosynthetic intermediate A40926 (a process achieved during recovery from the fermentation broth), protection of the carboxyl group present in the aminosugar, conversion of the C-terminal carboxyl group into a (3-dimethylamino)-1-propylamide, and final deprotection of the aminosugar carboxyl group [11]. The main components of the A40926 complex differ mainly in the acyl chain attached to the sugar, with B_0 and B_1 as the major representatives, characterized respectively by an iso-C12:0 and a n-C12:0 acyl moiety bound to the aminoglucuronic acid moiety [12]. The structures of A40926 and of dalbavancin are shown in Figure 1.

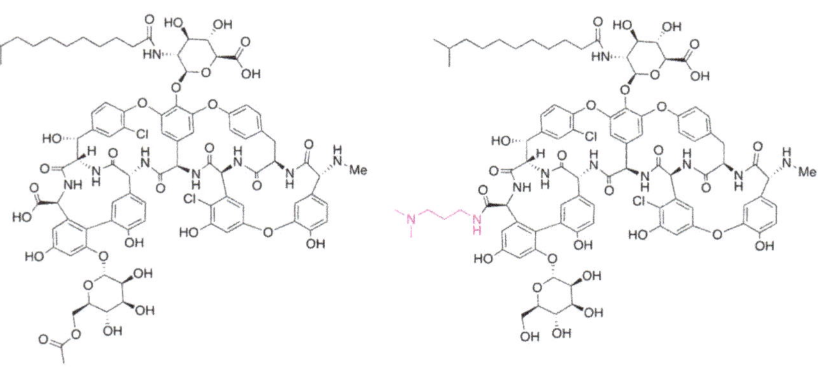

Figure 1. Chemical structures of *O*-acetyl A40926 and of dalbavancin. Only the component B0 is shown for simplicity. The chemical modification present in dalbavancin is indicated in red type.

Dalbavancin obtained market authorization in 2014 in the USA and the following year in Europe. This was a noteworthy success in view of the intricate history related to its development, which started back in the early 1990s and involved at least six different legal entities, as recently summarized [13].

In this review, we have organized the text into three separate sections: the first concerns the improvement of antibiotic yield by modifying the media components; the second describes the biosynthetic gene cluster and its transcriptional organization (Figure 2), along with the biosynthetic steps (Figure 3); and the last section deals with the cluster specific regulatory genes.

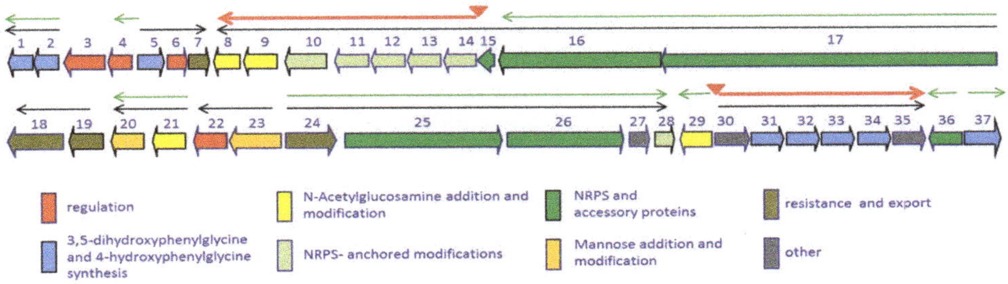

Figure 2. Genetic organization of the *dbv* cluster. The thin black arrows indicate experimentally determined operons. Red triangles indicate experimentally determined Dbv4 binding sites, with the corresponding transcripts as red thick arrows; the thin green arrows represent the transcriptional units controlled by Dbv3. The *dbv* genes are grouped by functional category as indicated. See also Table 1.

Table 1. Transcriptional units and biosynthetic roles of the corresponding proteins.

Transcriptional Unit	Function(s)
dbv1-dbv2	Hpg biosynthesis
dbv3	Regulation
dbv4	Regulation
dbv5-dbv7	Hpg biosynthesis; regulation; resistance
dbv8-dbv14	*N*-Acylation; Halogenation; glycosylation; cross-links
dbv15-dbv17	NRPS
dbv18-dbv19	Export
dbv20-dbv21	Mannose addition; *N*-sugar deacetylation
dbv22-dbv23	Regulation; mannose *O*-acetylation
dbv24-dbv28	Export; NRPS; Tyr β-hydroxylation
dbv29	*N*-sugar oxidation
dbv30-dbv35	Dpg biosynthesis
dbv36	NRPS accessory protein
dbv37	Hpg and Dpg biosynthesis

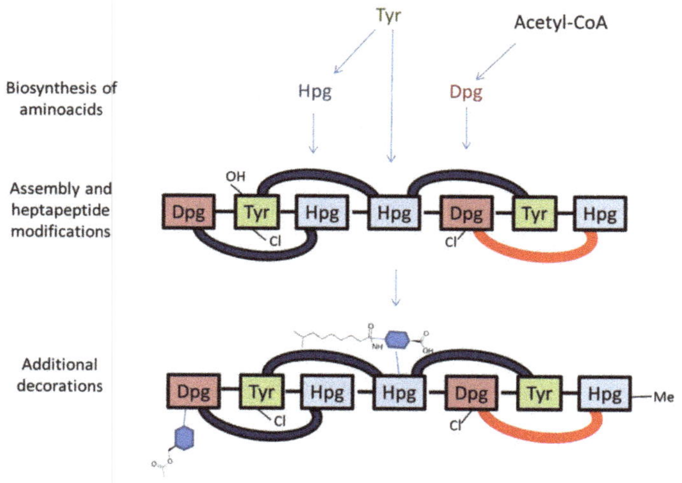

Figure 3. Simplified model of *O*-acetyl A40926 biosynthesis. Note that the heptapeptide is drawn right (N-terminus) to left (C-terminus), consistent with Figure 1. Cross-links are indicated by blue (C–O–C) or red (C–C) arcs. Sugars are represented as blue hexagons. Refer to Figure 2 and Table 1 for details.

3. Improvement of A40926 Production

Improvement of glycopeptide production has very likely been achieved through several rounds of mutagenesis and screening, leading to the current industrial strains producing vancomycin, teicoplanin, chloroeremomycin and A40926. However, most of this work has not surfaced in the scientific literature, and we will limit ourselves to published reports on the A40926 process.

Initial work established the influence of growth conditions on A40926 production by *Nonomuraea* sp. ATCC 39727, recently classified as *N. gerenzanensis* [14]. In a chemically defined medium, low initial concentrations of phosphate and ammonium led to increased A40926 production, while glucose limitation did not (Figure 4). In particular, the level of residual ammonium and phosphate strongly influenced A40926 production rates and final titers, but not the initiation of production [15]. In a similar medium, A40926 production was repressed by calcium, but supported when L-glutamine or L-asparagine were added as nitrogen sources instead of ammonium salts (Figure 4) [16]. Since the catabolic products of branched chain amino acids represent biosynthetic precursors for the formation of the branched chain acyl moieties of A40926 [17], studies were undertaken on the influence of valine supplementation. Addition of 1 to 3 g/L-valine to complex media improved both the relative and absolute production of the B_0 congener with decrease of the B_1 component in the A40926 complex [18]. A40926 yields were found to also be controlled by stringent response in both complex and chemically defined media (Figure 4) [19].

It has also been recently reported that a *Nonomuraea* strain producing high levels of A40926 in an optimized production medium was isolated after UV mutagenesis. This mutant strain was used to study the effect of carbon and nitrogen sources and of different ions on antibiotic productivity; addition of the scarcely assimilated carbon source maltodextrin and the nitrogen source soybean meal strongly affected A40926 production, which reached 1 g/L in a 10-L fermenter. Furthermore, Cu^{2+} stimulated A40926 biosynthesis while Co^{2+} showed an inhibitory effect. As shown for valine, even L-leucine addition led to an increased production of total A40926 and changed the complex toward the B_0 compound (Figure 4) [20]. While the shift in complex composition after amino acid addition can be easily rationalized, there are currently no clues as to why certain carbon sources and metal ions stimulate or inhibit growth and/or A40926 production.

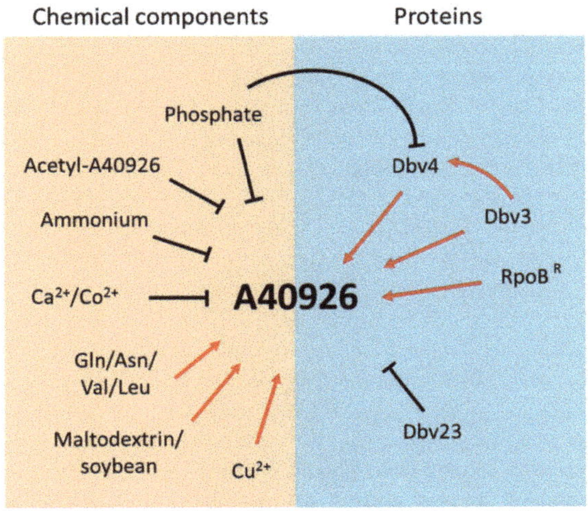

Figure 4. Nutrients, biosynthetic products and proteins regulating A40926 production in *Nonomuraea gerenzanensis*.

4. The *dbv* Gene Cluster: Main Features

The characterization of the gene cluster necessary for A40926 biosynthesis [21] laid the foundation for understanding the regulatory mechanisms working in the producer strain [22,23]. The *dbv* gene cluster is constituted by 37 protein coding sequences involved in antibiotic biosynthesis, regulation, immunity, and export [21] (Figure 2).

In particular, Dbv1, Dbv2, Dbv5, Dbv30-34 and Dbv37 are involved in biosynthesis of the two non proteinogenic amino acids Hpg and Dpg, while Dbv16-17 and Dbv25-26 constitute the NRPS that joins the amino acids Hpg, Tyr, Dpg, Hpg, Hpg, Tyr and Dpg in a ribosome-independent manner. The A40926 aryl groups are linked by three ether links and one C–C link through the action of Dbv11-14 P450s, while the single halogenase Dbv10 chlorinates Dpg-3 and Tyr-6. By analogy with other glycopeptides, halogenation should occur on an NRPS-bound substrate [5], while Tyr beta hydroxylation [24] might also involve interaction with an NRPS-bound substrate or intermediate. Additional modifications require the action of: Dbv27, for *N*-methylation of the terminal Hpg-1 residue; Dbv9, Dbv21, Dbv8 and Dbv29, for addition of *N*-acetyl glucosamine, deacetylation and acylation with long chain fatty acids, and sugar oxidation, respectively [25–27]; and Dbv20 and Dbv23, for mannosylation of Dpg-7 and its *O*-acetylation [28]. The different functions are illustrated in Figure 2 and summarized in Table 1, and a simplified model of *O*-acetyl A40926 biosynthesis is depicted in Figure 3.

The last biosynthetic step is possibly represented by acetylation at position 6 of the mannose moiety carried out by Dbv23 [28,29]. A strain deleted in *dbv23* produced only glycopeptides lacking the O-linked acetyl residue. Interestingly, antibiotic production in a complex medium by the mutant strain occurred at twice the levels of the wild type. The low amount of glycopeptide produced by the wild-type strain might be dependent upon an inhibitory effect exerted by the acetylated compound, the final pathway intermediate. Consistently, spiking the production medium with 1 μg/mL of the acetylated glycopeptide inhibited total glycopeptide production in the mutant strain, while the deacetylated glycopeptide had no effect [28]. It is thus tempting to speculate that A40926 production is regulated by its end product, ensuring that A40926 does not occur during growth of the strain. This might occur through a two-component signal transduction process, in which a specific receptor could activate a response regulator and repress A40926 biosynthesis. This might be relevant in industrial processes, in which a seed culture is eventually used to inoculate the production medium.

Any A40926 produced in the seed culture might be sufficient to inhibit A40926 production when the strain is inoculated in the production medium. This mechanism might be related to the inherent sensitivity of the strain to its own product, as described below.

Glycopeptides bind to the D-Ala-D-Ala portion of lipid II and thus inhibit the transpeptidation and transglycosylation reactions, thereby blocking peptidoglycan polymerization. The first and best characterized mechanism for glycopeptide resistance was established in enterococci, where glycopeptide action is avoided by deploying a modified target through a complex process that requires at least three biosynthetic genes (*vanHAX*) and a regulatory circuit (reviewed in [30–32]). Glycopeptide resistance in actinomycetes can also involve reprogramming of the peptidoglycan precursor by the action of VanHAX-related enzymes, as, for example, in *Amycolatopsis balhimycina* [33,34]. Instead, *Nonomuraea gerenzanensis* lacks the typical *vanHAX* cassette and the *dbv* cluster encodes the carboxypeptidase Dbv7, which has been shown to provide a modest but measurable resistance effect in the wild-type strain and in a heterologous background [35]. It should be noted that glycopeptide resistance in actinomycetes is still far from being completely understood, with there being a subtle interplay between glycopeptide resistance and glycopeptide tolerance [36]. Finally, the ABC transporters Dbv18, Dbv19, and Dbv24 and ion-dependent transmembrane transporter Dbv35 may contribute to glycopeptide resistance through active export from the cell, as observed for the Dbv24 homolog in the balhimycin producer [37].

The transcriptional organization of the *dbv* cluster was elucidated by RT-PCR targeting desired regions of the gene cluster [22]. The results, illustrated in Figure 2, denote a complex transcriptional organization, with at least 14 promoters, the two-gene operons *dbv1-dbv2*, *dbv19-dbv18*, *dbv21-dbv20*, *dbv23-dbv22*, the larger operons *dbv5-dbv7*, *dbv24-dbv28*, and *dbv30-dbv35*, and the largest operon *dbv17-dbv8*. Apparently, *dbv3*, *dbv4*, *dbv29*, *dbv36* and *dbv37* are transcribed as monocistronic units. The results are summarized in Table 1, which also lists the functions of the corresponding proteins. Real-time RT-PCR showed that a promoter is present upstream to *dbv14*, directing expression of the *dbv14-dbv8* operon through a leaderless transcript. However, a longer operon is likely to be transcribed from upstream promoter(s), since RT-PCR analysis showed the existence of a transcript spanning *dbv15* and *dbv14*. Some of the associated regulatory networks controlling A40926 biosynthesis are described below.

5. Cluster-Specific Regulatory Genes

The *dbv* cluster contains two regulatory genes, *dbv3* and *dbv4*, and the members of a putative two-component system, *dbv6* and *dbv22* [21–23]. Over a decade ago, a comparative analysis of the then-available five glycopeptide gene clusters—namely, those for chloroeremomycin, balhimycin, A47934, A40926 and teicoplanin—revealed that a StrR-like protein (i.e., Dbv4) was present in all clusters [38]. We previously demonstrated cross-binding among StrR-like regulators from glycopeptide clusters; specifically, Bbr (from the balhimycin cluster) can bind to the *dbv30* upstream region, while Dbv4 binds to the regions upstream of *bbr* and *oxyA* in the balhimycin cluster [22]. The target regions of Bbr and Dbv4 contain the highly conserved palindromic consensus sequence GTCCAR(N)$_{17}$TTGGAC. This sequence was considered to be the Dbv4 binding site and was found in two regions of the *dbv* cluster and in five regions of the balhimycin cluster [39]. Consistently, this conserved palindrome is part of a conserved intergenic region present in the five glycopeptide clusters mentioned above [38].

In addition to the common regulation of the oxygenase transcription through a Dbv4-type regulator, diverse regulatory schemes are apparently used in the other biosynthetic gene clusters. Actually, in the teicoplanin cluster, Dbv4-like protein also positively regulates the transcription of the gene operon involved in Dpg biosynthesis and, as a matter of fact, the Dbv4 target sequence was found upstream of this operon, suggesting a Dbv4-type dependent regulation. In contrast, while Bbr, through the binding of its upstream region [39], functions as an autoregulatory protein, Dbv4 did not. Similarly, since the conserved palindrome is apparently missing in the region upstream of the corresponding genes in the A47394 and teicoplanin, the Dbv4-like regulator is not expected to control its own expression in these clusters.

A40926 production is repressed by high initial concentrations of phosphate, and this repression was demonstrated to occur through Dbv4 [22]: phosphate depletion induces *dbv4* transcription in a defined medium, allowing Dbv4 to enhance expression of the operons *dbv14-dbv8* and *dbv30-dbv35*. However, phosphate did not influence the expression of most analyzed *dbv* genes [22]. The biosynthesis of many diverse secondary metabolites is controlled by phosphate [40]. Phosphate control of antibiotic biosynthesis in *Streptomyces coelicolor* and *S. lividans* is dependent upon the two-component system PhoR-PhoP [41], and PhoP was found to bind promoters of phosphate-regulated genes in *S. coelicolor* [42]. However, we were unable to identify Pho boxes in the region upstream of *dbv4* [22], suggesting divergence in phosphate control of antibiotic biosynthesis in different actinomycetes.

Another regulator from the *dbv* gene cluster has been experimentally characterized: Dbv3, a LuxR solo regulator belonging to the large ATP-binding regulators of the LuxR protein family. Dbv3 positively regulates A40926 production, since the Δ*dbv3* strain does not produce antibiotic and shows reduced transcription levels of *dbv4* and of many other *dbv* genes [23]. Thus, both the LuxR- and StrR-like regulators act as activators of A40926 biosynthesis.

The experimental evidence obtained for different glycopeptide pathways indicates that the StrR-like regulators Bbr, Tei15 and Dbv4 regulate balhimycin, teicoplanin and A40926 biosynthesis, respectively [22,39,43], whereas the LuxR-like regulators Dbv3 and Tei16 positively regulate A40926 and teicoplanin biosynthesis, respectively [23,43]. The balhimycin cluster does not encode a LuxR-like regulator. In A40926 biosynthesis, Dbv3 positively regulates Hpg biosynthesis, heptapeptide backbone biosynthesis, mannosylation, hexose oxidation and export. In addition, Dbv3 was found to hierarchically control *dbv*4 transcription in a cascade-like regulatory mechanism, so that Dpg biosynthesis and transcription of the *dbv14-dbv8* operon are also under indirect control of Dbv3. In addition, Dbv4 and Dbv3 expression seems to be differently modulated, since transcription of *dbv4* and Dbv4 target genes was found to be repressed by phosphate, while the Dbv3 target genes were not [22]. It should be noted that in teicoplanin biosynthesis, the expression of at least 17 genes is directly governed by Tei15, the Dbv4-like regulator, which directly controls transcription of *tei*16, the *luxR*-type regulator [43]. The targets of Tei16 have not been reported yet.

Notwithstanding the absence of obvious targets for yield improvements by gene knockouts (e.g., repressor genes), genetic manipulation of selected *dbv* genes has led to increased yields of A40926. Knockout of the acetyltransferase *dbv23* (see above) or overexpression of Dbv3 resulted in higher (2-fold) A40926 production than in the wild type strain in rich medium, providing useful examples of knowledge-based strain improvement [23,28]. Analysis of the additional regulators encoded by the *dbv* cluster, the sensor kinase Dbv22 and the response regulator Dbv6, has established their role in the regulation of A40926 and provided additional strategies for rational intervention.

6. Future Perspectives

This review summarizes the main achievements in understanding A40926 biosynthesis in *N. gerenzanensis* in relation to other glycopeptide producers and model *Streptomyces* strains. While many studies have addressed antibiotic production in model streptomycetes, like *S. coelicolor*, we continuously learn new mechanisms and pathways as we extend these analyses to industrially relevant antibiotics and, especially, to actinomycetes other than *Streptomyces* spp. In this respect, strains belonging to the genus *Nonomuraea* represent complex systems, with limited genetic tools available. Current results suggest an interplay between nutrients, resistance determinants and the end product. Even if many factors and proteins have been found to control A40926 biosynthesis (Figure 4), further studies are necessary to fill the many gaps present in our understanding of the strain's physiology and of the interplay between A40926 production and resistance before this information can be applied for A40926 yield improvement.

The recent availability of the *N. gerenzanensis* genome sequence [44] and of a large insert library [45] represent important assets for further work on the complex but intriguing regulatory network of the A40926-producing strain.

Acknowledgments: Most of the research described in this report was supported by Fondo per il finanziamento delle attività base di ricerca 2017 (FFABR 2017) 2017 to Rosa Alduina.

Author Contributions: Rosa Alduina, Margherita Sosio and Stefano Donadio wrote, read and approved the final manuscript.

Conflicts of Interest: The authors declare no conflict of interest.

References

1. Kahne, D.; Leimkuhler, C.; Lu, W.; Walsh, C. Glycopeptide and Lipoglycopeptide Antibiotics. *Chem. Rev.* **2005**, *105*, 425–448. [CrossRef] [PubMed]
2. Van Bambeke, F. Lipoglycopeptide antibacterial agents in gram-positive infections: A comparative review. *Drugs* **2015**, *75*, 2073–2095. [CrossRef] [PubMed]
3. Klinker, K.P.; Borgert, S.J. Beyond vancomycin: The tail of the lipoglycopeptides. *Clin. Ther.* **2015**, *37*, 2619–2636. [CrossRef] [PubMed]
4. Tailhades, J.; Schoppet, M.; Greule, A.; Peschke, M.; Brieke, C.; Cryle, M.J. A route to diastereomerically pure phenylglycine thioester peptides: Crucial intermediates for investigating glycopeptide antibiotic biosynthesis. *Chem. Commun. (Camb.)* **2018**, *54*, 2146–2149. [CrossRef] [PubMed]
5. Kittila, T.; Kittel, C.; Tailhades, J.; Butz, D.; Schoppet, M.; Buttner, A.; Goode, R.J.A.; Schittenhelm, R.B.; van Pee, K.H.; Sussmuth, R.D.; et al. Halogenation of glycopeptide antibiotics occurs at the amino acid level during non-ribosomal peptide synthesis. *Chem. Sci.* **2017**, *8*, 5992–6004. [CrossRef] [PubMed]
6. Peschke, M.; Brieke, C.; Goode, R.J.A.; Schittenhelm, R.B.; Cryle, M.J. Chlorinated glycopeptide antibiotic peptide precursors improve cytochrome P450-catalyzed cyclization cascade efficiency. *Biochemistry* **2017**, *56*, 1239–1247. [CrossRef] [PubMed]
7. Thaker, M.N.; Wang, W.; Spanogiannopoulos, P.; Waglechner, N.; King, A.M.; Medina, R.; Wright, G.D. Identifying producers of antibacterial compounds by screening for antibiotic resistance. *Nat. Biotechnol.* **2013**, *31*, 922–927. [CrossRef] [PubMed]
8. Yim, G.; Kalan, L.; Koteva, K.; Thaker, M.N.; Waglechner, N.; Tang, I.; Wright, G.D. Harnessing the synthetic capabilities of glycopeptide antibiotic tailoring enzymes: Characterization of the UK-68,597 biosynthetic cluster. *Chembiochem* **2014**, *15*, 2613–2623. [CrossRef] [PubMed]
9. Yim, G.; Thaker, M.N.; Koteva, K.; Wright, G. Glycopeptide antibiotic biosynthesis. *J. Antibiot. (Tokyo)* **2014**, *67*, 31–41. [CrossRef] [PubMed]
10. Stegmann, E.; Frasch, H.J.; Wohlleben, W. Glycopeptide biosynthesis in the context of basic cellular functions. *Curr. Opin. Microbiol.* **2010**, *13*, 595–602. [CrossRef] [PubMed]
11. Malabarba, A.; Goldstein, B.P. Origin, structure, and activity in vitro and in vivo of dalbavancin. *J. Antimicrob. Chemother.* **2005**, *55*, ii15–ii20. [CrossRef] [PubMed]
12. Zerilli, L.F.; Edwards, D.M.; Borghi, A.; Gallo, G.G.; Selva, E.; Denaro, M.; Lancini, G.C. Determination of the acyl moieties of the antibiotic complex A40926 and their relation with the membrane lipids of the producer strain. *Rapid Commun. Mass Spectrom.* **1992**, *6*, 109–114. [CrossRef] [PubMed]
13. Maffioli, S.I.; Cruz, J.C.S.; Monciardini, P.; Sosio, M.; Donadio, S. Advancing cell wall inhibitors towards clinical applications. *J. Ind. Microbiol. Biotechnol.* **2016**, *43*, 177–184. [CrossRef] [PubMed]
14. Dalmastri, C.; Gastaldo, L.; Marcone, G.L.; Binda, E.; Congiu, T.; Marinelli, F. Classification of *Nonomuraea* sp. ATCC 39727, an actinomycete that produces the glycopeptide antibiotic A40926, as *Nonomuraea gerenzanensis* sp. nov. *Int. J. Syst. Evol. Microbiol.* **2016**, *66*, 912–921. [CrossRef] [PubMed]
15. Gunnarsson, N.; Bruheim, P.; Nielsen, J. Glucose metabolism in the antibiotic producing actinomycete *Nonomuraea* sp. ATCC 39727. *Biotechnol. Bioeng.* **2004**, *88*, 652–663. [CrossRef] [PubMed]
16. Technikova-Dobrova, Z.; Damiano, F.; Tredici, S.M.; Vigliotta, G.; di Summa, R.; Palese, L.; Abbrescia, A.; Labonia, N.; Gnoni, G.V.; Alifano, P. Design of mineral medium for growth of *Actinomadura* sp. ATCC 39727, producer of the glycopeptide A40926: Effects of calcium ions and nitrogen sources. *Appl. Microbiol. Biotechnol.* **2004**, *65*, 671–677. [CrossRef] [PubMed]

17. Jovetic, S.; Feroggio, M.; Marinelli, F.; Lancini, G. Factors influencing cell fatty acid composition and A40926 antibiotic complex production in *Nonomuraea* sp. ATCC 39727. *J. Ind. Microbiol. Biotechnol.* **2008**, *35*, 1131–1138. [CrossRef] [PubMed]
18. Beltrametti, F.; Jovetic, S.; Feroggio, M.; Gastaldo, L.; Selva, E.; Marinelli, F. Valine influences production and complex composition of glycopeptide antibiotic A40926 in fermentations of *Nonomuraea* sp. ATCC 39727. *J. Antibiot. (Tokyo)* **2004**, *57*, 37–44. [CrossRef] [PubMed]
19. Vigliotta, G.; Tredici, S.M.; Damiano, F.; Montinaro, M.R.; Pulimeno, R.; di Summa, R.; Massardo, D.R.; Gnoni, G.V.; Alifano, P. Natural merodiploidy involving duplicated *rpob* alleles affects secondary metabolism in a producer actinomycete. *Mol. Microbiol.* **2005**, *55*, 396–412. [CrossRef] [PubMed]
20. Chen, M.; Xu, T.; Zhang, G.; Zhao, J.; Gao, Z.; Zhang, C. High-yield production of lipoglycopeptide antibiotic A40926 using a mutant strain *Nonomuraea* sp. Dp-13 in optimized medium. *Prep. Biochem. Biotechnol.* **2016**, *46*, 171–175. [CrossRef] [PubMed]
21. Sosio, M.; Stinchi, S.; Beltrametti, F.; Lazzarini, A.; Donadio, S. The gene cluster for the biosynthesis of the glycopeptide antibiotic A40926 by *Nonomuraea* species. *Chem. Biol.* **2003**, *10*, 541–549. [CrossRef]
22. Alduina, R.; Lo Piccolo, L.; D'Alia, D.; Ferraro, C.; Gunnarsson, N.; Donadio, S.; Puglia, A.M. Phosphate-controlled regulator for the biosynthesis of the dalbavancin precursor A40926. *J. Bacteriol.* **2007**, *189*, 8120–8129. [CrossRef] [PubMed]
23. Lo Grasso, L.; Maffioli, S.; Sosio, M.; Bibb, M.; Puglia, A.M.; Alduina, R. Two master switch regulators trigger A40926 biosynthesis in *Nonomuraea* sp. Strain ATCC 39727. *J. Bacterial.* **2015**, *197*, 2536–2544. [CrossRef] [PubMed]
24. Stinchi, S.; Carrano, L.; Lazzarini, A.; Feroggio, M.; Grigoletto, A.; Sosio, M.; Donadio, S. A derivative of the glycopeptide A40926 produced by inactivation of the beta-hydroxylase gene in *Nonomuraea* sp. ATCC 39727. *FEMS Microbiol. Lett.* **2006**, *256*, 229–235. [CrossRef] [PubMed]
25. Zou, Y.; Brunzelle, J.S.; Nair, S.K. Crystal structures of lipoglycopeptide antibiotic deacetylases: Implications for the biosynthesis of A40926 and teicoplanin. *Chem. Biol.* **2008**, *15*, 533–545. [CrossRef] [PubMed]
26. Kruger, R.G.; Lu, W.; Oberthur, M.; Tao, J.; Kahne, D.; Walsh, C.T. Tailoring of glycopeptide scaffolds by the acyltransferases from the teicoplanin and A-40,926 biosynthetic operons. *Chem. Biol.* **2005**, *12*, 131–140. [CrossRef] [PubMed]
27. Li, Y.S.; Ho, J.Y.; Huang, C.C.; Lyu, S.Y.; Lee, C.Y.; Huang, Y.T.; Wu, C.J.; Chan, H.C.; Huang, C.J.; Hsu, N.S.; et al. A unique flavin mononucleotide-linked primary alcohol oxidase for glycopeptide A40926 maturation. *J. Am. Chem. Soc.* **2007**, *129*, 13384–13385. [CrossRef] [PubMed]
28. Sosio, M.; Canavesi, A.; Stinchi, S.; Donadio, S. Improved production of A40926 by *Nonomuraea* sp. through deletion of a pathway-specific acetyltransferase. *Appl. Microbiol. Biotechnol.* **2010**, *87*, 1633–1638. [CrossRef] [PubMed]
29. Sosio, M.; Donadio, S. Understanding and manipulating glycopeptide pathways: The example of the dalbavancin precursor A40926. *J. Ind. Microbiol. Biotechnol.* **2006**, *33*, 569–576. [CrossRef] [PubMed]
30. Arthur, M.; Molinas, C.; Courvalin, P. The VanS-VanR two-component regulatory system controls synthesis of depsipeptide peptidoglycan precursors in *Enterococcus faecium* BM4147. *J. Bacteriol.* **1992**, *174*, 2582–2591. [CrossRef] [PubMed]
31. Arthur, M.; Reynolds, P.; Courvalin, P. Glycopeptide resistance in enterococci. *Trends Microbiol.* **1996**, *4*, 401–407. [CrossRef]
32. Arthur, M.; Depardieu, F.; Courvalin, P. Regulated interactions between partner and non-partner sensors and response regulators that control glycopeptide resistance gene expression in enterococci. *Microbiology* **1999**, *145*, 1849–1858. [CrossRef] [PubMed]
33. Stegmann, E.; Frasch, H.J.; Kilian, R.; Pozzi, R. Self-resistance mechanisms of actinomycetes producing lipid II-targeting antibiotics. *Int. J. Med. Microbiol.* **2015**, *305*, 190–195. [CrossRef] [PubMed]
34. Frasch, H.J.; Kalan, L.; Kilian, R.; Martin, T.; Wright, G.D.; Stegmann, E. Alternative pathway to a glycopeptide-resistant cell wall in the balhimycin producer *Amycolatopsis balhimycina*. *ACS Infect. Dis.* **2015**, *1*, 243–252. [CrossRef] [PubMed]
35. Marcone, G.L.; Beltrametti, F.; Binda, E.; Carrano, L.; Foulston, L.; Hesketh, A.; Bibb, M.; Marinelli, F. Novel mechanism of glycopeptide resistance in the A40926 producer *Nonomuraea* sp. ATCC 39727. *Antimicrob. Agents Chemother.* **2010**, *54*, 2465–2472. [CrossRef] [PubMed]

36. Hugonnet, J.E.; Haddache, N.; Veckerle, C.; Dubost, L.; Marie, A.; Shikura, N.; Mainardi, J.L.; Rice, L.B.; Arthur, M. Peptidoglycan cross-linking in glycopeptide-resistant actinomycetales. *Antimicrob. Agents Chemother.* **2014**, *58*, 1749–1756. [CrossRef] [PubMed]
37. Menges, R.; Muth, G.; Wohlleben, W.; Stegmann, E. The ABC transporter Tba of *Amycolatopsis balhimycina* is required for efficient export of the glycopeptide antibiotic balhimycin. *Appl. Microbiol. Biotechnol.* **2007**, *77*, 125–134. [CrossRef] [PubMed]
38. Donadio, S.; Sosio, M.; Stegmann, E.; Weber, T.; Wohlleben, W. Comparative analysis and insights into the evolution of gene clusters for glycopeptide antibiotic biosynthesis. *Mol. Genet. Genom.* **2005**, *274*, 40–50. [CrossRef] [PubMed]
39. Shawky, R.M.; Puk, O.; Wietzorrek, A.; Pelzer, S.; Takano, E.; Wohlleben, W.; Stegmann, E. The border sequence of the balhimycin biosynthesis gene cluster from *Amycolatopsis balhimycina* contains *bbr*, encoding a *strR*-like pathway-specific regulator. *J. Mol. Microbiol. Biotechnol.* **2007**, *13*, 76–88. [CrossRef] [PubMed]
40. Martin, J.F. Phosphate control of the biosynthesis of antibiotics and other secondary metabolites is mediated by the PhoR-PhoP system: An unfinished story. *J. Bacteriol.* **2004**, *186*, 5197–5201. [CrossRef] [PubMed]
41. Sola-Landa, A.; Moura, R.S.; Martin, J.F. The two-component PhoR-PhoP system controls both primary metabolism and secondary metabolite biosynthesis in *Streptomyces lividans*. *Proc. Natl. Acad. Sci. USA* **2003**, *100*, 6133–6138. [CrossRef] [PubMed]
42. Sola-Landa, A.; Rodriguez-Garcia, A.; Franco-Dominguez, E.; Martin, J.F. Binding of PhoP to promoters of phosphate-regulated genes in *Streptomyces coelicolor*: Identification of pho boxes. *Mol. Microbiol.* **2005**, *56*, 1373–1385. [CrossRef] [PubMed]
43. Horbal, L.; Kobylyanskyy, A.; Truman, A.W.; Zaburranyi, N.; Ostash, B.; Luzhetskyy, A.; Marinelli, F.; Fedorenko, V. The pathway-specific regulatory genes, *tei15** and *tei16**, are the master switches of teicoplanin production in *Actinoplanes teichomyceticus*. *Appl. Microbiol. Biotechnol.* **2014**, *98*, 9295–9309. [CrossRef] [PubMed]
44. D'Argenio, V.; Petrillo, M.; Pasanisi, D.; Pagliarulo, C.; Colicchio, R.; Tala, A.; de Biase, M.S.; Zanfardino, M.; Scolamiero, E.; Pagliuca, C.; et al. The complete 12 Mb genome and transcriptome of *Nonomuraea gerenzanensis* with new insights into its duplicated "magic" RNA polymerase. *Sci. Rep.* **2016**, *6*, 18. [CrossRef] [PubMed]
45. Alduina, R.; Giardina, A.; Gallo, G.; Renzone, G.; Ferraro, C.; Contino, A.; Scaloni, A.; Donadio, S.; Puglia, A.M. Expression in *Streptomyces lividans* of *Nonomuraea* genes cloned in an artificial chromosome. *Appl. Microbiol. Biotechnol.* **2005**, *68*, 656–662. [CrossRef] [PubMed]

antibiotics

MDPI

Review

Novel Aspects of Polynucleotide Phosphorylase Function in *Streptomyces*

George H. Jones

Department of Biology, Emory University, Atlanta, GA 30322, USA; george.h.jones@emory.edu

Received: 14 February 2018; Accepted: 16 March 2018; Published: 18 March 2018

Abstract: Polynucleotide phosphorylase (PNPase) is a 3′–5′-exoribnuclease that is found in most bacteria and in some eukaryotic organelles. The enzyme plays a key role in RNA decay in these systems. PNPase structure and function have been studied extensively in *Escherichia coli*, but there are several important aspects of PNPase function in *Streptomyces* that differ from what is observed in *E. coli* and other bacterial genera. This review highlights several of those differences: (1) the organization and expression of the PNPase gene in *Streptomyces*; (2) the possible function of PNPase as an RNA 3′-polyribonucleotide polymerase in *Streptomyces*; (3) the function of PNPase as both an exoribonuclease and as an RNA 3′-polyribonucleotide polymerase in *Streptomyces*; (4) the function of (p)ppGpp as a PNPase effector in *Streptomyces*. The review concludes with a consideration of a number of unanswered questions regarding the function of *Streptomyces* PNPase, which can be examined experimentally.

Keywords: polynucleotide phosphorylase; *Streptomyces*; ribonuclease; regulation; promoter; RNA decay; polyadenylation; (p)ppGpp; antibiotic

1. Introduction

Polynucleotide phosphorylase (PNPase, EC 2.7.7.8) was the first enzyme shown to synthesize polyribonucleotides [1], and for some time, it was thought to be the bacterial RNA polymerase. The enzyme was subsequently characterized in *Escherichia coli* and other bacteria, and was shown to catalyze the following reaction:

$$(p^{5'}N^{3'}OH)_X + Pi \leftrightarrows (p^{5'}N^{3'}OH)_{X-1} + pp^{5'}N$$

where N is any of the four bases found in RNA [2,3]. As written, the reaction depicts the phosphorolytic degradation of RNA chains, and this activity appears to reflect the major function of PNPase in vivo. The reaction is reversible, however, and PNPase will synthesize polyribonucleotide chains, using nucleoside diphosphates (NDPs), rather than triphosphates, as substrates. The polymerizing activity of PNPase played an important role in the synthesis of polyribonucleotides used to unravel the genetic code [4,5].

PNPase is found in all bacteria examined to date, except *Mycoplasma*, and is also present in eukaryotic organelles [6]. The enzyme has not been identified in Archaea [7]. In *E. coli*, PNPase and RNase II are the major 3′-exonucleases involved in RNA degradation [8]. In addition to its degradative function, PNPase plays a role in the bacterial response to environmental stresses, such as cold shock [9–11], is involved in biofilm formation [12,13] and virulence determination [14,15], and the activity of the enzyme is modulated by a number of small molecule effectors, at least in vitro [16–19].

Streptomyces are Gram-positive, soil-dwelling bacteria, notable for their ability to form spores and for their capacity to produce antibiotics [20,21]. Nearly 70% of all antibiotics used in clinical and veterinary medicine worldwide are synthesized as natural products by members of the genus [22]. Of particular relevance to this review, a number of biochemical and genetic features of *Streptomyces*

PNPase distinguish it from its counterparts in other bacteria. In what follows, the functions of PNPase in *Streptomyces* will be explored. The reader is referred to several excellent reviews as sources of additional information on PNPase from *E. coli* and other bacteria [2,3,23,24].

2. Organization and Expression of the PNPase Gene in *Streptomyces*

In *E. coli*, and other organisms that have been studied, the PNPase gene, *pnp*, is a part of an operon that also includes *rpsO*, the gene for ribosomal protein S15 [9,25,26]. That operon is transcribed from two promoters in *E. coli*, designated P*rpsO* and P*pnp* [25,27]. P*rpsO* is situated upstream of the *rpsO* gene, and P*pnp* is located in the intergenic region between the two genes. Transcription from P*rpsO* ends at a rho-independent terminator and produces the *rpsO* transcript, but transcription through this terminator occurs with significant frequency and produces a readthrough transcript, containing both *rpsO* and *pnp*. Transcription from the intergenic promoter, P*pnp*, produces a transcript containing *pnp* only [25,27]. In addition to the *rpsO* terminator, the *rpsO–pnp* intergenic region contains a second stem–loop that functions as a processing site for the double strand specific endoribonuclease, RNase III. RNase III processing plays an important role in *pnp* expression [25,28–30].

Our interest in the mechanisms of RNA decay in *Streptomyces* led to an examination of the transcriptional organization of the *rpsO–pnp* operon in *Streptomyces coelicolor*, the paradigm for biological studies in the genus. To our surprise, primer extension analysis of RNAs isolated from a parental strain of *S. coelicolor* and from an RNase III null mutant revealed not two, but four extension products, suggesting the presence of four promoters within the *rpsO–pnp* operon [31]. A visual inspection of 162 putative *Streptomyces* promoters from Strohl [32] and Yamazaki et al. [33], and a group of synthetic promoters generated by Seghezzi et al. [34], indicated that all four of the 5′-ends identified by primer extension in our studies were preceded by sequences similar to the −10 and −35 regions of characterized streptomycete promoters. Promoter probe cloning of DNA fragments containing these putative promoter sequences verified that the *rpsO–pnp* operon of *S. coelicolor* is transcribed from four promoters, two situated upstream of *rpsO* (P*rpsO*A and P*rpsO*B) and two situated in the intergenic region, upstream of *pnp* (P*pnp*A and P*pnp*B, Figure 1).

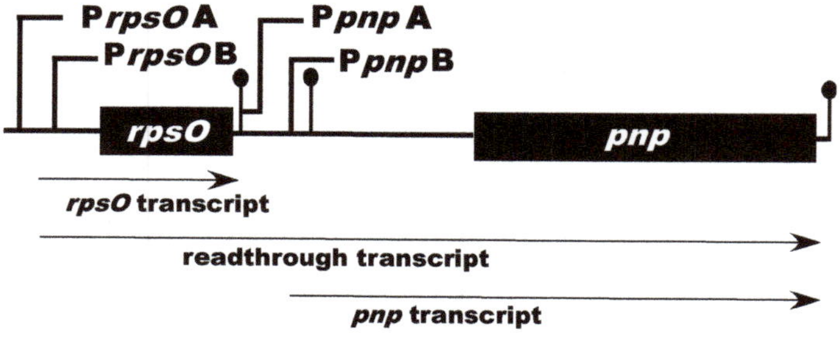

Figure 1. Schematic representation of the *Streptomyces coelicolor rpsO–pnp* operon. P*rpsO*A, B and P*pnp*A, B represent the upstream and intergenic promoters found in *S. coelicolor*, respectively. The ball-and-stick structures immediately following *rpsO* and *pnp* represent rho-independent transcription terminators. The ball-and-stick structure just upstream of *pnp* represents the intergenic hairpin which is cleaved by RNase III. The diagram is not drawn to scale.

Of particular interest in this analysis was the observation that the four promoters were temporally regulated. That is, the activity of the four promoters varied with time after inoculation of liquid cultures and the variations were promoter-specific, as shown in Figure 2. In terms of maximal activity, P*pnp*B was most active followed by P*rpsO*B, P*pnp*A, and P*rpsO*A.

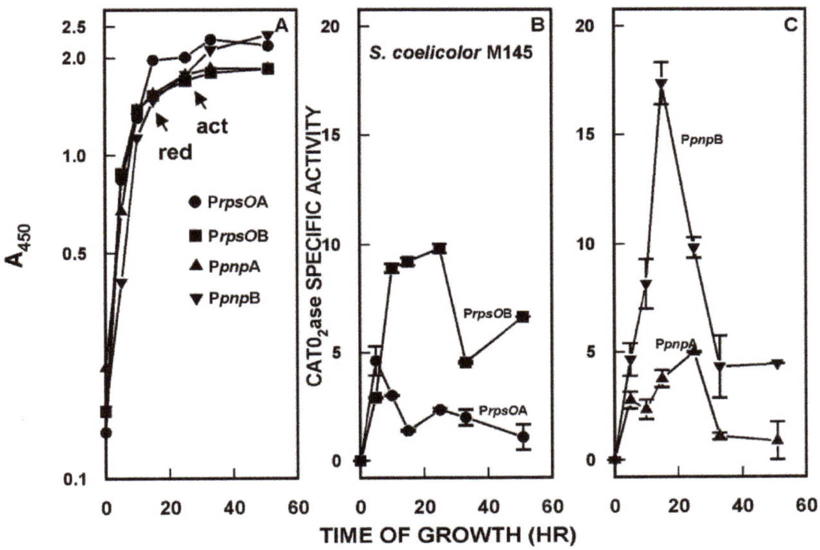

Figure 2. (**A**) Growth of the *S. coelicolor* strains containing promoter probe constructs. Growth was measured as the increase in optical density at 450 nm. The arrows in the figure indicate the onset of the production of two of the secondary metabolites synthesized by *S. coelicolor*, undecylprodigiosin (red) and actinorhodin (act). (**B**) Catechol dioxygenase ($CATO_2ase$) activity of mycelial extracts of *S. coelicolor* derivatives containing the putative *rpsO*A and *rpsO*B promoters, cloned in the promoter probe vector pIPP2 [35]. Mycelium was harvested at the indicated times, disrupted by sonication, and following centrifugation, supernatants were assayed for catechol dioxygenase, as described previously [31,35]. The catechol dioxygenase gene is the reporter in the promoter probe vector [35]. (**C**) $CATO_2ase$ activities of extracts of strains containing the putative *pnp*A and *pnp*B promoters. The results shown are the averages of duplicate assays from two independent experiments ± SEM. This figure is reprinted from *Gene*, 536, Patricia Bralley, Marcha L. Gatewood, George H. Jones, Transcription of the *rpsO–pnp* operon of *Streptomyces coelicolor* involves four temporally regulated, stress responsive promoters. 177–185, Copyright (2014), with permission from Elsevier [31].

A major mechanism for the modulation of promoter activity in bacterial systems involves the use of alternative sigma factors by RNA polymerase. The *S. coelicolor* genome encodes over sixty alternative sigma factors, many of which play roles in differentiation and responses to stress (reviewed in [36]). It was of considerable interest to determine whether the four *rpsO–pnp* promoters might require alternative sigma factors for transcription in *S. coelicolor*.

To this end, we obtained null mutants for a number of sigma factors, including σ^B, σ^H, and σ^L [36], and their corresponding parental strains. We transferred the *rpsO–pnp* promoter probe constructs to each strain and measured promoter activity. The results obtained for the σ^H and σ^L mutants were quite similar to those for the parental strain of *S. coelicolor*, that is, the same pattern of temporal regulation was observed in both of these sigma factor mutants as in the parental strain (cf. Figure 2). In marked contrast, the P*rpsO*A, P*rpsO*B, and P*pnp*B promoters were completely inactive in the σ^B mutant. P*pnp*A, however, was as active in the σ^B mutant as in the parental strain, and showed a similar pattern of temporal expression [31]. This result indicates that P*rpsO*A and B and P*pnp*B are dependent on σ^B for activity, and suggests that these promoters are transcribed by an RNA polymerase holoenzyme containing σ^B.

PNPase is a cold shock protein in many bacteria, and it has been shown that cold shock leads to an increase in PNPase levels in *E. coli* and other organisms [9–11]. It was of interest to determine whether PNPase levels increased in cold shock in *S. coelicolor*, and whether any such increase reflected changes in the activity of the *rpsO–pnp* promoters. As shown in Figure 3C, PNPase activity increased significantly (two-fold) in *S. coelicolor* over three hours of cold shock at 10 °C. This increase in activity was accompanied by an increase in the activities of all four of the *rpsO–pnp* promoters (Figure 3A,B) as compared with their activities at the normal growth temperature, 30 °C.

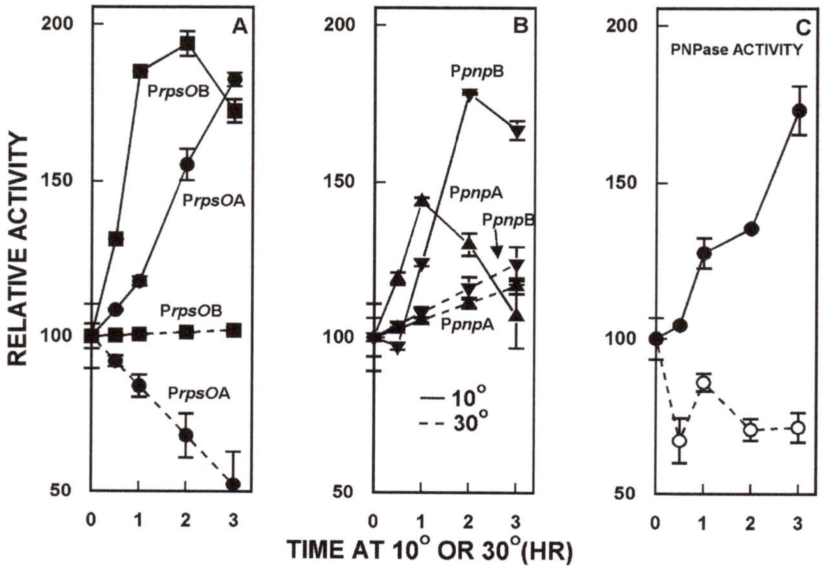

Figure 3. Cold shock responses of *S. coelicolor*. Derivatives containing the *rpsO–pnp* promoter probe constructs were grown and 30 °C, and half of each culture was then shifted to 10 °C. Mycelium was harvested at the indicated times, disrupted by sonication, and following centrifugation, supernatants were assayed for promoter activity, as described [31,35]. Panel **C** shows the results of PNPase polymerization assays. In Panels **A** and **B**, PNPase promoter activities are expressed relative to the activity measured at 30 °C at zero time, immediately before the shift to 10 °C. The results shown are the averages of duplicate assays from two independent experiments ± SEM. In the first experiment, PNPase levels were measured in *S. coelicolor* containing *Prps*OA and in the second, PNPase levels were measured in the derivative containing *Ppnp*B. This figure is reprinted from Gene, 536, Patricia Bralley, Marcha L. Gatewood, George H. Jones, Transcription of the *rpsO-pnp* operon of *Streptomyces coelicolor* involves four temporally regulated, stress responsive promoters. 177–185, Copyright (2014), with permission from Elsevier [31].

Thus, PNPase is a cold shock protein in *Streptomyces*, and the cold shock response involves changes in the activities of the promoters responsible for transcription of the *rpsO–pnp* operon [31].

3. PNPase Function as an RNA 3′-Polyribonucleotide Polymerase in *Streptomyces*

As is the case in eukaryotes, bacterial RNAs have oligo- and polyribonucleotide tails at their 3′-ends, and these tails are added post-transcriptionally (reviewed in [37]). In *E. coli*, the primary enzyme responsible for the synthesis of these tails is poly(A) polymerase I (PAP I), and the tails are composed primarily of A residues [38]. In bacteria, poly(A) tails function to facilitate RNA degradation as the major 3′-exoribonucleases, viz. RNase II and PNPase in *E. coli*, digest these tails processively in vitro and in vivo [8,39].

It was shown some years ago by Mohanty and Kushner that an *E. coli* mutant lacking PAP I still added tails to the 3′-ends of its RNAs. However, these tails were not composed exclusively of A residues; the tails contained G, C, and U residues as well, i.e., they were heteropolymeric [40]. Mohanty and Kushner demonstrated that the enzyme responsible for the synthesis of these heteropolymeric tails was none other than PNPase [40]. Thus, even in the absence of PAP I, PNPase can add 3′-tails to facilitate the degradation of cellular RNAs.

Streptomyces do not contain PAP I. Yet the 3′-ends of streptomycete RNAs do possess tails. Moreover, the tails are heteropolymeric in composition, like those synthesized by PNPase in *E. coli* in the absence of PAP I [41,42]. These observations, and the report that PNPase functions as the RNA 3′-polyribonucleotide polymerase in plant chloroplasts and in cyanobacteria [43,44], led to the hypothesis that PNPase played the same role in *Streptomyces* [45]. The straightforward way to test this hypothesis would be to create a *pnp* null mutant, e.g., in *S. coelicolor*, and to determine whether the mutant was still capable of adding 3′-tails to its RNAs. However, attempts to disrupt *pnp* in *S. coelicolor* and in the sister species, *Streptomyces antibioticus*, were only successful when a second copy of *pnp* was added to the genome. In other words, gene disruption attempts revealed that, unlike the situation in *E. coli* and in the Bacilli, *pnp* is an essential gene in *Streptomyces* [42].

A model for the function of PNPase as an RNA 3′-polyribonucleotide polymerase is presented in detail below.

4. Function of PNPase as Both an Exoribonuclease and as an RNA 3′-Polyribonucleotide Polymerase

PNPase activity is highly processive and the enzyme is impeded by stem–loop structures [46]. Streptomycete genomes are GC rich, so that enzymes involved in RNA decay may have evolved to degrade the RNAs derived from these genomes efficiently. A possible strategy for facilitating this degradation was suggested by the observation that PNPase appears to utilize its polymerizing activity to add 3′-tails to streptomycete RNAs [42,45]. It seemed possible that the enzyme might add such tails during phosphorolysis to create single stranded 3′-ends that would then function as the substrates for that phosphorolysis. If this were the case, it might be expected that nucleoside diphosphates, the substrates for polymerization, would stimulate phosphorolysis. To test this hypothesis, two model PNPase substrates were constructed from the sequence of the *rpsO–pnp* operon of *S. coelicolor*. Both substrates contained the *rpsO–pnp* terminator and the intergenic hairpin. Thus, both model substrates contained secondary structure that would be expected to impede phosphorolysis by PNPase. One substrate, designated 5601, also possessed a single stranded 3′-tail, 33 bases in length, while the other substrate, 5650, terminated at the base of the intergenic hairpin, and did not have a single stranded tail [47].

The phosphorolysis of these two substrates was studied in the absence and presence of a mixture of all four nucleoside diphosphates (NDPs) with the interesting result, predicted by our hypothesis, that the NDPs, normally the substrates for the polymerizing activity of PNPase, did stimulate RNA degradation by phosphorolysis. Figure 4 shows the results of phosphorolysis of the 5650 transcript (labeled RP3 in the figure).

Analysis of the results shown in Figure 4 revealed that NDPs at 20–30 μM in phosphorolysis mixtures stimulated that reaction by 2–3 fold as compared with controls, but only when the structured RNA, 5650 (RP3) was used as the substrate [47]. NDPs had no effect on the phosphorolysis of the 5601 substrate, possessing the single stranded tail. Kinetic analyses showed that NDPs affected the K_m for phosphorolysis. Thus, the K_m value for phosphorolysis of the 5650 substrate in the absence of NDPs was 3.1 μM. This value decreased to 0.65 μM in the presence of all four NDPs at 20 μM. This latter K_m was almost identical to that obtained in the absence of NDPs for the 5601 substrate, which has a single stranded 3′-tail (0.62 μM). NDPs did not further decrease the K_m for the 5601 substrate. It is noteworthy as well, that NDPs had no effect on the phosphorolysis of either substrate by *E. coli*

PNPase, and that the *E. coli* enzyme was intrinsically less active with the structured substrate than was its counterpart from *S. coelicolor* [47].

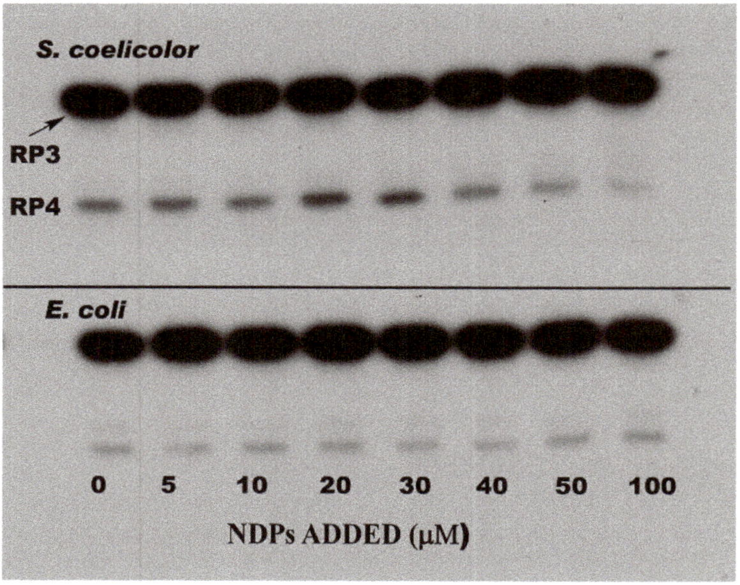

Figure 4. Effects of nucleoside diphosphates on the phosphorolysis of the 5650 transcript. Phosphorolysis reactions were performed as described in [47], and reaction products were separated by gel electrophoresis. The top panel shows the results obtained with *S. coelicolor* PNPase and the bottom panel results using *E. coli* PNPase. Reactions were conducted in the presence of increasing concentrations of a mixture of ADP, CDP, UDP, and GDP (nucleoside diphosphates (NDPs)) as indicated. RP3 is the 5650 transcript, and RP4 represents the product obtained by complete digestion of the intergenic hairpin in RP3 by PNPase. Note that as PNPase is highly processive [48], no intermediates with mobilities between those of RP3 and RP4 were observed. Copyright © American Society for Microbiology (*J. Bacteriol.* 190, 2008, 98–106, DOI:10.1128/JB.00327-07) [47].

Our model for the explanation of the foregoing observations is shown in Figure 5 [49].

We posit that the stem–loops of structured substrates, like 5650, block PNPase action. The addition of short 3′-tails during phosphorolysis, or the presence of naturally occurring tails on substrates like 5601, allows for the breathing of stems and thus permits PNPase, which might otherwise stall at the stem–loops in structured substrates, to continue phosphorolysis through those structures. As indicated above, this mechanism may represent an evolutionary adaptation occasioned by the high GC content of streptomycete genomes and their transcripts.

It must be noted, however, that the evidence for the function of PNPase as a 3′-polyribonucleotide polymerase in *Streptomyces* is indirect. In vivo evidence for this function remains to be uncovered.

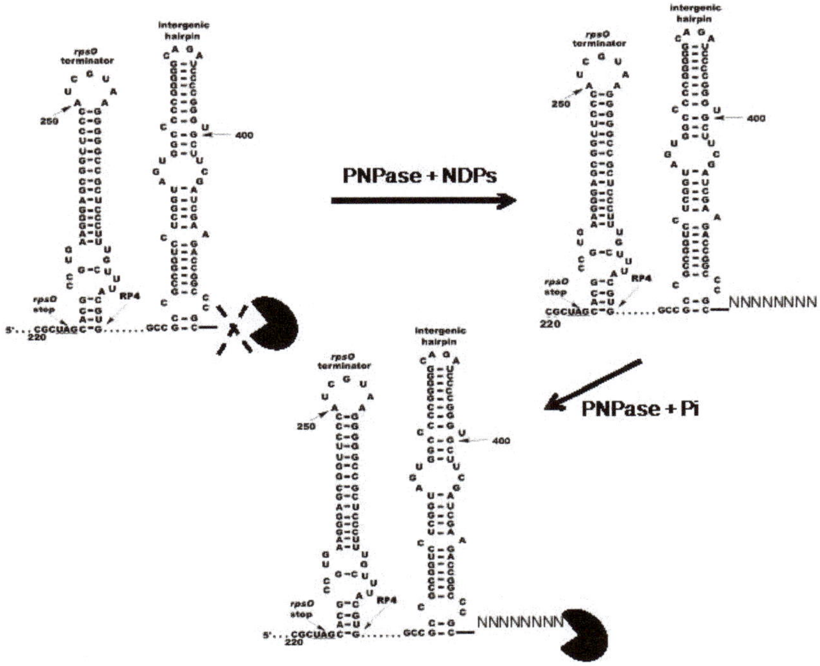

Figure 5. Model for the effects of NDPs on the activity *S. coelicolor* PNPase. The model posits that *S. coelicolor* PNPase (PacMan symbol) is able to phosphorolyze 5650 and other structured substrates to a limited extent in the absence of NDPs, as indicated by the dashed X. In the presence of NDPs, PNPase synthesizes unstructured 3′-tails in vivo, and these tails then provide an anchor for the enzyme, thus facilitating the digestion of structured substrates. Copyright © American Society for Microbiology (*J. Bacteriol.* 195, 2013, 5151–5159, DOI:10.1128/JB.00936-13) [49].

5. (p)ppGpp as a PNPase Effector

Highly phosphorylated guanine nucleotides, (p)ppGpp, guanosine pentaphosphate, and guanosine tetraphosphate, are alarmones that play a number of roles in the regulation of bacterial metabolism (reviewed in [50,51]). (p)ppGpp is synthesized by the product of the *relA* gene in *E. coli* [52], and that gene is found in *S. coelicolor* [53,54], *S. antibioticus* [55], and other streptomycetes [56]. In *Streptomyces*, ppGpp plays an important role in the regulation of antibiotic synthesis [53–56].

PNPase from *S. antibioticus* was shown to synthesize pppGpp in vitro [57,58]. While this activity may be an in vitro artifact, as no other PNPases are known to possess it, the observation suggested a possible relationship between (p)ppGpp and RNA decay in *Streptomyces*. To begin to examine this relationship, the effects of (p)ppGpp on polymerization and phosphorolysis by *S. coelicolor*, *S. antibioticus*, and *E. coli* PNPases were measured in the absence and presence of (p)ppGpp [59]. As shown in Figure 6, both guanosine penta- and tetraphosphates inhibited the activity of *S. coelicolor* PNPase, in both phosphorolysis and polymerization, though ppGpp was a more potent inhibitor than pppGpp.

Essentially identical results were obtained for *S. antibioticus* PNPase (not shown). By contrast, neither ppGpp nor pppGpp were effective inhibitors of the activity of *E. coli* PNPase. Indeed, at concentrations up to 1 mM, pppGpp actually stimulated the polymerizing activity of the *E. coli* enzyme.

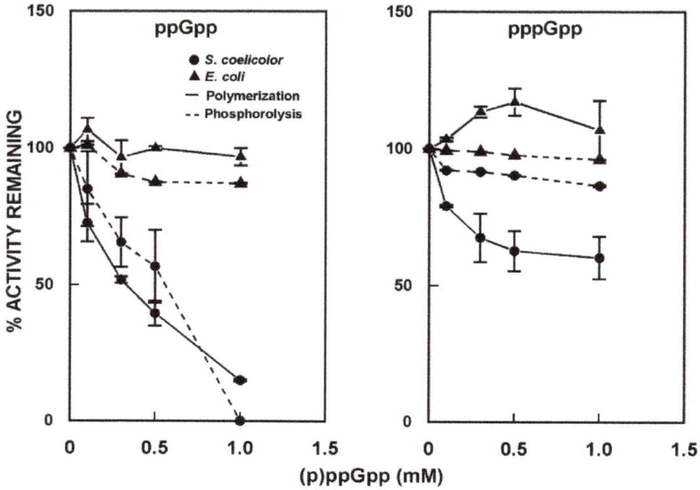

Figure 6. Effects of (p)ppGpp on the activity of PNPase. Polymerization and phosphorolysis reactions were performed in the absence and presence of guanosine tetraphosphate (ppGpp) or guanosine pentaphosphate (pppGpp), using purified PNPase from *S. coelicolor* and *E. coli* [59]. Results are expressed relative to the activities measured in the absence of (p)ppGpp, set arbitrarily to 100%.

In the same study, the effects of (p)ppGpp on the stability of bulk mRNA in *S. coelicolor* were examined [59]. It was initially observed that the half-life of bulk mRNA increased by 1.8-fold in stationary phase cultures as compared with exponential phase. That this increase might be related to the effects of (p)ppGpp was suggested by studies with an *S. coelicolor relA* mutant, and a strain containing an inducible *relA* gene. While the half-life of bulk mRNA was longer in the *relA* mutant than in the parental strain (e.g., 8.9 min vs 3.2 min in exponential phase), the half-life decreased slightly in the *relA* mutant in stationary phase (to 7.2 min). In the strain containing an inducible *relA* gene, producing increased levels of (p)ppGpp, induction occasioned a ca. two-fold increase in the half-life of bulk mRNA, from 6.6 to 11.8 min. Taken together, these observations suggest that (p)ppGpp may stabilize mRNAs in stationary phase *S. coelicolor* cells, as compared with cells growing exponentially.

Why and how might this stabilization occur? It is well established that although levels of RNA and protein synthesis decrease dramatically as *Streptomyces* cultures move from the exponential to the stationary phase of growth, a basal level of synthesis is maintained throughout stationary phase [60,61]. This basal level of macromolecular synthesis is presumably required to produce enzymes and other proteins involved in the synthesis of the secondary metabolites these organisms produce in stationary phase. Stabilization of the transcripts for these proteins would represent one strategy the organisms could employ to ensure the persistence of macromolecular synthesis to support secondary metabolite production. It is known that (p)ppGpp is present in significant amounts, even in stationary phase streptomycete cultures [62,63]. Thus, the inhibition of PNPase by (p)ppGpp might represent a strategy used by *Streptomyces* to stabilize essential mRNAs during stationary phase. It would be interesting to determine whether (p)ppGpp inhibits the activity of other exo- and endonucleases and while such analyses have yet to be performed, it is noteworthy that ppGpp inhibits PNPase from another actinomycete, *Nonomuraea* sp. [64].

6. Conclusions and Unanswered Questions

It is apparent from the brief analysis above that PNPase is a multitalented enzyme that plays a critical role in the metabolic activities of bacterial cells. Despite the wealth of information that has been accumulated about PNPase, a number of important biological questions remain unanswered, particularly as they relate to the functions of *Streptomyces* PNPase.

First, why is the *rpsO–pnp* operon of *S. coelicolor* transcribed from four promoters? The answer to this question may relate to the fact that the operon contains a ribosomal protein gene, as well as the gene for PNPase. It is possible that the promoters are not only critical to the regulation of *pnp* expression, but that they also play and important role in ribosome biogenesis via their regulation of the levels of the *rpsO* transcript. Mutation of the four promoter sequences may provide insight into these possibilities.

Second, what is the significance, if there is any, to the fact that the *S. coelicolor rpsO–pnp* operon produces six transcripts (two *rpsO* transcripts from P*rpsO*A and B, two readthrough transcripts from the same two promoters, and two *pnp* transcripts from P*pnp*A and B)? It should be noted that Northern blot analysis of the transcripts derived from the *S. coelicolor rpsO–pnp* operon did not reveal the presence of six separate transcripts [65]. It is possible that the different transcripts are not sufficiently different in size to have been resolved on the Northern blotting gels. Another intriguing possibility is that the longer transcripts, obtained from the upstream promoter in each case, might be processed at their 5′-ends by RNase J, which possesses 5′–3′-exoribonuclease activity. RNase J has recently been characterized in *S. coelicolor* [66,67].

Third, as described above, PNPase is a cold shock protein in *Streptomyces*. It is relevant to ask whether PNPase responds to other environmental stresses, such as heat, oxidative stress, metal ion stress, etc. It would be of interest, in particular, to examine the effects of various types of stress on the activities of the *rpsO–pnp* promoters. Mutational analyses again might reveal important aspects of promoter function in stress conditions in *Streptomyces*.

Fourth, a number of small effector molecules modulate the activity of *E. coli* PNPase, e.g., ATP, citrate, and cyclic-diGMP [16–19]. It has been proposed that these effectors connect RNA decay to other metabolic pathways in bacterial cells. It would be interesting to determine whether these effectors also affect the activity of *Streptomyces* PNPase. It is noteworthy, in this regard, that in silico molecular docking studies suggest that citrate, which inhibits the activity of *E. coli* PNPase, will bind to PNPase from *S. antibioticus* [18].

Finally, in *E. coli* and other organisms, PNPase is part of a larger macromolecular complex generally referred to as the degradosome ([68,69]. In *E. coli*, the components of the degradosome are organized around a scaffold provided by the single strand specific endoribonuclease, RNase E [70]. RNase E is present in *S. coelicolor*, and has been shown to interact with PNPase in vivo [71]. However, unlike the situation in *E. coli*, the identities of other proteins that might be involved in the degradative machine are unknown in *Streptomyces*.

It is fervently hoped that the foregoing and other important questions related to PNPase structure and function will continue to attract interest and experimentation to provide answers to them.

Acknowledgments: Much of the research described in this report was supported by grants MCB-0133520 and MCB-0817177 from the U.S. National Science Foundation to the author.

Conflicts of Interest: The author declares no conflicts of interest.

References

1. Grunberg-Manago, M.; Ochoa, S. Enzymatic synthesis and breakdown of polynucleotides: Polynucleotide phosphorylase. *J. Am. Chem. Soc.* **1955**, *77*, 3165–3166. [CrossRef]
2. Godefroy-Colburn, T.; Grunberg-Manago, M. Polynucleotide phosphorylase. *Enzymes* **1972**, *7*, 533–574.
3. Littauer, U.Z.; Soreq, H. Polynucleotide phosphorylase. *Enzymes* **1982**, *15*, 517–553.

4. Lengyel, P.; Speyer, J.F.; Ochoa, S. Synthetic polynucleotides and the amino acid code. *Proc. Natl. Acad. Sci. USA* **1962**, *47*, 1936–1942. [CrossRef]
5. Matthaei, J.H.; Jones, O.W.; Martin, R.G.; Nirenberg, M. Characteristics and composition of RNA coding units. *Proc. Natl. Acad. Sci. USA* **1962**, *48*, 666–676. [CrossRef] [PubMed]
6. Zuo, Y.; Deutscher, M. Exoribonuclease superfamilies: Structural analysis and phylogenetic distribution. *Nucleic Acids. Res.* **2001**, *29*, 1017–1026. [CrossRef] [PubMed]
7. Lin-Chao, S.; Chiou, N.T.; Schuster, G. The PNPase, exosome and RNA helicases as the building components of evolutionarily-conserved RNA degradation machines. *J. Biomed. Sci.* **2007**, *14*, 523–532. [CrossRef] [PubMed]
8. Donovan, W.P.; Kushner, S.R. Polynucleotide phosphorylase and ribonuclease II are required for cell viability and mRNA turnover in *Escherichia coli* k-12. *Proc. Natl. Acad. Sci. USA* **1986**, *83*, 120–124. [CrossRef] [PubMed]
9. Clarke, D.J.; Dowds, B.C. The gene coding for polynucleotide phosphorylase in *Photorhabdus* sp. Strain k122 is induced at low temperatures. *J. Bacteriol.* **1994**, *176*, 3775–3784. [CrossRef] [PubMed]
10. Goverde, R.L.J.; Huis in't Veld, J.H.J.; Kusters, J.G.; Mooi, F.R. The psychrotropic bacterium *Yersinia enterocolitica* requires expression of *pnp*, the gene for polynucleotide phosphorylase, for growth at low termperature (5 °C). *Mol. Microbiol.* **1998**, *28*, 555–569. [CrossRef] [PubMed]
11. Zangrossi, S.; Briani, F.; Ghisotti, D.; Regonesi, M.E.; Tortora, P.; Deho, G. Transcriptional and post-transcriptional control of polynucleotide phosphorylase during cold acclimation in *Escherichia coli*. *Mol. Microbiol.* **2000**, *36*, 1470–1480. [CrossRef] [PubMed]
12. Carzaniga, T.; Antoniani, D.; Deho, G.; Briani, F.; Landini, P. The RNA processing enzyme polynucleotide phosphorylase negatively controls biofilm formation by repressing poly-*N*-acetylglucosamine (PNAG) production in *Escherichia coli* C. *BMC Microbiol.* **2012**, *12*, 270. [CrossRef] [PubMed]
13. Pobre, V.; Arraiano, C.M. Next generation sequencing analysis reveals that the ribonucleases RNAse II, RNAse R and PNPase affect bacterial motility and biofilm formation in *E. coli*. *BMC Genomics* **2015**, *16*, 72. [CrossRef] [PubMed]
14. Rosenzweig, J.A.; Chopra, A.K. The exoribonuclease polynucleotide phosphorylase influences the virulence and stress responses of *Yersiniae* and many other pathogens. *Front. Cell. Infect. Microbiol.* **2013**, *3*, 81. [CrossRef] [PubMed]
15. Engman, J.; Negrea, A.; Sigurlasdottir, S.; Georg, M.; Eriksson, J.; Eriksson, O.S.; Kuwae, A.; Sjolinder, H.; Jonsson, A.B. *Neisseria meningitidis* polynucleotide phosphorylase affects aggregation, adhesion, and virulence. *Infect. Immun.* **2016**, *84*, 1501–1513. [CrossRef] [PubMed]
16. Del Favero, M.; Mazzantini, E.; Briani, F.; Zangrossi, S.; Tortora, P.; Deho, G. Regulation of *Escherichia coli* polynucleotide phosphorylase by ATP. *J. Biol. Chem.* **2008**, *283*, 27355–27359. [CrossRef] [PubMed]
17. Nurmohamed, S.; Vincent, H.A.; Titman, C.M.; Chandran, V.; Pears, M.R.; Du, D.; Griffin, J.L.; Callaghan, A.J.; Luisi, B.F. Polynucleotide phosphorylase activity may be modulated by metabolites in *Escherichia coli*. *J. Biol. Chem.* **2011**, *286*, 14315–14323. [CrossRef] [PubMed]
18. Tuckerman, J.R.; Gonzalez, G.; Gilles-Gonzalez, M.A. Cyclic di-GMP activation of polynucleotide phosphorylase signal-dependent RNA processing. *J. Mol. Biol.* **2011**, *407*, 633–639. [CrossRef] [PubMed]
19. Stone, C.M.; Butt, L.E.; Bufton, J.C.; Lourenco, D.C.; Gowers, D.M.; Pickford, A.R.; Cox, P.A.; Vincent, H.A.; Callaghan, A.J. Inhibition of homologous phosphorolytic ribonucleases by citrate may represent an evolutionarily conserved communicative link between RNA degradation and central metabolism. *Nucleic Acids Res.* **2017**, *45*, 4655–4666. [CrossRef] [PubMed]
20. Liu, G.; Chater, K.F.; Chandra, G.; Niu, G.; Tan, H. Molecular regulation of antibiotic biosynthesis in *Streptomyces*. *Microbiol. Mol. Biol. Rev.* **2013**, *77*, 112–143. [CrossRef] [PubMed]
21. Chandra, G.; Chater, K.F. Developmental biology of *Streptomyces* from the perspective of 100 actinobacterial genome sequences. *FEMS Microbiol. Rev.* **2014**, *38*, 345–379. [CrossRef] [PubMed]
22. Berdy, J. Recent advances in and prospects of antibiotic research. *Process. Biochem.* **1980**, *15*, 28–35.
23. Littauer, U.Z. From polynucleotide phosphorylase to neurobiology. *J. Biol. Chem.* **2005**, *280*, 38889–38897. [CrossRef] [PubMed]
24. Briani, F.; Carzaniga, T.; Deho, G. Regulation and functions of bacterial PNPase. *Wiley Interdiscip. Rev. RNA* **2016**, *7*, 241–258. [CrossRef] [PubMed]

25. Régnier, P.; Portier, C. Initiation, attenuation and RNase III processing of transcripts from the *Escherichia coli* operon encoding ribosomal protein S15 and polynucleotide phosphorylase. *J. Mol. Biol.* **1986**, *187*, 23–32. [CrossRef]
26. Luttinger, A.; Hahn, J.; Dubnau, D. Polynucleotide phosphorylase is necessary for competence development in *Bacillus subtilis*. *Mol. Microbiol.* **1996**, *19*, 343–356. [CrossRef] [PubMed]
27. Portier, C.; Regnier, P. Expression of the *rpsO* and *pnp* genes: Structural analysis of a DNA fragment carrying their control regions. *Nucleic Acids Res.* **1984**, *12*, 6091–6102. [CrossRef] [PubMed]
28. Robert-Le Meur, M.; Portier, C. *E. coli* polynucleotide phosphorylase expression is autoregulated through an RNase III-dependent mechanism. *EMBO J.* **1992**, *11*, 2633–2641. [PubMed]
29. Robert-Le Meur, M.; Portier, C. Polynucleotide phosphorylase of *Escherichia coli* induces degradation of its RNase III processed messenger by preventing its translation. *Nucleic Acids Res.* **1994**, *22*, 397–403. [CrossRef] [PubMed]
30. Jarrige, A.-C.; Mathy, N.; Portier, C. Pnpase autocontrols its expression by degrading a double-stranded structure in the *pnp* mRNA leader. *EMBO J.* **2001**, *20*, 6845–6855. [CrossRef] [PubMed]
31. Bralley, P.; Gatewood, M.L.; Jones, G.H. Transcription of the *rpsO-pnp* operon of *Streptomyces coelicolor* involves four temporally regulated, stress responsive promoters. *Gene* **2014**, *536*, 177–185. [CrossRef] [PubMed]
32. Strohl, W.R. Compilation and analysis of DNA sequences associated with apparent streptomycete promoters. *Nucleic Acids Res.* **1992**, *20*, 961–974. [CrossRef] [PubMed]
33. Yamazaki, H.; Ohnishi, Y.; Horinouchi, S. Transcriptional switch on of ssgA by a-factor, which is essential for spore septum formation in *Streptomyces griseus*. *J. Bacteriol.* **2003**, *185*, 1273–1283. [CrossRef] [PubMed]
34. Seghezzi, N.; Amar, P.; Koebmann, B.; Jensen, P.R.; Virolle, M.J. The construction of a library of synthetic promoters revealed some specific features of strong *Streptomyces* promoters. *Appl. Microbiol. Biotechnol.* **2011**, *90*, 615–623. [CrossRef] [PubMed]
35. Jones, G.H. Integrative, *xyle*-based promoter probe vectors for use in *Streptomyces*. *Plasmid* **2011**, *65*, 219–225. [CrossRef] [PubMed]
36. Sun, D.; Liu, C.; Zhu, J.; Liu, W. Connecting metabolic pathways: Sigma factors in *Streptomyces* spp. *Front. Microbiol.* **2017**, *8*, 2546. [CrossRef] [PubMed]
37. Mohanty, B.K.; Kushner, S.R. Bacterial/archaeal/organellar polyadenylation. *Wiley Interdiscip. Rev. RNA* **2011**, *2*, 256–276. [CrossRef] [PubMed]
38. Cao, G.J.; Sarkar, N. Identification of the gene for an *Escherichia coli* poly(A) polymerase. *Proc. Natl. Acad. Sci. USA* **1992**, *89*, 10380–10384. [CrossRef] [PubMed]
39. Cao, G.J.; Kalapos, M.P.; Sarkar, N. Polyadenylated mrna in *Escherichia coli*: Modulation of poly(A) RNA levels by polynucleotide phosphorylase and ribonuclease II. *Biochimie* **1997**, *79*, 211–220. [CrossRef]
40. Mohanty, B.K.; Kushner, S.R. Polynucleotide phosphorylase functions both as a 3′-5′ exonuclease and a poly(A) polymerase in *Escherichia coli*. *Proc. Natl. Acad. Sci. USA* **2000**, *97*, 11966–11971. [CrossRef] [PubMed]
41. Bralley, P.; Jones, G.H. cDNA cloning confirms the polyadenylation of RNA decay intermediates in *Streptomyces coelicolor*. *Microbiology* **2002**, *148*, 1421–1425. [CrossRef] [PubMed]
42. Bralley, P.; Gust, B.; Chang, S.A.; Chater, K.F.; Jones, G.H. RNA 3′-tail synthesis in *streptomyces*: In vitro and in vivo activities of RNase PH, the *SCO3896* gene product and pnpase. *Microbiology* **2006**, *152*, 627–636. [CrossRef] [PubMed]
43. Yehudai-Resheff, S.; Hirsh, M.; Schuster, G. Polynucleotide phosphorylase functions as both an exonuclease and a poly(A) polymerase in spinach chloroplasts. *Mol. Cell Biol.* **2001**, *21*, 5408–5416. [CrossRef] [PubMed]
44. Rott, R.; Zipor, G.; Portnoy, V.; Liveanu, V.; Schuster, G. RNA polyadenylation and degradation in cyanobacteria are similar to the chloroplast but different from *E. coli*. *J. Biol. Chem.* **2003**, *278*, 15771–15777. [CrossRef] [PubMed]
45. Sohlberg, B.; Huang, J.; Cohen, S.N. The *Streptomyces coelicolor* polynucleotide phosphorylase homologue, and not the putative poly(A) polymerase can polyadenylate RNA. *J. Bacteriol.* **2003**, *185*, 7273–7278. [CrossRef] [PubMed]
46. Spickler, C.; Mackie, G.A. Action of RNase II and polynucleotide phosphorylase against RNAs containing stem-loops of defined structure. *J. Bacteriol.* **2000**, *182*, 2422–2427. [CrossRef] [PubMed]

47. Chang, S.A.; Cozad, M.; Mackie, G.A.; Jones, G.H. Kinetics of polynucleotide phosphorylase: Comparison of enzymes from *Streptomyces* and *Escherichia coli* and effects of nucleoside diphosphates. *J. Bacteriol.* **2008**, *190*, 98–106. [CrossRef] [PubMed]
48. Symmons, M.; Jones, G.H.; Luisi, B. A duplicated fold is the structural basis for polynucleotide phosphorylase catalytic activity, processivity and regulation. *Structure* **2000**, *8*, 1215–1226. [CrossRef]
49. Jones, G.H.; Mackie, G.A. *Streptomyces coelicolor* polynucleotide phosphorylase can polymerize nucleoside diphosphates under phosphorolysis conditions, with implications for the degradation of structured rnas. *J. Bacteriol.* **2013**, *195*, 5151–5159. [CrossRef] [PubMed]
50. Potrykus, K.; Cashel, M. (p)ppgpp: Still magical? *Annu. Rev. Microbiol.* **2008**, *62*, 35–51. [CrossRef] [PubMed]
51. Srivatsan, A.; Wang, J.D. Control of bacterial transcription, translation and replication by (p)ppgpp. *Curr. Opin. Microbiol.* **2008**, *11*, 100–105. [CrossRef] [PubMed]
52. Petersen, F.S.; Kjelgaard, N.O. Analysis of the *relA* gene product of *Escherichia coli*. *Eur. J. Biochem.* **1977**, *76*, 91–97. [CrossRef]
53. Chakraburtty, R.; White, J.; Takano, E.; Bibb, M.J. Cloning and characterization and disruption of a (p)ppgpp synthetase gene (*relA*) of *Streptomyces coelicolor* a3(2). *Mol. Microbiol.* **1996**, *19*, 357–368. [CrossRef] [PubMed]
54. Chakraburtty, R.; Bibb, M.J. The ppgpp synthetase gene (*relA*) of *Streptomyces coelicolor* a3(2) plays a conditional role in antibiotic production and morphological differentiation. *J. Bacteriol.* **1997**, *179*, 5854–5864. [CrossRef] [PubMed]
55. Hoyt, S.; Jones, G.H. *RelA* is required for actinomycin production in *Streptomyces antibioticus*. *J. Bacteriol.* **1999**, *181*, 3824–3829. [PubMed]
56. Jin, W.; Ryu, Y.G.; Kang, S.G.; Kim, S.K.; Saito, N.; Ochi, K.; Lee, S.H.; Lee, K.J. Two *relA*/*spoT* homologous genes are involved in the morphological and physiological differentiation of *Streptomyces clavuligerus*. *Microbiology* **2004**, *150*, 1485–1493. [CrossRef] [PubMed]
57. Jones, G.H. Purification and properties of atp:Gtp 3′-pyrophosphotransferase (guanosine pentaphosphate synthetase) from *Streptomyces antibioticus*. *J. Bacteriol.* **1994**, *176*, 1475–1481. [CrossRef] [PubMed]
58. Jones, G.H. Activation of ATP-GTP 3′-pyrophosphotransferase (guanosine pentaphosphate synthetase) in *Streptomyces antibioticus*. *J. Bacteriol.* **1994**, *176*, 1482–1487. [CrossRef] [PubMed]
59. Gatewood, M.L.; Jones, G.H. (p)ppgpp inhibits polynucleotide phosphorylase from *Streptomyces* but not from *Escherichia coli* and increases the stability of bulk mRNA in *Streptomyces coelicolor*. *J. Bacteriol.* **2010**, *192*, 4275–4280. [CrossRef] [PubMed]
60. Jones, G.H. RNA synthesis in *Streptomyces antibioticus*: In vitro effects of actinomycin and transcriptional inhibitors from 48-h cells. *Biochemistry* **1976**, *15*, 3331–3341. [CrossRef] [PubMed]
61. Liras, P.; Villanueva, J.R.; Martin, J.F. Sequential expression of macromolecule biosynthesis and candicidin formation in *Streptomyces griseus*. *J. Gen. Microbiol.* **1977**, *102*, 269–277. [CrossRef] [PubMed]
62. Kelly, K.S.; Ochi, K.; Jones, G.H. Pleiotropic effects of a *relC* mutation in *Streptomyces antibioticus*. *J. Bacteriol.* **1991**, *173*, 2297–2300. [CrossRef] [PubMed]
63. Hesketh, A.; Sun, J.; Bibb, M.J. Induction of ppGpp synthesis in *Streptomyces coelicolor* A3(2) grown under conditions of nutritional sufficiency elicits *actII-ORF4* transcription and actinorhodin biosynthesis. *Mol. Microbiol.* **2001**, *39*, 136–141. [CrossRef] [PubMed]
64. Siculella, L.; Damiano, F.; di Summa, R.; Tredici, S.M.; Alduina, R.; Gnoni, G.V.; Alifano, P. Guanosine 5′-diphosphate 3′-diphosphate (ppgpp) as a negative modulator of polynucleotide phosphorylase activity in a 'rare' actinomycete. *Mol. Microbiol.* **2010**, *77*, 716–729. [CrossRef] [PubMed]
65. Gatewood, M.L.; Bralley, P.; Jones, G.H. RNase III-dependent expression of the *rpsO-pnp* Operon of *Streptomyces coelicolor*. *J. Bacteriol.* **2011**, *193*, 4371–4379. [CrossRef] [PubMed]
66. Bralley, P.; Aseem, M.; Jones, G.H. SCO5745, a bifunctional RNase J Ortholog, affects antibiotic production in *Streptomyces coelicolor*. *J. Bacteriol.* **2014**, *196*, 1197–1205. [CrossRef] [PubMed]
67. Pei, X.Y.; Bralley, P.; Jones, G.H.; Luisi, B.F. Linkage of catalysis and 5′ end recognition in ribonuclease RNase J. *Nucleic Acids Res.* **2015**, *43*, 8066–8076. [CrossRef] [PubMed]
68. Py, B.; Causton, C.F.; Mudd, E.A.; Higgins, C.F. A protein complex that mediates mRNA degradation in *Escherichia coli*. *Mol. Microbiol.* **1994**, *14*, 717–729. [CrossRef] [PubMed]
69. Ait-Bara, S.; Carpousis, A.J. RNA degradosomes in bacteria and chloroplasts: Classification, distribution and evolution of RNase E homologs. *Mol. Microbiol.* **2015**, *97*, 1021–1135. [CrossRef] [PubMed]

70. Vanzo, N.E.; Li, Y.S.; Py, B.; Blum, E.; Higgins, C.F.; Raynal, L.C.; Krisch, H.M.; Carpousis, A.J. Ribonuclease E organizes the protein interactions in the *Escherichia coli* degradosome. *Genes Dev.* **1998**, *12*, 2770–2781. [CrossRef] [PubMed]
71. Lee, K.; Cohen, S.N. A *Streptomyces coelicolor* functional orthologue of *Escherichia coli* RNAse e shows shuffling of catalytic and PNPase-binding domains. *Mol. Microbiol.* **2003**, *48*, 349–360. [CrossRef] [PubMed]

MDPI
St. Alban-Anlage 66
4052 Basel
Switzerland
Tel. +41 61 683 77 34
Fax +41 61 302 89 18
www.mdpi.com

Antibiotics Editorial Office
E-mail: antibiotics@mdpi.com
www.mdpi.com/journal/antibiotics

www.ingramcontent.com/pod-product-compliance
Lightning Source LLC
Chambersburg PA
CBHW061600220926
51850CB00005B/384
9783039369102